The Plant Hormone Ethylene

Editors

Autar K. Mattoo, Ph.D.
Research Leader
Plant Molecular Biology Laboratory
Beltsville Agricultural Research Center-West
United States Department of Agriculture
Agricultural Research Service
Beltsville, Maryland

Jeffrey C. Suttle, Ph.D.
Research Plant Physiologist
Biosciences Research Laboratory
United States Department of Agriculture
Agricultural Research Service
State University Station
Fargo, North Dakota

CRC Press
Taylor & Francis Group
Boca Raton London New York

CRC Press is an imprint of the
Taylor & Francis Group, an informa business

First published 1991 by CRC Press
Taylor & Francis Group
6000 Broken Sound Parkway NW, Suite 300
Boca Raton, FL 33487-2742

Reissued 2018 by CRC Press

Library of Congress Cataloging-in-Publication Data

The Plant hormone ethylene/editors, Autar K. Mattoo, Jeffrey C.
Suttle.
p. cm.
Includes bibliographical references and index.
ISBN 0-8493-4566-9
1. Ethylene. 2. Plant hormones I. Mattoo, Autar. II. Suttle,
Jeffrey Charles.
QK898.E8P57 1991
581.19'27--dc20 90-27594

A Library of Congress record exists under LC control number: 90027594

ISBN 13: 978-1-315-89666-3 (hbk)
ISBN 13: 978-1-351-07576-3 (ebk)

Visit the Taylor & Francis Web site at http://www.taylorandfrancis.com and the
CRC Press Web site at http://www.crcpress.com

PREFACE

The pace of scientific progress is inextricably linked to the introduction and utilization of new tools and techniques. Nowhere in plant biology is this fact more clearly seen than in the field of ethylene research. In the first third of this century, the pharmacological effects of chemically pure ethylene on plant development were studied in detail. These initial observations were useful from both an applied standpoint and a scientific perspective. By defining the biological actions of exogenous ethylene, these studies provided insight into the potential applications of applied ethylene as well as its possible involvement in various postharvest disorders and plant development in general. Analysis of ethylene evolution by plants could only be performed using bioassay techniques that were both tedious and relatively insensitive.

The late 1950s and early 1960s witnessed the introduction of gas chromatography for the analysis of ethylene produced by plant tissues. With the introduction of this new technology came a renewed interest in ethylene physiology. The rates of endogenous ethylene evolution were determined in a variety of plant tissues at various developmental stages or following environmental or chemical treatments.

In many cases, the experimental systems being studied were those previously identified as ethylene sensitive by the earlier workers using applied ethylene. Although extremely informative, these studies (at best) established a correlation between ethylene evolution and the particular process being studied. The only known inhibitor of ethylene action was CO_2 and its inhibitory effects were not only variable in nature, but also difficult to interpret owing to other known responses of plants to CO_2. The synthesis of ethylene could only be inhibited by fairly drastic and nonspecific agents or conditions such as dinitrophenol and anoxia. This situation was soon to change.

Beginning in the mid-1970s, several important discoveries by different research groups resulted in the identification of a series of biochemical tools that have had, and continue to have, a major impact in the field of ethylene physiology. These tools were the (reasonably) specific inhibitors of ethylene biosynthesis (rhizobitoxin, aminoethoxyvinylglycine) and actions (silver salts, cyclic olefins). Furthermore, it was found that the effects of the inhibitors of ethylene synthesis and those of the cyclic olefins (i.e., 2,5-norbornadiene) could be reversed by exogenous ethylene. The utility of an inhibitor as a probe in the biological sciences is generally dependent on the reversibility of its actions by physiologically appropriate agents. Thus, when carefully optimized for each system, these ethylene synthesis and action inhibitors provide invaluable information on the involvement and role of ethylene in a given process.

Two other major developments have occurred in the last 10 to 15 years. First, the entire pathway of ethylene biosynthesis in higher plants has been elucidated, followed by the characterization and molecular cloning of the messenger RNA for a key (generally rate-limiting) enzyme 1-aminocyclopropane-1-carboxylic acid synthase. Second, the powerful techniques of recombinant DNA technology have been brought to bear on the questions of ethylene action at the level of gene expression. Specific ethylene-affected gene products have been cloned and information concerning ethylene-responsive regulatory elements is currently being gathered.

More recently, various ethylene-response mutants have been identified. These should prove extremely useful in the elucidation of currently unknown aspects of ethylene action (i.e., signal transduction). Thus, in the seventeen years that have elapsed since the publicaton of the last monograph on ethylene, a major renaissance in ethylene research has occurred. The time was ripe for a fresh perspective on the entire field of ethylene physiology; hence the conception of this monograph.

The breadth and depth of knowledge concerning ethylene synthesis and action, coupled with the rapid pace of new progress makes a survey of the field a daunting task. We felt that experts actively engaged in different aspects of ethylene research would be in the best position to provide both insight and perspective to each topic.

In this monograph, we have been fortunate to enlist the able talents of a range of active ethylene researchers from different countries, spanning four continents. To be useful, a monograph of this nature should not only provide historical perspective but also suggest future trends and possibilities. We believe that this has been accomplished. We thank all contributors to this volume for their efforts and patience. We hope that readers will find this monograph informative and useful.

Autar K. Mattoo
Jeffrey C. Suttle

EDITORS

Autar K. Mattoo, Ph.D., is Research Leader and Supervisory Plant Physiologist, Plant Molecular Biology Laboratory, Beltsville Agricultural Research Center, United States Department of Agriculture, Agricultural Research Service, Beltsville, Maryland. He obtained his M.Sc. (1965) and Ph.D. (1969) degrees from the Maharaja Sayajirao University of Baroda (India). He was a postdoctoral fellow/visiting scientist at well-known research centers: Department of Biochemistry, University of Adelaide, Australia (1975-1976); Department of Botany, University of New South Wales, Kensington, Australia (1976); Beltsville Agricultural Research Center, Beltsville, Maryland, (1976-1977); Molecular Biology Unit, Tata Institute of Fundamental Research, Bombay, India (1978); and Department of Plant Genetics, The Weizmann Institute of Science, Rehovot, Israel (1979-1980; 1985; 1988). Dr. Mattoo was a Lecturer in the postgraduate Department of Microbiology, M. S. University of Baroda, India (1969-1979) and held a joint Research Associate positon at USDA, Plant Hormone Laboratory, Beltsville, Maryland and the Department of Botany, University of Maryland at College Park, Maryland (1980-1984). He has been with the USDA/ARS, Beltsville Agricultural Research Center, Maryland since October 1984, and in his present position since 1988.

Dr. Mattoo is a member of the American Association for Advancement of Science, American Chemical Society's Biological Section, American Society of Biochemistry and Molecular Biology, American Society of Plant Physiologists, Biochemical Society (U.K.), International Plant Molecular Biology Society, and Plant Growth Regulator Society of America. He served as Secretary-Treasurer of the Washington Section of the American Society of Plant Physiologists in 1987. He was the acting Chairman/Vice Chairman for the 1988 Gordon Research Conference on Plant Senescence and is the Chairman for the same conference in 1992.

Dr. Mattoo has presented over 40 invited lectures at national and international meetings, and about 35 guest lectures at various universities and research institutes. He has published more than 110 research papers. His current major research interests relate to regulation of ethylene biosynthesis and fruit ripening, stress biology, and elucidation of mechanisms involved in the assembly and turnover of photosystem II reaction center proteins.

Jeffrey C. Suttle, Ph.D., is a Research Plant Physiologist in the Plant Science Research Group, U.S. Department of Agriculture, Agricultural Research Service Biosciences Research Laboratory, Fargo, North Dakota.

Dr. Suttle received his B.A. degree from the University of Texas, Austin, Texas in 1974. He obtained his Ph.D. in 1979 from Michigan State University, East Lansing. He joined the USDA-ARS in 1979.

Dr. Suttle is a member of the American Association for the Advancement of Science, American Society of Plant Physiologists, Plant Growth Regulator Society of America, British Society for Plant Growth Regulation, and the honorary society Sigma Xi.

Dr. Suttle has published more than 45 research papers and presentations. He is on the editorial boards for the journals *Plant Physiology* and *Plant Growth Regulation.* His current research interests concern the interactions between synthetic plant growth regulators and endogenous phytohormone systems, especially phytohormone receptors.

CONTRIBUTORS

Thomas Boller, Ph.D.
Professor
Botanisches Institut der Universität Basel
Basel, Switzerland

Colin J. Brady, Ph.D.
CSIRO
Division of Horticulture
North Ryde, Australia

Richard M. Broglie, Ph.D.
Research Supervisor
Agricultural Products Department
E.I. duPont deNemours & Co.
Experimental Station
Wilmington, Delaware

Karen Broglie, Ph.D.
Research Biochemist
Agricultural Products Department
E.I. duPont deNemours & Co.
Experimental Station
Wilmington, Delaware

Yohji Esashi, Ph.D.
Department of Biological Science
Tohoku University
Kawauchi, Oabaku
Sendai, Japan

Hideo Fukuda, Ph.D.
Professor
Department of Applied Microbial Technology
Kumamoto Institute of Technology
Kumamoto, Japan

Professor Michael A. Hall, ARCS, FIBiol, Ph.D., D.Sc.
Department of Botany and Microbiology
University College of Wales
Penglais
Aberystwyth, Wales

Hiroshi Hyodo, Ph.D.
Professor
Department of Biological Science
Faculty of Agriculture
Shizuoka University
Ohya, Shizuoka, Japan

Hidemasa Imaseki, Ph.D.
Research Institute for Biochemical Regulation
School of Agricultural Sciences
Nagoya University
Chikusa
Nagoya, Japan

Michael B. Jackson, D.Sc.
Department of Agricultural Sciences
University of Bristol
Long Ashton Research Station
Long Ashton
Bristol, England

Klaus Lürssen, Ph.D.
Pflanzenschutz
Entwicklung Herbizide
Bayer AG
Leverkusen, Germany

Autar K. Mattoo, Ph.D.
Research Leader
Plant Molecular Biology Laboratory
U.S. Department of Agriculture
Agricultural Research Service
Beltsville, Maryland

Takahira Ogawa, Ph.D.
Associate Professor
Department of Applied Microbial Technology
Kumamoto Institute of Technology
Kumamoto, Japan

Daphne J. Osborne, D.Sc.
Department of Plant Sciences
University of Oxford
Oxford, England

William L. Pengelly, Ph.D.
Associate Professor
Chemical and Biological Sciences
Oregon Graduate Institute
Beaverton, Oregon

Ilya Raskin, Ph.D.
Associate Professor
Center for Agricultural Molecular Biology
Cook College, Rutgers University
New Brunswick, New Jersey
Wilmington, Delaware

Michael S. Reid, M.Sc., Ph.D.
Department of Environmental Horticulture
University of California, Davis
Davis, California

Edward C. Sisler, Ph.D.
Professor
Department of Biochemistry
North Carolina State University
Raleigh, North Carolina

Jim Speirs
Principal Research Scientist
CSIRO Horticulture
North Ryde, Australia

Ling-Yuan Su, Ph.D.
Postdoctoral Research Associate
Chemical and Biological Sciences
Oregon Graduate Institute
Beaverton, Oregon

Jeffrey C. Suttle, Ph.D.
Research Physiologist
Biosciences Research Laboratory
U.S. Department of Agriculture
Agricultural Research Service
Fargo, North Dakota

W. Bruce White, Ph.D.
Department of Anaerobic Microbiology
Virginia Polytechnic Institute
and State University
Blacksburg, Virginia

Meng-Jen Wu, Ph.D.
Department of Environmental Horticulture
University of California, Davis
Davis, California

This book is dedicated to the memory of
Morris Lieberman
a pioneer in the field of ethylene physiology,
and to
all past, present, and future students
of this fascinating plant hormone.

TABLE OF CONTENTS

Chapter 1

THE BIOCHEMISTRY OF ETHYLENE BIOSYNTHESIS

Hidemasa Imaseki

TABLE OF CONTENTS

I. INTRODUCTION

The history of studies on ethylene biosynthesis probably represents the agonies and frustration of the plant scientists who were involved with the topic. Because of the fact that homogenization of ethylene-producing tissues completely abolishes the ethylene-producing activity, studies of the biosynthetic pathway had to be done at the tissue level or with nonenzymatic model systems. These studies with many different plant tissues including ripe fruits, auxin-treated tissues, wounded and other stressed tissues, and senescing tissues clearly indicated that methionine was an intermediate and that ethylene was derived from C-3 and C-4 atoms of the amino acid, while the methylthio group was metabolized to nonvolatile substances. The progress of those early studies on ethylene biosynthesis is well described by Burg,[1] Abeles,[2] Lieberman,[3] and Yang and Hoffman.[4]

Elucidation of the pathway from methionine required another decade. During that period, however, the biochemical as well as physiological bases of ethylene biosynthesis were established, so that when 1-aminocyclopropane-1-carboxylic acid (ACC) was identified as an immediate precursor of ethylene in higher plants in 1979[5] and ACC synthase activity was detected in crude extracts of ethylene-producing tissues in the same year,[6,7] the major regulatory step of ethylene biosynthesis was soon identified at the biochemical level. In the following years, attention has been focused on the regulatory mechanism of synthesis of ACC synthase. Ethylene biosynthesis from methionine in higher plants includes three enzymes; methionine adenosyltransferase, *S*-adenosylmethionine methylthioadenosine-lyase (ACC synthase), and the commonly called ethylene-forming enzyme (EFE). Methionine adenosyltransferase is widely distributed among living organisms and supplies *S*-adenosyl-L-methionine (SAM), which serves as a donor of the methyl group to various cellular components including phenolic compounds and nucleic acids, and also, after decarboxylation, as a donor of the aminopropyl group to polyamines. By contrast, ACC synthase and EFE are unique to higher plants, although the enzymatic entity of EFE has not been identified.

Ethylene is also biosynthesized in large amounts by some bacteria and fungi, but biosynthetic pathways in microorganisms seem to be different from those in higher plants.[8] As the microbial system is described in Chapter 15 of this monograph, only biochemical aspects of ethylene biosynthesis in higher plants will be dealt with in this chapter.

II. BIOSYNTHETIC PATHWAY

A. METHIONINE CYCLE

In all organisms, L-methionine is, other than being incorporated into proteins, "activated" yielding SAM. It is reasonable to speculate that methionine may be "activated" before it is converted to ethylene. In fact, most of the labeled methionine supplied to apple fruit tissues is converted to SAM.[9] Since ethylene production requires oxygen and is inhibited by arsenite, that is restored by phosphate, Burg proposed that ethylene biosynthesis involves a reaction requiring a high energy phosphate bond, probably the formation of SAM.[10] The observation by Murr and Yang[11] that ethylene biosynthesis is inhibited by uncouplers of oxidative phosphorylation also supported the view that SAM is an intermediate. Indirect experimental evidence for this view was provided by Adams and Yang.[12] They assumed that if SAM served as an intermediate between methionine and ethylene, 5′-methylthioadenosine (MTA) would be produced concomitantly with ethylene formation. Using [^{35}S]methionine, they found that tissue plugs from climacteric apple fruits produced both radioactive MTA and 5′-methylthioribose (MTR) but tissue plugs from preclimacteric fruits did not. Further, the formation of MTA and MTR was strongly suppressed by aminoethoxyvinylglycine (AVG), which inhibits ethylene biosynthesis. Thus, MTA formation from methionine in apple tissue correlates with ethylene synthesis implicating SAM as an intermediate. Definite proof of this hypothesis was obtained when ACC synthase activity, utilizing SAM as a substrate, was detected in tomato extracts.

Adams and Yang also found that [^{35}S]MTA supplied to apple tissues was efficiently converted to MTR and methionine, and that [^{35}S]MTR was incorporated into methionine.[12] This indicates that MTA is hydrolyzed to MTR, the sulfur atom of which is recycled into methionine. The methylthio group of MTR is recycled into methionine without splitting between sulfur and methyl group. Based on these observations, they proposed a cyclic pathway of ethylene biosynthesis including methionine, SAM, MTA, and MTR in apple fruit. It was later shown that the methylthio group and the ribose moiety of MTR were also incorporated into methionine,[13,14] and that MTR is phosphorylated to MTR-1′-phosphate (MTR-P). A modified methionine cycle has been proposed (Figure 1).[15] Thus, the cyclic pathway continuously recycles the methylthio group of methionine for methionine resynthesis by utilizing the ribose moiety of input ATP as the source of the butyrate moiety of methionine as ethylene is synthesized. The physiological implications of the methionine cycle are significant. When a considerable amount of ethylene is produced, a co-product of ethylene from SAM is recycled to produce the ethylene precursor, methionine, and this allows a constant supply of methionine for ethylene biosynthesis, even in tissues containing a very small pool of methionine, such as apple fruits. MTA is also produced during polyamine biosynthesis, and it is possible that active polyamine biosynthesis is supported by the methionine cycle.

1. Methylthioadenosine Nucleosidase

In plants, MTA is hydrolyzed by a nucleosidase into MTR and adenine. MTA nucleosidase has been purified from yellow lupine seeds.[16] The native form of the nucleosidase has a relative molecular mass of 62 kDa and consists of two identical subunits of 31 kDa. The nucleosidase has a broad substrate specificity with the highest specificity toward MTA. The Km value for MTA of the lupine nucleosidase is as low as 0.41 μM. This high affinity for MTA agrees well with the observation that MTA formed from SAM is readily converted to MTR.[12] The nucleosidase is competitively inhibited by adenine, another product of the reaction, with a Ki value of about 11 μM. MTA nucleosidase in tomato suspension cultured cells appears to be cytosolic.[17] Conversion of MTA to MTR also occurs in *Lemna* and most of the adenine produced from MTA is metabolized to form adenine nucleotides.[18] It is therefore expected that the endogenous level of adenine is kept low and the significantly lower Km value for MTA than Ki value for adenine allows a constant flow of MTA to MTR, while MTR is accumulated.

2. Methylthioribose Kinase

Cell-free extracts from many plant tissues catalyze the ATP-dependent phosphorylation of MTR to MTR-P.[19,20] MTR kinase of yellow lupine seeds has been purified about 2000-fold and was characterized.[20] As judged by gel filtration, molecular weight of the native enzyme is 70,000. The reaction by lupine MTR kinase is absolutely dependent on the divalent cations, Mg^{++} and Mn^{++}, but Mn^{++} is only 12% as effective as Mg^{++}. The Km values for MTR and ATP are 4.3 and 8.3 μM, respectively, and deoxyATP also serves as a phosphate donor. The kinase is irreversibly inactivated after incubation with a SH reagent, *P*-chloromercuribenzoate.

3. Formation of 2-Keto-4-Methylthiobutyrate

Avocado extracts convert MTR to 2-keto-4-methylthiobutyrate (KMB) and 2-hydroxy-4-methylthiobutyrate (HMB) only in the presence of ATP, but conversion of MTR-P to KMB occurs in the absence of ATP.[19] This also indicates that MTR is first phosphorylated by MTR kinase, and MTR-P is then metabolized to KMB and HMB. Plant enzymes cata-

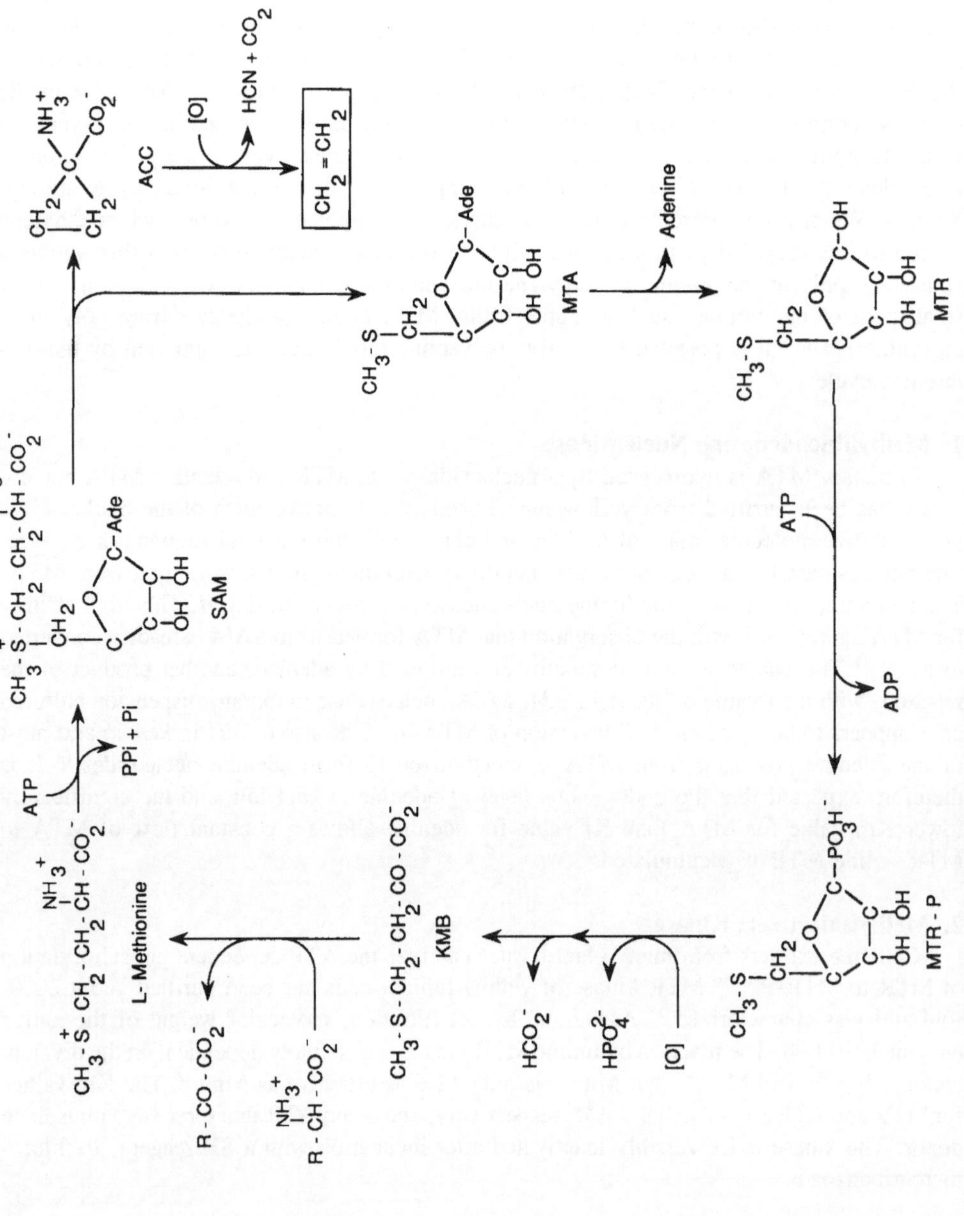

FIGURE 1. The methionine cycle. (Modified from Miyazaki, J. H. and Yang, S. F., *Physiol. Plant.*, 69, 366, 1987.)

lyzing the formation of KMB from MTR-P have not been characterized. However, in rat liver extracts, three enzymes are shown to be involved in the conversion.[21] The first enzyme isomerizes MTR-P to 5-methylthioribulose-1-phosphate, the second enzyme produces two unidentified compounds from 5-methylthioribulose-1-phosphate, and the third enzyme converts these two compounds to KMB with uptake of oxygen. As Wang et al.[14] demonstrated an oxygen requirement for conversion of MTR to methionine, it is likely that similar enzymes are involved in the conversion by plant extracts, but this should be demonstrated experimentally.

4. Formation of Methionine from KMB

The last step of the methionine cycle is conversion of KMB to methionine, possibly by transamination as found in rat liver.[22] KMB is converted to methionine and HMB in avocado extracts in the presence of arginine, but HMB is not converted to methionine.[19] That KMB serves as an ethylene precursor through methionine was shown in tomato pericarp tissue. [U-^{14}C]KMB supplied to tomato tissue produces labeled ethylene, and incorporation of the label is inhibited by AVG, which specifically inhibits conversion of methionine to ethylene.[19] Early in 1971, incorporation of KMB into methionine was found in apple tissue.[23] An enzyme responsible for conversion of KMB to methionine has not been characterized. Miyazaki and Yang[15] discussed the possibility that HCN, formed during the conversion of ACC to ethylene, is incorporated into the amide nitrogen of arginine. If the amide nitrogen is transferred to KMB to form methionine, then the amino nitrogen of methionine will be salvaged by the methionine cycle.

B. ACC PATHWAY

Ethylene biosynthesis requires oxygen, and ethylene production from plant tissues is rapidly arrested when tissues are placed under anaerobic conditions. When such tissues are returned to air, ethylene production immediately starts at a rate significantly higher than the control tissues never placed under anaerobic conditions. However, this increased rate of production is transient and the production rate resumes the control level after a few hours. Burg and Thimann[24] interpreted this phenomenon by proposing that an intermediate of ethylene biosynthesis accumulated during the anaerobic period and that the intermediate was rapidly converted to ethylene in the presence of oxygen. Adams and Yang,[5] using apple fruit tissue, examined the metabolism of methionine under these conditions. They found that a metabolite of [U-^{14}C]methionine accumulated only in tissue placed in anaerobic conditions, the accumulated metabolite disappeared when the tissue was returned to air, and that the label of the metabolite isolated from the tissue fed with [U-^{14}C]methionine was incorporated into ethylene. This metabolite was identified as ACC. Incorporation of [U-^{14}C]methionine into ethylene was suppressed by an exogenous supply of unlabeled ACC, but unlabeled methionine did not suppress incorporation of labeled ACC to ethylene. They also showed that labeled [methyl-^{14}C] or [^{35}S]methionine was incorporated into another metabolite identified as MTR, consistent with the formation of ACC via SAM from methionine. Identification of ACC as an immediate precursor derived from SAM facilitated the identification of the steps of AVG inhibition and of oxygen requirement as being conversion of methionine to ACC and that of ACC to ethylene, respectively. Independently, by screening of chemicals that affect ethylene production in soybean leaf discs, Lürssen et al.[25] found that tissue discs supplied with ACC produced a large amount of ethylene and the properties of this ACC-dependent ethylene production were similar to those of endogenous production in terms of inhibition by anaerobiosis, osmotic shock treatment, and Co^{++}. ACC-dependent ethylene production also occurs in many other plant tissues.

ACC was originally found in fruits of perry pear and cider apple by Burroughs,[26] and in ripe cowberries by Vahatalo and Virtanen,[27] as a nonprotein amino acid. It was not detected in unripe cowberries. Burroughs also observed that ACC increased during storage of pear fruits after harvest.[28] Research on ethylene biosynthesis has since progressed considerably since the recognition of ACC as an immediate precursor of ethylene. It should be mentioned that ACC was commercially available and this greatly helped the progress of research by other workers. Methods of quantitative determination of ACC have been established by Boller et al.[6] and Lizada and Yang.[29] Administration of exogenous ACC to many plant tissues was shown to produce ethylene in an oxygen-dependent manner, and an increase of ACC content that was inhibited by AVG was associated with an increased ethylene production. These results are considered evidence that most plant tissues produce ethylene primarily through the ACC pathway (see Chapter 11 by Osborne).

1. ACC Synthase [SAM MTA-Lyase]

a. Enzymological Properties

By establishing methods for the quantitative determination of ACC, Boller et al.[6] and Yu et al.[7] detected *in vitro* ACC synthase activity in extracts of wounded tomato fruits that utilized SAM as a substrate and pyridoxal-5′-phosphate as a cofactor. Pyridoxal-5′-phosphate is not tightly bound to the enzyme, since removal of the cofactor from an enzyme solution can be achieved by gel filtration, resulting in a significant reduction of the enzyme activity. Enzyme activity is fully restored after re-addition of the cofactor to the reaction mixture. The requirement of pyridoxal-5′-phosphate in ethylene biosynthesis had been proposed by Murr and Yang[11] based on inhibition of conversion of methionine to ethylene by L-canaline, an inhibitor of pyridoxal-5′-phosphate-linked enzymes. ACC synthase activity has been found in crude extracts from many other plant tissues, including hypocotyls of auxin-treated mungbean,[30] tomato and winter squash,[31] bean leaves,[32] and fruits.[33-35]

Most of the enzymological properties of ACC synthase have been determined with crude or partially purified extracts from tomato fruits,[5,6] auxin-treated mungbean,[30] and winter squash fruits.[36] The enzyme is present in cytosolic fractions of these tissues, and utilizes specifically SAM as the substrate. AVG and aminooxyacetic acid, both inhibitors of pyridoxal phosphate-linked enzymes, are competitive inhibitors of this enzyme. The properties determined with crude or partially purified preparations from various sources are summarized in Table 1. The Km values for SAM range from 12.1 to 60 μM and the Ki values for AVG vary from 0.12 to 10 μM. This variation may have resulted from different experimental conditions, particularly from slight changes in pH of the reaction mixtures. Satoh and Yang[37] found with a partially purified tomato ACC synthase that Km for SAM increased from 38 to 167 μM as pH of the reaction mixture increased from 7.5 to 9.5. In contract, the Ki for AVG decreased from 2.9 to 0.25 μM. The optimal pH of the reaction is in the alkaline range between pH 8.5 and 9.5.

ACC synthase has a rigid specificity for stereochemistry of SAM at both the sulfonium center and α-carbon of the methionine moiety. Naturally occurring SAM is (−)-L-SAM. Racemic SAM at the sulfonium center, (±)-L-SAM, shows half the activity as substrate and has a Km twice as high as (−)-L-SAM, indicating that unnatural (+)-L-SAM is inactive as the substrate.[38] Observation that (±)-D-SAM was totally inactive as substrate indicates that (−)-D-SAM does not serve as substrate. (+)-L-SAM[38] and a SAM analog, the antibiotic Sinefungin [5′-deoxy-5′-(1,4-diamino-4-carboxylbutyl)adenosine],[39,40] are strong inhibitors of ACC synthase.

b. Reaction Mechanisms

A possible reaction mechanism of ACC synthase has been proposed by analogy of 1,3-elimination in organic chemistry.[5,7] Schiff base formation between the 2-amino group of SAM and the pyridoxal phosphate-enzyme complex eliminates the proton at C-2, generating a carbanion. The positive sulfonium ion of SAM thus activates the methylene at C-4 and facilitates the intramolecular nucleophilic displacement reaction by the carbanion, generating ACC and MTA (Figure 2). The direct γ-displacement mechanism was experimentally sup-

TABLE 1
Enzymological Properties of Wound- and Auxin-Induced ACC Synthases

Source	Type	Km (SAM) μ*M*	Ki μ*M*	pH opt.	pI	$T^a_{1/2}$ min	*Mr* (Monomer) kDa	Ref.
Tomato								
Pericarp	Wound	13	0.2(AVG)	8.5			57	6
							55	59
Pericarp	Wound	20	0.8(AOA)	8.5				7
Pericarp	Wound					30		46
Pericarp	Wound	17	10(AVG)				50	49
Pericarp	Wound				5.8		45	51
Pericarp	Wound				7.0		67	50
					5.3			50
					9.0			50
Hypocotyl	Auxin	40						31
Winter squash								
Mesocarp	Wound	13.3	2.1(AVG)				58	99
Mesocarp	Wound	12.1	2.5(AVG)	9.5				36
Hypocotyl	Auxin	13	0.12(AVG)	9.0—9.5				31
Zucchini								
Fruit	W + A[b]						54	53
Mung bean								
Hypocotyl	Auxin	60		8.5				30
Hypocotyl	Auxin					25		44
Hypocotyl	Auxin	55		8.0			65	54

[a] *In vivo* half-life of the endogenous enzyme activity.
[b] Slices were treated with a mixture of IAA, BAP, AOA, and LiCl.

ported by Ramalingam et al.[41] They synthesized (±)-*S*-adenosyl-L-(3*R*;*·4*S**)-[3,4-$^2H_2$]methionine (*cis*-SAM) and (±)-*S*-adenosyl-L-(3*S**,4*R**)-[3,4-$^2H_2$]methionine (*trans*-SAM) as substrates for tomato fruit ACC synthase, and the configuration of the deuterium at C-2 and C-3 of the ACC formed was examined by 270 MHz ^1H-NMR. As *cis*-SAM and *trans*-SAM produced *cis*-ACC and *trans*-ACC, respectively, they proposed that there is no participation of β-hydrogen in conversion, and ACC is formed by direct γ-displacement. Thus, the reaction mechanism of ACC synthase is unique among the pyridoxal phosphate-mediated γ-replacement reactions, because most of the reactions involve participation of β-hydrogen whereby *cis,trans* isomerization occurs.

The amino hydrogen at C-6 of SAM seems important for interaction of the substrate with the enzyme, since substitution of the amino group with the hydroxyl group (*S*-inosyl- and *S*-guanosyl-methionine) or with the dimethylamino group abolishes substrate activity, whereas substitution with the monomethylamino group retains some activity.[42]

c. SAM-Induced Inactivation

As predicted by Sakai and Imaseki,[43] ACC synthase is rapidly inactivated within cells. The half-life of ACC synthase activity in auxin-treated mung bean hypocotyls[44,45] and in tomato fruits,[46] is estimated to be 25 and 30 min, respectively. In the presence of AVG, the half-life doubles.[44] Satoh and Esashi[47] reported that ACC synthase from auxin-treated mung bean hypocotyls was inactivated by its substrate, SAM, during the catalytic reaction. A similar inactivation of the enzyme was reported for the enzyme from tomato fruit[37] and winter squash.[36] Inactivation of ACC synthase partially purified from winter squash fruits is suppressed in the absence of pyridoxal phosphate even when SAM is present, or in the presence of *S*-adenosyl-homocysteine or homocysteine, which inhibit the ACC synthase reaction at high concentrations. These results suggest that inactivation is associated with the

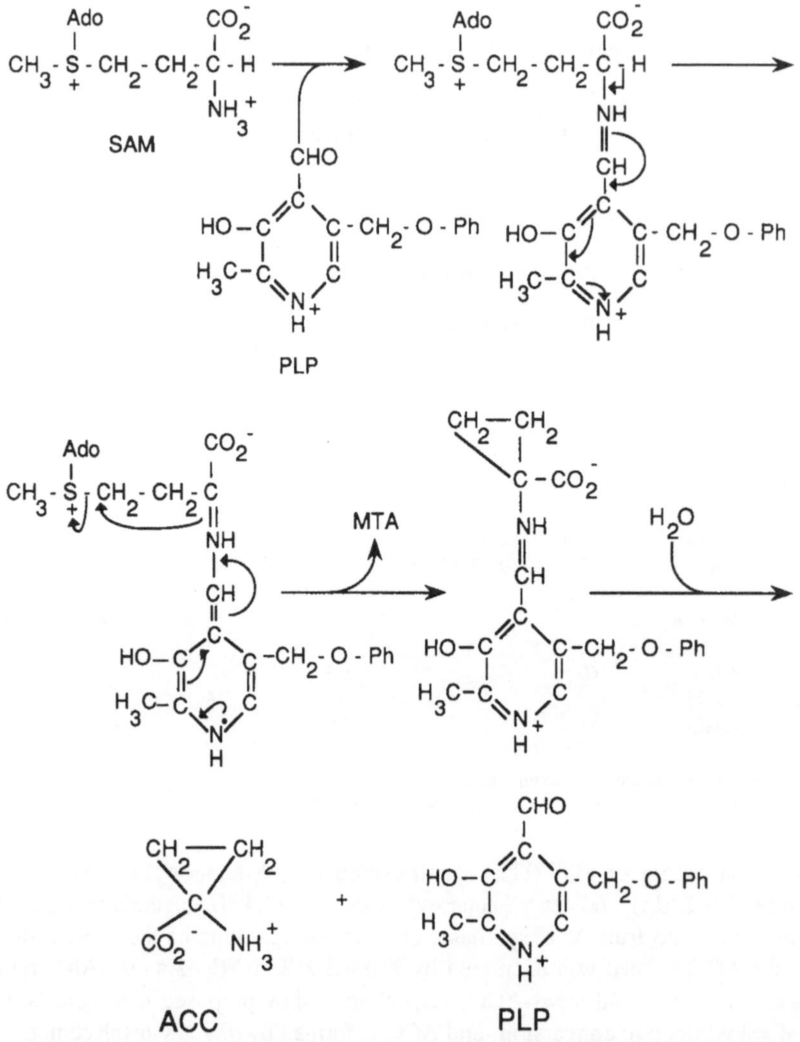

FIGURE 2. A proposed reaction mechanism of ACC synthase. (Modified from Adams, D. O. and Yang, S. F., *Proc. Natl. Acad. Sci. U.S.A.*, 76, 170, 1979.)

catalytic reaction. Based on the inactivation rates at different concentrations of SAM, Satoh and Esashi[47] suggested that the SAM-induced inactivation of ACC synthase is responsible for the rapid inactivation *in vivo*. During inactivation of the enzyme, labeled [3,4-^{14}C]SAM, but not [methyl-^{14}C]SAM[48] covalently binds to ACC synthase in the presence of pyridoxal phosphate. Although it is assumed that the aminobutyrate moiety binds to the enzyme, the precise portion is yet to be determined. Satoh and Yang[37] suggest that SAM binds at the active site of the enzyme, since AVG, a competitive inhibitor with respect to SAM, suppresses the SAM-induced inactivation. Many pyridoxal phosphate-linked enzymes are known to be inactivated during catalysis.

Covalent binding of a portion of SAM with the enzyme will allow for the purification of the peptide fragment containing the active site, leading to the amino acid sequence determination and to the analysis of the interaction of the enzyme with the substrate.

d. Purification and Molecular Weight

ACC synthase is very unstable after extraction and is present in a small quantity. This has hampered purification of the enzyme, but several attempts have been made to purify the enzyme for more precise characterization and for studies of induction mechanisms of the enzyme. Although enzymological properties of ACC synthases from various plant tissues induced by different stimuli are similar, there are many different estimates for the relative molecular mass of the enzyme, even from the same plant tissues. Bleecker et al.[49] reported partial purification of ACC synthase from tomato fruits by a combination of anion exchange, hydroxylapatite, and Affi-Gel Blue column chromatography and obtained a 6500-fold purified preparation with 5% recovery. The final preparation had a specific activity of 27,500 units (1 unit is defined as 1 nmol ACC formed per hour) per mg protein. In this preparation, ACC synthase consisted of a single subunit of 50 kDa. Bleecker et al. also noted that the immunoprecipitable protein migrated on SDS-PAGE as a diffuse band of 80 to 100 kDa if it was not treated with dithiothreitol before electrophoresis.[49] This may indicate that tomato ACC synthase contains at least one disulfide bond in the molecule, but the electrophoretic behavior is unusual if the enzyme consists of a single subunit.

Mehta et al.[50] reported a partial purification of tomato ACC synthase using buffers containing protease inhibitors and minimal salts and obtained 900-fold purified preparation having a specific activity of 805 units/mg. Gel filtration on Sephadex G-100 gave a molecular mass of about 65 kDa for the native enzyme. A monoclonal antibody was prepared for the preparation and was used for immunoprecipitation of the enzyme activity. It recognized a 67-kDa polypeptide on SDS-PAGE. However, when extracts were prepared with buffers containing NaCl, the 67-kDa polypeptide changed to smaller polypeptides which retained reactivity with the monoclonal antibody. Mehta et al. reported that crude extracts of tomato fruits contained at least three isoforms of the enzyme with different pI values, 5.3, 7, and 9, and the component with pI 7 contained a 67-kDa polypeptide which reacted with the monoclonal antibody.[50] A partial purification of tomato ACC synthase has also been reported by Van Der Straeten et al.[51] The final preparation is estimated to be between 25 and 50% pure and has a specific activity of 200,000 units/mg. The activity is associated with a 45-kDa polypeptide with pI value of 5.8. Nakajima et al.[52] purified ACC synthase from wounded mesocarp of *Cucurbita maxima* fruits to an apparent homogeneity by using an immunoaffinity column prepared from a polyclonal antibody raised against a partially purified preparation, obtained by a combination of chromatography with CM-, Q-Sepharose, Sephacryl S-300, MonoQ, and Superose 12.[36] The 33,600-fold purified preparation with a specific activity of 141,300 units per mg protein was obtained with 35% recovery. It contained a single polypeptide of 50 kDa on SDS-PAGE, and polyclonal antibody for this protein effectively immunoprecipitated the enzyme activity. However, immunostained Western blots of SDS-PAGE of crude extracts or fluorograms of immunoreactive proteins from mesocarp disks labeled with [^{35}S]methionine showed two positive polypeptides of 58 and 50 kDa, the 58-kDa polypeptide being more abundant than the 50-kDa polypeptide. They also noted that the 58-kDa polypeptide gradually converted to the 50-kDa polypeptide on incubation of a crude extract and concluded that the 50-kDa polypeptide they purified was a limited hydrolysis product of ACC synthase and that the subunit size of intact ACC synthase must be 58 kDa. The size of an *in vitro* translation product bound to this antibody was 58 kDa. Satoh and Theologis[53] reported that an antibody against ACC synthase from zucchini fruit recognized a 46-kDa polypeptide in partially purified preparations while a 53-kDa polypeptide was detected in the *in vitro* translation products.

The winter squash enzyme activity in a native form migrated at a 160-kDa position by gel filtration over Sephacryl S-300 or Superose 12. As reported for the tomato enzyme,[49] ACC synthase isolated from winter squash, if not treated with 2-mercaptoethanol, migrates at a 100- to 120-kDa position. It is thus highly likely that the enzyme consists of two subunits, probably linked by a disulfide bond, but an accurate subunit structure of the enzyme remains to be determined.

A purification of auxin-induced ACC synthase from IAA-treated mungbean hypocotyls has been achieved.[54] A high concentration (1.5 *M*) of phosphate in the extraction buffers stabilized the auxin-induced enzyme and only a four-step procedure including hydroxylapatite, phenyl-Sepharose, and MonoQ chromatography was required to obtain an active preparation containing a single polypeptide of 60 kDa on SDS-PAGE. The enzyme activity migrated at a position of 125 kDa on gel filtration over Bio-Gel A — 0.5 *M*, suggesting that the native enzyme is a dimer. However, the specific activity of this preparation was only 21,400 units/mg protein,[54] much lower than the wound-induced enzyme purified from squash fruits by Nakajima et al.[52] (147,330 units/mg), and an estimate for the tomato enzyme by Bleecker et al.[49] (400,000 units/mg). Whether this large difference in the specific activity is due to differences between the wound-induced enzyme and auxin-induced enzyme is not known. Nakajima et al. found that amino terminal sequencing of the wound-induced enzyme of squash did not yield any amino acid, suggesting blocking of the N terminal amino acid of the enzyme,[52] whereas Arteca[55] reported that N terminal sequence of auxin-induced mung bean ACC synthase has been obtained, although no amino acid sequence was presented.

The causes of different estimates for the relative molecular mass of ACC synthase from tomato fruits is not clear. All groups seem to work with the wound-induced enzymes. As Mehta et al. reported,[50] there may be isoforms with different pI values and possibly with different sizes although this should be rigorously proven, and different workers may have dealt with different isoforms. Another possibility is that as Nakajima et al.[52] observed with the wound-induced enzyme from winter squash mesocarp, tomato ACC synthases extracted by different workers may have been partially hydrolyzed during purification.

e. Immunochemical Difference Between the Wound- and Auxin-Induced Enzymes

A characteristic feature of ethylene production in higher plants is that the rate of production is dramatically increased by various stimuli. Auxin, wounding, and ripening induce production of large amounts of ethylene. In all cases, the limiting step is the endogenous level of ACC synthase. This raises an interesting and important question: are ACC synthases induced by those different stimuli the same enzyme protein or different proteins? During the study of cross-reactivity of the polyclonal antibody obtained for the wound-induced enzyme from winter squash mesocarp, Nakagawa et al.[31] found that the antibody cross-reacted with the wound-induced enzymes from tomato fruits and hypocotyls of winter squash seedlings, but not with the auxin-induced enzymes from hypocotyls of winter squash, tomato, or mung bean seedlings. The results show that the wound-induced enzymes are immunochemically different from the auxin-induced enzymes even in the same species. Moreover, the two types of the enzyme from winter squash were chromatographically separated.[56] It is thus apparent that the wound- and auxin-induced enzymes in winter squash or tomato plants are different in their primary structures and originate from expression of different genes (Figure 3). Although it remains to be proven by comparison of the base sequences of the respective cDNAs whether there are two different genes coding for wound- and auxin-induced ACC synthases, the suggestion opens a new avenue for studies on the regulatory mechanisms of gene expression during environmental and developmental responses of plants. The antibody did not cross-react with the ripening-induced ACC synthase from climacteric apple fruits. As specific antibodies against auxin- and ripening-induced ACC synthase are not available, it is not known whether the ripening-induced enzyme immunochemically belongs to the auxin-induced enzyme or constitutes a third group.

There have been few cases where isozymes of an enzyme such as ACC synthase are regulated differentially by different stimuli. Wounding is an event irregularly imposed on plants, whereas auxin is an endogenous growth regulator and regulates induction of ACC synthase during regular development of plants. The physiological significance of wound-induced ethylene has not been fully understood (see also Chapter 3), but it has been thought to be related to the defense mechanism of wounded plants.[57] Plants may have survived in

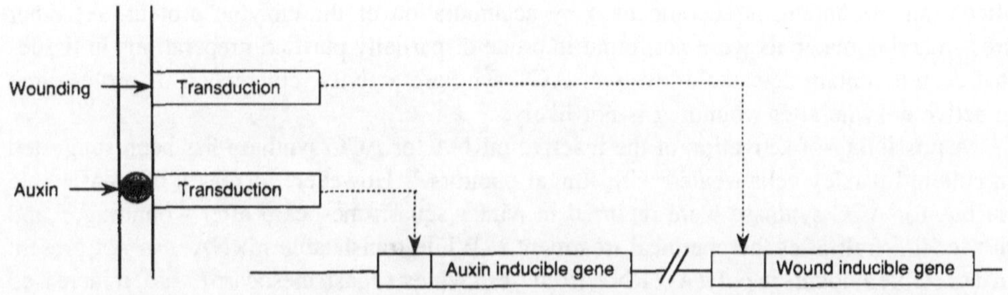

FIGURE 3. A two-gene hypothesis for the wound- and auxin-inducible enzymes.

the course of evolution by preparing a set of genes for emergency expression in addition to the regular developmentally regulated genes for enzymes that play important roles in both regular development and processes which must be activated by irregular stimuli.

f. Induction Mechanisms

Increases in the endogenous activity of ACC synthase may result from transcriptional activation of a gene coding for the enzyme, post-transcriptional modification of preformed mRNA, or post-translational activation of preexisting inactive enzyme. Early observation that increases of ACC synthase activity after auxin treatment or wounding are efficiently inhibited by inhibitors of protein synthesis (cycloheximide, MDMG) and RNA synthesis (α-amanitin, actinomycin D) suggested that the increase of enzyme activity requires RNA and protein synthesis. However, in cultured parsley cells, an increase of ACC synthase after fungal elicitor treatment was suppressed by inhibitors of protein synthesis but not by inhibitors of RNA synthesis, suggesting that RNA synthesis may not be required for the increase.[58] The mechanism of induction of ACC synthase has been examined at different levels.

Direct evidence of *de novo* synthesis of ACC synthase was obtained by the density labeling method for wound-induced tomato,[59] and fungal elicitor-induced parsley cell ACC synthase.[58] The tissues were labeled with ^2H_2O during the induction period, and crude extracts were centrifuged isopycnically in a CsCl gradient and the buoyant density of ACC synthase activity was compared with that of nonlabeled ACC synthase. In both cases, the buoyant densities of the induced enzymes were greater than that of the nonlabeled enzyme, indicating that ACC synthase protein is newly synthesized.

Whether the increase in ACC synthase activity in stimulated tissues is associated with an increase in the enzyme protein was examined after specific antibodies became available. Bleecker et al.,[60] using the partially purified wound-induced tomato enzyme, have prepared monoclonal antibodies. Two lines of the monoclonal antibodies against ACC synthase recognized two different epitopes of the enzyme. ELISA analysis with a monoclonal antibody revealed that an immunodetectable protein increased parallel with the increase of ACC synthase activity over time after wounding. They also showed that when radioactive methionine was supplied to tomato pericarp disks, an immunoprecipitable radioactive protein of 50 kDa changed in a similar pattern to the ELISA detectable protein. Western blot and immunostaining of crude extracts from squash mesocarp tissues at different times after wounding indicated that ACC synthase protein increased in quantity over time.[61] Satoh and Theologis[53] prepared polyclonal antiserum against a partially purified ACC synthase from zucchini fruit treated with a mixture of IAA, benzylaminopurine, LiCl, and AOA. They further purified the antiserum by treating with a protein extract from nontreated zucchini fruit. An immunodetectable polypeptide of 46 kDa was detected only in an extract from the treated fruit. These results demonstrate that the increase of endogenous ACC synthase activity after tissue wounding is accompanied by accumulation of the enzyme protein. As other cross-reacting materials were not found in crude or partially purified preparations in tissues that do not contain detectable levels of ACC synthase activity, conversion of pro-enzyme to active enzyme after wounding is not likely.

A possibility of activation of the inactive mRNA for ACC synthase has been suggested in cultured parsley cells treated with fungal elicitors.[58] However, increases of translatable mRNA for ACC synthase were reported in winter squash mesocarp after wounding,[52] and in zucchini fruit after the chemical treatment.[53] While translatable mRNA was not present in detectable amounts in poly(A)$^+$RNA from fresh winter squash mesocarp tissue, it increased after wounding parallel with an increase in ACC synthase activity.[52,61] Correlation between translatable mRNA for ACC synthase and the enzyme activity was also found during the effect of ethylene on increases of the enzyme activity. When wound-induced increase of ACC synthase was suppressed by ethylene or stimulated by 2,5-norbornadiene, translatable mRNA similarly decreased or increased, respectively.[52]

These results indicate that translatable mRNA for ACC synthase increases by tissue wounding, but do not imply that the increase is due to transcriptional activation of the gene. Whether or not synthesis of mRNA encoding ACC synthase is increased in response to wounding and other stimuli has been examined by isolation of complementary DNA for the enzyme. Imaseki et al.[62] and Satoh and Theologis[53] have isolated cDNA clones for ACC synthase from, respectively, wounded winter mesocarp and zucchini fruit treated with a mixture of indoleacetic acid, benzylaminopurine, LiCl, and AOA. Northern blot analysis of RNAs obtained from wounded or treated tissues with the respective cDNA as a probe showed that the mRNAs were not detected in fresh tissues but increased after wounding or the chemical treatment. The size of the mRNAs was about 1.9 kb in both tissues. These results demonstrate that treatment that increases endogenous activity of ACC synthase eventually activates the gene for the enzyme.

Recently, nucleotide sequences of cDNAs for ACC synthase from wounded tissues of winter squash[62a] and tomato fruit[62b] were reported. A direct alignment comparison of the nucleotide sequences of the two cDNAs indicates that there is only 30% homology; however, there is 64% homology at the maximum match comparison and also 65% homology in the amino acid sequences. Nakajima et al.[62a] showed that *Escherichia coli* harboring a plasmid that contained the cDNA insert expressed active ACC synthase under the inductive condition and accumulated ACC in the culture medium. They also reported that the amino acid sequence of winter squash ACC synthase contained a partial sequence which had a high homology to that proposed for the PLP-binding site of aspartate aminotransferase, another PLP-linked enzyme. Yip et al.[62c] isolated peptides containing the active site of ACC synthase from apple and tomato fruits, to which the aminobutyrate moiety was covalently bound when the enzyme was inactivated by the substrate, and their amino acid sequences were determined. The amino acid sequences of these peptides fully agree with the PLP-binding site of winter squash ACC synthase.

2. Ethylene-Forming Enzyme (EFE)

Conversion to ethylene of ACC exogenously supplied to plant tissues has been observed in a number of plant species that do not produce appreciable amounts of ethylene. This indicates that the enzyme catalyzing conversion of ACC to ethylene is constitutive in most plant tissues, although in some fruits EFE activity increases during fruit development. However, the biochemical entity of the enzyme has not been established, as the activity is completely abolished upon tissue homogenization, and a cell-free system representing the *in vivo* activity has not been obtained to date. Thus, EFE activity can only be determined by supplying ACC to tissues or cells, and several interesting properties of EFE activity have been determined.

a. Properties of EFE

Conversion of ACC to ethylene by plant tissues is absolutely dependent on oxygen, thus the enzyme responsible may be oxidase or hydroxylase. Ethylene is produced from C-3 and C-4 of methionine, that provide C-2 and C-3 of ACC. The carboxyl group of methionine, that is also the carboxyl group of ACC is released as CO_2 during ethylene biosynthesis. Thus, C-2 and C-3, and carboxyl carbon of ACC are fragmented to ethylene and CO_2, respectively, in the EFE reaction. Adams and Yang[63] have suggested that during EFE reaction, cyanoformic acid, which is readily fragmented to HCN and carbon dioxide, is formed. Experimental proof for formation of HCN has been presented by Peiser et al.[64] They found that when [1-^{14}C]ACC was administered to mung bean hypocotyls and *Vicia sativa* epicotyls, radioactivity was incorporated into asparagine and β-cyanoalanine plus γ-glutamyl-β-cyanoalanine, respectively, and there was a stoichiometric relation between ethylene formation and incorporation of radioactivity into these metabolities. As [^{14}C]HCN was incorporated into asparagine in mung bean hypocotyls and into β-cyanoalanine in *Vicia sativa* epicotyls, and as the radioactivity of [^{14}C]asparagine and [^{14}C]β-cyanoalanine isolated from tissues supplied with [1-^{14}C]ACC was found in the C-4 position of these amino acids, which is an expected carbon of HCN incorporation, they concluded that cyanide is formed concurrently with ethylene.

EFE is strongly inhibited by Co^{++},[65] uncouplers of oxidative phosphorylation (2,4-DNP and CCCP),[66] putative free radical scavengers (3,5-diiodo- or dibromo-4-hydroxybenzonitrile, 3,4,5-trichlorophenol, and *n*-propyl gallate),[67] polyamines,[68] and 2-aminobutyric acid.[69] Ethylene production is inhibited in tissues by treatments that chemically or physically change the membrane properties, such as dilute detergents or osmotic shock.[70,71] The step affected by these treatments was later found to be EFE,[67] suggesting that EFE may localize in cellular membranes and its structure is maintained by the highly structured membrane integrity.

The ethylene production rate of tissues is roughly proportional to the endogenous concentration of ACC,[30] with the exception of cell cultures (see Chapter 14). However, the rate is saturable with respect to the internal concentration of ACC. The apparent Km values for ACC are estimated to be 66 μM for pea epicotyls,[72] 16 μM for oat leaves,[15] and 200 μM for cocklebur cotyledons.[69]

The EFE seems to be activated by CO_2. Observed inhibition by light of the EFE activity in green leaves turned out to be due to decreased CO_2 concentration by photosynthesis.[73-75] Light inhibition of EFE is completely overcome by 1% CO_2 in the ambient air, and stimulation of EFE activity by CO_2 is reversible.[76,77] Results demonstrating alternate decrease and increase in EFE activity during alternate light and dark treatments of rice and tobacco leaves are reproduced by alternate removal and supply of CO_2 in the light; and indicate that response of EFE to CO_2 is quick, suggesting activation of EFE by the gas.

Stereospecific interaction of ACC with EFE has been assessed by Hoffman et al.[78] The two methylene groups of ACC are identical but lack rotational symmetry. Thus, if an ethyl group is introduced to each of the methylene hydrogens, the substituted molecule, 1-amino-2-ethylcyclopropane-1-carboxylic acid (AEC), creates two asymmetric carbon atoms at carbon 1 and 2, and yields four stereoisomers of AEC, having absolute configurations of (1*R*,2*R*), (1*R*,2*S*), (1*S*,2*R*), and (1*S*,2*S*). Among the four stereoisomers of AEC fed to apple fruit, cantaloupe fruit, and IAA-treated mung bean hypocotyls, only (1*R*,2*S*)-AEC is converted to 1-butene. The activity to convert (1*R*,2*S*)-AEC to 1-butene parallels the activity of ACC conversion to ethylene under the various conditions, and is inhibited by anaerobiosis and Co^{++}. The AEC isomer inhibits conversion of ACC to ethylene and vice versa. Therefore, the same enzyme catalyzes both reactions, thus ACC is stereospecifically converted to ethylene by EFE.[78] This has been explained by assuming that ACC interacts with EFE at four points: carboxyl, amino, pro(*S*)-methylene group, and the hydrogen *trans* to the amino group of pro(*R*)-methylene group. This stereospecific conversion of 2-substituted

ACC stereoisomers has now been employed as a critical criterion for evaluating the physiological relevance of cell-free enzyme systems to convert ACC to ethylene.

b. Cellular Localization of EFE

Many soluble *in vitro* enzyme systems that produce ethylene from ACC have been reported. These enzyme systems are oxygen-dependent and inhibited by free-radical scavengers, but require Mn^{++}, monophenol, and a reductant for the full activity. Moreover, the Km value for ACC is very high (millimolar or greater). Horse radish peroxidase also catalyzes conversion of ACC to ethylene in the presence of Mn^{++}, monophenol, and a reductant. These characteristics of the cell-free systems are similar to the natural EFE reaction to some extent but these soluble systems do not require the highly structured membrane integrity, which seems to be essential for EFE activity. As described in a preceding section, EFE interacts stereospecifically with ACC. However, all of the soluble cell-free systems convert all the stereoisomers to 1-butene at a similar rate, and thus, do not show stereospecificity toward AEC stereoisomers.[72,79,80]

A different cell-free system which may be associated with the membranes was reported by Bousquet and Thimann.[81] The cell-free system consists of lipoxygenase, linoleic acid, pyridoxal-5′-phosphate, and Mn^{++}. Based on several characteristics of this system including the membrane association of lipoxygenase, correlation between changes of lipoxygenase activity and ethylene production during leaf senescence, and natural presence of linoleic acid in the membrane lipids, Bousquet and Thimann claimed that the system causes *in vivo* ethylene production in senescent leaves. However, Pirrung,[82] using AMC (2-methyl-substituted ACC), and Wang and Yang,[83] using AEC, reported a complete lack of stereospecific conversion of AMC and AEC isomers in the lipoxygenase system, indicating that the system does not represent EFE.

As previously described, a possible association of EFE with the membranes has been postulated. Direct evidence that EFE is localized in vacuoles has been obtained by Guy and Kende,[79,84] who showed that vacuoles isolated from protoplasts of pea and broad bean mesophyll cells contained 85% of cellular ACC, retained 80% of the ethylene-producing activity of protoplasts, and converted AEC isomers stereospecifically as does EFE. They also observed that when isolated protoplasts were supplied with [U-^{14}C]methionine, the specific radioactivity of the C-2 and C-3 of ACC increased much faster than the specific radioactivity of ethylene. Erdmann et al.[85] reported that whereas protoplasts with central vacuoles, isolated from leaves of *Petunia hybrida*, or the vacuoles isolated from the protoplasts, produced ethylene from exogenously supplied ACC, evacuolated protoplasts lost EFE activity. However, when the evacuolated protoplasts were cultured, ACC-dependent ethylene production began after 20 h of incubation. The development of EFE activity was preceded by the regeneration of the central vacuoles in the evacuolated protoplasts. These data indicate that a major portion of cellular EFE is associated with the vacuoles. It appears that ACC is synthesized from methionine via SAM in the cytoplasm, then gradually transported to the vacuole, where ethylene is synthesized. It is still not clear whether EFE is localized in the tonoplast or in the soluble matrix of the vacuole. Complete loss of EFE activity was observed when vacuoles isolated from *Vicia faba* and *Petunia hybrida* leaves were disrupted.[85,86]

Membrane vesicles prepared from kiwi fruit have been reported to convert ACC to ethylene and AEC to 1-butene stereospecifically.[87] However, the membrane vesicles retained only 5% of EFE activity of the fruit tissue. It is unknown what caused this large loss of activity and what types of membranes are responsible for this activity. Isolated tonoplast vesicles from ruptured vacuoles of *Petunia hybrida* did not show EFE activity.[85]

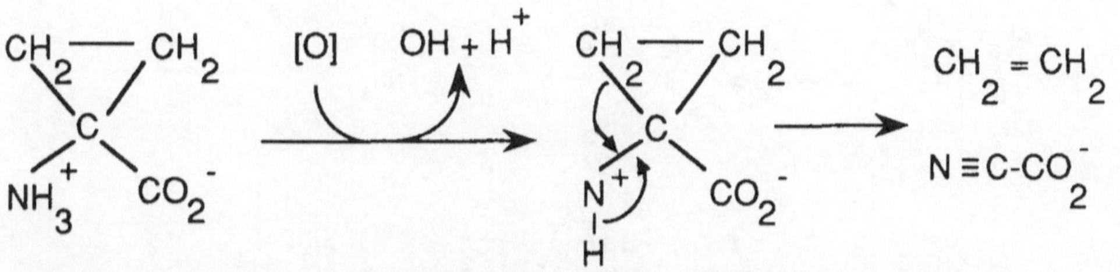

FIGURE 4. A reaction mechanism of EFE involving a niterenium intermediate. (Proposed by Adams, D. O. and Yang, S. F., *Proc. Natl. Acad. Sci. U.S.A.*, 76, 170, 1979.)

c. *Reaction Mechanisms*

EFE reaction produces ethylene from C-2 and C-3, CO_2 from carboxyl carbon and HCN from C-1 and amino nitrogen of ACC. As EFE has not been characterized *in vitro,* the mechanism of the EFE reaction is not known with certainty. However, plausible reaction mechanisms have been presented by analogy to the chemical oxidation mechanism of ACC. Based on the chemical oxidation of 1-phenylcyclopropylamine by lead tetraacetate and by sodium hypochlorite yielding benzonitrile and ethylene,[88] Adams and Yang[63] proposed a mechanism in which the amino group is first oxidized either by dehydrogenation or by hydroxylation followed by dehydration to form a niterenium intermediate, which is then fragmented to ethylene and cyanoformic acid (Figure 4). Cyanoformic acid is readily fragmented to HCN and CO_2. However, involvement of a niterene intermediate cannot be ruled out, because photochemical and acid-catalyzed oxidation of 1-azidocyclopropane-carboxylic acid, representing, respectively, the niterene and niterenium intermediate, yield ethylene, CO_2, and HCN at identical rates.[89]

The mechanism proposed by Adams and Yang depicts that the reaction, i.e., formation of the niterenium intermediate and fragmentation of the intermediate, is a concerted reaction. This predicts that the configuration of two methylene groups of ACC would be retained in ethylene formed. Adlington et al.[90] examined the stereochemistry of the *in vivo* and *in vitro* ethylene formation from ACC using synthesized *cis*-[2,3$^2H_2$]ACC and *trans*-[2,3-$^2H_2$]ACC. Oxidation of *cis*-[$^2H_2$]ACC and *trans*-[$^2H_2$]ACC by hypochlorite yielded *cis*-[$^2H_2$]ethylene and *trans*-[$^2H_2$]ethylene, respectively, indicating the chemical oxidation proceeds in a concerted reaction. However, feeding of *cis*-[$^2H_2$]ACC to apple tissue produced ethylene with a 1:1 mixture of *cis* form and *trans* form, indicating that during EFE reaction, scrambling of hydrogen at carbons 2 and 3 of ACC occurred. This scrambling was also confirmed by Pirrung with apple, cantaloupe, and mung bean tissues.[91] Pirrung also showed that electrochemical oxidation of *cis*-[$^2H_2$]ACC produced ethylene which is stereochemically scrambled.[91] These results, and the fact that oxidation of ACC to ethylene is a two-electron oxidation, suggest that ethylene biosynthesis proceeds in a stepwise mechanism involving an intermediate that allows scrambling of the methylene hydrogens of ACC (Figure 5). The first single-electron oxidation yields a free-radical intermediate corresponding to cyclopropylaminyl radicals, which undergo rapid ring opening yielding the second intermediate that allows free rotation of the methylene groups resulting in loss of stereochemistry. A second single electron oxidation produces the third intermediate which by β-fragmentation gives ethylene and cyanoformic acid. The loss of stereochemistry of methylene groups of ACC can only be explained by formation of a ring-opened intermediate as postulated by Pirrung. Elucidation of the precise mechanism of EFE reaction must await characterization of a bonafide enzyme which converts ACC to ethylene *in vivo*.

3. Metabolism of ACC

In certain microorganisms, ACC is degraded to α-aminobutyric acid and ammonia by ACC ammonia-lyase.[92] In plants, however, ACC is *N*-malonylated to *N*-malonylACC

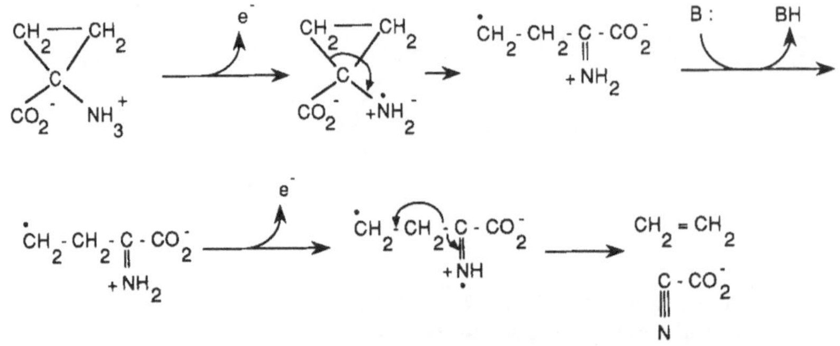

FIGURE 5. Two-step oxidation of ACC by EFE. (Proposed by Pirrung, M. C., *J. Am. Chem. Soc.*, 105, 7207, 1983.)

(MACC).[93-95] Thus, in addition to its conversion to ethylene, ACC is metabolized to MACC. Physiological studies of the relationship among endogenous levels of ACC, MACC, and ethylene production suggest that MACC is not a buffering form of ACC for regulation of ethylene production, but an end-product of ACC metabolism. It should be noted that although conversion of ACC to ethylene has a great physiological significance, the major metabolite of ACC, MACC, may be physiologically inactive. As the rate of formation of MACC in plant tissues affects the endogenous level of ACC in tissues which have a low activity of ACC synthase, ACC malonyltransferase has some relevance to regulation of ethylene biosynthesis.

An enzyme activity which catalyzes formation of MACC from ACC and malonyl-CoA was first detected in a cell-free extract from light-grown mung bean hypocotyls.[96] The malonyltransferase precipitates between 30 and 60% saturation of ammonium sulfate, has pH optimum at 8.0, and shows Km values of 0.17 m*M* for ACC and 0.25 m*M* for malonyl-CoA. It is inhibited by CoA, and higher concentrations of malonyl-CoA. Although many D-amino acids inhibit MACC formation to various degrees, L-amino acids have no such inhibitory activity. D-Phenylalanine, which is an effective inhibitor of MACC formation, was shown to be a competitive inhibitor with respect to ACC with a Ki value of 1.2 m*M*. Malonyl-transferase is activated by monovalent anions; NO_3 being most effective. The enzyme from etiolated mung bean hypocotyls is strongly inhibited by SH reagents such as PCMB and NEM, indicating that SH groups are involved in the active site of the enzyme.[97] Enzyme preparations from green or etiolated mung bean hypocotyls also catalyze *N*-malonylation of various D-amino acids. Because the activity ratios of MACC formation and D-amino acid malonylation did not change during purification of the enzyme and because D-amino acid malonylation was competitively inhibited by ACC and vice versa, it has been proposed that the same enzyme catalyzes both reactons. Stereospecific recognition of ACC by malonyltransferase has supported the above view.[98]

III. CONCLUDING REMARKS

Studies on ethylene biosynthesis have now reached a stage where a molecular approach is possible for one aspect. The basic structure of cDNA and genes coding for ACC synthase should soon be deciphered and it will be possible to assess the structure of the enzyme based on the deduced primary structure. Thus, one of the key steps at present is to isolate cDNA clones for differently induced ACC synthases. The regulatory mechanisms of expression of ACC synthase genes by different stimuli will then be explained based on DNA structure, though it will not be easy.

A number of problems have remained unsolved at the biochemical level. Particularly important is to elucidate the enzymological entity that catalyzes the terminal step in ethylene biosynthesis, viz., that of EFE. The fact that EFE loses its activity upon disruption of cellular membranes may suggest that EFE is an enzyme complex that is active only in a particular conformation delineated by the membrane integrity or the correct topological arrangement with the membranes. Alternatively, EFE reaction may simply require as yet unknown factor(s) that are diluted or inactivated during disruption of cells.

Not all the enzymes involved in the methionine cycle and the ACC pathway have been isolated and characterized. As biosyntheses of ethylene and polyamines, that are sources of MTA in the cycle, do not occur at constant rates all the time, operation of the methionine cycle must be regulated at the substrate level. This regulatory mechanism should also be elucidated in order to understand regulation of ethylene biosynthesis.

REFERENCES

1. **Burg, S. P.,** Physiology of ethylene formation, *Annu. Rev. Plant Physiol.*, 13, 265, 1962.
2. **Abeles, F. B.,** *Ethylene in Plant Biology*, Academic Press, New York, 1973.
3. **Lieberman, M.,** Biosynthesis and action of ethylene, *Annu. Rev. Plant Physiol.*, 30, 533, 1979.
4. **Yang, S. F. and Hoffman, N. E.,** Ethylene biosynthesis and its regulation in higher plants, *Annu. Rev. Plant Physiol.*, 35, 155, 1984.
5. **Adams, D. O. and Yang, S. F.,** Ethylene biosynthesis: identification of 1-aminocyclopropane-1-carboxylic acid as an intermediate in the conversion of methionine to ethylene, *Proc. Natl. Acad. Sci. U.S.A.*, 76, 170, 1979.
6. **Boller, T., Herner, R. C., and Kende, H.,** Assay for and enzymatic formation of an ethylene precursor, 1-aminocyclopropane-1-carboxylic acid, *Planta*, 145, 293, 1979.
7. **Yu, Y. B., Adams, D. O., and Yang, S. F.,** 1-Aminocyclopropanecarboxylate synthase, a key enzyme in ethylene biosynthesis, *Arch. Biochem. Biophys.*, 198, 280, 1979.
8. **Chou, T. W. and Yang, S. F.,** The biogenesis of ethylene in *Penicillium digitatum*, *Arch. Biochem. Biophys.*, 157, 73, 1973.
9. **Burg, S. P. and Clagett, C. O.,** Conversion of methionine to ethylene in vegetative tissues and fruits, *Biochem. Biophys. Res. Commun.*, 27, 125, 1967.
10. **Burg, S. P.,** Ethylene in plant growth, *Proc. Natl. Acad. Sci. U.S.A.*, 70, 591, 1973.
11. **Murr, D. P. and Yang, S. F.,** Inhibition of in vivo conversion of methionine to ethylene by L-canaline and 2,4-dinitrophenol, *Plant Physiol.*, 55, 79, 1975.
12. **Adams, D. O. and Yang, S. F.,** Methionine metabolism in apple tissue. Implication of S-adenosylmethionine as an intermediate in the conversion of methionine to ethylene, *Plant Physiol.*, 60, 892, 1977.
13. **Yung, K. H., Yang, S. F., and Schlenk, F.,** Methionine synthesis from 5-methylthioribose in apple tissue, *Biochem. Biophys. Res. Commun.*, 104, 771, 1982.
14. **Wang, S. Y., Adams, D. O., and Lieberman, M.,** Recycling of 5′-methylthioadenosine ribose carbon into methionine in tomato tissue in relation to ethylene production, *Plant Physiol.*, 70, 117, 1982.
15. **Miyazaki, J. H. and Yang, S. F.,** The methionine salvage pathway in relation to ethylene and polyamine biosynthesis, *Physiol. Plant.*, 69, 366, 1987.
16. **Guranowski, A. B., Chiang, P. K., and Cantoni, G. L.,** 5′-Methylthioadenosine nucleosidase. Purification and characterization of the enzyme from *Lupinus luteus* seeds, *Eur. J. Biochem.*, 114, 293, 1981.
17. **Wasternack, C., Guranowski, A., Glund, K., Tewas, A., and Walther, R.,** Subcellular localization of some purine-metabolizing enzymes in tomato *(Lycopersicon esculentum)* cells grown in suspension, *J. Plant Physiol.*, 120, 19, 1985.
18. **Giovanelli, J., Datko, A. H., Mudd, S. H., and Thompson, G. A.,** In vivo metabolism of 5′-methylthioadenosine in *Lemna*, *Plant Physiol.*, 71, 319, 1983.
19. **Kushad, M. M., Richardson, D. G., and Ferro, A. J.,** Intermediates in the recycling of 5-methylthioribose to methionine in fruits, *Plant Physiol.*, 73, 257, 1983.
20. **Guranowski, A.,** Plant 5-methylthioribose kinase: properties of the partially purified enzyme from yellow lupin *(Lupinus luteus)* seeds, *Plant Physiol.*, 71, 932, 1983.
21. **Trackman, P. C. and Abeles, R. H.,** Methionine synthesis from 5′-S-methylthioadenosine, *J. Biol. Chem.*, 258, 6717, 1983.

22. **Backlund, P. S., Jr., Chang, C. P., and Smith, R. A.,** Identification of 2-keto-4-methylthiobutyrate as an intermediate compound in methionine synthesis from 5′-methylthioadenosine, *J. Biol. Chem.*, 257, 4196, 1982.
23. **Baur, A. H., Yang, S. F., Pratt, H. K., and Biale, J. B.,** Ethylene biosynthesis in fruit tissues, *Plant Physiol.*, 47, 696, 1971.
24. **Burg, S. P. and Thimann, K. V.,** The physiology of ethylene formation in apple tissue, *Proc. Natl. Acad. Sci. U.S.A.*, 45, 335, 1959.
25. **Lürssen, K., Naumann, K., and Schroder, R.,** 1-Aminocyclopropane-1-carboxylic acid—an intermediate of the ethylene biosynthesis in higher plants, *Z. Pflanzenphysiol.*, 92, 285, 1979.
26. **Burroughs, L. F.,** 1-Aminocyclopropane-1-carboxylic acid: a new amino acid in perry pears and cider apples, *Nature*, 179, 360, 1957.
27. **Vahatalo, M. I. and Virtanen, A. I.,** A new cyclic α-aminocarboxylic acid in berries of cowberry, *Acta Chim. Scand.*, 11, 741, 1957.
28. **Burroughs, L. F.,** The free amino acids of certain British fruits, *J. Sci. Food Agric.*, 11, 14, 1960.
29. **Lizada, M. A. C. and Yang, S. F.,** A simple and sensitive assay for 1-aminocyclopropane-1-carboxylic acid, *Anal. Biochem.*, 100, 140, 1979.
30. **Yoshii, H. and Imaseki, H.,** Biosynthesis of auxin-induced ethylene. Effects of IAA, benzyladenine and abscisic acid on endogenous levels of 1-aminocyclopropane-1-carboxylic acid (ACC) and ACC synthase, *Plant Cell Physiol.*, 22, 369, 1981.
31. **Nakagawa, N., Nakajima, N., and Imaseki, H.,** Immunochemical difference of wound-induced 1-aminocyclopropane-1-carboxylate synthase from the auxin-induced enzyme, *Plant Cell Physiol.*, 29, 1255, 1988.
32. **Führer, J.,** Ethylene biosynthesis and cadmium toxicity in leaf tissue of beans *(Phaseolus vulgaris L.)*, *Plant Physiol.*, 70, 162, 1982.
33. **Hoffman, N. E. and Yang, S. F.,** Enhancement of wound-induced ethylene synthesis by ethylene in preclimacteric cantaloupe, *Plant Physiol.*, 69, 317, 1982.
34. **Riov, J. and Yang, S. F.,** Autoinhibition of ethylene production in citrus peel discs. Suppression of 1-aminocyclopropane-1-carboxylic acid synthesis, *Plant Physiol.*, 69, 687, 1982.
35. **Wang, C. Y. and Adams, D. O.,** Chilling-induced ethylene production in cucumbers *(Cucumis sativus L.)*, *Plant Physiol.*, 69, 424, 1982.
36. **Nakajima, N. and Imaseki, H.,** purification and properties of 1-aminocyclopropane-1-carboxylate synthase of mesocarp of *Cucurbita maixma* Duch. fruits, *Plant Cell Physiol.*, 27, 969, 1986.
37. **Satoh, S. and Yang, S. F.,** S-Adenosylmethionine-dependent inactivation and radiolabeling of 1-aminocyclopropane-1-carboxylate synthase isolated from tomato fruits, *Plant Physiol.*, 88, 109, 1988.
38. **Khani-Oskouee, S., Jones, J. P., and Woodard, R. W.,** Stereochemical course of the biosynthesis of 1-aminocyclopropane-1-carboxylic acid. I. Role of the asymmetric sulfonium pole and the -amino acid center, *Biochem. Biophys. Res. Commun.*, 121, 181, 1984.
39. **Icekson, I. and Apelbaum, A.,** Antifungal antibiotics and SIBA inhibit 1-aminocyclopropane-1-carboxylic acid synthase activity, *Biochem. Biophys. Res. Commun.*, 113, 586, 1983.
40. **Miura, G. A. and Chiang, P. K.,** A radioisotope assay for 1-aminocyclopropane-1-carboxylic acid synthase: S-adenosylhomocysteine analogs as inhibitors of the enzyme involved in plant senescence, *Anal. Biochem.*, 147, 217, 1985.
41. **Ramalingam, K., Lee, K. M., Woodard, R. W., Bleecker, A. B., and Kende, H.,** Stereochemical course of the reaction catalyzed by the pyridoxal phosphate-dependent enzyme 1-aminocyclopropane-1-carboxylate synthase, *Proc. Natl. Acad. Sci. U.S.A.*, 82, 7820, 1985.
42. **Khani-Oskouee, S., Ramalingam, K., Kalvin, D., and Woodard, R. W.,** Alternate substrates and inhibitors of 1-aminocyclopropane-1-carboxylic acid synthase, *Bioorganic Chem.*, 15, 92, 1987.
43. **Sakai, S. and Imaseki, H.,** Auxin-induced ethylene production by mung bean hypocotyl segments, *Plant Cell Physiol.*, 12, 349, 1971.
44. **Yoshii, H. and Imaseki, H.,** Regulation of auxin-induced ethylene biosynthesis. Repression of inductive formation of ACC synthase by ethylene, *Plant Cell Physiol.*, 23, 639, 1982.
45. **Imaseki, H., Yoshii, H., and Todaka, I.,** Regulation of auxin-induced ethylene biosynthesis in plants, in *Plant Growth Substances 1982*, Wareing, P. F., Ed., Academic Press, London, 1982, 259.
46. **Kende, H. and Boller, T.,** Wound ethylene and 1-aminocyclopropane-1-carboxylate synthase in ripening tomato fruit, *Planta*, 151, 476, 1981.
47. **Satoh, S. and Esashi, Y.,** Inactivation of 1-aminocyclopropane-1-carboxylic acid synthase of etiolated mung bean hypocotyl segments by its substrate, S-adenosyl-L-methionine, *Plant Cell Physiol.*, 27, 285, 1986.
48. **Satoh, S. and Yang, S. F.,** Specificity of S-adenosyl-L-methionine in the inactivation and the labeling of 1-aminocyclopropane-1-carboxylate synthase isolated from tomato fruits, *Arch. Biochem. Biophys.*, 271, 107, 1989.

49. **Bleecker, A. B., Kenyon, W. H., Sommerville, S. C., and Kende, H.,** Use of monoclonal antibodies in the purification and characterization of 1-aminocyclopropane-1-carboxylate synthase, an enzyme in ethylene biosynthesis, *Proc. Natl. Acad. Sci. U.S.A.*, 83, 7755, 1986.
50. **Nehta, A. M., Jordan, R. L., Anderson, J. D., and Mattoo, A. K.,** Identification of a unique isoform of 1-aminocyclopropane-1-carboxylic acid synthase by monoclonal antibody, *Proc. Natl. Acad. Sci. U.S.A.*, 85, 8810, 1988.
51. **Van Der Straeten, D., Van Wiemeersch, L., Goodman, H. M., and Van Montagu, M.,** Purification and partial characterization of 1-aminocyclopropane-1-carboxylate synthase from tomato pericarp, *Eur. J. Biochem.*, 182, 639, 1989.
52. **Nakajima, N., Nakagawa, N., and Imaseki, H.,** Molecular size of wound-induced 1-aminocyclopropane-1-carboxylate synthase from Cucurbita maxima Duch. and change of translatable mRNA of the enzyme after wounding, *Plant Cell Physiol.*, 29, 989, 1988.
53. **Sato, T. and Theologis, A.,** Cloning the mRNA encoding 1-aminocyclopropane-1-carboxylate synthase, the key enzyme for ethylene biosynthesis in plants, *Proc. Natl. Acad. Sci. U.S.A.*, 86, 6621, 1989.
54. **Tsai, E. S., Arteca, R. N., Bachman, J. M., and Phillips, A. T.,** Purification and characterization of 1-aminocyclopropane-1-carboxylate synthase from etiolated mung bean hypocotyls, *Arch. Biochem. Biophys.*, 264, 632, 1988.
55. **Arteca, R. N.,** Hormonal inducton of ACC synthase, in *Biochemical and Physiological Aspects of Ethylene*, Clijsters, H., De Proft, M., Marcelle, R., and Van Pouck, Eds., Kluwer Academic, Dordrecht, 1989, 119.
56. **Nakagawa, N. and Imaseki, H.,** unpublished data, 1989.
57. **Imaseki, H.,** Hormonal control of wound-induced response, in *Hormonal Regulation of Development III*, Pharis, R. P. and Reid, D. M., Eds., Springer-Verlag, Berlin, 1985, 485.
58. **Chappel, J., Hahlbröck, K., and Boller, T.,** Rapid induction of ethylene biosynthesis in cultured parsley cells by fungal elicitor and its relationship to the induction of phenylalanine ammonia-lyase, *Planta*, 161, 475, 1984.
59. **Acaster, M. A. and Kende, H.,** Properties and partial purification of ACC synthase, *Plant Physiol.*, 72, 139, 1982.
60. **Bleecker, A. B., Robinson, G., and Kende, H.,** Studies on the regulation of 1-aminocyclopropane-1-carboxylate synthase in tomato using monoclonal antibodies, *Planta*, 173, 385, 1988.
61. **Imaseki, H., Nakajima, N., and Todaka, I.,** Biosynthesis of ethylene and its regulation in plants, in *Biomechanisms Regulating Growth and Development*, Steffens, G. L. and Ramsey, T. S., Eds., Kluwer Academic, Dordrecht, 1988, 205.
62. **Imaseki, H., Nakajima, N., and Nakagawa, N.,** Mechanism of ACC synthase induction in plants, presented at 5th FAOB Congress, Seoul, August 13-19, 1989, S-12.
62a. **Nakajima, N., Mori, H., Yamazaki, K., and Imaseki, H.,** Molecular cloning and sequence of a complementary DNA encoding 1-aminocyclopropane-1-carboxylate synthase induced by tissue wounding, *Plant Cell Physiol.*, 31, 1021, 1990.
62b. **Van Der Straeten, D., Van Wiemeersch, L., Goodman, H. M., and Van Montagu, M.,** Cloning and sequence of two different cDNAs encoding 1-aminocyclopropane-1-carboxylate synthase in tomato, *Proc. Natl. Acad. Sci. U.S.A.*, 87, 4859, 1990.
62c. **Yip, W., Dong, J., Kenny, J. W., Thompson, G. A., and Yang, S. F.,** Characterization and sequencing of the active site of 1-aminocyclopropane-1-carboxylate synthase, *Proc. Natl. Acad. Sci. U.S.A.*, 87, 7930, 1990.
63. **Adams, D. O. and Yang, S. F.,** Ethylene the gaseous plant hormone: mechanism and regulation of biosynthesis, *Trends Biochem. Sci.*, 6, 161, 1981.
64. **Peiser, G. D., Wang, T. T., Hoffman, N. E., Yang, S. F., Liu, H. W., and Walsh, C. T.,** Formation of cyanide from carbon 1 of 1-aminocyclopropane-1-carboxylic acid during its conversion to ethylene, *Proc. Natl. Acad. Sci. U.S.A.*, 81, 3059, 1984.
65. **Yu, Y. B. and Yang, S. F.,** Auxin-induced ethylene production and its inhibition by aminoethoxyvinylglycine and cobalt ion, *Plant Physiol.*, 64, 1074, 1979.
66. **Apelbaum, A., Wang, S. Y., Burgoon, A. C., Baker, J. E., and Lieberman, M.,** Inhibition of the conversion of 1-aminocyclopropane-1-carboxylic acid to ethylene by structural analogs, inhibitors of electron transfer, uncouplers of oxidative phosphorylation, and free radical scavengers, *Plant Physiol.*, 67, 74, 1981.
67. **Apelbaum, A., Burgoon, A. C., Anderson, J. D., Solomos, T., and Lieberman, M.,** Some characteristics of the system converting 1-aminocyclopropane-1-carboxylic acid to ethylene, *Plant Physiol.*, 67, 80, 1981.
68. **Hyodo, H. and Tanaka, K.,** Inhibition of 1-aminocyclopropane-1-carboxylic acid synthase by polyamines, their related compounds and metabolites of S-adenosylmethionine, *Plant Cell Physiol.*, 27, 391, 1986.
69. **Satoh, S. and Esashi, Y.,** α-Aminoisobutyric acid, propyl gallate and cobalt ion and the mode of inhibition of ethylene production by cotyledonary segments of cocklebur seeds, *Physiol. Plant.*, 57, 521, 1983.
70. **Odawara, S., Watanabe, A., and Imaseki, H.,** Involvement of cellular membrane in regulation of ethylene production, *Plant Cell Physiol.*, 18, 569, 1977.
71. **Imaseki, H. and Watanabe, A.,** Inhibition of ethylene production by osmotic shock. A further evidence for control of ethylene production by membrane, *Plant Cell Physiol.*, 19, 345, 1978.

72. **McKeon, T. A. and Yang, S. F.**, A comparison of the conversion of 1-amino-2-ethylcyclopropane-1-carboxylic acid stereoisomers to 1-butene by pea epicotyls and by a cell-free system, *Planta*, 160, 84, 1984.
73. **Gepstein, S. and Thimann, K. V.**, The effect of light on the production of ethylene from 1-aminocyclopropane-1-carboxylic acid by leaves, *Planta*, 149, 196, 1980.
74. **De Laat, M. M., Brandenburg, D. C. C., and Van Loon, L. C.**, The modulation of conversion of 1-aminocyclopropane-1-carboxylic acid to ethylene by light, *Planta*, 153, 193, 1981.
75. **Grodzinski, B., Boesel, I., and Horton, R. F.**, The effect of light intensity on the release of ethylene from leaves, *J. Exp. Bot.*, 33, 1185, 1982.
76. **Kao, C. H. and Yang, S. F.**, Light inhibition of the conversion of 1-aminocyclopropane-1-carboxylic acid to ethylene in leaves is mediated through carbon dioxide, *Planta*, 155, 261, 1982.
77. **Philosoph-Hadas, S., Aharoni, N., and Yang, S. F.**, Carbon dioxide enhances the development of the ethylene forming enzyme in tobacco leaf discs, *Plant Physiol.*, 82, 925, 1986.
78. **Hoffman, N. E., Yang, S. F., Ichihara, A., and Sakamura, S.**, Stereospecific conversion of 1-aminocyclopropanecarboxylic acid to ethylene by plant tissues. Conversion of stereoisomers of 1-amino-2-ethylcyclopropanecarboxylic acid to 1-butene, *Plant Physiol.*, 70, 195, 1982.
79. **Guy, M. and Kende, H.**, Conversion of 1-aminocyclopropane-1-carboxylic acid to ethylene by isolated vacuoles of *Pisum sativum* L., *Planta*, 160, 281, 1984.
80. **Venis, M. A.**, Cell-free ethylene-forming systems lack stereochemical fidelity, *Planta*, 162, 85, 1984.
81. **Bousquet, J. F. and Thimann, K. V.**, Lipid peroxidation forms ethylene from 1-aminocyclopropane-1-carboxylic acid and may operate in leaf senescence, *Proc. Natl. Acad. Sci. U.S.A.*, 81, 1724, 1984.
82. **Pirrung, M. C.**, Mechanism of a lipoxygenase model for ethylene biosynthesis, *Biochemistry*, 25, 114, 1986.
83. **Wang, T. T. and Yang, S. F.**, The physiological role of lipoxygenase in ethylene formation from 1-aminocyclopropane-1-carboxylic acid in oat leaves, *Planta*, 170, 190, 1987.
84. **Guy, M. and Kende, H.**, Ethylene formation in *Pisum sativum* and *Vicia faba* protoplasts, *Planta*, 160, 276, 1984.
85. **Erdmann, H., Griesbach, R. J., Lawson, R. H., and Mattoo, A. K.**, 1-Aminocyclopropane-1-carboxylic acid-dependent ethylene production during re-formation of vacuoles in evacuolated protoplasts of *Petunia hybrida*, *Planta*, 179, 196, 1989.
86. **Mayne, R. G. and Kende, H.**, Ethylene biosynthesis in isolated vacuoles of *Vicia faba* L. — requirement for membrane integrity, *Planta*, 167, 159, 1986.
87. **Mitchell, T., Porter, A. J. R., and John, P.**, Authentic activity of the ethylene-forming enzyme observed in membranes obtained from kiwifruit *(Actinidia deliciosa)*, *New Phytol.*, 109, 313, 1988.
88. **Hiyama, T., Koide, H., and Nozaki, H.**, Oxidation of cyclopropylamines and aziridines with lead tetraacetate, *Bull. Chem. Soc. Jpn.*, 48, 2918, 1975.
89. **Pirrung, M. C. and McGeehan, G. M.**, Ethylene biosynthesis. 1. A model for two reactive intermediates, *J. Org. Chem.*, 48, 5143, 1983.
90. **Adlington, R. M., Aplin, R. T., Baldwin, J. E., Rawlings, B. J., and Osborne, D.**, On the biosynthesis of ethylene, *J. Chem. Soc. Chem. Commun.*, 1086, 1982.
91. **Pirrung, M. C.**, Ethylene biosynthesis. 2. Stereochemistry of ripening, stress, and model reactions., *J. Am. Chem. Soc.*, 105, 7207, 1983.
92. **Honma, M. and Shimomura, T.**, Metabolism of 1-aminocyclopropane-1-carboxylic acid, *Agric. Biol. Chem.*, 42, 1825, 1974.
93. **Amrhein, N., Schneebeck, D., Skorupka, H., Tophof, S., and Stockigt, J.**, Identification of a major metabolite of the ethylene precursor 1-aminocyclopropane-1-carboxylic acid in higher plants, *Naturwissenschaften*, 68, 619, 1981.
94. **Hoffman, N. E. and Yang, S. F.**, Identification of 1-(malonylamino)cyclopropane-1-carboxylic acid as a major conjugate of 1-aminocyclopropane-1-carboxylic acid, an ethylene precursor in higher plants, *Biochem. Biophys. Res. Commun.*, 104, 765, 1982.
95. **Amrhein, N., Breuing, F., Eberle, J., Skorupka, H., and Tophoff, S.**, The metabolism of 1-aminocyclopropane-1-carboxylic acid, in *Plant Growth Substances 1982*, Wareing, P. F., Ed., Academic Press, London, 1982, 249.
96. **Kionka, C. and Amrhein, N.**, The enzymatic malonylation of 1-aminocyclopropane-1-carboxylic acid in homogenates of mung-bean hypocotyls, *Planta*, 162, 226, 1984.
97. **Su, L. Y., Liu, Y., and Yang, S. F.**, Relationship between 1-aminocyclopropanecarboxylate malonyltransferase and D-amino acid malonyltransferase, *Phytochemistry*, 24, 1141, 1985.
98. **Liu, Y., Su. L. Y., and Yang, S. F.**, Stereospecificity of 1-aminocyclopropanecarboxylate malonyltransferase toward stereoisomers of 1-amino-2-ethylcyclopropanecarboxylic acid, *Arch. Biochem. Biophys.*, 235, 319, 1984.
99. **Hyodo, H., Tanaka, K., and Watanabe, K.**, Wound-induced ethylene production and 1-aminocyclopropane-1-carboxylic acid synthase in mesocarp tissue of winter squash fruit, *Plant Cell Physiol.*, 24, 963, 1983.

Chapter 2

REGULATION OF ETHYLENE BIOSYNTHESIS

Autar K. Mattoo and W. Bruce White

TABLE OF CONTENTS

I. INTRODUCTION

Ethylene is produced by most higher plants that have been studied and, in trace amounts, influences many aspects of plant growth and development. Ethylene production is highly regulated during plant development, being induced during seed germination, ripening of fruits, abscission of leaves, and senescence of flowers. A variety of extrinsic factors including mechanical wounding, various environmental factors, and certain chemicals also induce ethylene production (see Table 1). The salient features of regulation specific to each of these processes of induction or inhibition of ethylene biosynthesis are dealt with in subsequent chapters. A major stimulation of research in this area was catalyzed by the elucidation of the primary ethylene biosynthetic pathway in higher plants (see the previous chapter).[1] It has thus become possible to identify the target site(s) in ethylene biosynthesis that is (are) regulated during plant development or by an environmental stimulus. In this chapter we attempt to integrate the current knowledge on the physiology and regulation of ethylene biosynthesis, focusing upon factors studied in the context of the primary biosynthetic pathway.

II. REGULATION BY ETHYLENE

Ethylene regulates its own biosynthesis, both positively and negatively. In certain ripening fruits and senescing tissues a massive increase in ethylene production is triggered by exposure to ethylene. This autoenhancement is known as the *autocatalytic effect*. The autocatalytic role of ethylene is complex, and probably an indirect one. It has been suggested that endogenous ripening inhibitors exist in preclimacteric fruits, and these ripening inhibitors are destroyed following ethylene exposure.[2] Upon destruction of these inhibitors, ripening ensues, and, as a consequence, an increase in ethylene production (the climacteric) occurs.

McMurchie et al.[3] have proposed a model of two regulatory systems that control ethylene biosynthesis. System I controls basal ethylene levels and stress-related ethylene production. It is present in both climacteric and nonclimacteric fruits. System II regulates autostimulation (autocatalysis) of ethylene and operates only in climacteric fruit. In this hypothetical model, the System I receptor responds to ethylene by promoting the activity of ethylene-forming enzyme (EFE) that converts 1-aminocyclopropane-1-carboxylic acid (ACC) to ethylene, and ACC malonylation. The System II receptor, on the other hand, responds temporally to ethylene and induces ACC synthase activity. Several reports are consistent with such a regulatory scheme.

Ethylene also increases the capacity of excised tissues of preclimacteric cantaloupe fruit,[5] citrus leaves,[6] tobacco leaves,[7] intact preclimacteric tomato and cantaloupe fruits,[8] as well as carnation petals detached from freshly cut flowers,[9] to convert exogenous ACC to ethylene. Autocatalysis involves the stimulation of EFE, as well as the synthesis of ACC. In studies on the effect of exogenous ethylene on the biosynthesis of ethylene in intact preclimacteric cantaloupe and tomato fruit, Liu et al.[8] observed that the increase in EFE activity preceded that of ACC synthase activity. During the initial 18 h of ethylene treatment there was no increase in ethylene production or ACC content, but the tissue's ability to convert exogenous ACC to ethylene increased markedly. Whether or not the increase in EFE precedes that of ACC synthase during the natural ripening of fruits is unclear. Chalutz et al.[7] suggested that the increased conversion of ACC to ethylene in ethylene-treated tissue may be mediated through a change in the membrane milieu of EFE (a membrane-associated enzyme system). In carnation petals detached from freshly cut flowers, Manning[9] examined the stimulation of EFE activity by ethylene in the presence of aminoethoxyvinylglycine (AVG), which inhibits ACC synthesis. He found that EFE activity was stimulated 27-fold, approaching the rate for control petals given ACC alone.

TABLE 1
Effectors and Their Probable Control Site(s) in Ethylene Biosynthesis Pathway

Effectors	Reaction affected: SAM → ACC	Reaction affected: ACC → MACC	Reaction affected: ACC → C_2H_4	Net result on C_2H_4	Ref.
Endogenous					
Metabolites					
Carbohydrates					
Galactose	+ ve	0	+ ve	+ ve	83-89,91,95
Glucose	+ ve	n.t.	+ ve	+ ve	90
Mannitol	n.t.	n.t.	n.t.	+ ve	87
Mannose	n.t.	n.t.	n.t.	− ve	97
Orthophosphate	+ ve	n.t.	− ve	− ve	52-54
Polyamines	− ve	n.t.	− ve	− ve	32,44,74-77
Hormonal					
Auxin (IAA)	+ ve	n.t.	0	+ ve (vegetative) − ve (fruits)	58,115-119,121-123 120
IAA-L-alanine	n.t.	n.t.	n.t.	+ ve	86,130
ABA (low conc.)	− ve	n.t.	+ ve ?	− ve	123,137,138,142,143
ABA (high conc.)	+ ve	n.t.	+ ve	+ ve	144
Brassinosteroids + IAA	+ ve	n.t.	n.t.	+ ve	155
Cytokinin—CK (low conc)	+ ve	n.t.	+ ve	+ ve	24,113,140,147-150
CK (high conc.)	n.t.	n.t.	− ve	− ve	132,145,146
CK + ABA	+ ve	n.t.	n.t.	+ ve	142,151,152
CK + IAA	+ ve	n.t.	n.t.	+ ve	24,123,149
Ethylene	0	+ ve	+ ve	+ ve	5-9,11,13,14,25
Ethylene	− ve	0	0	− ve	18-24
Methyl jasmonate	n.t.	n.t.	n.t.	+ ve	154
Environmental					
Light	− ve	+ ve	− ve	− ve	99-104,114
CO_2	n.t.	n.t.	+ ve	+ ve	24,103,105,109,113
Light + CO_2	0	n.t.	+ ve	+ ve	103,105-108
Dehydration	n.t.	n.t.	n.t.	+ ve	26,27

Note: + ve, stimulation; − ve, inhibition; 0, no effect; n.t., not tested.

Protein synthesis is necessary for the enhancement of EFE activity, since ethylene formation is substantially inhibited by cycloheximide, an inhibitor of nucleocytoplasmic protein synthesis. Woodson and Lawton[10] have examined mRNA populations following exposure of carnation flowers to ethylene. A number of changes were noted as determined by *in vitro* translation of petal polyadenylated RNA. Their results indicate that changes in petal physiology resulting from exposure to ethylene may be the result of rapid changes in gene expression.

Ethylene can also regulate its own synthesis by modifying the plant's ability to conjugate ACC to malonyl-ACC (MACC) (malonylation). Ethylene treatment of preclimacteric tomato pericarp tissue greatly promotes the capacity of the tissue to form MACC from exogenous ACC. In addition, the level of malonyl transferase activity in ethylene-treated fruit was increased approximately sixfold as compared to fruit tissue incubated in air. These observations complement other findings that showed a progressive increase in the level of MACC in tomatoes as the ethylene production rate increased during maturation and ripening.[12] Ethylene has also been shown to promote ACC malonyltransferase activity in flavedo tissue of grapefruit[13] and tobacco leaf discs.[14]

For some time it was assumed that MACC was functionally inactive, i.e., it did not serve as a storage pool for ACC and hence for ethylene production. However, hydrolysis of exogenous MACC to ACC leading subsequently to ethylene production has been demonstrated in watercress stem sections,[15] tobacco leaf discs,[15] and petal tissue.[16] The MACC-induced ethylene production was not inhibited by AVG (an inhibitor of ACC synthase), but was inhibited by $CoCl_2$ (an inhibitor of EFE). This eliminated the possibility that MACC stimulates *de novo* ACC synthesis, and suggested that the ACC formed was derived from the MACC applied. Direct confirmation of this was obtained from the observation that radioactive ACC was formed from labeled MACC applied to the tissues. The physiological importance of the MACC hydrolase reaction is not clear because the endogenous MACC levels are normally far below the apparent K_m for MACC hydrolase. Preliminary characterization of the MACC hydrolase indicated that it is a metal-requiring enzyme, whose activity, like other aminopeptidases,[17] is enhanced by Mn^{2+} ions. It is not known if the MACC-hydrolase is capable of hydrolyzing other *N*-malonyl-D-amino acids.

Ethylene also inhibits its own biosynthesis, a phenomenon known as *autoinhibition.* Autoinhibition has been recognized in a number of fruit and vegetative tissues.[1,18-20] Vendrell and McGlasson[19] demonstrated quite some time ago that ethylene treatment significantly inhibited wound ethylene production in banana pulp slices. Later, it was shown that storage of avocado fruit in an ethylene atmosphere at 6°C irreversibly suppressed ethylene production by these fruit after removal from cold storage.[20] When unripe fig fruit were enclosed in a jar, ethylene emission ceased when the concentration of ethylene in the jar reached 0.6 μl/l.[21] Upon removal of ethylene by aeration, ethylene production in the figs was renewed. Treatment with exogenous ethylene also caused an immediate and complete cessation of ethylene production in fig fruit. Autoinhibition of ethylene production also occurs in vegetative tissues.[18] Exogenous ethylene or propylene (an ethylene analog) inhibited wound ethylene synthesis in excised pea segments during application, but ethylene synthesis was rapidly restored following removal of ethylene. Similarly, propylene suppresses ethylene evolution in intact green bananas.[3]

The target site of ethylene-mediated autoinhibition was demonstrated only after the primary ethylene biosynthesis pathway was elucidated. Wound ethylene production by flavedo tissue of citrus fruits was greatly inhibited by exogenous ethylene.[22] In the absence of exogenous ethylene, ethylene production rate by the flavedo discs increased more than twentyfold, parallel with an increase in ACC content. However, in ethylene-treated discs there was little increase in ACC content or ethylene production. Upon application of ACC, ethylene production in ethylene-pretreated discs was increased to the level produced in the untreated tissue. These data indicate that autoinhibition of ethylene synthesis results largely from a limitation in the availability of ACC but not by impairing the tissues' ability to convert ACC to ethylene.[22] How ethylene exerts this indirect effect on ACC synthase activity is not fully clear.

Yoshii and Imaseki[23] have suggested that autoinhibition by ethylene occurs partly via feedback repression of the synthesis of ACC synthase. In mung bean hypocotyls treated with IAA, extractable ACC synthase activity increased fourfold when AVG was administered to the tissue. This stimulatory effect of AVG can be partially nullified by exogenous ACC or ethylene. In this system, it was not possible to examine whether or not ethylene also exerts additional autoinhibition by inhibiting ACC synthase activity. Treatment of tobacco leaf discs with ethylene (25 μl/l) in the presence of IAA inhibited the conversion of labeled methionine to labeled ethylene by about 90%.[24] The system eventually produced ethylene at a higher rate in response to IAA treatment. Apparently, the concentration of ethylene required for autoinhibition in this system was high, or autoinhibition occurred before the system was induced to produce ethylene at a high rate. The conjugation of ACC to MACC by malonyl transferase is markedly increased when fruits[13] or leaves[25] are exposed to ethylene, indicating an additional mechanism by which ethylene can regulate its own production.

It should be noted that ethylene does not cause the above-mentioned effects in cell-free systems. Therefore, ethylene appears to act indirectly, and the definition of autoinhibition or autocatalysis should be taken in a physiological rather than a biochemical sense.

III. MEMBRANE ASSOCIATION AND INVOLVEMENT OF MEMBRANE FUNCTION

An early indication that membrane integrity may be an important factor regulating ethylene biosynthesis was the finding that in apple tissue ethylene production was sensitive to solute concentration of the medium. Ethylene production continued during dehydration, but swelling inhibited ethylene production.[26,27] Further support for membrane involvement was provided by experiments with excised apple tissue,[28,29] apple protoplasts,[30] and mung bean hypocotyls[31] where ethylene production was inhibited by lipophilic membrane probes. Ethylene production by apple fruit protoplasts was greatly inhibited by the nonionic detergent Triton X-100, and osmotic shock that caused lysis of the protoplasts.[30] Similarly, ethylene production in IAA-treated mung bean hypocotyl segments was inhibited by lecithin, Tween 20, Triton X-100, and sodium dodecyl sulfate.[31] In the latter study, the inhibition by Triton X-100 and sodium dodecyl sulfate was partially reversed by washing these materials off the tissue segments. It was concluded that inhibition resulted from structural changes of cell membranes caused by lipophilic substances, and that ethylene production was regulated by membrane function. Discontinuities in Arrhenius plots have been commonly found in the activities of membrane enzymes[33,34] and have been shown to parallel a physical change in the membrane lipids from liquid-crystalline to gel state. In apple and tomato fruit slices, discontinuity in Arrhenius plots of incubation temperature against ethylene production has been used as suggestive evidence for membrane involvement.[28,32] These observations suggest that ethylene formation involves highly structured and regulated enzymes, and requires membrane integrity.[35]

Growth retardants, such as the triazole (LAB 150 978) and norbornenodiazetine (tetcyclacis) can alter the membrane composition of cell suspensions[36] and also inhibit the conversion of ACC to ethylene.[37] Thus, the inhibitory effect of these chemicals on ethylene production may be mediated via an effect on cellular membranes; although a direct effect on EFE is also possible.

It is now apparent that EFE is membrane associated. The conversion of ACC to ethylene by apple protoplasts is severely inhibited by osmotic shock or Triton X-100.[38] Also, the rate of conversion of ACC to ethylene is inhibited by temperatures above 29°C and below 12°C.[38] Inhibitor studies suggested the involvement of oxygen or free radicals in the conversion of ACC to ethylene.[39] In addition, the potent inhibition of this reaction by uncouplers of oxidative phosphorylation, viz. dinitrophenol (DNP) and carbonyl cyanide *m*-chlorophenylhydrazone (CCCP), suggested the necessity for a functional, energy-requiring membrane system for the conversion of ACC to ethylene. The effects of calcium and magnesium ions[30,40,41] and of silver ions[42] on ethylene production may be due to stabilization of membrane-associated EFE.[43-45] Calcium ions can inhibit lipid peroxidation and protein leakage in apple slices, and this interaction with cellular membranes may stabilize EFE activity. Infiltration of calcium into fruits[46-48] reduces respiration and ethylene production, and results in the maintenance of fruit firmness and in lengthening shelf-life. Calcium ions may also facilitate translocation of ACC to the tonoplast where EFE may be localized, thus promoting ACC conversion to ethylene.[49,50] Long-term treatment of mature-green tomato fruit with silver ions resulted in a stimulation of both the conversion of ACC to ethylene and the synthesis of ACC.[51] Silver ions are also known to increase accumulation of ACC in leaf tissue.[25]

Orthophosphate ions act as reversible inhibitors of ethylene biosynthesis in higher plants.[52] This effect of phosphate was subsequently shown to be directed at EFE when internal phosphate concentrations reached 6.5 to 10 m*M*.[53] Conversely, phosphate ions enhance the stability of ACC synthase activity *in vitro*.[53] Indeed, the use of phosphate buffers stabilized the activity of IAA-induced ACC synthase from mung bean hypocotyl tissue, aiding in the purification of the enzyme.[54]

Much indirect evidence has been generated supporting a membrane association of the EFE. The vacuole fraction isolated from pea (*Pisum sativum* L.) leaf protoplasts accounted for 80% of the ethylene production by the protoplasts.[55] Over 80% of the ACC was found to be localized in the vacuole. Furthermore, it was shown that ethylene synthesis by isolated vacuoles saturates at 1 m*M* ACC, as does ACC-dependent ethylene synthesis in pea leaf discs. The apparent K_m for ACC-dependent ethylene synthesis by isolated pea[55] and petunia[56] vacuoles was found to be 61 μM and that by pea epicotyls had a similar K_m of 66 μM. In the vacuoles, ethylene formation from endogenous ACC was inhibited by Co^{2+}, *n*-propyl gallate, and N_2 as is the case with intact plant tissue.[39,57,58] The EFE in pea vacuoles exhibited stereospecificity inasmuch as it catalyzed the conversion of (+)-allocoronamic acid to 1-butene but not that of (+)-coronamic acid, as had been previously demonstrated in intact leaf tissue.[55] Thus, the EFE of isolated vacuoles resembles the enzyme in intact tissue. It was suggested that the EFE activity is associated with the inner face of the vacuolar membrane, although experimental proof of this hypothesis is lacking. The EFE does not appear to be readily accessible to externally applied ACC. In vacuoles that were isolated from AVG-treated protoplasts and required exogenous ACC for the formation of ethylene, a lag time of 1 to 2 hours was observed between the addition of ACC and the accumulation of ethylene.[55,56] Furthermore, a critical concentration of substrate had to be reached inside the vacuole before ethylene biosynthesis commenced. Ethylene synthesis ceased when the vacuoles were lysed. Isolated vacuoles from bean *(Vicia faba* L.)[59] and petunia *(Petunia hybrida)*[56] leaves behave similarly to those from pea[55] with respect to EFE. Ethylene synthesis in the vacuolar fraction was inhibited by CCCP and DNP, and by ionophores valinomycin, nigericin, and A23187. These observations support the hypothesis that ethylene biosynthesis in vacuolar preparations is, in some way, dependent upon membrane integrity. In cells induced to accumulate ACC, ethylene synthesis may occur primarily in the vacuole. In those cells where ACC turnover is rapid and which thus have a small ACC pool, such as in apples[57] or tomato fruits,[60] ethylene formation may also take place at the plasma membrane.[27,61] In any case, the intact vacuole appears to be the simplest organelle that possesses properties characteristic of EFE.

EFE activity has been compared in leaf discs, isolated cells, protoplasts, and vacuoles derived from leaves of pea and bean.[62] It was found that 95% of the activity observed with leaf tissue was lost when cells were isolated from the tissue, suggesting that total EFE activity requires tissue integrity in addition to a previously noted requirement for cell membrane integrity. These findings support the view that in addition to the vacuoles, ethylene synthesis from ACC may also occur in other subcellular membranes.[27,55,61]

Ethylene biosynthesis is also regulated by various environmental stimuli. These stimuli may generate signals at the cellular membrane that in turn regulate the synthesis or activity of ACC synthase.[63] Wound-induced ACC synthase activity in tomato fruit slices is inhibited or stimulated by agents that affect membrane potential (e.g., osmotic shock, uncouplers),[1,32] or are highly charged (e.g., spermine).[32] Calcium ions that largely accumulate in the cell

TABLE 2
Effect of Protease Inhibitors on Cellulysin-Mediated Ethylene Production in Tobacco Leaf Discs (A) and on Wound-Induced ACC Synthase in Tomato Pericarp Slices (B)

Addition to the incubation medium	Ethylene production or ACC synthase (% of control)
A. Control,[a] cellulysin (2%)	100.0
Plus phenylmethanesulfonic acid (1 m*M*)	2.7
Plus soybean trypsin inhibitor (1 mg)	27.2
Plus pepstatin A (0.1 mg)	75.5
B. Control[b]	100.0
Plus phenylmethanesulfonic acid (1 m*M*)	33.2
Plus soybean trypsin inhibitor (5 mg/ml)	50.5

[a] Tobacco leaf discs (6 per 25 ml Erlenmeyer flask) were floated on 1 ml of incubation medium containing 600 m*M* sorbitol, 10 m*M* $CaCl_2$, 10 m*M* MES, pH 6.0, 50 μg/ml streptomycin sulfate, 50 U/ml penicillin G and 2% cellulysin. Ethylene accumulation in the gaseous phase between 0 and 2 h was quantified by gas chromatography. 100% control represents 3.1 nmol ethylene formed per h per g fresh wt of tobacco tissue (from Reference 67).

[b] Over-ripe tomato fruit pericarp slices were incubated with 600 m*M* sorbitol and 10 m*M* MES, pH 6.0, with and without the indicated protease inhibitors. Homogenates were prepared, gel filtered, and ACC synthase activity determined.[65] ACC synthase activity in control was 4.97 nmol; ACC formed per h per mg protein (from Reference 32).

wall/cell membrane[64] may also increase ACC production. Since ACC synthase is not directly affected by calcium ions *in vitro*,[43,65] the increase in ACC level must be regulated indirectly or by an increase in the concentration of active ACC synthase.[43,66] The induction of ACC synthase activity in tomato fruit upon wounding is inhibited by certain protease inhibitors, viz. phenylmethanesulfonyl fluoride (PMSF) and soybean trypsin inhibitor, while other protease inhibitors, trasylol and acetylaminobenzamide, stimulate the wound-induced rise in ACC synthase activity.[32] Soybean trypsin inhibitor and PMSF also inhibit cellulysin-mediated increases in ACC level and ethylene production in tobacco leaf discs (Table 2).[67] The ethylene-inducing factor from cellulysin has been purified and found to consist of a family of proteins with endoxylanase activity that is resistant to protease attack, except in the presence of urea.[68] The concentrations of PMSF and soybean trypsin inhibitor that inhibited the cellulysin-mediated effect did not inhibit ACC synthase activity *in vitro*, or the chemical oxidation of ACC to ethylene. These results suggest that some proteolytic activity *in vivo* is associated with the ethylene induction processes, possibly with the activation and/or inactivation of ACC synthase, or with the induction signal itself.

There is evidence that suggests some degree of association of ACC synthase with membranes. A small but significant amount of ACC synthase activity in particulate fractions from tomato fruit extracts could be modulated by effectors such as tetrabutylammonium bromide that is known to abolish electrical gradients across membranes.[32] Other characteristics of ACC synthase favor its association with cellular membranes. These include its hydrophobic nature,[65,69] the presence of a highly charged form with a pI of 9[32,70] that can interact with negative charges on membrane phospholipids, and the ease of its isolation from some tissues using membrane-dissolving detergents.[71] The availability of poly- and monoclonal antibodies (see previous chapter) specific for ACC synthase will be useful in future studies aimed at localizing this enzyme within the cell.

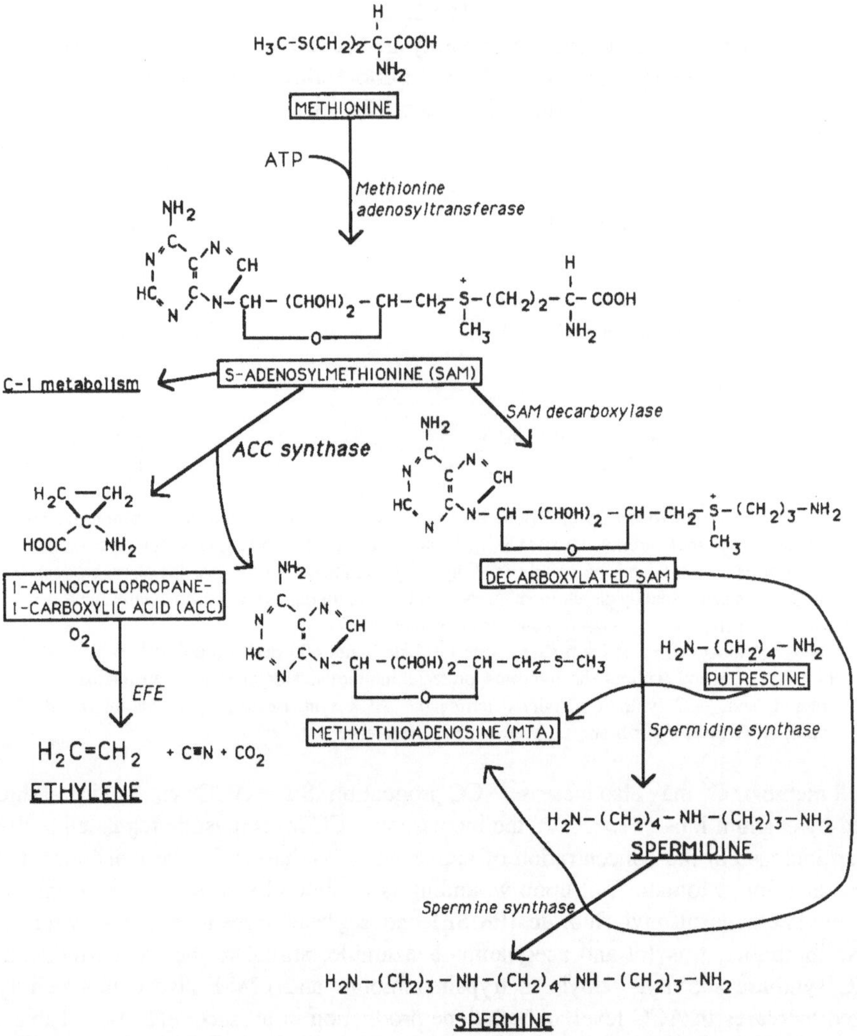

FIGURE 1. A schematic representation of the linkage between the biosynthetic pathways for ethylene and polyamines in higher plants.

IV. LINKAGE WITH POLYAMINE BIOSYNTHESIS

The demonstration (see previous chapter) that *S*-adenosylmethionine (SAM) is an intermediate in ethylene biosynthesis revealed a possible linkage of several metabolic pathways (Figure 1). In addition to its formation during the conversion of SAM to ACC, methylthioadenosine (MTA) is also generated by the aminopropyl transferase that utilizes decarboxylated SAM as a substrate in the biosynthesis of the polyamines spermidine and spermine. Both ethylene and these polyamines share SAM as a common precursor and MTA as a common byproduct. The fact that polyamines exhibit marked antisenescence properties[72,73] and directly antagonize many ethylene-mediated responses prompted an examination of the ability of polyamines to control SAM levels and/or influence ethylene biosynthesis. In senescing *Tradescantia* petals and in IAA-treated mung bean hypocotyls, the addition of exogenous polyamines reduced ethylene production.[74] Similar results were independently reported for senescing apple fruit, leaf tissues of tobacco and pinto bean,[75] and senescing orange peels.[76] Inhibition of ethylene was also observed in excised oat leaves.[77] In fruit tissues the inhibition of ethylene biosynthesis by polyamines was related to polyvalent cationic progression.[75] Spermine, the strongest base, was the most effective inhibitor of ethylene biosynthesis, followed by spermidine, and then putrescine.

Both the formation of ACC[76,77] and its subsequent conversion to ethylene[74,75] are inhibited by polyamines. In aged orange peel slices, when the conversion of SAM to ACC was blocked by polyamines or AVG, an increased incorporation of SAM into spermidine was observed.[76] Thus, the antisenescence properties of exogenous polyamines may result from both an inhibition of ethylene biosynthesis and promotion of endogenous polyamine biosynthesis. Support for this view was provided by Roberts et al.,[78] who used specific inhibitors of both ethylene and polyamine biosynthesis. They found that in senescing carnation flowers, the inhibition of ACC synthase by α-aminooxyacetic acid decreased ethylene production and increased polyamine levels, while the onset of senescence was delayed. Conversely, inhibition of polyamine biosynthesis by α-difluoromethylarginine (DFMA) or methylglyoxalbis-[guanylhydrazone] (MGBG) led to increased rates of ethylene production and hastened senescence. However, in this case, the inhibition of ethylene biosynthesis by α-aminoisobutyric acid, which blocks the conversion of ACC to ethylene, had no significant effect on polyamine levels in the tissue, suggesting that ACC accumulation itself did not influence polyamine biosynthesis.

Calcium can partially antagonize the polyamine-induced inhibition of ethylene synthesis in senescing leaf tissues.[75,77] This suggests that a part of the polyamine effect is mediated through ionic interactions with membranes possibly by modulating the activity of the EFE. A strong correlation between the change in the membrane microviscosity as affected by spermine and temperature and ethylene production by apple slices, enabled Ben-Arie et al.[44] to suggest that inhibition of ethylene biosynthesis by spermine may be mediated via its interaction with cellular membranes. The mechanism of this regulation appears to be indirect. Polyamines have been shown also to restrict ACC uptake into tomato pericarp vacuoles.[50] Thus, polyamines may limit ethylene production by limiting the availability of the substrate ACC to EFE located in the tonoplast.

Another aspect of polyamine-mediated regulation of ethylene biosynthesis is the effect of spermine on the development of ACC synthase activity. This phenomenon was apparent when it was shown that increase in ACC synthase activity during wounding of tomato fruit tissue was inhibited by spermine while the activity of the enzyme in cell-free extracts remained unaffected.[32] These studies have indicated possible sites of regulation in the metabolism of SAM that serve to link the biosynthesis of ethylene with that of polyamines. Interestingly, ethylene inhibits SAM decarboxylase activity in etiolated pea seedlings.[79] The mechanism by which ethylene inhibits the formation of decarboxylated SAM is unknown. Further work is required to fully elucidate the mechanisms that control the balance between ethylene and polyamine levels in plants. Kushad et al.[80] found little ethylene production during maximum polyamine synthesis and vice versa in ripening avocado fruits, indicating that although the two pathways are closely linked, they perhaps do not actively compete for their mutual substrates at any given stage of fruit development *in situ*.

The relation between endogenous polyamine levels and ethylene production rates has been deciphered by studies on tomatoes that ripen more slowly than other horticultural varieties and have long shelf lives. One is a fruit of landrace *Alcobaca*, containing the recessive allele *alc*,[81] and the other a slow ripening cultivar (*L. esculentum* Mill. cv. Liberty). The 'Liberty' cultivar produces 16% and 38% of the ethylene produced by 'Pik Red' and 'Rutgers' cultivars, respectively. The level of polyamines, particularly that of putrescine, was 6-fold higher in the 'Liberty' cultivar. The putrescine level of *alc* fruit increased during ripening and was as high as that in the immature green fruit.[81] These data, together with others described above, suggest a linkage between the polyamine levels and the plants' ability to produce ethylene, and thus, may account for ripening and storability properties of different tomato cultivars.

V. REGULATION BY CARBOHYDRATES

Several mono-, di-, and trisaccharides have been shown to stimulate ethylene production in a number of plant tissues.[83] For example, different carbohydrates stimulate ethylene production in petiole and lamina of excised mustard cotyledons,[84] mung bean hypocotyl segments,[85] leaf discs of tobacco, tomato, bean, and cotton, as well as in attached cotton leaves,[86] citrus leaf discs,[87] and mature-green tomato fruit.[88] In mung bean hypocotyls,[85] of the many carbohydrates tested, galactose alone elicited high rates of ethylene evolution. Galactose and IAA were found to act synergistically with respect to ethylene induction. In most cases, carbohydrate-mediated induction of ethylene production takes a day or two to develop, and thus is considerably slower than induction by auxin alone. In tobacco leaf discs, galactose produced a larger, transient stimulation of ethylene production than did other sugars.[83,89] Although the stimulation of ethylene production by galactose has been ascribed to toxicity, it is not the case, nor does the effect appear to be mediated by osmotic effects. In tobacco leaf discs, sucrose or glucose (5 m*M)* induced an increase in ethylene production lasting several days.[90] Neither L-glucose[90] nor 3-0-methylglucose[84] (both non-catabolizable carbohydrates) elicited this effect. Therefore, the glucose and sucrose effects may be mediated via their metabolism, rather than toxicity. Although the precise mechanisms by which these sugars stimulate ethylene production are unknown, prolonged incubation (5 to 6 d) of tobacco leaf discs with sucrose or glucose results in increased ethylene production via an enhancement of both ACC formation and its conversion to ethylene.[90] This was also observed to be the case for the stimulatory effect of mannitol on ethylene production in citrus leaf discs.[87] The pattern of galactose-stimulated ethylene production in tobacco leaf discs was accompanied by increased formation of ACC, which accumulated without being metabolized to ethylene or to MACC.[91]

The promotion of ethylene evolution by galactose may have physiological significance during fruit ripening. A substantial loss of cell wall galactosyl residues occurs during ripening of tomato fruit[92] and other fruits.[93,94] Infiltration of galactose into mature-green tomatoes not only stimulates ethylene production but also promotes their ripening.[88] These results suggest that free galactose, solubilized from tomato cell wall polysaccharides during ripening, may possibly be involved in fruit ripening through the stimulation of ethylene biosynthesis. Concurrently with the stimulation of ethylene production by galactose, ACC content and ACC synthase activity also increased.[95] However, galactose did not stimulate ACC synthase activity *in vitro,* suggesting an indirect effect on enzyme production. Galactose infiltrated into pericarp tissue was rapidly metabolized, decreasing to endogenous levels within 50 h. Of the variety of sugars tested, galacturonic acid, dulcitol, and mannose stimulated transient increases in ethylene production similar to that of galactose. The relationship between the release of soluble uronides, ethylene production, and fruit softening was tested in the elegant studies of Giovannoni et al.[96] A chimeric polygalacturonase gene was expressed in *rin* (ripening inhibitor) tomato fruit, and the activity of polygalacturonase, soluble uronides, ethylene production, and softening of the transgenic *rin* tomato were analyzed. While polygalacturonase activity was correlated with cell wall polyuronide degradation, it was not sufficient for increasing ethylene production or for fruit softening.

Mannose has been shown to inhibit ethylene evolution in ripening pear fruit.[97] This effect is associated with a decrease in inorganic phosphate and the accumulation of presumably toxic mannose-6-phosphate. In cut flowers as well, various sugars exhibit an inhibitory effect on ethylene production, suggesting a different mechanism of ethylene regulation in these systems than in those where carbohydrates stimulate ethylene production. In microsomal membrane preparations from carnation petals, the conversion of ACC to ethylene is inhibited by sucrose, glucose, fructose, and several other sugars.[98] Presumably this inhibition does not appear to be related to osmotic effects. It was suggested that the inhibition of EFE activity in this system may possibly be the result of a direct interaction with the enzyme, or, alternatively, with free radicals that might also participate in the enzymatic conversion of ACC to ethylene. However, in the absence of validation of this system as a reflection of the *in vivo* system (due to lack of stereospecificity of the *in vitro* system), the physiological relevance of these observations is in doubt.

VI. REGULATION BY LIGHT AND CARBON DIOXIDE

Both light and CO_2 can promote or inhibit ethylene production, depending on the tissue studied. Exposure of etiolated pea seedlings to a single dose of red light resulted in a decrease in ethylene production.[99] Far-red irradiation following the red light treatment reversed the red effect, suggesting that the ethylene production is under phytochrome control. Subsequently, the red light effect was also observed in the hypocotyl hook of bean seedlings,[100] where an increase in CO_2 evolution was observed concurrent with a decrease in ethylene production.

The effect of light on ethylene production has been studied most extensively in various green-leaf tissues where light strongly inhibits ethylene production.[101-104] Gepstein and Thimann[101] first observed that white light inhibits the conversion of ACC to ethylene. Their experiments were conducted with segments of oat leaves (a monocotyledon) and discs isolated from green leaves of tobacco (a dicotyledon), demonstrating the general nature of this phenomenon. Since leaves were grown and preincubated in the light, yet could produce from 2 to 9 times as much ethylene in the dark as in the light, it followed that the light effect is fully reversible. The inhibition by light was approximately equal to the inhibition exerted in the dark by $CoCl_2$, an inhibitor of EFE. The light-inhibition effect was partially reversed by dithiothreitol and fully reversed by mercaptoethanol. Thus, it appears light acts probably via the photosynthetic system, by oxidizing essential SH groups of the enzyme system converting ACC to ethylene. The suggestion that the light effect involves reversible oxidation of essential sulfhydryl groups[101] has been challenged.[102] In the latter study, it was observed that the inhibitory effect was not reversible by red/far-red light treatments, and was inhibited by dichlorophenyldimethylurea (DCMU), an inhibitor of photosynthetic electron transport.[102] These results provided experimental support for the suggestion that the light inhibition of the conversion of ACC to ethylene was in some manner related to the photosynthetic system. Light-mediated inhibition has a requirement for *de novo* protein synthesis. This was shown by experiments in which leaf discs were shifted from light to darkness in the absence and presence of protein synthesis inhibitors, cycloheximide and 2-(4-methyl-2,6-dinitroanilo-*N*-methylpropioamide) (D-MDMP). Both these inhibitors inhibited dark-induced increase in endogenous EFE activity.[102] Wright,[104] on the other hand, proposed that ethylene biosynthesis requires an obligatory dark stage. This conclusion was based on the observation that, when water-stressed leaves were kept in dark chambers for several hours and then transferred to the light, ethylene production continued at the *dark* rate for over 2 h before beginning to decline, whereas leaves that had been in the light, when transferred to darkness, immediately produced ethylene at the *dark* rate. In all of these studies[101,102,104] the leaf tissues were incubated in enclosed containers under light or dark conditions, where the concentrations of carbon dioxide were not controlled.

Observing that CO_2 promotes ethylene production in leaf tissues, Grodzinski et al.[103] and Kao and Yang[105] reasoned that the inhibition of ethylene evolution by light might result from a decrease in internal CO_2 concentrations. It was observed that when an external source of CO_2 is provided in the light, the rate of ethylene production increased markedly to a level approaching[103,106] or exceeding that observed in the dark.[103,105,107,108] However, CO_2 alone did not stimulate ethylene production in the dark. Kao and Yang[105] found that, in excised rice leaves under light, CO_2 markedly promoted ethylene production without significantly influencing the endogenous level of ACC. The rate of conversion of exogenous ACC to ethylene was inhibited by light, the inhibition being relieved in the presence of CO_2. Similar results were obtained with tobacco leaf discs. Thus, it appears that CO_2 exerts its effect by promoting the conversion of ACC to ethylene.

The CO_2 concentrations that give half-maximal promotion of ACC-dependent ethylene production were found to be only severalfold higher than the ambient level of CO_2 in the atmosphere (0.06% and 0.18% for rice and tobacco leaf tissue, respectively). Dhawan et al.[109] used a continuous-flow system to study the interactions between carbon dioxide and ethylene in intact sunflower plants in continuous light. They observed that as the concentration of carbon dioxide increased above the ambient level (0.033%) in the atmosphere, the rate of ethylene production in plants increased, while a decrease in carbon dioxide concentration resulted in a decrease in the rate of ethylene production. All these data put together support the idea that it is CO_2 metabolism rather than light per se that regulates the evolution of ethylene.[103,105] When sufficient CO_2 is present, light becomes a stimulatory factor in many green tissues, suggesting that the amount of ethylene generated by photosynthetic tissues is controlled directly by the availability of CO_2.[1,103,105,108] The CO_2 effect on ethylene production does not seem to be a direct result of stomatal control.[110]

A possibility that promotion of ethylene evolution may occur by CO_2 displacing ethylene from its endogenous binding sites, resulting in more ethylene being released from the leaves,[103] does not seem tenable since the total number of ethylene binding sites present in tobacco leaf tissue has been estimated to be 3.5 pmol g^{-1} tissue,[111] whereas CO_2-dependent ethylene production rate in the presence of ACC can be as high as 3 nmol g^{-1} h^{-1} in the same tissue.[105] Thus, the maximal amount of ethylene possibly bound to the tissue is far too small to account for the increase in ethylene evolution caused by CO_2. Also, CO_2 is ineffective in altering the amount of bound ethylene in the tissue.[103]

The stimulatory effect of CO_2 on EFE might involve not only an enhancement of the activity but also its turnover (net synthesis and degradation).[112] CO_2 (5 to 15%) enhances ethylene production in tobacco leaf discs during dark incubation for several days.[24,113] When ACC-dependent ethylene production was examined in similar tissues during dark incubation under 10% CO_2, a marked increase in ethylene production rate was observed with increasing incubation time, up to 24 h.[112] If CO_2 promoted only EFE activity, the ACC-dependent ethylene production rate, measured under conditions that give maximal activity, would not be expected to change during incubation, unless the enzyme underwent a rapid turnover in the tissue. The results indicated that under prolonged incubation, CO_2 can also induce enzyme synthesis and accumulation in addition to its promotive effect on the EFE activity, thereby resulting in higher ACC-dependent ethylene production rates. Promotion of EFE development was found to be dependent on CO_2 concentration (saturated at 2% CO_2) and duration (maximum at 24 h in the dark), and was abolished by cycloheximide. Application of exogenous ethylene or light treatment further increased the CO_2-enhanced development of EFE, implying that these two factors can also effect EFE development via an interaction with CO_2. The mechanisms by which ethylene and light interact to affect EFE development are presently unknown.

In addition to an effect on EFE, light may also affect ethylene biosynthesis by stimulating the conjugation of ACC, thus limiting its availability as a substrate for the EFE. Light promotes malonylation in etiolated wheat seedling leaves.[114] It was observed that etiolated wheat leaves have high ACC levels and high rates of ethylene production. Illumination resulted in reduced ACC levels, and consequently in reduced rates of ethylene production. From the following observations the authors concluded that promotion of ACC malonylation by light accounted for the rapid decline in ACC level: (1) a net decrease in ACC content was accompanied by a corresponding increase in MACC content; (2) there was little or no difference in ethylene production during 0 to 4 or 0 to 6 h between light and dark incubation, indicating that the rapid decline in ACC content under light could not be attributed to the increased amount of ACC converted to ethylene, and (3) the rapid decline in ACC content under light could not be due to the decrease in ACC synthesis. Interestingly, the extractable ACC-malonyltransferase activity from light-treated tissue was not significantly higher than that of the dark control. Therefore, the mechanism for the promotion of ACC malonylation by light is still unknown, although the effect seems mediated through phytochrome as shown by red/far-red reversibility.[114] The mechanism by which phytochrome modulates ACC malonylation is also not known.

VII. INTERACTIONS WITH OTHER PLANT GROWTH REGULATORS

Plant hormones interact with one another to influence the metabolism in plant cells. Physiological responses to plant hormones vary greatly and depend on the species. Among the plant hormones that affect the biosynthesis of ethylene, the best studied is auxin (IAA). Other plant growth regulators that markedly influence ethylene biosynthesis include abscisic acid (ABA), cytokinins, methyl jasmonate, and brassinosteroids.

In vegetative tissues, which normally produce quite low amounts of ethylene, auxin, such as IAA, markedly induces ethylene biosynthesis[115-119] whereas in fruit tissues auxin generally inhibits ethylene production.[120] This auxin effect has been studied mainly in etiolated mung bean and pea seedlings, where IAA stimulates ethylene production several hundredfold.[1,121] In these tissues, the conversion of methionine to ethylene requires the presence of IAA, but the conversion of methionine to SAM and the conversion of ACC to ethylene are not dependent upon IAA.[58] These observations indicated that IAA stimulates ethylene production by inducing the synthesis of ACC from SAM, and that this step is rate limiting. Several studies have confirmed this view and shown that ethylene production rates induced by IAA closely parallel the level of endogenous ACC and the activity of ACC synthase.[58,122,123] Yoshii and Imaseki[23] observed that ACC synthase activity induced by IAA treatment is severely inhibited by the application of cycloheximide, actinomycin D, and α-amanitin, suggesting that IAA exerts its effect by inducing the *de novo* synthesis of ACC synthase. The regulation of ACC synthase level by IAA is discussed in detail in the previous chapter.

IAA conjugates play an important role in the physiology and metabolism of auxin. They are thought to function as components for transport, storage, and subsequent reuse of IAA within plant tissue.[124] In mung bean hypocotyls[118] and in pea epicotyls,[116] a positive correlation was obtained between the rate of ethylene production and the internal level of IAA. In these two tissues, IAA-induced ethylene production continued as long as free IAA was present in the medium. In tobacco leaves, however, a 4-hour pulse of IAA stimulated ethylene production for several days.[125] This sustained increase in ethylene production by IAA occurred only when sucrose, or some other carbohydrate was present in the incubation medium.[125,126] Together, these observations led Aharoni et al.[86] to hypothesize that the natural regulation of ethylene biosynthesis in the leaf is primarily dependent on the type and levels of the endogenous IAA conjugates formed in the tissue and their susceptibility to hydrolytic enzymes.

Amino acid conjugates of IAA support the growth of plant tissue cultures[127,128] and have a variety of effects on tissue differentiation.[129] Some of these conjugates, such as IAA-L-alanine (IAAla), mimic a frequently renewed supply of free IAA.[129] Applied IAAla stimulates ethylene production in pea and tomato hypocotyl segments[130] and in leaf discs of tobacco, tomato, cotton, and bean.[86] In leaf discs, the stimulatory effect of IAAla on ethylene production is observed 3 to 4 d after the initial treatment and corresponds with a climacteric-like peak of ethylene production that is also observed in control discs. In the presence of

50 m*M* sucrose, however, IAAla stimulates ethylene production within a few hours and the effect persists for several days. During this period, an acceleration of IAAla hydrolysis was observed which was accompanied by an increase in the activity of IAAla hydrolase.[83,131]

Three different IAA esters have been isolated from IAA-treated tobacco leaves, two of which were identified by GC-MS techniques as IAA-glucose and IAA-myo-inositol.[63] Although both of these IAA conjugates were more effective than free IAA in stimulating ethylene production in the presence of sucrose, it is still questionable whether endogenous IAA conjugates have a role in ethylene production in senescing leaves. In detached tobacco leaf, an increase in free IAA was associated with a rise in ethylene production, with both processes being suppressed by the senescence-retarding hormones, gibberellin and kinetin.[132] A decrease in the extractable level of IAA during leaf senescence may occur by rapid turnover of free IAA in the senescent leaf.[133-135] In this situation, IAA conjugates could serve as a source of free IAA that, in turn, may increase ethylene production.[83] The role of IAA conjugates in regulating ethylene production in other developmental processes is not well understood and needs to be further explored.

ABA participates in the regulation of embryogenesis, seed maturation, environmental stresses, and senescence.[136,139] Its effect on ethylene biosynthesis is complex. Some plants respond to ABA by producing more ethylene. In other tissues, ABA inhibits ethylene production induced by senescence,[140] IAA,[123,141] or water stress.[123,142] The ABA-mediated suppression of ethylene production induced by water stress mainly occurs via an inhibition of ACC synthase activity[142] although EFE activity is stimulated in some cases.[143] Tan and Thimann[143] have suggested that inhibition of ethylene production by ABA occurs because ACC precursors are diverted to other pathways. Inhibition of ethylene production by ABA may be achieved indirectly, i.e., via induction of senescence.[137,138] However, high concentrations of ABA promote ethylene production, by both increasing ACC synthesis and accelerating the conversion of ACC to ethylene.[144] Aging itself amplifies the ABA effect on ethylene production, by increasing the responsiveness of the tissue to the hormone. These varied observations could have arisen because the tissues tested were at different physiological ages or were previously stressed.

Cytokinins, like auxins, are important for cell division, growth, and organogenesis. The effects of cytokinins on ethylene biosynthesis are diverse. In carnation petals, cytokinins inhibit ethylene synthesis.[145,146] However, at low concentrations (such as 10^{-7} *M*), they can promote ethylene synthesis in several tissues.[24,113,147-150] In the latter situation, cytokinins that normally act as antisenescence agents actually promote senescence. This is a rare case, however.

In detached rice leaves cytokinins promote ethylene synthesis but, in this case, senescence is also delayed.[140] Cytokinins promote ACC and/or ethylene production in stressed or ABA-treated plant tissues.[142,151,152] Water-stressed wheat leaves treated with benzyladenine (BA) produced more ACC and ethylene compared to that produced by stress alone.[142,151] Kinetin antagonizes the inhibition of ACC accumulation by ABA in rice seedlings during submergence.[152] The relief of salt stress and thermoinhibition of lettuce seed germination by kinetin (see Chapter 8) is accompanied by enhanced pregermination ethylene production and the addition of ACC with kinetin elicits a synergistic increase in ethylene production.[153] In the absence of salt stress this synergism between ACC and kinetin was not observed.

In combination with IAA, cytokinin stimulates ethylene production in excised tobacco leaf discs[24] and pea seedlings.[149] The kinetin effect is dependent upon seedling age. BA stimulates ethylene production in IAA-treated mung bean hypocotyls by stimulating ACC synthase activity.[123] As discussed above, kinetin stimulates ethylene production in the absence of IAA but the degree of stimulation is much less than that observed with kinetin and IAA. The stimulatory effect of kinetin may be limited to young vegetative tissues, and may be species specific. Kinetin suppresses ethylene production in detached senescing tobacco leaves,[132] but not nonsenescing cotton leaves, and exhibits only marginal effects on leaf tissues from bean, sunflower, and corn. In mung bean hypocotyls, both BA and ABA regulate the synthesis or accumulation of ACC primarily by interacting with IAA. However, in the absence of IAA, BA significantly promotes ethylene production, whereas the ACC content is only slightly affected. These results suggest that BA may also stimulate ethylene production via an auxin-independent mechanism in this system.

Methyl jasmonate belongs to the cyclopentanone family of compounds that produce effects similar to ABA. Application of methyl jasmonate to detached unripe tomato fruits causes a surge in ethylene production and inhibition of color development.[154] However, the mechanism of this stimulation remains to be elucidated.

Brassinosteroids, originally isolated from *Brassica* pollen, are growth stimulators that seem to act synergistically with auxin and stimulate ethylene biosynthesis in mung bean segments.[155] In the absence of auxin, this plant steroid did not stimulate ethylene production in squash hypocotyl segments.[156] The brassin-mediated enhancement in IAA-induced ethylene production seems to occur at the level of ACC synthase.

VIII. ACC-INDEPENDENT ETHYLENE PRODUCTION

That ACC is an intermediate in the pathway of ethylene biosynthesis from methionine is now well established.[1] However, the existence of alternative routes for ethylene formation in certain aquatic plants, and under stress conditions in some higher plants, has been documented. For example, methionine does not serve as an ethylene precursor in the aquatic fern *Regnellidium* (see Chapter 11).[157] In sweet potato root tissues infected by *Ceratocystis*, ethylene production does not seem to involve ACC as an intermediate,[158] as is also the case with ethylene production in Norway spruce needles induced by acid stress.[159] Similarly, during chemical-induced senescence of the aquatic angiosperm, *Spirodela oligorrhiza*, ethylene production occurs via non-ACC pathways,[160] involving linolenic acid and/or methionine as precursors in a free-radical-mediated pathway. Ethylene production in these cases might occur via membrane damage. The simple criteria that distinguish the non-ACC pathway from the ACC-dependent pathway include: (1) inhibition of the former by selenomethionine,[160] which conversely stimulates the latter pathway,[122] and (2) ineffectiveness of AVG to inhibit the non-ACC pathway. Although free radicals have been implicated in mediating chemically produced ethylene, the latter observations provide support for the *in vivo* significance of such reactions in plants, particularly under biotic and abiotic stress conditions. Further studies will be required to determine the precise mechanism(s) that regulate non-ACC ethylene production in plants.

IX. CONCLUSIONS

Considerable information is now available on the physiology and regulation of ethylene biosynthesis. Following the elucidation of the primary biosynthetic pathway for ethylene formation, it has become clear that, in most plant tissues, the level of active ACC synthase governs the rate of ethylene production. There is already some evidence that the level of ACC synthase in the cell is regulated primarily at the transcriptional level, and future studies will no doubt be directed at determining the mechanism(s) that underly the transcriptional regulation of ACC synthase gene(s) during plant development. Interestingly, it now appears that there are more than one, or perhaps several different genes that encode ACC synthase. Characterization of these genes will enable determination and delineation of ACC synthase genes that are regulated during plant development and by environmental stimuli. Post-transcriptional regulation appears to be an additional mechanism that may possibly serve to regulate ACC synthase activity. The biochemical signal(s) that regulate the level/activity of

ACC synthase and the cellular mechanism through which the signal transduces this physiological effect are unknown.

Further work needs to be done to purify and study the regulation of enzymes associated with the recycling of MTA to methionine, since it appears that the regulation of this pathway allows for maximum utilization of methionine for ethylene biosynthesis. Additional efforts should be directed at the regulation, purification, and characterization of ACC:malonyl transferase since the conjugation of ACC by this enzyme is a major factor limiting the availability of ACC.

Vacuoles appear to be the simplest biological units that exhibit an enzyme activity characteristic of *in vivo* EFE. However, it is not clear that the vacuole is the sole site of ethylene biosynthesis in the cell.[56,61] This enzyme system is labile, membrane-associated, and regulated by CO_2, ethylene, and also, probably, by other hormones. The preparation of monoclonal or polyclonal antibodies specific for the protein(s) associated with EFE activity would be an extremely significant development. Perhaps future studies of EFE function and regulation will allow unambiguous identification of the protein(s) associated with ethylene-forming activity, making it possible to develop tools for the study of this very important enzyme system at the molecular level. Isolated vacuoles appear most promising in this regard.

The identification and characterization of the appropriate genes will provide better handles for studying structure-function relationships of the corresponding enzymes and in bioengineering the senescence and ripening processes in plants. Such probes should also help in identifying the subcellular sites where other plant hormones such as IAA and its conjugates, cytokinins, gibberellins, ABA, and molecules such as carbohydrates, polyamines, cations, brassinosteroids, and methyl jasmonate interact to regulate ethylene biosynthesis.

The induction of ethylene biosynthesis by specific oligossaccharides of either plant or fungal origin promises to be a relevant model of pathogenesis, in addition to fruit ripening. Identification and purification of these specific oligosaccharins should enable identification of receptors that coordinate cell wall degradation and ethylene biosynthesis during fruit ripening and senescence. Characterizing these receptors might facilitate discovery of specific, reversible blocking agents capable of controlling postharvest ripening and senescence in a more precise manner than is currently possible.

X. ACKNOWLEDGMENTS

We thank Edo Chalutz, Ken Gross, Roshni Mehta, Barbara Parsons, Robert Saftner, Jeffrey Suttle, and Mark Tucker for valuable suggestions and a critical reading of the first draft of this chapter, and Roshni Mehta for compiling Figure 1. We thank Jane Wooster for patient and careful word processing.

REFERENCES

1. **Yang, S. F. and Hoffman, N. E.,** Ethylene biosynthesis and its regulation in higher plants, *Annu. Rev. Plant Physiol.*, 35, 155, 1984.
2. **Peacock, B. C.,** Role of ethylene in the initiation of fruit ripening, *Queensl. J. Agric. Anim. Sci.*, 29, 137, 1972.
3. **McMurchie, E. J., McGlasson, W. B., and Eaks, I. L.,** Treatment of fruit with propylene gives information about the biogenesis of ethylene, *Nature*, 237, 235, 1972.
4. **Yang, S. F.,** The role of ethylene and ethylene synthesis in fruit ripening, in *Plant Senescence: Its Biochemistry and Physiology*, Thompson, W. W., Nothnagel, E. A., and Huffaker, R. C., Eds., The American Society of Plant Physiologists, Maryland, 1987, 156.
5. **Hoffman, N. E. and Yang, S. F.,** Enhancement of wound-induced ethylene synthesis in preclimacteric cantaloupe, *Plant Physiol.*, 69, 317, 1982.
6. **Riov, J. and Yang, S. F.,** Effects of exogenous ethylene on ethylene production in citrus leaf tissue, *Plant Physiol.*, 70, 136, 1982.

7. **Chalutz, E., Mattoo, A. K., Solomos, T., and Anderson, J. D.,** Enhancement by ethylene of Cellulysin-induced ethylene production by tobacco leaf discs, *Plant Physiol.*, 52, 38, 1984.
8. **Liu, Y., Hoffman, N. E., and Yang, S. F.,** Promotion by ethylene of the capability to convert 1-aminocyclopropane-1-carboxylic acid to ethylene in preclimacteric tomato and cantaloupe fruits, *Plant Physiol.*, 77, 407, 1985.
9. **Manning, K.,** The ethylene forming enzyme system in carnation flowers, in *Ethylene and Plant Development*, Roberts, J. A. and Tucker, G. A., Eds., Butterworths, London, 1986, 83.
10. **Woodson, W. R. and Lawton, K. A.,** Ethylene-induced gene expression in carnation petals: relationship to autocatalytic ethylene production and senescence, *Plant Physiol.*, 87, 498, 1987.
11. **Liu, Y., Su, L. Y., and Yang, S. F.,** Ethylene promotes the capability to malonylate 1-aminocyclopropane-1-carboxylic acid and D-amino acids in preclimacteric tomato fruits, *Plant Physiol.*, 77, 891, 1985.
12. **Su, L. Y., McKeon, T., Grierson, G., Cantwell, M., and Yang, S. F.,** Development of 1-aminocyclopropane-1-carboxylic acid synthase and poly-galacturonase activities during the maturation and ripening of tomato fruits, *HortScience*, 19, 576, 1984.
13. **Liu, Y., Hoffman, N. E., and Yang, S. F.,** Ethylene promoted malonylation of 1-aminocyclopropane-1-carboxylic acid participates in autoinhibition of ethylene synthesis in grape fruit flavedo discs, *Planta*, 164, 565, 1985.
14. **Amrhein, N., Dorzok, Y., Kionka, C., Kondiziolka, U., Skorupka, H., and Tophof, S.,** The biochemistry and physiology of 1-aminocyclopropane-1-carboxylic acid conjugation, in *Ethylene: Biochemical, Physiological, and Applied Aspects*, Fuchs, Y. and Chalutz, E., Eds., Martinus Nijhoff/Dr. W. Junk, The Hague, 1984, 11.
15. **Jiao, X., Philosoph-Hadas, S., Su, L., and Yang, S. F.,** The conversion of 1-(malonylamino)cyclopropane-1-carboxylic acid to 1-aminocyclopropane-1-carboxylic acid in plant tissues, *Plant Physiol.*, 81, 637, 1986.
16. **Hanley, K. M., Meir, S., and Bramlage, W. J.,** Activity of aging carnation flower parts and the effects of 1-(malonylamino)cyclopropane-1-carboxylic acid-induced ethylene, *Plant Physiol.*, 91, 1126, 1989.
17. **Bryce, G. F. and Rabin, B. R.,** The function of the metal ion in leucine aminopeptidase and the mechanism of action of the enzyme, *Biochem. J.*, 90, 513, 1964.
18. **Saltveit, M. E., Jr. and Dilley, D. R.,** Rapidly induced wound ethylene from excised segments of etiolated *Pisum sativum* L., cv. Alaska. II. Oxygen and temperature dependency, *Plant Physiol.*, 61, 675, 1978.
19. **Vendrell, M. and McGlasson, W. B.,** Inhibition of ethylene production in banana fruit tissue by ethylene treatment, *Aust. J. Biol. Sci.*, 24, 885, 1971.
20. **Zauberman, G. and Fuchs, Y.,** Ripening process in avocado stored in ethylene atmosphere in cold storage, *J. Am. Soc. Hortic. Sci.*, 98, 477, 1973.
21. **Zeroni, M., Galil, J., and Ben-Yehoshua, S.,** Autoinhibition of ethylene formation in nonripening stages of the fruit of sycamore fig *(Ficus sycomorus* L.), *Plant Physiol.*, 57, 647, 1976.
22. **Riov, J. and Yang, S. F.,** Autoinhibition of ethylene production in citrus peel discs. Suppression of 1-aminocyclopropane-1-carboxylic acid synthesis, *Plant Physiol.*, 69, 687, 1982.
23. **Yoshii, H. and Imaseki, H.,** Regulation of auxin-induced ethylene biosynthesis. Repression of inductive formation of ACC synthase by ethylene, *Plant Cell Physiol.*, 23, 639, 1982.
24. **Aharoni, N., Anderson, J. D., and Lieberman, M.,** Production and action of ethylene in senescing leaf discs: effect of indoleacetic acid, kinetin, silver ion, and carbon dioxide, *Plant Physiol.*, 64, 805, 1979.
25. **Philosoph-Hadas, S., Meir, S., and Aharoni, N.,** Autoinhibition of ethylene production in tobacco leaf discs: enhancement of 1-aminocyclopropane-1-carboxylic acid conjugation, *Plant Physiol.*, 63, 431, 1985.
26. **Burg, S. P. and Thimann, K. V.,** Studies on the ethylene production of apple tissue, *Plant Physiol.*, 35, 24, 1960.
27. **Mattoo, A. K. and Lieberman, M.,** Localization of the ethylene-synthesizing system in apple tissue, *Plant Physiol.*, 60, 794, 1977.
28. **Mattoo, A. K., Baker, J. E., Chalutz, E., and Lieberman, M.,** Effect of temperature on the ethylene-synthesizing systems in apple, tomato and *Penicillium digitatum*, *Plant Cell Physiol.*, 18, 715, 1977.
29. **Mattoo, A. K., Chalutz, E., and Lieberman, M.,** Effects of lipophilic and water-soluble membrane probes on ethylene synthesis in apple and *Penicillium digitatum*, *Plant Cell Physiol.*, 20, 1097, 1979.
30. **Anderson, J. D., Lieberman, M., and Stewart, R. N.,** Ethylene production by apple protoplasts, *Plant Physiol.*, 63, 931, 1979.
31. **Odawara, S., Watanabe, A., and Imaseki, H.,** Involvement of cellular membrane in regulation of ethylene production, *Plant Cell Physiol.*, 18, 569, 1977.
32. **Mattoo, A. K. and Anderson, J. D.,** Wound-induced increase in 1-aminocyclopropane-1-carboxylate synthase activity: regulatory aspects and membrane association of the enzyme, in *Ethylene: Biochemical, Physiological, and Applied Aspects*, Fuchs, Y. and Chalutz, E., Eds., Martinus Nijhoff/Dr. W. Junk, The Hague, 1984, 139.
33. **Bertoli, E., Parenti-Castelli, G., Landi, L., Sechi, A. M., and Lenaz, G.,** Activation energies of mitochondrial adenosine triphosphatase under different conditions, *Bioenergetics*, 4, 591, 1973.

34. **Raison, J. K.**, Temperature-induced phase changes in membrane lipids and their influence on metabolic regulation, *Symp. Soc. Exp. Biol.*, 27, 485, 1973.
35. **Lieberman, M.**, Biosynthesis and action of ethylene, *Annu. Rev. Plant Physiol.*, 30, 533, 1979.
36. **Grossman, K., Schmidt, H. O., and Jung, J.**, Changes in membrane permeability and mineral, phytohormone and polypeptide composition in rice suspension cells during growth and under the influence of the growth retardant tetcyclacis, *Plant Cell Rep.*, 5, 315, 1986.
37. **Sauerbrey, E., Grossmann, K., and Jung, J.**, Ethylene production by sunflower cell suspensions. Effects of plant growth retardants, *Plant Physiol.*, 87, 510, 1988.
38. **Apelbaum, A., Burgoon, A. C., Anderson, J. D., Solomos, T., and Lieberman, M.**, Some characteristics of the system converting 1-aminocyclopropane-1-carboxylic acid to ethylene, *Plant Physiol.*, 67, 80, 1981.
39. **Apelbaum, A., Wang, S. Y., Burgoon, A. C., Baker, J. E., and Lieberman, M.**, Inhibition of the conversion of 1-aminocyclopropane-1-carboxylic acid to ethylene by structural analogs, inhibitors of electron transfer, uncouplers of oxidative phosphorylation, and free radical scavengers, *Plant Physiol.*, 67, 74, 1981.
40. **Lau, O. L. and Yang, S. F.**, Synergistic effect of calcium and kinetin on ethylene production by mung bean hypocotyl, *Planta*, 118, 1, 1974.
41. **Lieberman, M. and Wang, S. Y.**, Influence of calcium and magnesium on ethylene production by apple tissue slices, *Plant Physiol.*, 69, 1150, 1982.
42. **Mattoo, A. K. and Lieberman, M.**, Role of silver ions in controlling senescence and conversion of 1-aminocyclopropane-1-carboxylic acid to ethylene, *Plant Physiol. Suppl.*, 69, 198, 1982.
43. **Burns, J. K. and Evensen, K. B.**, Ca^{2+} effects on ethylene, carbon dioxide and 1-aminocyclopropane-1-carboxylic acid synthase activity, *Physiol. Plant.*, 66, 609, 1986.
44. **Ben-Arie, R., Lurie, S., and Mattoo, A. K.**, Temperature dependent inhibitory effects of calcium and spermine on ethylene biosynthesis in apple discs correlate with changes in microsomal membrane viscosity, *Plant Sci. Lett.*, 24, 239, 1982.
45. **Legge, R. L., Thompson, J. E., Baker, J. E., and Lieberman, M.**, The effect of calcium on the fluidity and phase properties of microsomal membranes isolated from post-climacteric golden delicious apple, *Plant Cell Physiol.*, 23, 161, 1982.
46. **Paliyath, G. and Poovaiah, B. W.**, Calmodulin inhibitor in senescing apples and its physiological and pharmacological significance, *Proc. Natl. Acad. Sci. U.S.A.*, 81, 2065, 1984.
47. **Eaks, I. L.**, Effect of calcium on ripening, respiratory rate, ethylene production, and quality of avocado fruit, *J. Am. Soc. Hortic. Sci.*, 110, 145, 1985.
48. **Conway, W. S. and Sams, C. E.**, The effects of postharvest infiltration of calcium, magnesium, or strontium on decay, firmness, respiration, and ethylene production in apples, *J. Am. Soc. Hortic. Sci.*, 112, 300, 1987.
49. **Cheverry, J. L., Pouliquen, J., Le Guyader, H., and Marcellin, P.**, Calcium regulation of exogenous and endogenous 1-aminocyclopropane-1-carboxylic acid bioconversion to ethylene, *Physiol. Plant.*, 74, 53, 1988.
50. **Saftner, R. A.**, Effects of organic amines on α-aminoisobutyric acid uptake into the vacuole and on ethylene production by tomato pericarp slices, *Physiol. Plant.*, 75, 485, 1989.
51. **Atta-Aly, M. A., Saltveit, M. E., and Hobson, G. E.**, Effect of silver ions on ethylene biosynthesis by tomato fruit tissue, *Plant Physiol.*, 83, 44, 1987.
52. **Chalutz, E., Mattoo, A. K., and Fuchs, Y.**, Biosynthesis of ethylene: the effect of phosphate, *Plant Cell Environ.*, 3, 349, 1980.
53. **Fuchs, Y., Mattoo, A. K., Chalutz, E., and Rot, I.**, Biosynthesis of ethylene in higher plants: the metabolic site of inhibition by phosphate, *Plant Cell Environ.*, 4, 291, 1981.
54. **Tsai, D.-S., Arteca, R. N., Backman, J. M., and Phillips, A. T.**, Purification and characterization of 1-aminocyclopropane-1-carboxylate synthase from etiolated mung bean hypocotyls, *Arch. Biochem. Biophys.*, 264, 632, 1988.
55. **Guy, M. and Kende, H.**, Conversion of 1-aminocyclopropane-1-carboxylic acid to ethylene by isolated vacuoles of *Pisum sativum L.*, *Planta*, 160, 281, 1984.
56. **Erdmann, H., Griesbach, R. J., Lawson, R. H., and Mattoo, A. K.**, 1-Aminocyclopropane-1-carboxylic-acid-dependent ethylene production during re-formation of vacuoles in evacuolated protoplasts of *Petunia hybrida*, *Planta*, 179, 196, 1989.
57. **Adams, D. O. and Yang, S. F.**, Ethylene biosynthesis: identification of 1-aminocyclopropane-1-carboxylic acid as an intermediate in the conversion of methionine to ethylene, *Proc. Natl. Acad. Sci. U.S.A.*, 76, 170, 1979.
58. **Yu, Y. B. and Yang, S. F.**, Auxin-induced ethylene production and its inhibition by aminoethoxyvinylglycine and cobalt ion, *Plant Physiol.*, 64, 1074, 1979.
59. **Mayne, R. G. and Kende, H.**, Ethylene biosynthesis in isolated vacuoles of *Vicia faba* L.—requirement for membrane integrity, *Planta*, 167, 159, 1986.
60. **Kende, H. and Boller, T.**, Wound ethylene and 1-aminocyclopropane-1-carboxylate synthase in ripening tomato fruit, *Planta*, 151, 476, 1981.

61. **Bouzayen, M., Latche, A., and Pech, J.-C.,** Subcellular localization of the sites of conversion of 1-aminocyclopropane-1-carboxylic acid into ethylene in plant cells, *Planta,* 180, 175, 1990.
62. **Porter, A. J. R., Borlakoglu, J. T., and John, P.,** Activity of the ethylene-forming enzyme in relation to plant cell structure and organization, *J. Plant Physiol.*, 125, 207, 1986.
63. **Mattoo, A. K. and Aharoni, N.,** Ethylene and plant senescence, in *Senescence and Aging in Plants,* Nooden, L. and Leopold, A. C., Eds., Academic Press, London, 1988, 241.
64. **Ferguson, I. B.,** Calcium in plant senescence and fruit ripening, *Plant Cell Environ.*, 7, 477, 1983.
65. **Mattoo, A. K., Adams, D. O., Patterson, G. W., and Lieberman, M.,** Inhibition of 1-aminocyclopropane-1-carboxylic acid synthase by phenothiazines, *Plant Sci. Lett.*, 28, 173, 1982/1983.
66. **Evensen, K. B.,** Calcium effects on ethylene and ethane production and 1-aminocyclopropane-1-carboxylic acid content in potato discs, *Physiol. Plant.*, 60, 125, 1984.
67. **Anderson, J. D., Mattoo, A. K., and Lieberman, M.,** Induction of ethylene in tobacco leaf discs by cell wall digesting enzymes, *Biochem. Biophys. Res. Commun.*, 107, 588, 1982.
68. **Fuchs, Y. and Anderson, J. D.,** Purification and characterization of ethylene inducing proteins from cellulysin, *Plant Physiol.*, 84, 732, 1987.
69. **Acaster, M. A. and Kende, H.,** Properties and partial purification of 1-aminocyclopropane-1-carboxylate synthase, *Plant Physiol.*, 72, 139, 1983.
70. **Mehta, A. M., Jordan, R. L., Anderson, J. D., and Mattoo, A. K.,** Identification of a unique isoform of 1-aminocyclopropane-1-carboxylic acid synthase by monoclonal antibody, *Proc. Natl. Acad. Sci. U.S.A.*, 85, 8810, 1988.
71. **Bufler, G. and Bangerth, F.,** Effects of propylene and oxygen on the ethylene producing system of apples, *Physiol. Plant.*, 58, 486, 1983.
72. **Smith, T. A.,** Polyamines, *Annu. Rev. Plant Physiol.*, 36, 117, 1985.
73. **Slocum, R. D., Kaur-Sawhney, R., and Galston, A. W.,** The physiology and biochemistry of polyamines in plants, *Arch. Biochem. Biophys.*, 235, 283, 1984.
74. **Suttle, J. C.,** Effect of polyamines on ethylene production, *Phytochemistry,* 20, 1477, 1981.
75. **Apelbaum, A., Burgoon, A. C., Anderson, J. D., Lieberman, M., Ben-Arie, R., and Mattoo, A. K.,** Polyamines inhibit biosynthesis of ethylene in higher plant tissue and fruit protoplasts, *Plant Physiol.*, 68, 453, 1981.
76. **Even-Chen, Z., Mattoo, A. K., and Goren, R.,** Inhibition of ethylene biosynthesis by aminoethoxyvinylglycine and by polyamines shunts label from 3,4-(^{14}C) methionine into spermidine in aged orange peel discs, *Plant Physiol.*, 69, 385, 1982.
77. **Fuhrer, J., Kaur-Sawhney, R., Shih, L. M., and Galston, A. W.,** Effects of exogenous 1,3-diaminopropane and spermidine on senescence of oat leaves. II. Inhibition of ethylene biosynthesis and possible mode of action, *Plant Physiol.*, 70, 1597, 1982.
78. **Roberts, D. R., Walker, M. A., Thompson, J. E., and Dumbroff, E. B.,** The effects of inhibitors of polyamine and ethylene biosynthesis on senescence, ethylene production, and polyamine levels in cut carnation flowers, *Plant Cell Physiol.*, 25, 315, 1984.
79. **Icekson, I., Goldlust, A., and Apelbaum, A.,** Influence of ethylene on S-adenosylmethionine decarboxylase activity in etiolated pea seedlings, *J. Plant Physiol.*, 119, 335, 1986.
80. **Kushad, M. M., Yelenosky, G., and Knight, R.,** Interrelationship of polyamine and ethylene biosynthesis during avocado fruit development and ripening, *Plant Physiol.*, 87, 463, 1988.
81. **Dibble, A. R. G., Davies, P. J., and Mutschler, M. A.,** Polyamine content of long-keeping Alcobaca tomato fruit, *Plant Physiol.*, 86, 338, 1988.
82. **Saftner, R. A. and Baldi, B. G.,** Polyamine levels and tomato fruit development: possible interaction with ethylene, *Plant Physiol.*, 92, 547, 1990.
83. **Meir, S., Philosoph-Hadas, S., Epstein, E., and Aharoni, N.,** Carbohydrates stimulate ethylene production in tobacco leaf discs. I. Interaction with auxin and the relation to auxin metabolism, *Plant Physiol.*, 78, 131, 1985.
84. **Moore, K. D.,** Effects of sugars on ethylene production by excised *Sinapis* cotyledons, *Ann. Bot.*, 40, 543, 1976.
85. **Colclasure, G. C. and Yopp, J. H.,** Galactose-induced ethylene evolution in mung bean hypocotyls: a possible mechanism for galactose retardation of plant growth, *Physiol. Plant.*, 37, 298, 1976.
86. **Aharoni, N., Philosoph-Hadas, S., and Meir, S.,** Control of the biosynthesis of ethylene in senescing tissues, in *Ethylene: Biochemical, Physiological, and Applied Aspects,* Fuchs, Y. and Chalutz, E., Eds., Martinus Nijhoff/Dr. W. Junk, The Hague, 1984, 129.
87. **Riov, J. and Yang, S. F.,** Stimulation of ethylene production in citrus leaf discs by mannitol, *Plant Physiol.*, 70, 142, 1982.
88. **Gross, K. C.,** Promotion of ethylene evolution and ripening of tomato fruit by galactose, *Plant Physiol.*, 79, 306, 1985.
89. **Philosoph-Hadas, S., Meir, S., and Aharoni, N.,** Carbohydrates stimulate ethylene production in tobacco leaf discs. II. Sites of stimulation in the ethylene biosynthesis pathway, *Plant Physiol.*, 78, 139, 1985.

90. **Meir, S., Philosoph-Hadas, S., and Aharoni, N.,** Role of IAA conjugates in inducing ethylene production by tobacco leaf discs, *J. Plant Growth Regul.*, 3, 169, 1984.
91. **Philosoph-Hadas, S. and Aharoni, N.,** Galactose inhibits the conversion of 1-aminocyclopropane-1-carboxylic acid to ethylene in aged tobacco leaf discs, *Plant Physiol.*, 83, 8, 1987.
92. **Gross, K. C. and Wallner, S. J.,** Degradation of cell wall polysaccharides during tomato fruit ripening, *Plant Physiol.*, 63, 117, 1979.
93. **Gross, K. C. and Sams, C. E.,** Changes in cell wall neutral sugar composition during fruit ripening: a species survey, *Phytochemistry*, 23, 2457, 1984.
94. **Labavitch, J. M.,** Cell wall turnover in plant development, *Annu. Rev. Plant Physiol.*, 32, 385, 1981.
95. **Kim, J., Gross, K. C., and Solomos, T.,** Characterization of the stimulation of ethylene production by galactose in tomato *(Lycopersicon esculentum* Mill.) fruit, *Plant Physiol.*, 85, 804, 1987.
96. **Giovannoni, J. J., Della Penna, D., Bennett, A. B., and Fischer, R. L.,** Expression of a chimeric polygalacturonase gene in transgenic *rin* (ripening inhibitor) tomato fruit results in polyuronide degradation but not fruit softening, *Plant Cell*, 1, 53, 1989.
97. **Watkins, C. B. and Frenkel, C.,** Inhibition of pear fruit ripening by mannose, *Plant Physiol.*, 85, 56, 1987.
98. **Mayak, S. and Borochov, A.,** Nonosmotic inhibition by sugars of the ethylene-forming activity associated with microsomal membranes from carnation petals, *Plant Physiol.*, 76, 191, 1984.
99. **Goeschl, J. D., Pratt, H. K., and Bonner, B. A.,** An effect of light on the production of ethylene and the growth of the plumular portion of etiolated pea seedlings, *Plant Physiol.*, 42, 1077, 1967.
100. **Kang, B. G. and Ray, P. M.,** Ethylene and carbon dioxide as mediators in the response of the bean hypocotyl hook to light and auxins, *Planta*, 87, 206, 1969.
101. **Gepstein, S. and Thimann, K. V.,** The effect of light on the production of ethylene from 1-aminocyclopropane-1-carboxylic acid by leaves, *Planta*, 149, 196, 1980.
102. **De Laat, A. M. M., Brandenburg, D. C. C., and Van Loon, L. C.,** The modulation of the conversion of 1-aminocyclopropane-1-carboxylic acid to ethylene by light, *Planta*, 153, 193, 1981.
103. **Grodzinski, B., Boesel, I., and Horton, R. F.,** Ethylene release from leaves of *Xanthium strumarium* L. and *Zea mays* L., *J. Exp. Bot.*, 33, 344, 1982.
104. **Wright, S. T. C.,** The effect of light and dark periods on the production of ethylene from water-stressed wheat leaves, *Planta*, 153, 172, 1981.
105. **Kao, C. H. and Yang, S. F.,** Light inhibition of the conversion of 1-aminocyclopropane-1-carboxylic acid to ethylene in leaves is mediated through carbon dioxide, *Planta*, 155, 261, 1982.
106. **Grodzinski, B., Boesel, I., and Horton, R. F.,** The effect of light intensity on the release of ethylene from leaves, *J. Exp. Bot.*, 33, 1185, 1982.
107. **Bassi, P. K. and Spencer, M. S.,** Effect of carbon dioxide and light on ethylene production in intact sunflower plants, *Plant Physiol.*, 69, 1222, 1982.
108. **Grodzinski, B., Boesel, I., and Horton, R. F.,** Light stimulation of ethylene release from leaves of *Gomphrena globosa* L., *Plant Physiol.*, 71, 588, 1983.
109. **Dhawan, K. R., Bassi, P. K., and Spencer, M. S.,** Effects of carbon dioxide on ethylene production and action in intact sunflower plants, *Plant Physiol.*, 68, 831, 1981.
110. **Horton, R. F. and Saville, B. J.,** Carbon dioxide enrichment, transpiration and 1-aminocyclopropane-1-carboxylic acid-dependent ethylene release from oat leaves, *Plant Sci. Lett.*, 36, 131, 1984.
111. **Sisler, E. C.,** Measurement of ethylene binding in plant tissue, *Plant Physiol.*, 64, 538, 1979.
112. **Philosoph-Hadas, S., Aharoni, N., and Yang, S. F.,** Carbon dioxide enhances the development of the ethylene forming enzyme in tobacco leaf discs, *Plant Physiol.*, 82, 925, 1986.
113. **Aharoni, N., Lieberman, M., and Sisler, H. D.,** Patterns of ethylene production in senescing leaves, *Plant Physiol.*, 64, 796, 1979.
114. **Jiao, X., Yip, W., and Yang, S. F.,** The effect of light and phytochrome on 1-aminocyclopropane-1-carboxylic acid metabolism in etiolated wheat seedling leaves, *Plant Physiol.*, 85, 643, 1987.
115. **Abeles, F. B. and Rubinstein, B.,** Regulation of ethylene evolution and leaf abscission by auxin, *Plant Physiol.*, 39, 963, 1964.
116. **Kang, B. G., Newcomb, W., and Burg, S. P.,** Mechanisms of auxin-induced ethylene production, *Plant Physiol.*, 47, 504, 1971.
117. **Abeles, F. B.,** *Ethylene in Plant Biology*, Academic Press, New York, 1973.
118. **Lau, O. L. and Yang, S. F.,** Mechanisms of a synergistic effect of kinetin on auxin-induced ethylene suppression: suppression of auxin conjugation, *Plant Physiol.*, 51, 1011, 1973.
119. **Imaseki, H., Kondo, K., and Watanabe, A.,** Mechanisms of cytokinin action on auxin-induced ethylene production, *Plant Cell Physiol.*, 16, 777, 1975.
120. **Rhodes, M. J. C.,** The maturation and ripening of fruits, in ***Senescence in Plants***, Thimann, K. V., Ed., CRC Press, Boca Raton, FL, 1980, 157.

121. **Imaseki, H.,** Regulation of ethylene biosynthesis in auxin-treated plant tissues, in *The New Frontiers in Plant Biochemistry,* Akazawa, T., Asashi, T., and Imaseki, H., Eds., Martinus Nijhoff/Dr. W. Junk, The Hague, 1983, 133.
122. **Jones, J. F. and Kende, H.,** Auxin-induced ethylene biosynthesis in subapical stem sections of etiolated seedlings of *Pisum sativum* L., *Planta,* 146, 649, 1979.
123. **Yoshii, H. and Imaseki, H.,** Biosynthesis of auxin-induced ethylene, effects of indole-3-acetic acid, benzyladenine and abscisic acid on endogenous levels of 1-aminocyclopropane-1-carboxylic acid (ACC) and ACC synthase, *Plant Cell Physiol.,* 22, 369, 1982.
124. **Cohen, J. D. and Bandurski, R. S.,** Chemistry and physiology of the bound auxins, *Annu. Rev. Plant Physiol.,* 33, 403, 1982.
125. **Aharoni, N. and Yang, S. F.,** Auxin-induced ethylene production as related to auxin metabolism in leaf discs of tobacco and sugar beet, *Plant Physiol.,* 73, 598, 1983.
126. **Philosoph-Hadas, S., Meir, S., and Aharoni, N.,** Carbohydrates stimulate ethylene biosynthesis in tobacco leaf discs, *Plant Physiol. Suppl.,* 72, 121, 1983.
127. **Feung, C.-S., Hamilton, R. H., and Mumma, R. O.,** Metabolism of indole-3-acetic acid. III. Biological properties of amino acid conjugates, *Plant Physiol.,* 59, 91, 1977.
128. **Rekoslavskaya, N. I. and Gamburg, K. Z.,** On the biological activity of IAA conjugates, *Biochem. Physiol. Pflanz.,* 169, 299, 1976.
129. **Hangarter, R. P., Peterson, M. D., and Good, N. E.,** Biological activities of indoleacetyl amino acids and their use as auxins in tissue culture, *Plant Physiol.,* 65, 761, 1980.
130. **Hangarter, R. P. and Good, N. E.,** Evidence that IAA conjugates are slow-release sources of free IAA in plant tissues, *Plant Physiol.,* 68, 1424, 1981.
131. **Meir, S., Riov, J., Philosoph-Hadas, S., and Aharoni, N.,** Carbohydrates stimulate ethylene production in tobacco leaf discs. III. Stimulation of enzymic hydrolysis of indole-3-acetyl-L-alanine, *Plant Physiol.,* 90, 1246, 1989.
132. **Even-Chen, Z., Atsmon, D., and Itai, C.,** Hormonal aspects of senescence in detached tobacco leaves, *Plant Physiol.,* 44, 377, 1978.
133. **Shoji, K., Addicott, F. T., and Swets, W. A.,** Auxin in relation to leaf blade abscission, *Plant Physiol.,* 26, 1989, 1951.
134. **Sweetser, P. B. and Swartzfager, D. G.,** Indole-3-acetic acid levels of plant tissue as determined by a new high performance liquid chromatographic method, *Plant Physiol.,* 61, 254, 1978.
135. **Roberts, J. A. and Osborne, D. J.,** Auxin and the control of ethylene production during the development and senescence of leaves and fruits, *J. Exp. Bot.,* 32, 875, 1981.
136. **Edgerton, L. J.,** Apple abscission, *HortScience,* 6, 378, 1971.
137. **Jackson, M. B. and Osborne, D. J.,** Abscisic acid, auxin, and ethylene in explant abscission, *J. Exp. Bot.,* 23, 849, 1972.
138. **Lieberman, M., Baker, J. E., and Sloger, M.,** Influence of plant hormones on ethylene production in apple, tomato, and avocado slices during maturation and senescence, *Plant Physiol.,* 60, 214, 1977.
139. **Sagee, O., Goren, R., and Riov, J.,** Abscission of citrus leaf explants: interrelationships of abscisic acid, ethylene, and hydrolytic enzymes, *Plant Physiol.,* 66, 750, 1980.
140. **Kao, C. H. and Yang, S. F.,** Role of ethylene in senescence of detached rice leaves, *Plant Physiol.,* 73, 881, 1983.
141. **Gertman, E. and Fuchs, Y.,** Effects of abscisic acid and its interactions with other plant hormones on ethylene production in two plant systems, *Plant Physiol.,* 50, 194, 1972.
142. **McKeon, T. A., Hoffman, N. E., and Yang, S. F.,** The effect of plant hormone pre-treatments on ethylene production and synthesis of 1-aminocyclopropane-1-carboxylic acid in water-stressed wheat leaves, *Planta,* 155, 437, 1982.
143. **Tan, Z.-Y. and Thimann, K. V.,** The roles of carbon dioxide and abscisic acid in the production of ethylene, *Physiol. Plant.,* 75, 13, 1989.
144. **Riov, J., Dagan, E., Goren, R., and Yang, S. F.,** Characterization of abscisic acid-induced ethylene production in citrus leaf and tomato fruit tissues, *Plant Physiol.,* 92, 48, 1990.
145. **Mor, Y., Spiegelstein, H., and Halevy, A. H.,** Inhibition of ethylene biosynthesis in carnation petals by cytokinin, *Plant Physiol.,* 71, 541, 1983.
146. **Cook, D., Rasche, M., and Eisinger, W.,** Regulation of ethylene biosynthesis and action in cut carnation flower senescence by cytokinins, *J. Am. Soc. Hortic. Sci.,* 110, 24, 1985.
147. **Abeles, F. B., Holm, R. E., and Gahagan, H. E.,** Abscission: the role of aging, *Plant Physiol.,* 42, 1351, 1967.
148. **Yu, Y., Yang, S. F., Corse, J., Kuhnle, J. A., and Hua, S.,** Structures of cytokinins influence synergistic production of ethylene, *Phytochemistry,* 20, 1191, 1981.
149. **Fuchs, Y. and Lieberman, M.,** Effects of kinetin, IAA, and gibberellin on ethylene production and their interaction in growth of seedlings, *Plant Physiol.,* 43, 2029, 1968.

150. **Suttle, J. C.,** Cytokinin-induced ethylene biosynthesis in non-senescing cotton leaves, *Plant Physiol.*, 82, 930, 1986.
151. **Wright, S. T. C.,** The effect of plant growth regulator treatments on the level of ethylene emanating from excised turgid and wilted leaves, *Planta*, 148, 381, 1980.
152. **Khan, A. A., Thakur, R., Akbar, M., HilleRisLambers, D., and Seshu, D. V.,** Hormonal regulation of elongation in floating rice during submergence, *Crop Sci.*, 28, 121, 1988.
153. **Khan, A. A. and Huang, X.-L.,** Synergistic enhancement of ethylene production and germination with kinetin and 1-aminocyclopropane-1-carboxylic acid in lettuce seeds exposed to salinity stress, *Plant Physiol.*, 87, 847, 1988.
154. **Saniewski, M. and Czapski, J.,** Stimulatory effect of methyl jasmonate on the ethylene production in tomato fruits, *Experientia*, 41, 256, 1985.
155. **Mandava, N. B.,** Plant growth-promoting brassinosteriods, *Annu. Rev. Plant Physiol.*, 39, 23, 1988.
156. **Eun, J.-S., Kuraishi, S., and Sakurai, N.,** Changes in levels of auxin and abscisic acid and the evolution of ethylene is squash hypocotyls after treatment with brassinolide, *Plant Cell Physiol.*, 30, 807, 1989.
157. **Cookson, C. and Osborne, D. J.,** The stimulation of cell extension by ethylene and auxin in aquatic plants, *Planta*, 144, 39, 1978.
158. **Hyodo, H. and Uritani, I.,** Ethylene production in sweet potato root tissue infected by *Ceratocystis fimbriata*, *Plant Cell Physiol.*, 25, 1147, 1984.
159. **Chen, Y.-M. and Wellburn, A. R.,** Enhanced ethylene emissions from red and Norway spruce exposed to acidic mists, *Plant Physiol.*, 91, 357, 1989.
160. **Mattoo, A. K., Baker, J. E., and Moline, H. E.,** Induction by copper ions of ethylene production in *Spirodela oligorrhiza:* evidence for a pathway independent of 1-aminocyclopropane-1-carboxylic acid, *J. Plant Physiol.*, 123, 193, 1986.

Chapter 3

STRESS/WOUND ETHYLENE

Hiroshi Hyodo

TABLE OF CONTENTS

I. INTRODUCTION

Ethylene production in plants is induced by various environmental factors such as wounding, physical load, disease, drought, waterlogging, chilling temperature, and exposure to various chemicals. Abeles[1] collectively called these kinds of ethylene production stress-induced ethylene production, or more simply, stress ethylene. The term stress is used for an external factor capable of inducing a potentially injurious strain in living organisms.[2] The stress may cause changes in metabolism, leading, in severe cases, to injury. Stress ethylene is one of the general phenomena observed in plant tissues subjected to various unfavorable conditions. Yang and Pratt[3] stated that stress ethylene is produced under conditions when plants are coping with such crises as the shedding of injured parts or growing in a difficult situation. Jaffe[4] suggested that the response to mechanical stress, so called thigmomorphogenesis, represents an adaptation designed to protect plants from stress. Environmental stresses can be divided into three groups: physical stress, chemical stress, and biological stress.[3] Ethylene production is induced by many forms of physical stress. Among them are mechanical wounding, physical load, or pressure, water deficit, flooding, low temperature, and irradiation. Also, ethylene is induced in plant tissues by chemical stimuli such as heavy metal ions, herbicides, and defoliants, and gases such as sulfur dioxide and ozone. Biological stresses are caused by organisms invading and attacking the plant. It often results from infection by fungi, bacteria, or viruses and also from invasion of insects. A large amount of ethylene produced in the diseased tissue as a result of host-parasite relationship is mostly of host origin. It was suggested by Uritani[5] that ethylene may be produced as a primary response that acts as a stimulating agent for the secondary response in the tissue infected by a pathogen. Stress ethylene has been covered by a number of reviews and readers can refer to the articles by Pratt and Goeschl,[6] Abeles,[1] Yang and Pratt,[3] Lieberman,[7] Tingey,[8] Yang and Hoffman,[9] and Imaseki.[10]

Since the discovery by Adams and Yang[11] that 1-aminocyclopropane-1-carboxylic acid (ACC) serves as an intermediate of ethylene biosynthesis, the methionine-ACC pathway has been established to operate in plants for ethylene biosynthesis.[9] Much progress has been made and various aspects of ethylene biosynthesis and its regulation in plants have been clarified.[9]

This chapter deals mainly with wound-induced ethylene production as well as general aspects of stress-induced ethylene production. The biosynthesis, regulation, and physiological role of stress ethylene and the ethylene-inducing factors will also be discussed.

II. STRESS ETHYLENE

Physical load, which restrains elongation or growth, is a common cause of ethylene production in plants. Plants often grow under various kinds of physical stress, which vary in strength from weak touch to strong pressures like bending forced by wind.[12,13] These may cause reduction of elongation, increased radial expansion, and change in the shape of plants. These changes in morphology are thought to be favorable for plants to cope with the burden applied by physical stress.

Goeschl et al.[14] studied the growth of pea epicotyls subjected to nonwounding physical stress. Pea epicotyls were enclosed in chambers where their elongation was restricted by means of a foam neoprene stopper or under glass beads. The obstructed epicotyls which touched the upper stopper produced large amounts of ethylene (Figure 1). These treatments resulted in reduced elongation and increased diameter of the internodes. Goeschl et al. suggested that the increase in diameter would increase both the total emergence force and the ability of the plant axes to support the resulting load under restrictive conditions.

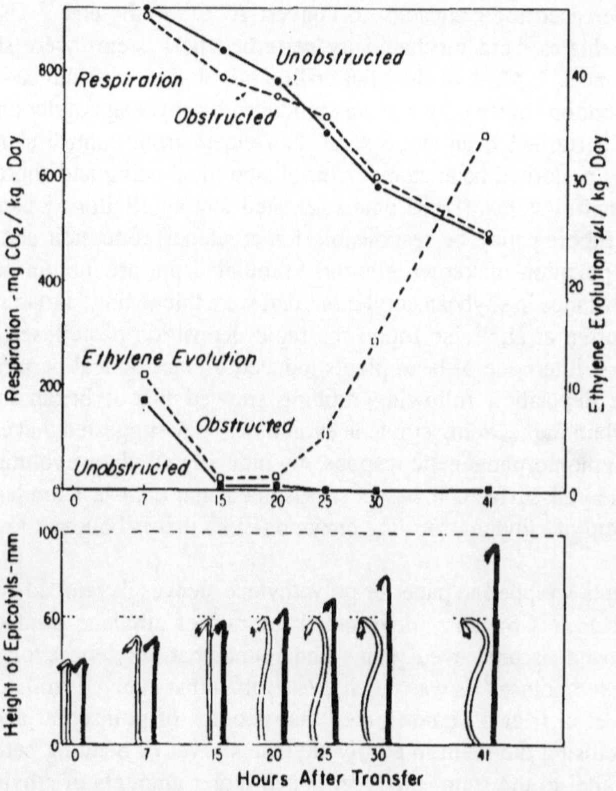

FIGURE 1. Growth pattern and gas exchange of pea epicotyls with and without obstructions. Unshaded figures in the lower panel represent growth habit of epicotyls that encountered a foam neoprene stopper about 18 h after transfer. Black figures represent the unobstructed or control epicotyls. Upper figure shows ethylene evolution and CO_2 evolution by the same epicotyls. (From Goeschl, J. D. et al., *Plant Physiol.*, 41, 877, 1966. With permission.)

Plants also respond to a gentle mechanical stimulus such as touch or rubbing, resulting in changes in plant growth and development. This phenomenon was called "thigmomorphogenesis" by Jaffe.[4,15] He studied morphological changes in young plants subjected to mechanical stimulation (perturbation) by rubbing. The rubbing stimulus was given by holding the internode between thumb and forefinger and gently rubbing it.[4] This treatment significantly retarded elongation in several plants tested. In tomato plants epinasty occurred in response to rubbing. The rubbing did not produce an injury, and when the stimulus was removed, the plant resumed normal growth.

Leopold et al.[16] found that physical stress on branches of white pine, apple, and peach trees caused by tying them in arcs (90° curvature) resulted in an increase in the ethylene content of the stressed branches. It was demonstrated by Hiraki and Ota[17] that the rate of ethylene production by *Lilium longiflorum* was enhanced by lightly stroking the leaf blades with a dusting brush daily or covering the plant top with a net, which resulted in reduced plant height.

Rubbing internodes of bean plant seedlings induced ethylene evolution.[18] After a lag period of about 1 h, a burst of ethylene production occurred in response to rubbing and lasted for about 3 h. ACC levels in perturbed tissues rose two to three times within the first 30 min. In the case of *Bryonia dioica*, however, rubbing internodes reduced their ethylene production but increased their capacity to convert ACC to ethylene.[19] Using the soybean cotyledon assay, eluates from mechanically perturbed bean stems were shown to contain elicitor-like substances.[20] Most of the elicitor-like production was induced within 1 h after mechanical perturbation. More ethylene was produced by soybean cotyledons after treatment with eluate from perturbed bean stems than with eluate from control stems. Also, eluate from mechanically perturbed bean stems induced stem thickening and shortening more than the eluate from control stems. These data suggested that an elicitor-like substance induced in internodes by rubbing may be responsible for ethylene production and thigmomorphogenesis.[20] The application of known elicitors stimulated the production of ethylene and a glyceollin-like substance in soybean cotyledons and stem thickening, and inhibited elongation of bean plants. Jaffee et al.[21] also found the rapid deposition of callose (β-1,3-glucan) in the cells of the first internode of bean plants induced by mechanical perturbation. The time course of callose deposition following rubbing showed that it began immediately after mechanical perturbation preceding ethylene evolution. They suggested that callose deposition may cause the thigmomorphogenetic response by inducing ethylene evolution. Since callose is deposited at the cell surface, it has been assumed that callose stimulates ethylene biosynthesis at the step of ethylene-forming enzyme (EFE) that is believed to be located at the cell membrane.[22]

Poinsettia plants wrapped in paper or polyethylene sleeves developed petiolar curvatures of leaves and bracts.[23,24] Sacalis[23] determined the rate of ethylene production by petioles excised from sleeved or unsleeved plants and found that ethylene evolved from excised petioles from sleeved poinsettias was much greater than that evolved from unsleeved excised petioles. Saltveit et al.[25] also demonstrated that petioles of poinsettia plants mechanically bent by either enclosing the plant in a polyethylene sleeve, or bending petioles upward and holding them parallel to the stem, produced much larger amounts of ethylene than petioles from unstressed plants. The cultivars susceptible to petiolar curvature (leaf epinasty) showed a greater enhancement of ethylene evolution by mechanical stress than cultivars showing more resistance. Foliar sprays of inhibitors of ethylene action (Ag^+) or ethylene biosynthesis (aminoethoxyvinylglycine [AVG]) before sleeving prevented development of leaf epinasty after sleeves were removed.[25,26] These data indicated that ethylene is closely related to leaf epinasty of sleeved plants.

There are a number of reports that water stress, either water deficit or water extreme, induces ethylene production in plants. As water deficit develops due to drought, a large increase in ethylene production occurs. This has been shown in intact cotton petioles,[27] detached Valencia orange leaves,[28] picked avocado fruits,[29] intact broad bean leaves,[30] young cotton bolls,[31] detached leaves of lettuce, sugar beet, squash, cabbage, tobacco, kohlrabi, bean, pepper, castor bean, and Shamouti orange,[32] excised wheat leaves,[33,34] and stored plantain fruits.[35] Stress ethylene induced by water deficit may play a regulatory role in inducing leaf abscission under the stressed condition.[32,26] In the cases of harvested avocado and plantain fruits, the enhanced rate of water loss accelerated their ripening.[29,35] It is evident that stress ethylene induced by water deficit is synthesized via the methionine and ACC pathway, as the ethylene production was inhibited by AVG[32,24] and the increase in ethylene production rate was accompanied by a parallel increase in ACC level in the stressed tissue.[34]

Plants are often exposed to flooding or waterlogging unless the drainage system is satisfactory. As a consequence, enhanced ethylene production and increased internal ethylene concentration can occur in those flooded plants, leading to leaf epinasty, leaf chlorosis, reduction in plant height, stem thickening, adventitious root formation, and wilting.[37]

Kawase[38] found that when basal portions (roots and stems) of intact plants of sunflower, chrysanthemum, and tomato were submerged in water, ethylene concentration increased not only in the submerged parts but also in the nonsubmerged upper portions. From these observations it was thought that ethylene formed in the submerged portion moved into the nonsubmerged portion or that local submersion increased ethylene concentration in other nonsubmerged parts of the intact plants. Likewise El-Beltagy and Hall[30] reported that flooding to the level of the soil surface resulted in a rise in ethylene concentration in the stem and leaves of broad bean. Jackson and Campbell[39] also demonstrated that waterlogging increased ethylene concentration in all parts of the shoot and caused leaf epinasty, leaf yellowing, and adventitious root formation in tomato plants. Further, they found that in tomato plants grown in nutrient solution supplied with a low concentration of oxygen, the ethylene concentration increased in the shoot and epinastic growth was promoted. These effects were not observed when detached shoots were used. They came to the conclusion that anaerobisis of the root system is a major factor for increasing ethylene concentration in the leaves and stems, and in stimulating petiole epinasty. It was proposed by them that the anaerobic root system could export some substances that may act as precursors or promotors of ethylene synthesis in the aerated shoot system. However, the cause of the increase in ethylene production in the flooded plants was not elucidated until Bradford and Yang[40] discovered that under waterlogging ACC accumulated in the anaerobic root and then was transported via the xylem to the shoot where it was converted to ethylene under aerobic conditon, resulting in petiole epinasty and other morphological responses. They pointed out that ACC can be synthesized in one organ and transported to another site where it exerts its hormonal activity through conversion to ethylene.

In relation to waterlogging it is interesting to note that submergence of deepwater rice *(Oryza sativa)* results in accumulation of ethylene, which in turn causes internode elongation. Kende and associates[41-44] have studied the growth response of submerged deepwater rice plants in relation to ethylene. Internodal growth was due to the accelerated cell division in the intercalary meristem and subsequent elongation of the newly formed cells. Low concentrations of O_2 resulted from submergence-enhanced ethylene synthesis. ACC synthase activity determined *in vivo* increased in the intercalary meristem and in the elongating region above it following submergence or under low levels of O_2.[45] Since GA_3 has the same effect as ethylene in enhancing internodal elongation and since GA_3 but not ethylene is effective in overcoming the growth inhibition exerted by tetcyclacis, an inhibitor of GA biosynthesis, Raskin and Kende[44] have concluded that ethylene enhances growth by increasing the activity of endogenous gibberellins. From the experiments with ancymidol, an inhibitor of gibberellin biosynthesis, and the determination of endogenous gibberellin levels in the submerged rice plants, Suge[44a] came to the same conclusion, that the presence of gibberellin is necessary for ethylene response in internodal elongation.

Exposure to low temperatures at which plants are susceptible to chilling injury causes plants to stimulate ethylene production. Ethylene production is enhanced when plants are transferred to warmer temperatures after exposure to the chilling temperature. This trend of ethylene production may be a good indicator of chilling sensitivity.[46]

Wang and Adams[47] found that chilling stress induced ACC synthase activity in cucumbers, followed by the increase in ACC level and ethylene production (Figure 2). ACC synthase activity was very low in cucumbers held at the chilling temperature (2.5°C) but upon warming (25°C), it increased rapidly and reached a plateau in 3 h. The increase in ACC synthase activity was inhibited by cycloheximide, but was not suppressed by cordycepin or α-amanitin, known inhibitors of mRNA synthesis. This implies that induction of mRNA synthesis has been completed during exposure to the chilling temperature. The ability to produce ethylene upon warming decreased after reaching a maximum with prolonged exposure to chilling temperature, even though the ACC level remained high. This suggests that the EFE system is impaired with severe chilling injury.[47,48]

Plants are sensitive to environmental chemicals. Exposure to air pollutants such as sulfur dioxide[49] and ozone[50] and to mercury in the atmosphere[51] enhanced the rate of ethylene production in plants. Peiser and Yang[49] showed that following SO_2 fumigation, the ethylene

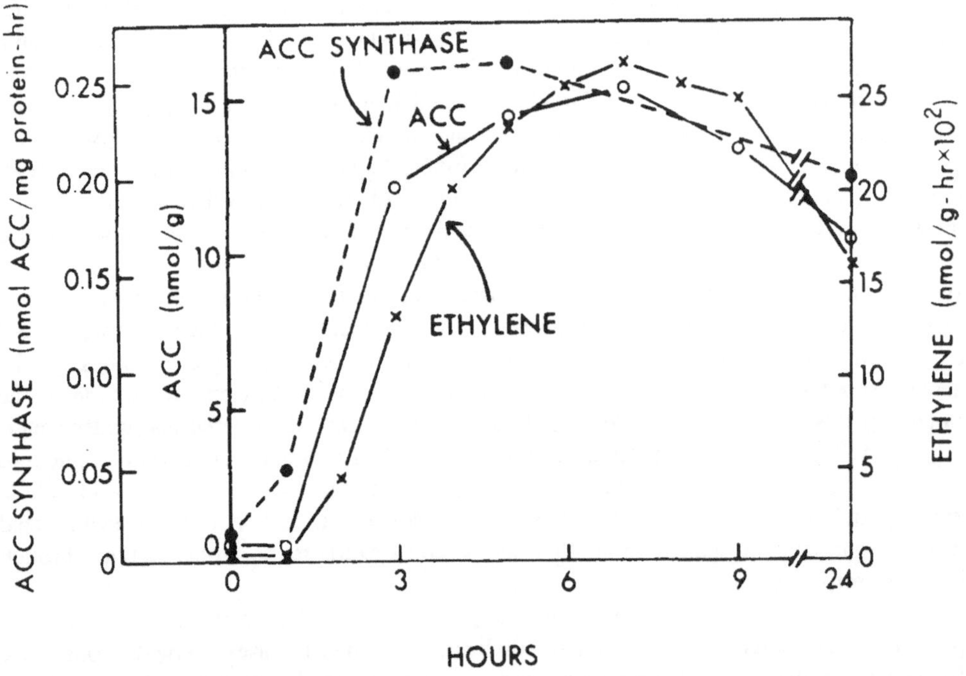

FIGURE 2. Changes in ACC synthase, ACC level, and ethylene production in cucumbers after transfer from a chilling temperature (2.5°C) to a warmer temperature (25°C). Cucumbers were exposed to chilling temperature for 4 d. ACC synthase and ACC level were measured in the skin tissues but intact fruit were used for the determination of ethylene production. (From Wang, C. Y. and Adams, D. O., *Plant Physiol.*, 69, 424, 1982. With permission.)

production rate rapidly increased in alfalfa plants, reaching a peak within 4 to 8 h, thereafter ethylene production declined to the basal level.

Heavy metal ions such as Cu^{2+} and Cd^{2+} are capable of inducing ethylene synthesis in plants. Abeles and Abeles[52] indicated that stress ethylene induced by application of $CuSO_4$, endothal, and ozone was synthesized from methionine in bean and tobacco leaves based on the evidence that [U-^{14}C]methionine was converted to radioactive ethylene. Yu and Yang[53] established that stress-ethylene induced by Cu^{2+} was synthesized via the methionine and ACC pathway by showing that the increase in ethylene evolution was preceded by and paralleled the rise in ACC formation in mung bean hypocotyls. Fuhrer[54] compared the stimulating effect of several heavy metals including Cd^{2+}, Cu^{2+}, Hg^{2+}, and Zn^{2+} on ethylene production in bean leaf tissue. Cadmium ions were shown to be the most effective. It has been likewise established that the methionine and ACC pathway is operating in Cd^{2+}-induced ethylene production. ACC synthase was rapidly induced in bean leaf tissue by exposing to Cd^{2+} solution. This activity reached a peak at 8 h and declined afterward. It should be noted that Cd^{2+} induces ACC synthase *in vivo,* while it inhibits ACC synthase activity *in vitro.* Thus, the decline in ACC synthase activity was considered to have resulted from direct interaction between Cd^{2+} and the enzyme. The rise and fall of ACC synthase are found in all systems, not limited to the Cd system. Ethylene production in *Spriodela oligorrhiza,* an aquatic angiosperm, was greatly stimulated by Cu^{2+}. This Cu^{2+}-induced ethylene production was not inhibited by AVG and the intracellular levels of ACC or its conjugates were below the detection limits, suggesting the presence of an ethylene biosynthetic pathway different from the methionine and ACC pathway.[54a]

It has been known that chemicals that cause abscission induce ethylene production in plants.[55-57] Suttle[56-58] has studied the effect of the defoliant, thidiazuron (*N*-phenyl-*N*′-1,2,3-

thiadiazol-5-ylurea) on ethylene production in mung bean hypocotyl segments, cotton leaf blades, and cotton explants containing an abscission zone. Treatment with thidiazuron resulted in a marked stimulation of ethylene evolution, parallel with a large increase in ACC formation in these tissues. He has concluded that the methionine and ACC pathway is operative in ethylene biosynthesis in thidiazuron-treated tissues and its ability to promote abscission may be accounted for by its ability to stimulate ethylene synthesis. Of several plant species tested, the cotton plant exhibited the greatest sensitivity to this chemical, while bean and corn plants showed only slight response.[58]

In response to infection by pathogens, ethylene production in plants is also stimulated. In addition to microbial penetration, invasion of insects and worms causes plants to enhance ethylene production.

Infection with root-knot nematode inhibited shoot growth in tomato plants, which was relieved by foliar spray with AVG and STS.[59] This implies that growth inhibition and other characteristics like petiole epinasty and chlorosis observed in the infected tomato plants is caused by the increased ethylene production. Although the nematode penetrated into roots, an increased ACC level was observed in both roots and leaves, and an increased ethylene production was observed in leaves.

Ethylene production rate rises in the tissue infected by fungi,[60,61] bacteria,[62,63] or viruses.[64,65] An early study was conducted by Williamson[66] using the triple response of etiolated pea seedlings for ethylene bioassay. He reported that of 12 species of plants infected with 19 pathogens, large quantities of ethylene were produced by rose leaves infected with *Diplocarpon rosae,* cherry leaves infected with *Cocomyces hiemalis,* and chrysanthemum flowers infected with *Ascochyta chrysanthemi.* It should be noted that in addition to infection with *Diplocarpon rosae,* ethylene production was induced in rose leaves by shredding or infestation with spider mite. He concluded that ethylene is evolved by the living plant tissue as a response to injury produced by invasion of certain pathogens and insect pests, and by mechanical wounding.

The increased ethylene production in sweet potato root tissue in response to infection by black rot fungus *(Ceratocystis fimbriata)* was studied by Stahmann and associates,[67] Uritani and associates,[68-71] and Chalutz and DeVay.[72] The rate of ethylene production by sweet potato root tissue inoculated with *Ceratocystis fimbriata* markedly increased with the progress of fungal invasion into the tissue. Both ACC synthase activity and ACC content in the infected tissue did not seem to be large enough to account for the high rate of ethylene production.[71] The inhibition of ethylene production by AVG in the inoculated surface layer was slight, whereas it was much greater in the inner healthy layer. The addition of ACC to the inoculated tissue did not stimulate ethylene production to the extent found in the inner healthy tissue. These lines of evidence indicate that the methionine and ACC pathway may operate in the healthy tissue adjacent to the fungus-growing area, whereas in the fungus-containing necrotic region ethylene biosynthesis may proceed partly along a pathway that may differ from the established methionine-ACC pathway.[71]

Ethylene is produced mainly by plant tissues as a result of host-parasite interaction. However, it is likely that in some cases ethylene originates from microorganisms invading plant tissues.[60] *Penicillium digitatum* has been known to be a fungus capable of producing ethylene.[73,74] Recently Goto et al.[75,76] have shown that the Kudzu strain of *Pseudomonas syringae* pv. *phaseolicola,* the halo blight pathogen, produces a large amount of ethylene. Microbial ethylene production is discussed elsewhere in this monograph (Chapter 15).

III. WOUND-INDUCED ETHYLENE PRODUCTION

Mechanical wounding refers to wounding caused by cutting, gashing, abrasion, stabbing, bruising, or intruding which may cause injury on the surface or to the inner tissues. In the host-parasite relationship, host cells may be mechanically injured as a consequence of invasion or intrusion of pathogens.

Meigh et al.[77] showed that ethylene production by tomato fruit was greatly stimulated by cutting the pericarp wall into slices of 4-mm cubes. The rate of ethylene production in slices was more than 23 times that in whole fruit. Wounding induced ethylene production also in *rin* mutant tomato fruits that produce little or no ethylene when intact.[78] McGlasson and Pratt[79] found striking differences in the rate of ethylene production between whole fruits and tissue slices (4 and 6 mm thick) of cantaloupe melon. The rate of ethylene production by the slices prepared from 15-d fruits was at least 10 times greater that that produced by whole fruits. Similarly, the rapid rate of ethylene production was induced in the discs of sweet potato root,[80] green banana fruit,[81] and grapefruit flavedo.[82]

According to Galil,[83] gashing of sycamore fig *(Ficus sycomorus)* fruits to cause the fruit to ripen has been practised over thousands of years in East-Mediterranean countries. This gashing technique has been known to induce ripening within a few days in parthenocarpic fruit, which otherwise would drop at an earlier stage. Galil describes "that the ancient practice of fig-gashing is indeed a unique method of ripening figs through traumatization of the undetached fruit". Later he and his associates[84] determined the amount of ethylene produced by gashed fruit and found that indeed there was a marked increase (50-fold) in the rate of ethylene production within 1 h of gashing, followed by a 2- to 3-fold increase in diameter and weight, and ripening within 3 d. They have therefore concluded that the ripening of the treated figs resulted from the increased ethylene production brought about by gashing. Application of ethylene, ethephon, or auxins caused the same effects as wounding.

Saltveit and Dilley[85] reported that excising epicotyl segments from etiolated pea seedlings into 9-mm length induced an increase in ethylene production. Wound-induced ethylene production increased after a lag of 26 min to reach a maximum at 56 min. It then declined to a minimum at 90 min, which was followed by a second lower maximum. A wound-induced increase in ethylene production was also observed in other tissues excised from different species of plants such as cucumber, barley, corn, wheat, and oat. Cutting subapical sections into smaller segments (2 mm) decreased the maximum rate of wound ethylene production about 30%.[86] Interestingly, the rate of wound ethylene synthesis was markedly reduced by incubating 9-mm subapical pea sections in water during the incubation (lag) period prior to wound ethylene synthesis, suggesting that a water-soluble factor necessary for inducing ethylene production may have been lost while in contact with water.[87] Working with rib segments excised from petals of morning glory, Hanson and Kende[88] found that ethylene production by rib segments was markedly increased by pressing them with a light-weight file. The rate of ethylene production rose within 1 h and then reached a maximum after an additional 1 to 2 h.

Bruising resulting from dropping impact causes fruits to produce ethylene at a greater rate. Apples bruised by dropping from a height of about 87 cm to a concrete floor had higher ethylene levels within the core cavity than unbruised controls.[89] The stimulation of ethylene evolution in the bruised apples was observed with freshly harvested apples but not with stored apples. Dropping from 40 cm onto a hard table top caused mature-green tomatoes to produce ethylene at an increased rate.[90] Both the magnitude of ethylene production and the severity of internal damage increased with number of dropping impacts. As a result ripening of bruised fruit tended to be accelerated. Fruit and vegetables are subjected to mechanical wounding and stress during harvesting, sorting, packing, and transportation, which may affect ripening and senescence processes. The effect of wounding on ripening of avocado fruit was studied by Zauberman and Fuchs.[91] Avocado fruit were wounded by excising 1 cm^3 of tissue including the peel and pulp. The rate of ethylene production was promoted at 20°C and the time to reach a peak of ethylene production became shorter at 14°C, though no increase in ethylene evolution was observed immediately after wounding at any temperature. Wounded fruit ripened faster than nonwounded fruit.

Citrus fruits are classified as nonclimacteric fruit, since mature fruits do not undergo a climacteric rise in respiration nor evolve a significantly high amount of ethylene. However, Aharoni[92] observed that young, immature oranges or grapefruit underwent a marked increase in respiration and in ethylene production after harvest. Similar observations have been made with young, immature Satsuma mandarin fruit *(Citrus unshiu)*.[93] The young fruit evolved a small amount of ethylene shortly after picking, followed by a marked increase. By cutting the fruit into halves, quarters, or eighths the rate of ethylene production was progressively enhanced. The wound-induced ethylene production was also observed in the flavedo and albedo tissues excised from the mature and fully developed fruit that did not produce ethylene at a significant level when intact.[82,94,95] The rate of ethylene production by freshly prepared albedo discs excised from Satsuma mandarin fruit was very low, but increased markedly during incubation at 26°C. It was demonstrated that L-[U-^{14}C]methionine was converted to [^{14}C]ethylene in the excised albedo tissue. Both ethylene evolution and conversion of methionine to ethylene were markedly reduced in the presence of cycloheximide.[94] Yu and Yang[53] determined ACC content in the albedo tissue excised from Valencia orange and found that both the ACC level and the rate of ethylene production increased progressively during the incubation period, and there was a close relationship between them, suggesting that the step of ACC synthesis may regulate the ethylene synthesis in the wounded albedo tissue. That ethylene was produced via the methionine and ACC pathway was supported by the observation that ethylene production in the albedo tissue was strongly inhibited by AVG. The existence of EFE activity was implicated by the observation that application of ACC to the tissue resulted in an appreciable increase in ethylene production.[96] EFE activity was, however, much higher in the aged albedo discs than in the freshly prepared ones, indicating that EFE was also inducible by wounding.[96] Riov and Yang[97] found that wound-induced ACC formation and ethylene production in flavedo tissue excised from grapefruit were almost completely arrested in the presence of applied ethylene (Figure 3). They considered that autoinhibition of ethylene synthesis was attributable to the repression of ACC synthase by ethylene.

Boller and Kende[98] investigated time courses for wound-induced ACC synthase, ACC, and ethylene production in excised pericarp tissue from green tomato fruit. The ACC synthase activity was low in freshly cut tissue, but increased 10-fold within 40 min, which was followed by a progressive increase in ACC content after a lag of about 20 min and by a progressive increase in ethylene production after a lag of 30 min (Figure 4). Wound induction of ACC synthesis in the pericarp tissue was suppressed by cycloheximide.[99] Yu and Yang[53] also found with pericarp tissue of green tomato fruit that ACC synthase activity, ACC content, and rate of ethylene production were increased by wounding. A close correlation between ACC synthase activity, ACC content, and the rate of ethylene production supported the view that the induction of ACC synthase is a key step for wound ethylene production. Enzymatic properties of ACC synthase extracted from tomato fruits have been studied.[100,101] ACC synthase induced upon wounding has similar kinetic characteristics as the enzyme that is induced during the natural development of the plant. Based on density labeling studies in which tomato pericarp tissue was incubated in D_2O, Acaster and Kende[102] concluded that ACC synthase was synthesized *de novo* following wounding.

Wound-induced ethylene synthesis in excised discs (10 mm in diameter and 3 mm thick) or plugs (1 cm in diameter and 2 cm in length) of preclimacteric cantaloupe fruit was studied by Hoffman and Yang.[103] Intact preclimacteric cantaloupe produced very little ethylene and contained little ACC, but cutting caused a massive increase in ACC and ethylene production and an increase in EFE activity. It is interesting that ethylene enhanced the development of EFE in excised tissue (Figure 5). They showed that the autocatalytic effect of ethylene in

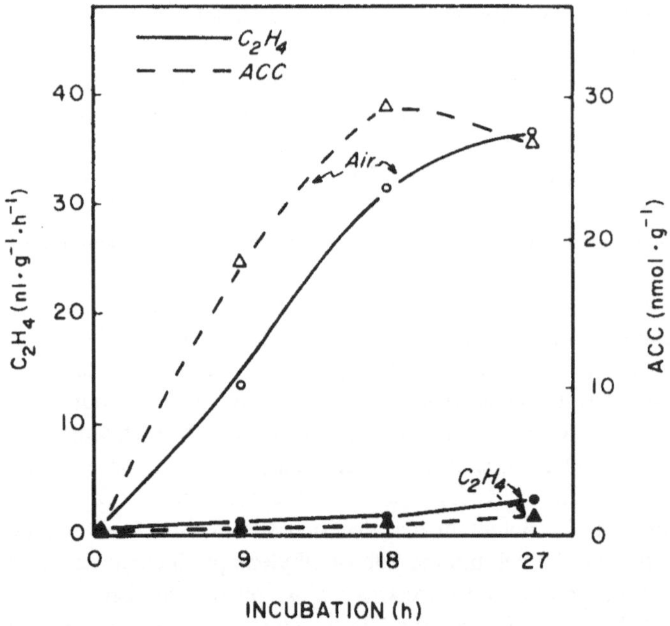

FIGURE 3. Effect of exogenous ethylene on inhibition of ethylene production rates and ACC accumulation in grapefruit flavedo discs. Ethylene production rates in the discs were determined under air after incubation for various periods under air or ethylene (20 μl/l). Immediately after completion of ethylene measurements, the same discs were used for assay of ACC. (From Riov, J. and Yang, S. F., *Plant Physiol.*, 69, 687, 1982. With permission.)

ethylene biosynthesis resulted from the fact that ethylene enhanced ethylene production by promoting the step of ACC conversion to ethylene. Recently Bufler[104,105] found in apple fruits that there were two steps in ethylene biosynthesis which were enhanced by ethylene, that is, formation of ACC by ACC synthase and formation of ethylene by EFE. He suggested that autocatalytic ethylene production may be brought about by promoting development of both ACC synthase activity and EFE activity by means of endogenous ethylene. Apple peel tissue was capable of producing significant amounts of wound ethylene after cutting.[105] Chalutz et al.[105a] observed that Cellulysin, a fungal cell wall digesting preparation, induced ethylene production in tobacco leaf tissue, which was further enhanced by pretreatment in ethylene. They indicated that the enhancement by ethylene of ethylene production resulted from a higher rate of conversion of ACC to ethylene.[105a]

Ethylene production also increased in segments excised from apple flesh.[106] The increased ethylene production reached a maximum 24 h after cutting. Both cyanide content and β-cyanoalanine synthase activity increased along with ethylene production in the wounded tissue and they peaked 36 h after cutting, later than that of ethylene production. Peiser et al.[107] have shown that cyanide released in the reaction of ACC oxidation by EFE was further metabolized to β-cyanoalanine and asparagine. In line with this, Manning[108] reported that β-cyanoalanine synthase activity increased with the rise in ethylene production in petals of carnation flowers.

A whole intact fruit of winter squash *(Cucurbita maxima)* produces only a trace amount of ethylene (less than 0.2 μl kg^{-1} h^{-1}). However, if a fruit is cut into halves or quarters, a marked increase in ethylene production occurs (more than 100 μl kg^{-1} h^{-1}). Small discs (9 mm in diameter and 2 mm thick) excised from mesocarp of winter squash fruit have also been studied with respect to wound-induced ethylene synthesis in response to cutting.[109,110,110a] A typical time course of changes in ACC synthase activity, ACC content,

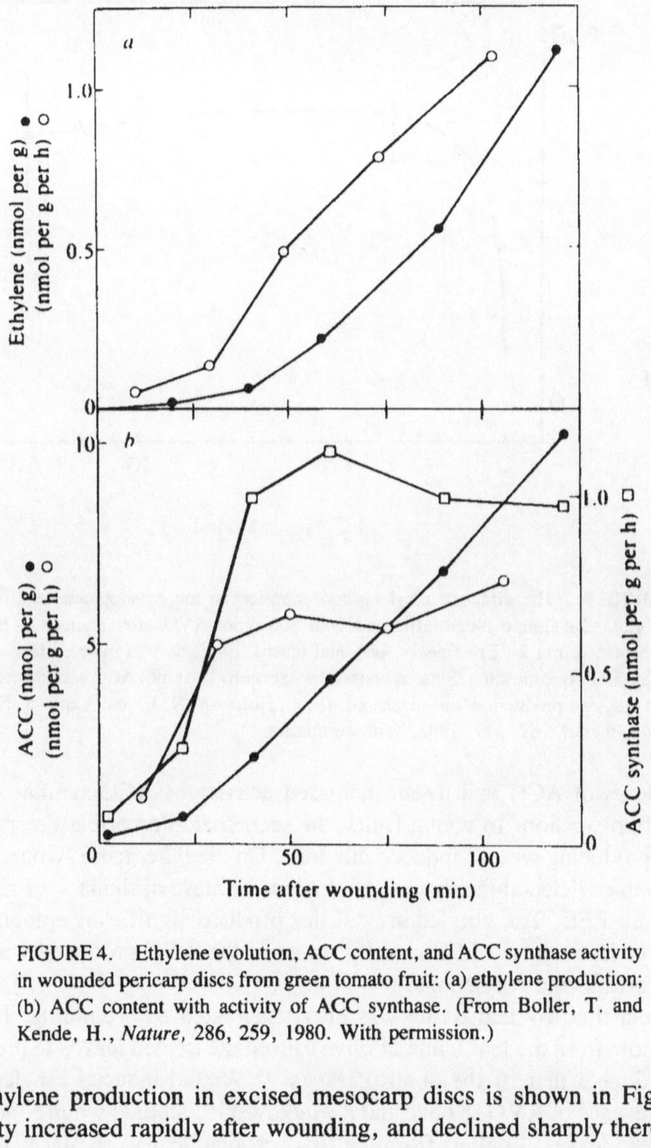

FIGURE 4. Ethylene evolution, ACC content, and ACC synthase activity in wounded pericarp discs from green tomato fruit: (a) ethylene production; (b) ACC content with activity of ACC synthase. (From Boller, T. and Kende, H., *Nature*, 286, 259, 1980. With permission.)

and rate of ethylene production in excised mesocarp discs is shown in Figure 6.[110] ACC synthase activity increased rapidly after wounding, and declined sharply thereafter. The rise in ACC synthase activity was followed by an increase in ACC content and in the rate of ethylene production. Most of the ACC found in the excised tissue was not conjugated and the bound form of ACC was hardly detectable. EFE activity was also found to increase rapidly in the excised tissue. The induction of ACC synthase was suppressed by ethylene exogenously applied to discs. The degree of inhibition increased with increasing concentration of ethylene until it reached a maximum at 92 μl/l ethylene, where the induction of ACC synthase was suppressed by more than 50%.[110] The enzyme activity was inhibited by polyamines such as putrescine, spermidine, and spermine.[111] ACC synthase in the wounded mesocarp tissue of *Cucurbita maxima* was highly purified by Nakajima and Imaseki[112] and Nakajima et al.[112a] The apparent molecular size of the purified ACC synthase was 50 kDa as estimated by SDS-polyacrylamide gel electrophoresis.[112a] The enzyme was synthesized *de novo* in the wounded mesocarp tissue. The labeling of enzyme with [^{35}S]methionine was maximal 6 h after wounding.[112a]

The fruit tissues of young and mature preclimacteric avocadoes produced a large amount of ethylene 24 h after excision.[113] The increased rate of ethylene production was accompanied

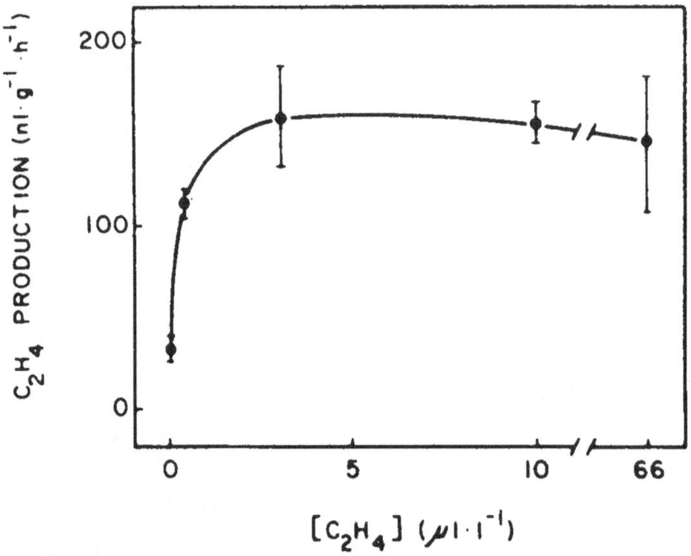

FIGURE 5. The effect of ethylene concentration on the development of EFE activity. Cantaloupe plugs infiltrated with 100 nmol AVG after excision were each placed in a 2-l Erlenmeyer flask and sealed. Ethylene was injected to give the desired concentration. Sixteen hours after excision, 100 nmol ACC was injected and ethylene production was measured. (From Hoffman, N. E. and Yang, S. F., *Plant Physiol.*, 69, 317, 1982. With permission.)

by an elevated level of ACC and by an increased activity of ACC synthase. EFE activity also increased after excision. In young fruits, the seed coat and pericarp responded actively to wounding by producing wound-induced ethylene. On the other hand, wounded endosperm tissue did not produce detectable amounts of ethylene because of the lack of ability to induce ACC synthase and EFE. The cotyledons did not produce significant amounts of ethylene either. Because of its very low EFE activity, most of the ACC produced was converted to the conjugated form in the cotyledons.

Ethylene production by leaf tissue was also accelerated by wounding. Punching out a 6-mm diameter hole from the leaf blade of olive caused the treated leaves to produce ethylene at a rate about double that of the control leaves.[114] Wound-induced ethylene was totally arrested in the presence of AVG. Konze and Kwiatkowski[115] studied wound-induced ethylene production by bean leaves. Leaflets from the first trifoliate leaf were placed on tissue paper and were pressed firmly with a light-weight metal file. When leaves were thus wounded, they began to produce ethylene at a rapid rate after a lag phase of 15 to 20 min at 20°C. Induction of wound ethylene and increase in ACC content in the leaf tissue were almost completely prevented by the addition of AVG, whereas ethylene formation from ACC was not inhibited by AVG. These results indicate that ethylene is produced in the wounded leaf tissue via the methionine and ACC pathway as already shown in other tissues.

Peeling (stripping) the epidermis from oat leaves induced a 4-fold increase in ethylene synthesis.[116] This wound ethylene induced the development of resistance to cellulolytic digestion by Cellulysin of mesophyll cell walls. In addition to mechanical wounding, ethylene production was stimulated by point freezing of sugar beet leaf tissue.[117] Ethylene formation occurred in the leaf area surrounding the frozen parts, showing a wound response of intact cells adjacent to the frozen, killed cells.

It has been known that ethylene production occurs dramatically during flower senescence. The initial observation that ethylene is closely associated with flower wilting goes back to the findings by Crocker and Knight[118] that exposing open carnation flowers to 0.5 µl/l

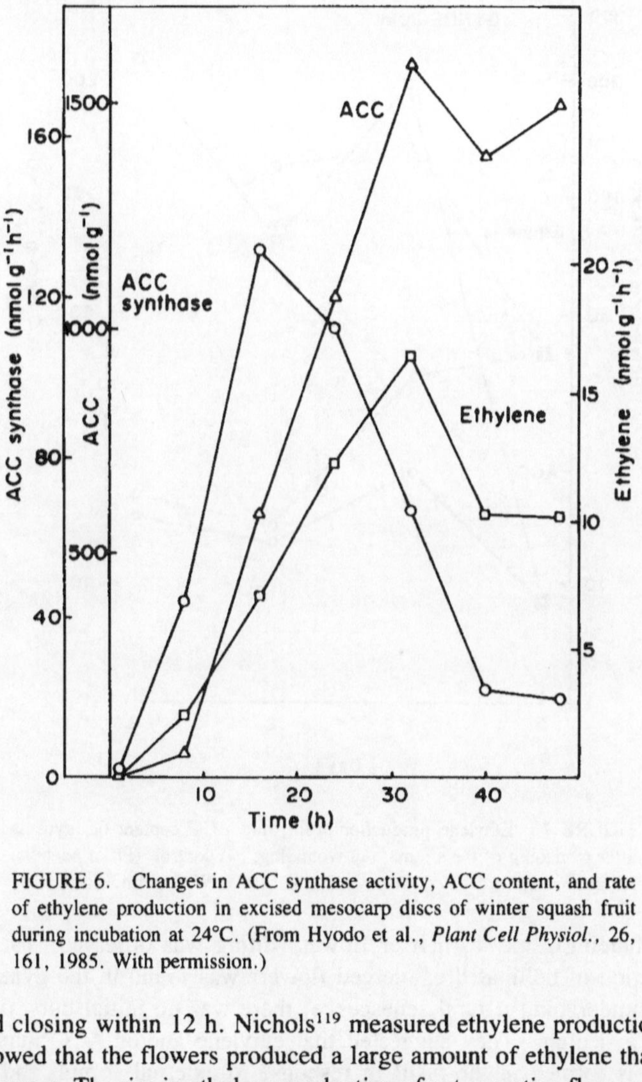

FIGURE 6. Changes in ACC synthase activity, ACC content, and rate of ethylene production in excised mesocarp discs of winter squash fruit during incubation at 24°C. (From Hyodo et al., *Plant Cell Physiol.*, 26, 161, 1985. With permission.)

ethylene caused closing within 12 h. Nichols[119] measured ethylene production by carnation flowers and showed that the flowers produced a large amount of ethylene that accompanied the wilting of flowers. The rise in ethylene production of cut carnation flowers was associated with a 30-fold increase in ACC content in the petals.[120]

Petunia flowers show a similar behavior to carnations in vase life and flower senescence. Gilissen[121] reported that wilting of the corolla of the petunia flower was accelerated by pollen tube growth in the style, while it was not affected by only pollination and germination of the stigmatic surface. It is interesting that wilting of the corolla was also accelerated by cutting off the stigma or cutting the style half-way down.[121] He suggested that the style functions as a sense organ that perceives a specific external stimulus and in turn induces metabolic changes in other parts of the flowers. It appeared that the stress given to the stigma and style by pollen tube growth or wounding may induce ethylene production in the pistil and corolla and thus accelerate the wilting of the corolla. Whitehead et al.[122] studied pollination- and wound-induced ethylene production in petunia flowers. They showed that there was a rapid rise in ethylene production by the flowers shortly after application of pollen to the stigma or wounding of the pistil. Wounding of pistils was carried out by repeated piercing (25 times) with a copper needle. The rate of ethylene production peaked about 3 h after pollination and wounding and then declined to a low level. This was followed by a more gradual rise reaching a second peak at 70 h in the pollinated flowers and at 100 h in flowers with wounded pistils. At the second peak of ethylene production, color change was visible

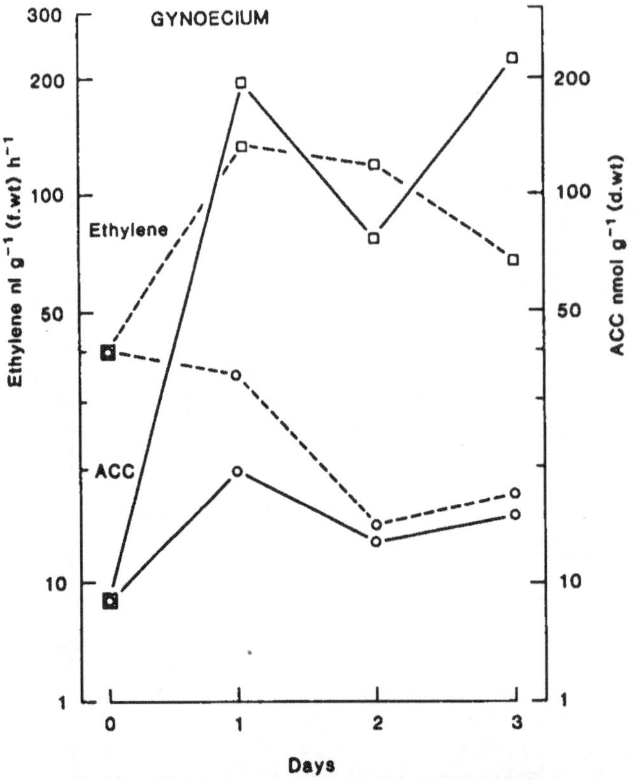

FIGURE 7. Ethylene production from, and ACC content of, gynecia after wounding of the stigma: (□) wounding; (○) control. (From Nichols, R. and Frost, C. E., *Sci. Hortic.*, 26, 47, 1985. With permission.)

in the corolla (blueing), and 40 h later flower wilting was observed. The early burst in ethylene production of pollinated or pierced flowers was found in the gynecium, while in control flowers undergoing natural senescence, there was no initial burst of ethylene production by the gynecium. They suggested that ethylene and/or ACC may be a possible signal(s) which is formed in the pistil in response to external stimuli and may stimulate ethylene production in the corolla. Nichols and Frost[123] determined the rate of ethylene production and ACC content in the wounded stigma of petunia flowers. To wound the tissue, the stigma was squeezed between forcep tips, which resulted in accelerated senescence of the corolla. The data indicated that increased production of ACC was induced in the gynecium within 1 day from stigma wounding (Figure 7). The ethylene production by the gynecium also increased coincident with an increase in ACC (Figure 7). The ACC content in the corolla significantly increased by the second day after wounding. It was assumed that ACC may be a possible candidate for the mobile signal translocating from the style to the corolla and inducing ethylene production and corolla wilting. However, Gilissen and Hoekstra[124] reported that a wilting factor found in eluates collected from the ovarian end of the pollinated or wounded styles was different from ACC.

IV. INDUCTION OF WOUND AND STRESS ETHYLENE

Wound and stress ethylene has been shown to be synthesized via the methionine and ACC pathway as mentioned previously. Information on the induction and regulation of enzymes involved in ethylene biosynthesis in response to environmental stimuli is essential to understanding the mechanism of wound- or stress-induced ethylene biosynthesis. Induction of ACC synthase and control of its activity are of particular interest, since it has been shown that the formation of ACC is a crucial step for inducing ethylene synthesis.[9] Induction of EFE is also essential in some cases.

Recent studies have shown that ethylene production is induced in plant tissues by fungal cell wall preparations (fungal elicitors),[125-128] by cell wall fragments following treatment with macerase,[129] and by treatment with cell wall digesting enzymes.[62,130-132] Takahashi and Jaffe[20] reported that an elicitor-like substance was formed in response to mechanical stress in bean internodes, which was capable of inducing ethylene synthesis in soybean cotyledons. The nature of elicitors released either from microorganisms or from plant cells due to microbial infection has been established.[133] There is considerable evidence indicating that these elicitors consist of oligosaccharides.[134-140]

The fungal elicitor isolated from *Phytophthora megasperma* was shown to induce ACC synthase in suspension-cultured cells of parsley.[127] That the effect of elicitors in stimulating ethylene production was reduced by AVG[126,127] supports the view that these active substances promote the synthesis of ACC by inducing ACC synthase. The induction of ACC synthase by elicitor in suspension-cultured cells was very rapid with a lag period of less than 20 min.[127] It resulted from *de novo* synthesis of the enzyme protein as demonstrated by density labeling with D_2O, although there was a lack of effect of inhibitors of RNA synthesis. The lack of inhibition by RNA synthesis inhibitors was also observed in the induction of ACC synthase by chilling temperature stress.[47] These results may suggest that the mRNA for ACC synthase is preexisting in suspension-cultured cells or its synthesis has been completed under chilling condition before new enzyme protein is synthesized during warming. On the other hand, induction of ACC synthase by wounding was suppressed by actinomycin D and cordycepin as well as by cycloheximide.[109] The lag periods for the inducton of ACC synthase by wounding were less than 20 min[98] and 2 h[110] in tomato pericarp tissue and winter squash mesocarp tissue, respectively. The difference in time for ACC synthase to be induced between these tissues may depend on how fast and effectively a trigger substance is formed and responds to stress.

With regard to the inducing substance it is interesting to note the studies by Ryan and associates[139] and by Ryan[141] on proteinase inhibitor inducing factor (PIIF). Their studies indicate that PIIF that initiates synthesis and accumulation of proteinase inhibitors is produced and released from wounded leaves of tomato and transported to unwounded leaves. A severe wound, for example crushing the leaf between a wooden dowel and a flat file, was quite effective in releasing PIIF. The wound-induced accumulation of proteinase inhibitors is considered to contribute to a defense reaction of plants against attacking pests. It has been shown that PIIF consists of pectic fragments with an average degree of polymerization of about 30. The smaller fragments obtained by enzyme hydrolysis with polygalacturonase were also found to have substantial PIIF activities.

Mattoo and Anderson[142] investigated the intracellular localization of ACC synthase induced in tomato fruit tissue by wounding. They found that a part of the ACC synthase was membrane-associated, the relative distribution in particulate fraction depending on the maturity of fruit used. The membrane-associated ACC synthase showed a higher affinity for SAM (Km $= 8\ \mu M$) than the soluble enzyme (Km $= 18\ \mu M$). They speculate that the newly synthesized ACC synthase may be associated with membrane and there may be a close interaction between ACC synthase and EFE in the vacuolar membrane, where the latter seems to reside.[143]

V. CONCLUSIONS

Ethylene producton is induced by various forms of stress. These include physical load, physical pressure, mechanical perturbation, mechanical wounding, water deficit, waterlogging, chilling temperature, contact with chemicals such as air pollutants, defoliants, and heavy metal ions, and invasion by microorganisms and insects. The stress ethylene has been shown to be synthesized via the methionine and ACC pathway. Induction of ACC synthase and EFE and control of their activities are of particular significance to elucidate the mechanism of stress- or wound-induced ethylene production. It seems possible to consider that some kinds of oligosaccharides originating from plant cell wall and fungal elicitors are associated with the enzyme induction involved in ethylene production.

Stress ethylene has been shown to be responsible for the alterations of growth, development, and differentiation in plants, which are exemplified by inhibition of growth, promotion of internode growth in aquatic plants, stem thickening, epinasty, abscission, acceleration of senescence and ripening, and development of disease symptoms. One of the striking features is that in response to physical stress varying from gentle touch to strong bending force, plant shoots or stems tend to be stunted and thickened, leading to a compact form of plants. This will make plants physically stronger and more resistant to subsequent severe stress.[22,144]

It is known that ethylene increases respiration, induces certain enzymes, and modifies certain metabolic activities. Although stress ethylene may be associated with wound healing[116] and defense reaction,[67] its biochemical role and its mechanism remain to be clarified. Since stress ethylene induces phenylalanine ammonia-lyase,[82] it is possible that ethylene stimulates lignification by enhancing phenylpropanoid synthesis, thus favoring the protection of wounded or diseased tissue. An exogenous application of ethylene is very effective in inducing various enzymes in plants,[145,146] but it remains to be shown whether or not endogenous ethylene formed in response to stress or wounding is acting as a signal triggering the subsequent enzyme induction in plant tissues.[147]

ACKNOWLEDGMENTS

I wish to thank Dr. Shang Fa Yang and Dr. Ikuzo Uritani for reading the manuscript and for their encouragement during its preparation.

REFERENCES

1. **Abeles, F. B.,** Stress ethylene, in *Ethylene in Plant Biology,* Academic Press, New York, 1973, chap. 5.
2. **Levitt, J.,** Stress and strain terminology, in *Response of Plants to Environmental Stresses,* Academic Press, New York, 1972, chap. 1.
3. **Yang, S. F. and Pratt, H. K.,** The physiology of ethylene in wounded plant tissues, in *Biochemistry of Wounded Plant Tissues,* Kahl, G., Ed., Walter de Gruyter, Berlin, 1978, 595.
4. **Jaffe, M. J.,** Thigmomorphogenesis: the response of plant growth and development to mechanical stimulation. With special reference to *Bryonia dioica, Planta,* 114, 143, 1973.
5. **Uritani, I.,** Protein changes in diseased plants, *Annu. Rev. Phytopathol.,* 9, 211, 1971.
6. **Pratt, H. K. and Goeschl, J. D.,** Physiological roles of ethylene in plants, *Annu. Rev. Plant Physiol.,* 20, 541, 1969.
7. **Lieberman, M.,** Biosynthesis and action of ethylene, *Annu. Rev. Plant Physiol.,* 30, 533, 1979.
8. **Tingey, D. T.,** Stress ethylene production—a measure of plant response to stress, *HortScience,* 15, 630, 1980.
9. **Yang, S. F. and Hoffman, N. E.,** Ethylene biosynthesis and its regulation in higher plants, *Annu. Rev. Plant Physiol.,* 35, 155, 1984.
10. **Imaseki, H.,** Hormonal control of wound-induced responses, *Encycl. Plant Physiol., New Ser.,* 11 (III), 485, 1985.
11. **Adams, D. O. and Yang, S. F.,** Ethylene biosynthesis: identification of 1-aminocyclopropane-1-carboxylic acid as an intermediate in the conversion of methionine to ethylene, *Proc. Natl. Acad. Sci. U.S.A.,* 76, 170, 1979.

12. **Telewski, F. W. and Jaffe, M. J.,** Thigmomorphogenesis: field and laboratory studies of *Abies fraseri* in response to wind or mechanical perturbation, *Physiol. Plant.*, 66, 211, 1986.
13. **Telewski, F. W. and Jaffe, M. J.,** Thigmomorphogenesis: the role of ethylene in the response of *Pinus taeda* and *Abies fraseri* to mechanical perturbation, *Physiol. Plant.*, 66, 227, 1986.
14. **Goeschl, J. D., Rappaport, L., and Pratt, H. K.,** Ethylene as a factor regulating the growth of pea epicotyls subjected to physical stress, *Plant Physiol.*, 41, 877, 1966.
15. **Jaffe, M. J.,** Morphogenetic responses of plants to mechanical stimuli or stress, *BioScience*, 30, 239, 1980.
16. **Leopold, A. C., Brown, K. M., and Emerson, F. H.,** Ethylene in the wood of stressed trees, *HortScience*, 7, 175, 1972.
17. **Hiraki, Y. and Ota, Y.,** The relationship between growth inhibition and ethylene production by mechanical stimulation in *Lilium longiflorum*, *Plant Cell Physiol.*, 16, 185, 1975.
18. **Biro, R. L. and Jaffe, M. J.,** Thigmomorphogenesis: ethylene evolution and its role in the changes observed in mechanically perturbed bean plants, *Physiol. Plant.*, 62, 289, 1984.
19. **Boyer, N., Desbiez, M. O., Hofinger, M., and Gasper, T.,** Effect of lithium on thigmomorphogenesis in *Bryonia dioica* ethylene production and sensitivity, *Plant Physiol.*, 72, 522, 1983.
20. **Takahashi, H. and Jaffe, M. J.,** Thigmomorphogenesis: the relationship of mechanical perturbation to elicitor-like activity and ethylene production, *Physiol. Plant.*, 61, 405, 1984.
21. **Jaffe, M. J., Huberman, M., Johnson, J., and Telewski, F. W.,** Thigmomorphogenesis: the induction of callose formation and ethylene evolution by mechanical perturbation in bean stems, *Physiol. Plant.*, 64, 271, 1985.
22. **Jaffe, M. J.,** The involvement of callose and elicitors in ethylene production caused by mechanical perturbation, in *Ethylene: Biochemical, Physiological and Applied Aspects*, Fuchs, Y. and Chalutz, E., Eds., Martinus Nijhoff/Dr. W. Junk, The Hague, 1984, 199.
23. **Sacalis, J. N.,** Ethylene evolution by petioles of sleeved poinsettia plants, *HortScience*, 13, 594, 1978.
24. **Staby, G. L., Thompson, J. F., and Kofranek, A. M.,** Postharvest characteristics of poinsettias as influenced by handling and storage procedures, *J. Am. Soc. Hortic. Sci.*, 103, 712, 1978.
25. **Saltveit, M. E., Jr., Pharr, D. M., and Larson, R. A.,** Mechanical stress induces ethylene production and epinasty in poinsettia cultivars, *J. Am. Soc. Hortic. Sci.*, 104, 452, 1979.
26. **Saltveit, M. E., Jr. and Larson, R. A.,** Reducing leaf epinasty in mechanically stressed poinsettia plants, *J. Am. Soc. Hortic. Sci.*, 106, 156, 1981.
27. **McMichael, B. L., Jordan, W. R., and Powell, R. D.,** An effect of water stress on ethylene production by intact cotton petioles, *Plant Physiol.*, 49, 658, 1972.
28. **Ben-Yehoshua, S. and Aloni, B.,** Effect of water stress on ethylene production by detached leaves of Valencia orange (*Citrus sinensis* Osbeck), *Plant Physiol.*, 53, 863, 1974.
29. **Adato, I. and Gazit, S.,** Water-deficit stress, ethylene production, and ripening in avocado fruits, *Plant Physiol.*, 53, 45, 1974.
30. **El-Beltagy, A. S. and Hall, M. A.,** Effect of water stress upon endogenous ethylene levels in *Vicia faba*, *New Phytol.*, 73, 47, 1974.
31. **Guinn, G.,** Water deficit and ethylene evolution by young cotton bolls, *Plant Physiol.*, 57, 403, 1976.
32. **Aharoni, N.,** Relationship between leaf water status and endogenous ethylene in detached leaves, *Plant Physiol.*, 61, 658, 1978.
33. **Wright, S. T. C.,** The relationship between leaf water potential and the levels of abscisic acid and ethylene in excised wheat leaves, *Planta*, 134, 183, 1977.
34. **Apelbaum, A. and Yang, S. F.,** Biosynthesis of stress ethylene induced by water deficit, *Plant Physiol.*, 68, 594, 1981.
35. **George, J. B., Marriott, J., Palmer, J. M., and Karikari, S. K.,** Sensitivity to water stress and ethylene of stored plantain fruits, *J. Exp. Bot.*, 33, 1194, 1982.
36. **Jordan, W. R., Morgan, P. W., and Davenport, T. L.,** Water stress enhances ethylene-mediated leaf abscission in cotton, *Plant Physiol.*, 50, 756, 1972.
37. **Kawase, M.,** Role of ethylene in induction of flooding damage in sunflower, *Physiol. Plant.*, 31, 29, 1974.
38. **Kawase, M.,** Effect of flooding on ethylene concentration in horticultural plants, *J. Am. Soc. Hortic. Sci.*, 97, 584, 1972.
39. **Jackson, M. B. and Campbell, D. J.,** Waterlogging and petiole epinasty in tomato: the role of ethylene and low oxygen, *New Phytol.*, 76, 21, 1976.
40. **Bradford, K. J. and Yang, S. F.,** Xylem transport of 1-aminocyclopropane-1-carboxylic acid, an ethylene precursor, in waterlogged tomato plants, *Plant Physiol.*, 65, 322, 1980.
41. **Métraux, J. P. and Kende, H.,** The role of ethylene in the growth response of submerged deep water rice, *Plant Physiol.*, 72, 441, 1983.
42. **Métraux, J. P. and Kende, H.,** The cellular basis of the elongation response in submerged deep-water rice, *Planta*, 160, 73, 1984.

43. **Raskin, I. and Kende, H.**, Regulation of growth in stem sections of deep-water rice, *Planta*, 160, 66, 1984.
44. **Raskin, I. and Kende, H.**, Role of gibberellin in the growth response of submerged deep water rice, *Plant Physiol.*, 76, 947, 1984.
44a. **Suge, H.**, Ethylene and gibberellin: regulation of internodal elongation and nodal root development in floating rice, *Plant Cell Physiol.*, 26, 607, 1985.
45. **Cohen, E. and Kende, H.**, In vivo 1-aminocyclopropane-1-carboxylate synthase activity in internodes of deepwater rice. Enhancement by submergence and low oxygen levels, *Plant Physiol.*, 84, 282, 1987.
46. **Chen, Y. Z. and Patterson, B. D.**, Ethylene and 1-aminocyclopropane-1-carboxylic acid as indicators of chilling sensitivity in various plant species, *Aust. J. Plant Physiol.*, 12, 377, 1985.
47. **Wang, C. Y. and Adams, D. O.**, Chilling-induced ethylene production in cucumbers (*Cucumis sativus* L.), *Plant Physiol.*, 69, 424, 1982.
48. **Etani, S. and Yoshida, S.**, Reversible and irreversible reduction of ACC-dependent ethylene formation in mung bean (*Vigna radiata* [L] Wilczek) hypocotyls caused by chilling, *Plant Cell Physiol.*, 28, 83, 1987.
49. **Peiser, G. D. and Yang, S. F.**, Ethylene and ethane production from sulfur dioxide-injured plants, *Plant Physiol.*, 63, 142, 1979.
50. **Hogsett, W. E., Raba, R. M., and Tingey, D. T.**, Biosynthesis of stress ethylene in soybean seedlings: similarities to endogenous ethylene biosynthesis, *Physiol. Plant.*, 53, 307, 1981.
51. **Goren, R. and Siegel, S. M.**, Mercury-induced ethylene formation and abscission in *Citrus* and *Coleus* explants, *Plant Physiol.*, 57, 628, 1976.
52. **Abeles, A. L. and Abeles, F. B.**, Biochemical pathway of stress-induced ethylene, *Plant Physiol.*, 50, 496, 1972.
53. **Yu, Y. B. and Yang, S. F.**, Biosynthesis of wound ethylene, *Plant Physiol.*, 66, 281, 1980.
54. **Fuhrer, J.**, Ethylene biosynthesis and cadmium toxicity in leaf tissue of beans (*Phaseolus vulgaris* L.), *Plant Physiol.*, 70, 162, 1982.
54a. **Mattoo, A. K., Baker, J. E., and Moline, H. E.**, Induction by copper ions of ethylene production in *Spirodela oligorrhiza:* evidence for a pathway independent of 1-aminocyclopropane-1-carboxylic acid, *J. Plant Physiol.*, 123, 193, 1986.
55. **Riov, J. and Yang, S. F.**, Effect of cycloheximide on ethylene production in intact and excised citrus fruit tissues, *J. Plant Growth Regul.*, 1, 95, 1982.
56. **Suttle, J. C.**, Effect of the defoliant thidiazuron on ethylene evolution from mung bean hypocotyl segments, *Plant Physiol.*, 75, 902, 1984.
57. **Suttle, J. C.**, Involvement of ethylene in the action of the cotton defoliant thidiazuron, *Plant Physiol.*, 78, 272, 1985.
58. **Suttle, J. C.**, Cytokinin-induced ethylene biosynthesis in nonsenescing cotton leaves, *Plant Physiol.*, 82, 930, 1986.
59. **Glazer, I., Apelbaum, A., and Orion, D.**, Reversal of nematode-induced growth retardation in tomato plants by inhibition of ethylene action, *J. Am. Soc. Hortic. Sci.*, 109, 886, 1984.
60. **Achilea, O., Chalutz, E., Fuchs, Y., and Rot, I.**, Ethylene biosynthesis and related physiological changes in *Penicillium digitatum*-infected grapefruit *(Citrus paradisi), Physiol. Plant Pathol.*, 26, 125, 1985.
61. **Achilea, O., Fuchs, Y., Chalutz, E., and Rot, I.**, The contribution of host and pathogen to ethylene biosynthesis in *Penicillium digitatum*-infected citrus fruit, *Physiol. Plant Pathol.*, 27, 55, 1985.
62. **Lund, B. M.**, The effect of certain bacteria on ethylene production by plant tissue, in *Fungal Pathogenicity and Plant's Response*, Byrde, R. J. W. and Cutting, C. V., Eds., Academic Press, London, 1973, 69.
63. **Goto, M., Yaguchi, Y., and Hyodo, H.**, Ethylene production in citrus leaves infected with *Xanthomonas citri* and its relation to defoliation, *Physiol. Plant Pathol.*, 16, 343, 1980.
64. **De Laat, A. M. M. and Van Loon, L. C.**, Regulation of ethylene biosynthesis in virus-infected tobacco leaves. I. Determination of the role of methionine as the precursor of ethylene, *Plant Physiol.*, 68, 256, 1981.
65. **De Laat, A. M. M. and Van Loon, L. C.**, Regulation of ethylene biosynthesis in virus-infected tobacco leaves. II. Time course of levels of intermediates and in vivo conversion rates, *Plant Physiol.*, 69, 240, 1982.
66. **Williamson, C. E.**, Ethylene, a metabolic product of diseased or injured plants, *Phytopathology*, 40, 205, 1950.
67. **Stahmann, M. A., Clare, B. G., and Woodbury, W.**, Increased disease resistance and enzyme activity induced by ethylene and ethylene production by black rot infected sweet potato tissue, *Plant Physiol.*, 41, 1505, 1966.
68. **Imaseki, H., Teranishi, T., and Uritani, I.**, Production of ethylene by sweet potato roots infected by black rot fungus, *Plant Cell Physiol.*, 9, 769, 1968.
69. **Sakai, S., Imaseki, H., and Uritani, I.**, Biosynthesis of ethylene in sweet potato root tissue, *Plant Cell Physiol.*, 11, 737, 1970.

70. **Kato, Y. and Uritani, I.**, Ethylene biosynthesis in diseased sweet potato root tissue with special reference to the methionine system, *Agric. Biol. Chem.*, 36, 2601, 1972.
71. **Hyodo, H. and Uritani, I.**, Ethylene production in sweet potato root tissue infected by *Ceratocystis fimbriata, Plant Cell Physiol.*, 25, 1147, 1984.
72. **Chalutz, E. and DeVay, J. E.**, Production of ethylene in vitro and in vivo by *Ceratocystis fimbriata* in relation to disease development, *Phytopathology*, 59, 750, 1969.
73. **Chou, T. M. and Yang, S. F.**, The biosynthesis of ethylene in *Penicillium digitatum, Arch. Biochem. Biophys.*, 157, 73, 1973.
74. **Fukuda, H., Fujii, T., and Ogawa, T.**, Preparation of a cell-free ethylene-forming system from *Penicillium digitatum, Agric. Biol. Chem.*, 50, 977, 1986.
75. **Goto, M., Ishida, Y., Takikawa, Y., and Hyodo, H.**, Ethylene production by the Kudzu strains of *Pseudomonas syringae* pv. *phaseolicola* causing halo blight in *Pueraria lobata* (Willd) Ohwi, *Plant Cell Physiol.*, 26, 141, 1985.
76. **Goto, M. and Hyodo, H.**, Ethylene production by cell-free extract of the Kudzu strain of *Pseudomonas syringae* pv. *phaseolicola, Plant Cell Physiol.*, 28, 405, 1987.
77. **Meigh, D. F., Norris, K. H., Craft, C. C., and Lieberman, M.**, Ethylene production by tomato and apple fruits, *Nature*, 186, 902, 1960.
78. **Herner, R. C. and Sink, K. C., Jr.**, Ethylene production and respiratory behavior of the *rin* tomato mutant, *Plant Physiol.*, 52, 38, 1973.
79. **McGlasson, W. B. and Pratt, H. K.**, Effects of wounding on respiration and ethylene production by cantaloupe fruit tissue, *Plant Physiol.*, 39, 128, 1964.
80. **Imaseki, H., Uritani, I., and Stahmann, M. A.**, Production of ethylene by injured sweet potato root tissue, *Plant Cell Physiol.*, 9, 757, 1968.
81. **McGlasson, W. B.**, Ethylene production by slices of green banana fruit and potato tuber tissue during the development of induced respiration, *Aust. J. Biol. Sci.*, 22, 489, 1969.
82. **Riov, J., Monselise, S. P., and Kahan, R. S.**, Ethylene-controlled induction of phenylalanine ammonia-lyase in citrus fruit peel, *Plant Physiol.*, 44, 631, 1969.
83. **Galil, J.**, An ancient technique for ripening sycamore fruit in East-Mediterranean countries, *Econ. Bot.*, 22, 178, 1968.
84. **Zeroni, M., Ben-Yehoshua, S., and Galil, J.**, Relationship between ethylene and growth of *Ficus sycomorus, Plant Physiol.*, 50, 378, 1972.
85. **Saltveit, M. E., Jr. and Dilley, D. R.**, Rapidly induced wound ethylene from excised segments of etiolated *Pisum sativum* L., cv. Alaska. I. Characterization of the response, *Plant Physiol.*, 61, 447, 1978.
86. **Saltveit, M. E., Jr. and Dilley, D. R.**, Rapidly induced wound ethylene from excised segments of etiolated *Pisum sativum* L., cv. Alaska. III. Induction and transmission of the response, *Plant Physiol.*, 62, 710, 1978.
87. **Saltveit, M. E., Jr. and Dilley, D. R.**, Studies of rapidly induced wound ethylene synthesis by excised sections of etiolated *Pisum sativum* L., cv. Alaska. IV. Requirement of a water-soluble, heat-stable factor, *Plant Physiol.*, 64, 417, 1979.
88. **Hanson, A. D. and Kende, H.**, Biosynthesis of wound ethylene in morning-glory flower tissue, *Plant Physiol.*, 57, 538, 1976.
89. **Lougheed, E. C. and Franklin, E. W.**, Ethylene production increased by bruising of apples, *HortScience*, 9, 192, 1974.
90. **McLeod, R. F., Kader, A. A., and Morris, L. L.**, Stimulation of ethylene and CO_2 production of mature-green tomatoes by impact bruising, *HortScience*, 11, 604, 1976.
91. **Zauberman, G. and Fuchs, Y.**, Effect of wounding on 'Fuerte' avocado ripening, *HortScience*, 16, 496, 1981.
92. **Aharoni, Y.**, Respiration of oranges and grapefruits harvested at different stages of development, *Plant Physiol.*, 43, 99, 1968.
93. **Hyodo, H.**, Ethylene production and respiration of Satsuma mandarin (*Citrus unshiu* Marc.) fruit harvested at different stages of development, *J. Jpn. Soc. Hortic. Sci.*, 45, 427, 1977.
94. **Hyodo, H.**, Ethylene production by albedo tissue of Satsuma mandarin (*Citrus unshiu* Marc.) fruit, *Plant Physiol.*, 59, 111, 1977.
95. **Evensen, K. B., Bausher, M. G., and Biggs, R. H.**, Wound-induced ethylene production in peel explants of 'Valencia' orange fruit, *HortScience*, 16, 43, 1981.
96. **Hyodo, H. and Nishino, T.**, Wound-induced ethylene formation in albedo tissue of citrus fruit, *Plant Physiol.*, 67, 421, 1981.
97. **Riov, J. and Yang, S. F.**, Autoinhibition of ethylene production in citrus peel discs. Suppression of 1-aminocyclopropane-1-carboxylic acid synthesis, *Plant Physiol.*, 69, 687, 1982.
98. **Boller, T. and Kende, H.**, Regulation of wound ethylene synthesis in plants, *Nature*, 286, 259, 1980.
99. **Kende, H. and Boller, T.**, Wound ethylene and 1-aminocyclopropane-1-carboxylate synthase in ripening tomato fruit, *Planta*, 151, 476, 1981.

100. **Boller, T., Herner, R. C., and Kende, H.,** Assay for and enzymatic formation of an ethylene precursor, 1-aminocyclopropane-1-carboxylic acid, *Planta,* 145, 293, 1979.
101. **Yu, Y. B., Adams, D. O., and Yang, S. F.,** 1-Aminocyclopropanecarboxylate synthase, a key enzyme in ethylene biosynthesis, *Arch. Biochem. Biophys.,* 198, 280, 1979.
102. **Acaster, M. A. and Kende, H.,** Properties and partial purification of 1-aminocyclopropane-1-carboxylate synthase, *Plant Physiol.,* 72, 139, 1983.
103. **Hoffman, N. E. and Yang, S. F.,** Enhancement of wound-induced ethylene synthesis by ethylene in preclimacteric cantaloupe, *Plant Physiol.,* 69, 317, 1982.
104. **Bufler, G.,** Ethylene-enhanced 1-aminocyclopropane-1-carboxylic acid synthase activity in ripening apples, *Plant Physiol.,* 75, 192, 1984.
105. **Bufler, G.,** Ethylene-promoted conversion of 1-aminocyclopropane-1-carboxylic acid to ethylene in peel of apple at various stages of fruit development, *Plant Physiol.,* 80, 539, 1986.
105a. **Chalutz, E., Mattoo, A. K., Solomos, T., and Anderson, J. D.,** Enhancement by ethylene of Cellulysin-induced ethylene production by tobacco leaf discs, *Plant Physiol.,* 74, 99, 1984.
106. **Mizutani, F., Hirota, R., and Kadoya, K.,** Cyanide metabolism involved in ethylene formation of ripening or wounded fruit tissues, *J. Jpn. Soc. Hortic. Sci.,* 55, 273, 1986.
107. **Peiser, G. D., Wang, T. T., Hoffman, N. E., Yang, S. F., Liu, H. W., and Walsh, C. T.,** Formation of cyanide from carbon 1 of 1-aminocyclopropane-1-carboxylic acid during its conversion to ethylene, *Proc. Natl. Acad. Sci. U.S.A.,* 81, 3059, 1984.
108. **Manning, K.,** Ethylene production and β-cyanoalanine synthase activity in carnation flowers, *Planta,* 168, 61, 1986.
109. **Hyodo, H., Tanaka, K., and Watanabe, K.,** Wound-induced ethylene production and 1-aminocyclopropane-1-carboxylic acid synthase in mesocarp tissue of winter squash fruit, *Plant Cell Physiol.,* 24, 963, 1983.
110. **Hyodo, H., Tanaka, K., and Yoshisaka, J.,** Induction of 1-aminocyclopropane-1-carboxylic acid synthase in wounded mesocarp tissue of winter squash fruit and the effects of ethylene, *Plant Cell Physiol.,* 26, 161, 1985.
110a. **Hyodo, H., Fujinami, H., Okada, E., and Mochizuki, T.,** Wound-induced ethylene production and 1-aminocyclopropane-1-carboxylic acid synthase in mesocarp tissue of *Cucurbita maxima,* in *Biochemical and Physiological Aspects of Ethylene Production in Lower and Higher Plants,* Clijsters, H., De Proft, M., Marcelle, R., and Van Poucke, M., Eds., Kluwer Academic Publishers, Dordrecht, 1989, 229.
111. **Hyodo, H. and Tanaka, K.,** Inhibition of 1-aminocyclopropane-1-carboxylic acid synthase activity by polyamines, their related compounds and metabolites of S-adenosylmethionine, *Plant Cell Physiol.,* 27, 391, 1986.
112. **Nakajima, N. and Imaseki, H.,** Purification and properties of 1-aminocyclopropane-1-carboxylate synthase of mesocarp of *Cucurbita maxima* Duch. fruits, *Plant Cell Physiol.,* 27, 969, 1986.
112a. **Nakajima, N., Nakagawa, N., and Imaseki, H.,** Molecular size of wound-induced 1-aminocyclopropane-1-carboxylate synthase from *Cucurbita maxima* Duch. and change of translatable mRNA of the enzyme after wounding, *Plant Cell Physiol.,* 29, 989, 1988.
113. **Sitrit, Y., Blumenfeld, A., and Riov, J.,** Ethylene biosynthesis in tissues of young and mature avocado fruits, *Physiol. Plant.,* 69, 504, 1987.
114. **Lavee, S. and Martin, G. C.,** Ethylene evolution from various developing organs of olive (*Olea europaea*) after excision, *Physiol. Plant.,* 51, 33, 1981.
115. **Konze, J. R. and Kwiatkowski, G. M. K.,** Rapidly induced ethylene formation after wounding is controlled by the regulation of 1-aminocyclopropane-1-carboxylic acid synthesis, *Planta,* 151, 327, 1981.
116. **Geballe, G. T. and Galston, A. W.,** Ethylene as an effector of wound-induced resistance to cellulase in oat leaves, *Plant Physiol.,* 70, 788, 1982.
117. **Elstner, E. F. and Konze, J. R.,** Effect of point freezing on ethylene and ethane production by sugar beet leaf disks, *Nature,* 263, 351, 1976.
118. **Crocker, W. and Knight, L. I.,** Effect of illuminating gas and ethylene upon flowering carnations, *Bot. Gaz. (Chicago),* 46, 259, 1908.
119. **Nichols, R.,** Ethylene production during senescence of flowers, *J. Hortic. Sci.,* 41, 279, 1966.
120. **Bufler, G., Mor, Y., Reid, M. S., and Yang, S. F.,** Changes in 1-aminocyclopropane-1-carboxylic-acid content of cut carnation flowers in relation to their senescence, *Planta,* 150, 439, 1980.
121. **Gilissen, L. J. W.,** The role of the style as a sense-organ in relation to wilting of the flower, *Planta,* 131, 201, 1976.
122. **Whitehead, C. S., Halevy, A. H., and Reid, M. S.,** Roles of ethylene and 1-aminocyclopropane-1-carboxylic acid in pollination and wound-induced senescence of *Petunia hybrida* flowers, *Physiol. Plant.,* 61, 643, 1984.
123. **Nichols, R. and Frost, C. E.,** Wound-induced production of 1-aminocyclopropane-1-carboxylic acid and accelerated senescence of *Petunia* corollas, *Sci. Hortic.,* 26, 47, 1985.

124. **Gilissen, L. J. W. and Hoekstra, F. A.,** Pollination-induced corolla wilting in *Petunia hybrida:* rapid transfer through the style of a wilting-inducing substance, *Plant Physiol.*, 75, 496, 1984.
125. **Paradies, I., Hümme, B., Hoppe, H. H., Heitefuss, R., and Elstner, E. F.,** Induction of ethylene formation in bean *(Phaseolus vulgaris)* hypocotyl segments by preparation isolated from germ tube cell walls of *Uromyces phaseoli, Planta,* 146, 193, 1979.
126. **Paradies, I., Konze, J. R., Elstner, E. F., and Paxton, J.,** Ethylene: indicator but not inducer of phytoalexin synthesis in soybean, *Plant Physiol.*, 66, 1106, 1980.
127. **Chappell, J., Hahlbrock, K., and Boller, T.,** Rapid induction of ethylene biosynthesis in cultured parsley cells by fungal elicitors and its relationship to the induction of phenylalanine ammonia-lyase, *Planta,* 161, 475, 1984.
128. **Toppan, A. and Esquerré-Tugayé, M. T.,** Cell surfaces in plant-microorganism interactions. IV. Fungal glycopeptides which elicit the synthesis of ethylene in plants, *Plant Physiol.*, 75, 1133, 1984.
129. **Tong, C. B., Labavitch, J. M., and Yang, S. F.,** The induction of ethylene production from pear cell culture by cell wall fragments, *Plant Physiol.*, 81, 929, 1986.
130. **Sato, K., Uritani, I., and Saito, T.,** Properties of terpene-inducing factor extracted from adults of the sweet potato weevil, *Cylas formicarius* Fabricius (Coleoptera:Brenthidae), *Appl. Entomol. Zool.*, 17, 368, 1982.
131. **Anderson, J. D., Chalutz, E., and Mattoo, A. K.,** Induction of ethylene biosynthesis by cell wall digesting enzymes, in *Cellular and Molecular Biology of Plant Stress,* Key, J. L. and Kosuge, T., Eds., Alan R. Liss, New York, 1985, 263.
132. **Fuchs, Y. and Anderson, J. D.,** Purification and characterization of ethylene inducing proteins from Cellulysin, *Plant Physiol.*, 84, 732, 1987.
133. **Darvill, A. G. and Albersheim, P.,** Phytoalexins and their elicitors—a defense against microbial infection in plants, *Annu. Rev. Plant Physiol.*, 35, 243, 1984.
134. **Nothnagel, E. A., McNeil, M., Albersheim, P., and Dell, A.,** Host-pathogen interactions. XXII. A galacturonic acid oligosaccharide from plant cell walls elicits phytoalexins, *Plant Physiol.*, 71, 916, 1983.
135. **Davis, K. R., Lyon, G. D., Darvill, A. G., and Albersheim, P.,** Host-parasite interactions. XXV. Endopolygalacturonic acid lyase from *Erwinia carotovora* elicits phytoalexin accumulation by releasing plant cell wall fragments, *Plant Physiol.*, 74, 52, 1984.
136. **Sharp, J. K., Valent, B., and Albersheim, P.,** Purification and partial characterization of a β-glucan fragment that elicits phytoalexin accumulation in soybean, *J. Biol. Chem.*, 259, 11312, 1984.
137. **Sharp, J. K., McNeil, M., and Albersheim, P.,** The primary structures of one elicitor-active and seven elicitor inactive hexa (β-D-glucopyranosyl)-D-glucitols isolated from the mycelial walls of *Phytophthora megasperma* f. sp. *glycinea, J. Biol. Chem.*, 259, 11321, 1984.
138. **Kendra, D. F. and Hadwiger, L. A.,** Characterization of the smallest chitosan oligomer that is maximally antifungal to *Fusarium solani* and elicits pisatin formation in *Pisum sativum, Exp. Mycol.*, 8, 276, 1984.
139. **Ryan, C. A., Bishop, P. D., Walker-Simmons, M., Brown, W. E., and Graham, J. S.,** Pectic fragments regulate the expression of proteinase inhibitor genes in plants, in *Cellular and Molecular Biology of Plant Stress,* Key, J. L. and Kosuge, T., Eds., Alan R. Liss, New York, 1985, 319.
140. **West, C. A., Moesta, P., Jin, D. F., Lois, A. F., and Wickham, K. A.,** The role of pectic fragments of the plant cell wall in the response to biological stresses, in *Cellular and Molecular Biology of Plant Stress,* Key, J. L. and Kosuge, T., Eds., Alan R. Liss, New York, 1985, 335.
141. **Ryan, C. A.,** Wound-regulated synthesis and vacuolar compartmentation of proteinase inhibitors in plant leaves, *Curr. Top. Cell. Regul.*, 17, 1, 1980.
142. **Mattoo, A. K. and Anderson, J. D.,** Wound-induced increase in 1-aminocyclopropane-1-carboxylate synthase activity: regulatory aspects and membrane association of the enzyme, in *Ethylene: Biochemical, Physiological and Applied Aspects,* Fuchs, Y. and Chalutz, E., Eds., Martinus Nijhoff/Dr. W. Junk, The Hague, 1984, 139.
143. **Guy, M. and Kende, H.,** Conversion of 1-aminocyclopropane-1-carboxylic acid to ethylene by isolated vacuoles of *Pisum sativum* L., *Planta,* 160, 281, 1984.
144. **Biddington, N. L.,** The effects of mechanically-induced stress in plants—a review, *Plant Growth Regul.*, 4, 103, 1986.
145. **Hyodo, H. and Yang, S. F.,** Ethylene-enhanced synthesis of phenylalanine ammonia-lyase in pea seedlings, *Plant Physiol.*, 47, 765, 1971.
146. **Boller, T., Gehri, A., Mauch, F., and Vögeli, U.,** Chitinase in bean leaves: induction by ethylene, purification, properties, and possible function, *Planta,* 157, 22, 1983.
147. **Mauch, F., Hadwiger, L. A., and Boller, T.,** Ethylene: symptom, not signal for the induction of chitinase and β-1,3-glucanase in pea pods by pathogens and elicitors, *Plant Physiol.*, 76, 607, 1984.

Chapter 4

ETHYLENE METABOLISM

Michael A. Hall

TABLE OF CONTENTS

I. INTRODUCTION

Studies on the metabolism (or more strictly, catabolism) of endogenous plant growth regulators have been a feature of work on these substances for many years. The basic premise of developmental physiology has been that there is a dose-response relationship between the endogenous concentration of plant growth regulator and developmental events. While this notion has been challenged,[1] work with mutants[2] and progress with receptors[3] tend to reinforce the validity of the hypothesis in all its essentials. As a consequence, any process involved in the control of endogenous concentrations of growth regulators must form a key part of any attempt to understand the way in which development is mediated.

In theory, this should apply as much to ethylene as to other growth regulators. The fact that this has not in general proved to be the case is a simple consequence of the fact that ethylene is a gas which can diffuse out of plant tissue with great rapidity. Thus, for many years in the case of ethylene, diffusion was seen as a substitute for metabolism and the latter process remained unexplored.

A countervailing consideration which served to promote work on the subject was the proposition that, in contrast to the situation with other plant growth regulators, ethylene metabolism had a direct role in the mechanism of action of the gas.[4]

Thus, there has been a large number of publications on this subject, although only a few groups have been involved.

II. EARLY STUDIES

The first reports on ethylene metabolism by higher plants began to appear in the late 1950s[5-8] indicating that the process did take place, albeit at very low rates at physiological concentrations of the growth regulator.

These early studies suffered from two major defects. First, no attempt was made to ensure that incubation of plant tissue with ethylene was carried out under aseptic conditions and second, the radioactive ethylene used was usually released from mercuric perchlorate complexes. The subsequent demonstration that at least some microorganisms can metabolize ethylene (see e.g., References 9 to 11) and that release of ethylene from perchlorate complexes leads to the formation of other substances, some of which are metabolized by, or incorporated into, higher plants[4,12,13] has confirmed that such reservations are justified. Indeed, even samples of "pure" ^{14}C-ethylene direct from the manufacturers contain impurities that higher plants can metabolize.[4,13] Of course, in most systems small amounts of impurities within a substrate would be unimportant because at reasonable rates of metabolism most of the products would be derived from the main component of a mixture. However, as noted above, ethylene metabolism in higher plants appeared to occur at low rates such that the proportion of substrate incorporated was of the same order as the proportion of impurities present or generated as described.

It is worth observing at this juncture that despite the well-established facts referred to, there still appear publications where no attention is paid to the purity of the ^{14}C-ethylene used.

As a result of the experimental deficiencies described above, most workers concluded that higher plants did not in fact metabolize ethylene (e.g., see Reference 14). This was not considered to be a difficulty since, as noted above, ethylene diffuses out of plants rapidly (e.g., see Reference 15) and this process was seen as a substitute for metabolism in maintaining a particular endogenous concentration of the growth regulator.

III. THE PATTERN AND FUNCTION OF ETHYLENE METABOLISM

In spite of the observations noted above, in 1975 Beyer[16] showed unambiguously that higher plants *could* metabolize ethylene. Using aseptically grown peas and highly purified ethylene Beyer demonstrated that the gas was metabolized to two separate components, one of which was trapped in NaOH and was deemed to be CO_2 and the other, termed "tissue incorporation", which consisted of at that time uncharacterized compounds. Subsequently, Blomstrom and Beyer[17] investigated the "tissue" component and identified ethylene glycol and its glucose conjugate [β-(2-hydroxy-ethyl)D-glucoside] as products of ^{14}C-ethylene metabolism in peas. Earlier, Jerie and Hall[18] had demonstrated that in developing cotyledons of *Vicia faba* the major product of ethylene metabolism was ethylene oxide and because there exist hydrolases capable of converting epoxides to glycols,[19] it seems probable that the products identified by Blomstrom and Beyer are in fact derived from ethylene oxide as the primary product, although this has never been rigorously proven.

While Beyer and co-workers had unequivocally demonstrated that ethylene metabolism was a real capability of higher plants, there remained serious questions regarding function. With the exception of the two examples noted below, the rates of ethylene metabolism observed in higher plants could not reasonably be expected to have a major effect on endogenous ethylene concentrations. Usually less than 10% of the ethylene produced by plant tissues is metabolized and in many cases, such as morning glory flowers, the figure falls to 0.2%.[20]

If control of endogenous concentration was not the role of metabolism, then what was? An answer to this question was sought by Beyer and co-workers by studies on ethylene metabolism in relation to a number of developmental events in a variety of species.

Thus it was shown that the inhibitors of ethylene action Ag(I) and CO_2 also inhibit metabolism, the former having its effect on tissue incorporation and the latter on metabolism of ethylene to CO_2. Strikingly, the extent of the inhibition of metabolism closely paralleled the effect of the inhibitor in antagonizing ethylene effects upon development[21] (Figures 1, 2, and 3).

In cut carnations[22] ethylene metabolism showed a complex pattern during petal expansion and senescence (Figure 4). For metabolism to CO_2, activity was low except during the period of petal wilt and expansion of the ovary, when a major peak of activity was detected. For tissue incorporation the pattern was similar in the petals, sepals, and bracts but the major peak of activity occurred just before full bloom. The peak of metabolism to CO_2 occurs at the point where the petals are most sensitive to ethylene.

The pattern in morning glory flowers *(Ipomea tricolor)* was somewhat similar to that in carnations in that for two to three days prior to flower opening, metabolism of ethylene occurs at a low rate. At about one day before full bloom, when the flowers become sensitive to ethylene, there is a dramatic increase in metabolism to CO_2, which declines shortly thereafter.[20]

In cotton leaf petioles metabolism of ethylene to CO_2 increases in the abscission zone just prior to the onset of abscission. The increase does not occur in adjacent petiolar tissue and changes in the rate of abscission are paralleled by changes in the rate of metabolism. Treatments with auxin or abscisic acid decreasing or increasing the rate of abscission respectively, correspondingly inhibited or promoted metabolism[23] (Figure 5). These correlations taken together seemed to indicate an association between ethylene metabolism and sensitivity to ethylene.

In effect this might be taken to imply that the metabolizing site(s) were somehow acting as receptor(s). Although examples are lacking there is nothing to suppose that such a mechanism is impossible or even improbable. The essence of a receptor-ligand interaction

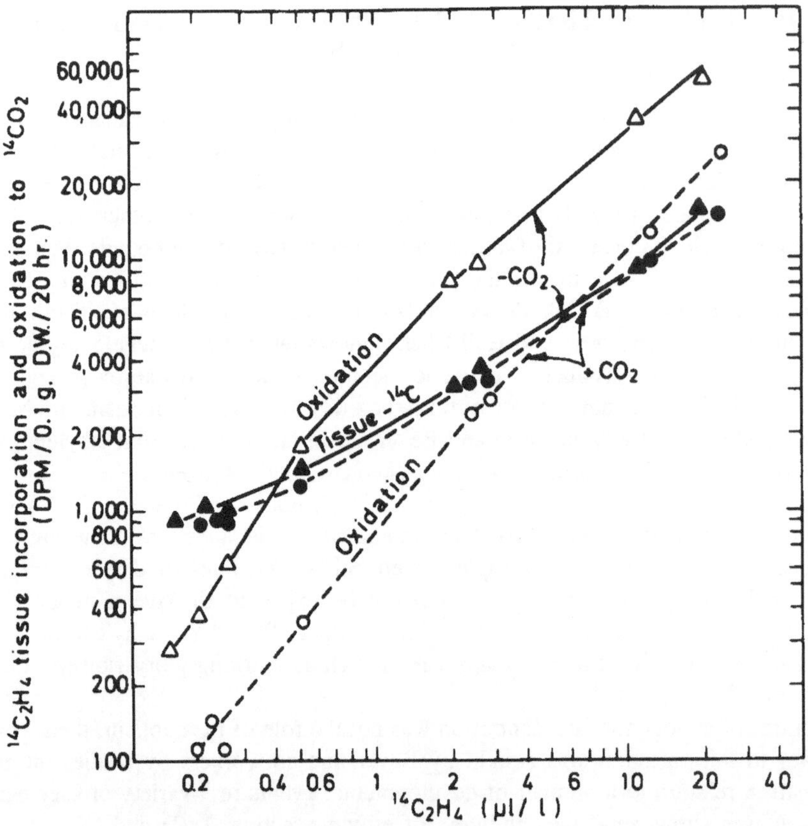

FIGURE 1. Effect of 7% CO_2 upon $^{14}C_2H_4$ metabolism in excised pea tips. (From Beyer, E. M., *Plant Physiol.*, 63, 169, 1979. With permission.)

is that as a consequence of binding of the ligand, the receptor undergoes a conformational change which in turn leads to biochemical event(s) characteristic of the effect of the growth regulator. The general concept is that the ligand is then released unchanged but there is no reason to suppose that it cannot simultaneously be metabolized. It should also be borne in mind that most of the observations and proposals on metabolism referred to antedated the characterization of candidate ethylene receptors, i.e., binding sites, in higher plants.[24-26]

IV. ETHYLENE METABOLISM IN *VICIA FABA*

A somewhat different perspective on the field was provided by the observations of Jerie and Hall[18] on ethylene metabolism in developing cotyledons of *Vicia faba.* This system differed in two important respects from those described by Beyer and co-workers. First, metabolism occurred at high rates at physiological concentrations of ethylene (up to two to three orders of magnitude higher than in peas) and second there was only one major product, namely ethylene oxide, although a small amount of CO_2 was produced and a number of minor metabolites such as ethylene glycol, glycerol, oxalate, glycollate, and ethanolamine were observed in the tissue.[27]

Subsequent work on cell-free preparations from this system[28] indicated that the enzyme involved was a monooxygenase catalysing the reaction

$$C_2H_4 + O_2 + NADPH + H^+ \rightarrow C_2H_4O + H_2O + NADP^+$$

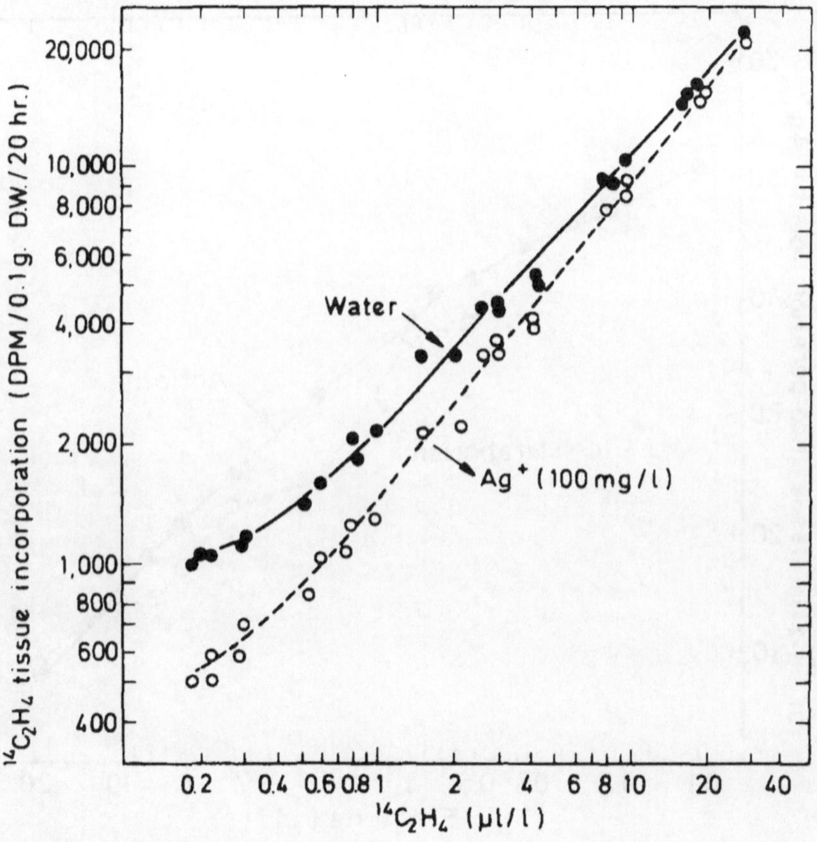

FIGURE 2. Effect of Ag^+ upon the rate of $^{14}C_2H_4$ tissue incorporation in excised pea tips. (From Beyer, E. M., *Plant Physiol.*, 63, 169, 1979. With permission.)

The observation reported by Beyer et al.[29] that ^{18}O is incorporated into the ethylene oxide apparent is further evidence that such a mechanism is likely. The enzyme has a high affinity for ethylene ($K_m = 2.5 \times 10^{-8}$ *M*) and an affinity for NADPH of 3.7×10^{-5} *M*. Similar enzymes are known in microorganisms[30,31] but so far only one other plant, namely alfalfa,[32] has been shown to behave like *Vicia faba;* the remainder of the species investigated resemble the pea system.

The ethylene-metabolizing system was detectable in other tissues of *Vicia faba* and changes in activity during leaf ontogeny were consistent with the hypothesis of Beyer, in that there appeared to be a relationship between ethylene metabolism and sensitivity to ethylene in terms of leaf senescence.[33]

V. METABOLIZING SYSTEM OR RECEPTOR?

During the same period that these investigations were proceeding, a number of problems, both technical and conceptual, presented themselves. The first of these,[4] subsequently confirmed by the work of Smith et al.,[28] demonstrated that ethylene oxide was absorbed by sodium hydroxide. Thus, in Beyer's earlier experiments radioactivity attributed to CO_2 (because it was detected in NaOH traps) could in fact have been due both to CO_2 and ethylene oxide. Equally, in order for the "metabolizing system/receptor" hypothesis to be valid, it would be expected that the affinity of the metabolizing system for ethylene would correspond with the concentration of ethylene giving a half-maximal effect in initiating developmental

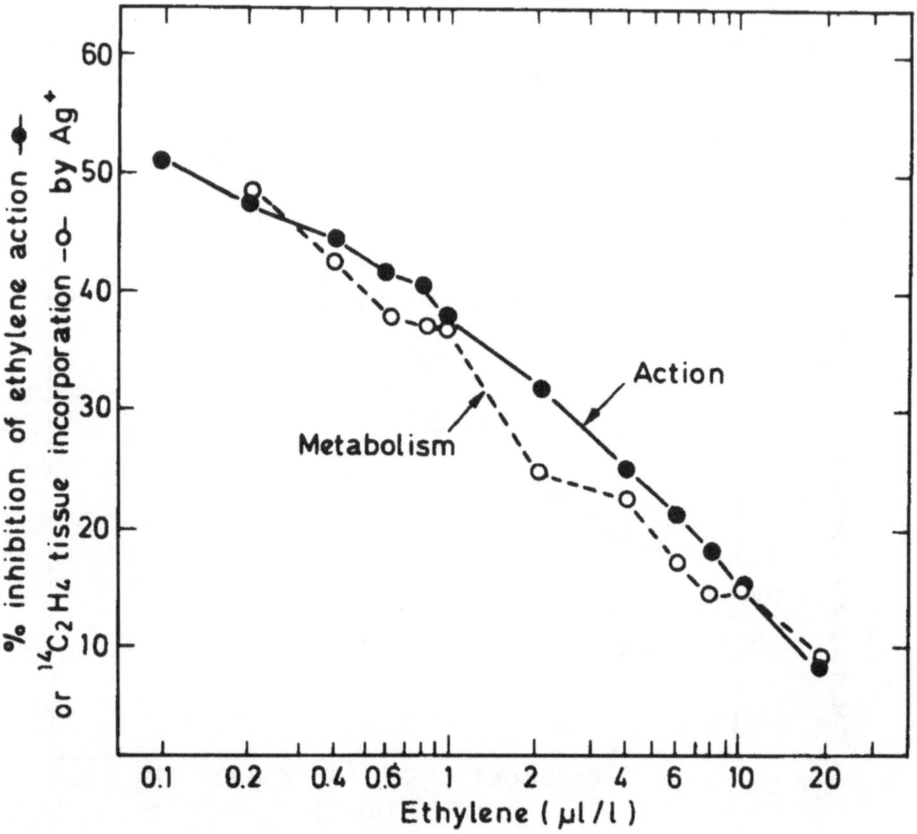

FIGURE 3. Comparison of the inibition of ethylene action and metabolism in peas by Ag^+. (From Beyer, E. M., *Plant Physiol.*, 63, 169, 1979. With permission.)

processes. Similarly, the relative affinities of the enzyme for physiologically active analogs of ethylene should parallel their relative physiological effectiveness. In the case of *Vicia faba* this was indeed the case[34] but the same could not be said for peas.

Thus Beyer[35] observed that peas could metabolize propylene, apparently more readily than ethylene. These findings were confirmed by Evans et al.,[36] who modified the experimental system to take account of ethylene oxide absorbed by the NaOH trap.

Moreover, this latter work demonstrated that the affinity of the pea system for ethylene was low — with an approximate K_m for both tissue incorporation and metabolism to CO_2 of about 10^{-6} *M*. This corresponds to a gas phase concentration of 220 µl l^{-1} — three to four orders of magnitude higher than the concentration at which ethylene induces half-maximal developmental responses in most tissues, including peas, and far in excess of naturally occurring ethylene concentrations in plants. This finding also explained the low rates of ethylene metabolism observed in peas at physiological ethylene concentrations. The same work demonstrated that the system had a higher affinity for propylene ($K_m = 1.5 \times 10^{-7}$ *M*) than for ethylene.

Clearly, therefore, these findings presented a problem for the metabolizing system/receptor hypothesis, especially since at about the same time binding sites for ethylene having the appropriate properties expected of receptors were being isolated.[24-26]

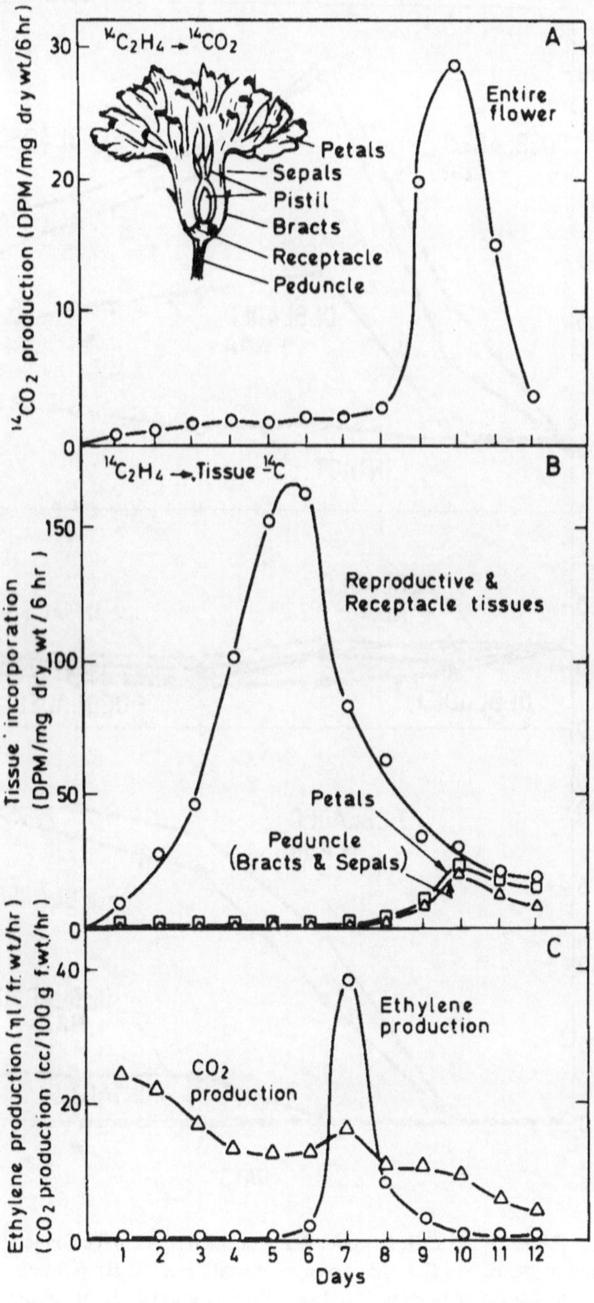

FIGURE 4. Ethylene metabolism in carnation flowers: (A) conversion of $^{14}C_2H_4$ to $^{14}CO_2$ by the entire flower; (B) tissue incorporation of $^{14}C_2H_4$; (C) production of CO_2 by $^{14}C_2H_4$-treated flowers and C_2H_4 production by untreated flowers. (From Beyer, E. M., *Plant Physiol.*, 60, 203, 1977. With permission.)

VI. THE ROLE OF ETHYLENE OXIDE

Beyer[4] observed that ethylene oxide could synergize with ethylene in inducing various developmental responses, namely the "triple response" in peas, leaf abscission in cotton, and growth stimulation in rice. He therefore modified his previous hypothesis and suggested

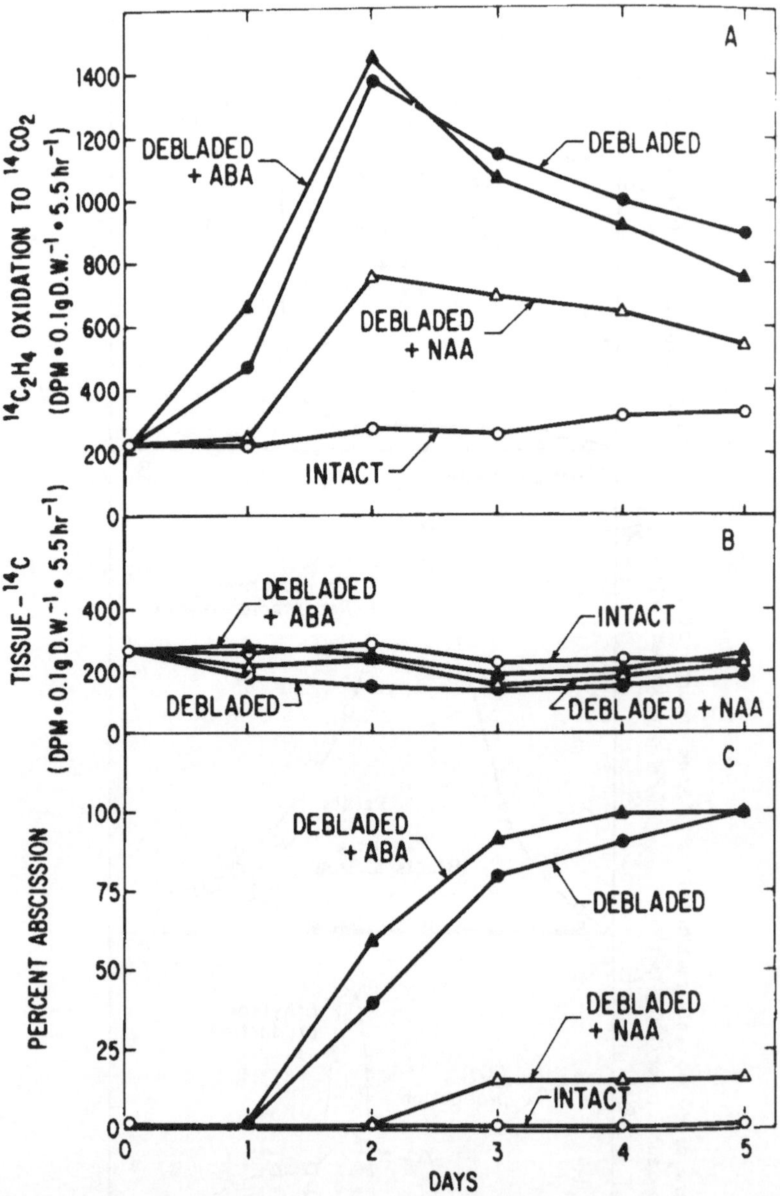

FIGURE 5. Oxidation of $^{14}C_2H_4$ by cotton leaf abscission zones in relation to abscission: (A) oxidation of $^{14}C_2H_4$ to $^{14}CO_2$; (B) tissue incorporation of $^{14}C_2H_4$; (C) time course of abscission in similar leaves to those in (A) and (B). (From Beyer, E. M., *Plant Physiol.*, 64, 971, 1979. With permission.)

that ethylene oxide somehow modulated tissue sensitivity to ethylene,[4,32] a proposition which was further discussed by Sanders et al.[37]

Abeles[38,39] challenged Beyer's hypothesis. In experiments with *Vicia faba* his evidence was (1) that silver nitrate partially blocked ethylene action but not the metabolizing activity, (2) that there was a lack of correlation between the effects of ethylene and its analogs on inhibition of epicotyl growth and their relative rates of metabolism, and (3) carbon disulfide markedly inhibited metabolism without affecting responsiveness to ethylene.

This work has been criticized on a number of grounds, principal among which was that the assay for the metabolizing system was insufficiently sensitive in that it relied solely on

TABLE 1
$^{14}C_2H_4$ Binding and Metabolism in Pea Epicotyl Tips Using Different Assay Procedures

Component	Procedure 1 Modified metabolism assay of Evans et al[36] (Bq g^{-1} F.W.)	Procedure 2 Binding & metabolism assay (Bq g^{-1} F.W.)	Procedure 3 Metabolism assay (Bq g^{-1} F.W.)	Procedure 4 Binding & metabolism assay + 100 μl/l CS_2 (Bq g^{-1} F.W.)
^{14}C in tissue	29.3 ± 1.8	7.9 ± 0.4	8.0 ± 0.3	1.9 ± 0.2
Volatile metabolites absorbed during $^{14}C_2H_4$ incubation period:				
$^{14}CO_2$	7.0 ± 0.7	8.2 ± 1.0	10.8 ± 1.3	2.8 ± 1.1
$^{14}C_2H_4O$				
Acidified 3 *M* NaOH	11.4 ± 0.9	5.6 ± 0.9	7.2 ± 0.4	0.5 ± 0.2
1 *M* H_2SO_4	—	42.3 ± 2.7	57.7 ± 9.2	0.9 ± 0.2
^{14}C released from tissue after $^{14}C_2H_4$ incubation period:				
$^{14}CO_2$	—	0.3 ± 0.1	—	0
$^{14}C_2H_4O$				
Acidified 3 *M* NaOH	—	0.3 ± 0.1	—	0.3 ± 0.1
0.25 *M* $Hg(ClO_4)_2$ + 4 *M* HCl	—	4.3 ± 0.4	—	0.7 ± 0.2
Bound $^{14}C_2H_4$	—	32.5 ± 4.1	—	34.1 ± 3.5
Total	47.7	100.5	83.7	41.1

Note: 0.7 g of pea epicotyl tips were placed in each flask and exposed to $^{14}C_2H_4$ (1 μl/l, 11,000 dpm cm^{-3}) for 20 h. Each value is the mean of 5 replicates ± s.e. minus the values for controls containing 10% $^{12}C_2H_4$.

From Sanders, I.O., Smith, A. R., and Hall, M. A., *Planta*, 174, 97, 1989. With permission.

measurement of consumption of ethylene or its analogs by the tissue and did not take possible effects on ethylene biosynthesis into account. The findings nevertheless did raise questions which necessitated answers.

VII. RECENT TECHNICAL AND CONCEPTUAL DEVELOPMENTS

For some years our group had been looking closely at the methods used to measure ethylene metabolism *in vivo*. This initially led to a modification of the technique to take into account ethylene oxide present in NaOH traps[28,36] but as the studies proceeded it emerged that factors not previously considered could lead to misapprehensions in relation both to the magnitude of metabolism and to the relative distribution of metabolites.[13] The first of these factors was that "tissue incorporation" which had previously been assumed to correspond to dissolved ethylene oxide and also its metabolites, could in fact in part be accounted for by "bound" ethylene. The second and more important factor was that NaOH traps are rather poor scavengers of ethylene oxide vapor even over quite long periods and as a consequence total metabolism to this product was being underestimated. New methods were therefore devised to measure ethylene metabolism and binding, simultaneously and separately. The effect of using these methods upon the magnitude and distribution of metabolism is shown (Tables 1 and 2). Procedure 1 corresponds to that used by Evans et al.[36] and is similar to that used previously by Beyer and co-workers but takes into account ethylene oxide trapped by NaOH; Procedure 2 measures both metabolism and binding; Procedure 3 measures metabolism only; and Procedure 4 measures binding alone (both Procedures 2 and 3 incorporate acid traps that adequately scavenge ethylene oxide vapor).

TABLE 2
Distribution of Radioactivity in Various Components as Measured by Different Assay Procedures

Component	Procedure 1	Procedure 2	Procedure 3	Procedure 2 + 100 μl l^{-1} CS_2
Tissue	29.3 (61.4)	7.9 (7.9)	8.0 (9.5)	1.9 (4.6)
CO_2	7.0 (14.7)	8.5 (8.5)	10.8 (12.9)	2.8 (6.8)
C_2H_4O	11.4 (23.9)	51.6 (51.3)	64.3 (77.6)	2.3 (5.7)
Bound C_2H_4	— (—)	32.5 (32.3)	— (—)	34.1 (82.9)

Note: Each column represents the total radioactivity in each fraction as derived from Table 1 (Bq g^{-1} F.W.) with percent distribution in parentheses.

From Sanders, I. O., Smith, A. R., and Hall, M. A., *Planta*, 174, 97, 1989. With permission.

Several important lessons emerge from these data. First, some of the tissue counts in previous work are due to ''bound'' ethylene (cf. Procedures 1 and 2). Second, metabolism to ethylene oxide (and hence total metabolism) has been seriously underestimated in the past (cf. Procedure 1 with Procedures 2 and 3). Last but not least, if the majority of the tissue counts in Procedures 2 and 3 are in fact ethylene oxide and its further metabolites, then metabolism to ethylene oxide represents about 85% of the total and metabolism to CO_2 the remainder. This is in contrast to the earlier work, where metabolism appeared to be more or less equally divided between each component. This also eliminates one of the major differences thought to exist between *Vicia faba* and most other tissues, including peas.

Further work was then carried out using these new techniques to reexamine some of the conclusions reached on ethylene metabolism.[41] Direct comparisons with previous work, both that of Beyer and of Evans et al. is not possible for the reasons set out above.

The studies examined ethylene metabolism in peas from three viewpoints, namely (1) the effects of inhibitors of ethylene action upon ethylene metabolism, (2) the effects of elicitors of ethylene-effects upon ethylene metabolism, and (3) the effects of inhibitors of ethylene metabolism upon ethylene action. The rationale behind this approach was that if ethylene metabolism or its products was involved in the mediation of ethylene effects then there should be parallel responses in the three types of experiment.

When the effects of CO_2 and Ag upon ethylene metabolism were reexamined it was clear that while both substances inhibited metabolism — as reported previously[21,36] — the relative and absolute magnitudes of the inhibitions were different from those previously described. Thus, 7% CO_2 inhibits ethylene action in peas in terms of epicotyl elongation by >90%,[40,21] however only a 23% reduction in total ^{14}C-ethylene metabolism was observed at this concentration; the effect was somewhat greater for metabolism to CO_2 than to ethylene oxide. The effect of Ag was generally similar; 0.6 m*M* $AgNO_3$ sprayed onto the seedlings inhibited metabolism by 22%. The same concentration of Ag inhibits ethylene action by about 50%.[21,41]

Sisler and Yang[42] demonstrated that 2,5-norbornadiene and cyclopentene are potent inhibitors of ethylene action. When the effects of these inhibitors on ethylene metabolism were assessed, they did indeed inhibit the process very effectively. However, while 2,5-norbornadiene is about seven times more effective than cyclopentene in antagonizing the effect of ethylene upon peas, cyclopentene is much more effective in inhibiting metabolism (see Table 3 for I_{50} values).

In similar experiments the effects of various structural analogs of ethylene were studied. To this end various concentrations of each analog were included during incubations with epicotyls which contained a fixed concentration of ^{14}C-ethylene. The results in Figure 6

TABLE 3

A Comparison of the Affinities of the Ethylene-Metabolizing and Ethylene-Binding Systems in *Pisum sativum* for Ethylene and its Analogs with the Effects of the Same Substances on Epicotyl Extension in the Same Species

	Approx. K_m or K_i for ethylene metabolism ($\mu l\ l^{-1}$)[a]	Approx. K_m or K_i relative to ethylene	K_D or K_i for ethylene binding ($\mu l\ l^{-1}$)[b]	K_D or K_i relative to ethylene	Conc. giving half-maximal induction or inhibition of ethylene responses in pea epicotyls ($\mu l\ l^{-1}$)	Conc. relative to ethylene
Ethylene	191	1	0.0205	1	0.1[c]	1
Propylene	17	0.09	7	341	10[c]	100
Vinyl chloride	17	0.09	12	586	140[c]	1400
Carbon monoxide	23442	123	62	3024	270[c]	2700
Acetylene	117	0.6	86	4195	280[c]	2800
Methyl acetylene	10	0.05	95	4634	800[c]	8000
1-Butene	9	0.047	2010	98050	27000[c]	270000
cis-2-Butene	6	0.03	5800	282926	7100[b]	71000
trans-2-Butene	3	0.016	—	—	—	—
2,5 Norbornadiene	33	0.17	57	2780	170[b]	1700
Cyclopentene	2	0.01	560	27317	1100[b]	11000
Ethane	15800	83	NE	—	NE	—
Propane	1200	6.3	NE	—	NE	—
Pentane	25	0.13	NE	—	NE	—
Carbon disulfide	0.0027	$1.4 - 10^{-5}$	NE	—	NE	—

Note: NE signifies that no effect is observed.

[a] Data from Reference 41.
[b] Data from Reference 56.
[c] Data from Reference 40.

From Sanders, I. O., Smith, A. R., and Hall, M. A., *Planta*, 174, 104, 1989. With permission.

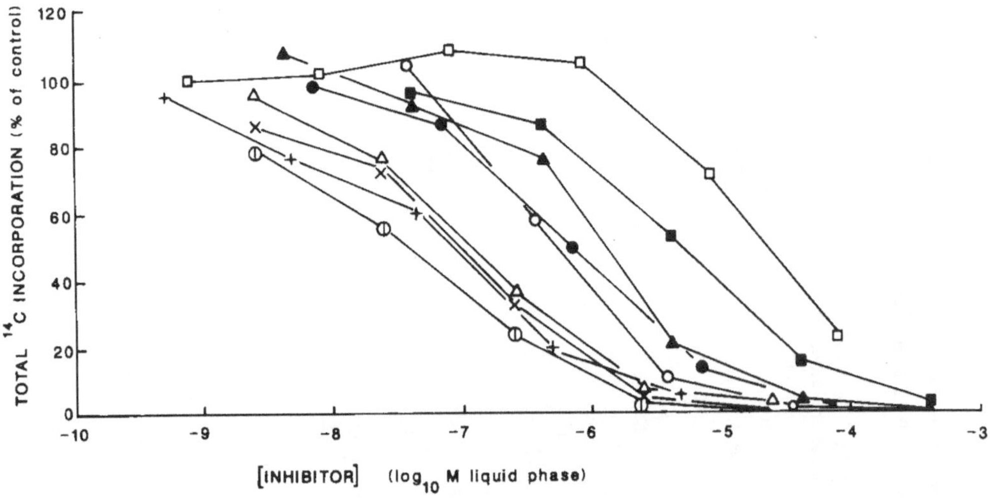

FIGURE 6. The effect of aliphatic olefins and carbon monoxide upon $^{14}C_2H_4$ metabolism in pea epicotyl tips. Ethylene (▲), carbon monoxide (□), acetylene (■), methyl acetylene (●), vinyl chloride (○), 1-butene (△), *cis*-2-butene (×), propylene (+), *trans*-2-butene (⦶). (From Sanders, I. O., Smith, A. R., and Hall, M. A., *Planta*, 174, 104, 1989. With permission.)

show that all of the analogs tested inhibited metabolism. The I_{50} values are given in Table 3. (Since the I_{50} for a competitive inhibitor is given by K_i $(1 + S/K_m)$ these values will approximate to the K_i in each case because in this work S was 10 to 100 times less than the K_ms for the three aspects of metabolism.)

With the exception of carbon monoxide and acetylene, all of the compounds used were more effective in reducing metabolism than ethylene itself. A comparison between the I_{50} values for the analogs and the concentrations of these substances bringing about half-maximal developmental responses (Table 3) shows that there is no relationship between the two. Indeed the analog most effective in inhibiting metabolism, namely *trans*-2-butene, does not elicit ethylene-induced responses. In general, an increase in molecular mass paralleled an increase in the ability to inhibit ethylene metabolism.

Further work indicated that the pea tissue could actually metabolize many of the analogs, including *cis*- and *trans*-2-butene. The ability of the tissue to metabolize an analog was roughly in proportion to its ability to inhibit ethylene metabolism as shown in previous experiments.

CS_2 is a very effective inhibitor of ethylene metabolism in peas[22] yet, as noted above, it does not appear to inhibit ethylene responses in *Vicia* epicotyls.[38,41] We were able to confirm the inhibitory effect of CS_2 upon metabolism; an observable response can be obtained at 1 nl l^{-1} gas phase (3.5×10^{-11} *M* liquid phase) and total inhibition occurs at 100 nl l^{-1}. Even at concentrations of CS_2 three orders of magnitude higher than these, there was no observable effect on ethylene-induced inhibition of epicotyl extension. The only apparently anomalous result in this work was given by dark-grown seedlings incubated in the absence of ethylene or at 0.01 $\mu l\ l^{-1}$ ethylene, where CS_2-treated epicotyls grew more than controls. However, as observed previously,[43] CS_2 is an effective inhibitor of ethylene biosynthesis, hence the flasks with epicotyls grown in the absence of CS_2 had somewhat higher ethylene concentrations than those where it was present. This had an effect where exogenous ethylene was not added or where the concentration applied was very low. In other work[41] it has been shown that even alkanes can inhibit ethylene metabolism, although these substances do not elicit ethylene responses (Table 3).

These experiments taken together indicate that there is no correlation between the characteristics of ethylene metabolism and ethylene action in peas. Equally, since the effects of the various treatments outlined above have a more or less parallel effect on the production of ethylene oxide a natural role for this metabolite seems unlikely.

Recent developments in the study of ethylene binding sites in peas[44-46] (and see the next chapter) provide negative evidence that metabolism is not a factor in responsiveness to ethylene if only because it seems unlikely that two different types of system would control the same response. Thus, the binding sites in pea show appropriate affinities for ethylene and its analogs, have the expected rate constants of association and dissociation, are inhibited by Ag at appropriate concentrations, and are unaffected by CS_2 (see Table 3 for a comparison of the effects of analogs in metabolism and binding).

VIII. THE NATURE OF ETHYLENE METABOLISM?

The type of metabolism in peas appears to be the norm for plant tissue and apart from the examples investigated by Beyer and co-workers, similar observations have been made in tomatoes, apples, pears, and medlars.[47] As noted above there now seems no reason to suppose that these processes are involved in the responses of plants to ethylene and as discussed earlier they cannot be deemed to have a role in controlling endogenous ethylene concentrations. The problem has always been that it is difficult to envisage that a metabolic process — especially one involving an important regulatory substance — has no function, especially since the existence of the phenomenon is not in doubt.

However, this presupposes that the substrate in question is indeed the natural substrate and here legitimate doubts can be raised. In none of the work described here have possible alternative naturally occurring substrates been used. Indeed this would be difficult or impossible since the enzyme(s) has not been successfully isolated and purified. Nevertheless, NADPH- or NADH-requiring monooxygenases capable of metabolizing alkenes and other substances containing double bonds e.g., zeaxanthin, 18-hydroxyoleic acid, and squalene, are known in higher plants[48,49] and animals epoxidize a wide range of alkenes including ethylene and its derivatives such as trichloroethylene.[50-52] In many of these cases cytochrome P-450 enzymes appear to be involved.[53] In this connection it is perhaps significant that the concentration of carbon monoxide necessary to inhibit ethylene metabolism in peas is close to that observed for inhibition of cytochrome P-450 monooxygenases.

It seems possible therefore that olefins, including ethylene, may be epoxidized to some extent by plant cytochrome P-450 enzymes which normally do not function in this way and whose true substrates have not yet been elucidated.

There remain the two other systems so far discovered, which do not fit into the general mold, namely those in *Vicia* and in alfalfa.

It must first be said, that in *Vicia* at least the only feature in common with the pea system is that the product is ethylene oxide. The affinity of the *Vicia* enzyme for ethylene is, as noted above, much higher than that in peas and whereas in peas the K_i for carbon monoxide is about 1.8×10^{-5} *M*,[38] in *Vicia* the corresponding figure is 6×10^{-7} *M*.[28] The first is consistent with carbon monoxide acting as an inhibitor of a cytochrome P-450 system whereas the latter is the figure expected if the inhibitor was acting as an ethylene analog.

Subsequent work has indicated that the situation is not as clear-cut as first appeared. The studies described above were carried out on developing cotyledons or on cell-free preparations from them. In the cell-free preparations most of the activity is membrane-bound. However, rehydrated mature seeds also metabolize ethylene,[54] albeit at a lower rate than developing cotyledons. Nevertheless, incubation of rehydrated seeds in 100 $\mu l\ l^{-1}$ ethylene for 24 h leads to a fivefold increase in activity to around the level found in developing cotyledons. However, most of this activity is soluble. Preliminary experiments on such rehydrated seeds indicate a change in the properties of the metabolizing system with a much lower affinity for ethylene (albeit still higher than in peas) and a decreased sensitivity to carbon monoxide.

At this stage it is unclear whether the results derive from a change in the properties of one enzyme system or whether there are in fact two such systems, one membrane-bound with a high affinity for ethylene and the other(s) with a low affinity and corresponding to the situation in peas. This problem may be resolved by further studies on purified preparations of the systems for *Vicia*.

Nevertheless, the *in vivo* metabolizing systems in *Vicia* and alfalfa species do have a high affinity for ethylene and do have a major impact on endogenous ethylene concentrations[15,54,55] whether or not ethylene is the true endogenous substrate.

It would seem worthwhile therefore to discover whether or not inhibition of metabolism has observable effects on ontogeny. If it does, then in these systems at least, there may be a real role for ethylene metabolism analogous to that perceived for other hormone metabolizing enzymes.

IX. CONCLUSIONS

We must conclude therefore that in most plants ethylene metabolism does not have a significant effect on endogenous ethylene concentration nor is it related to the mode of action of ethylene. The possible role of ethylene metabolism in systems such as *Vicia faba* and alfalfa remains to be elucidated.

REFERENCES

1. **Trewavas, A. J.,** Sensitivity and sensory adaptation in growth substance responses, in *Hormone Action in Plant Development,* Hoad, G. V., Lenton, J. R., Jackson, M. B., and Atkin, R. K., Eds., Butterworths, London, 1987, 19.
2. **Macmillan, J.,** Gibberellin deficient mutants of maize and pea and the molecular basis of gibberellin action, in *Hormone Action in Plant Development,* Hoad, G. V., Lenton, J. R., Jackson, M. B., and Atkin, R. K., Eds., Butterworths, London, 1987, 73.
3. **Venis, M. A.,** Hormone receptor sites and the study of plant development, in *Hormone Action in Plant Development,* Hoad, G. V., Lenton, J. R., Jackson, M. B., and Atkin, R. K., Eds., Butterworths, London, 1987, 53.
4. **Beyer, E. M.,** Recent advances in ethylene metabolism, *Dutch Plant Growth Regulator Group/British Plant Growth Regulator Group Monograph,* 6, 35, 1980.
5. **Buhler, D. R., Hansen, E., and Wang, C. H.,** Incorporation of ethylene into fruits, *Nature,* 179, 48, 1957.
6. **Jansen, E. F.,** Metabolism of labelled ethylene in the avocado—appearance of tritium in the methyl group of toluene, *J. Biol. Chem.,* 238, 1552, 1963.
7. **Jansen, E. F.,** Metabolism of labelled ethylene in the avocado. II. Benzene and toluene from ethylene-^{14}C; benzene from ethylene ^{3}H, *J. Biol. Chem.,* 239, 1664, 1964.
8. **Jansen, E. F.,** Ethylene and polyacetylenes: F. Metabolism of ethylene, in *Plant Biochemistry,* Bonner, J. and Varner, J. E., Eds., Academic Press, New York, 1965, 641.
9. **de Bont, J. A. M.,** Oxidation of ethylene by soil bacteria, *Antonie van Leeuwenhoek,* 42, 59, 1976a.
10. **de Bont, J. A. M. and Albers, R. A. J. M.,** Microbial metabolism of ethylene, *Antonie van Leeuwenhoek,* 42, 73, 1976b.
11. **de Bont, J. A. M. and Harder, W.,** Metabolism of ethylene by *Mycobacterium* E20, *FEMS Microbiol. Lett.,* 3, 89, 1978.
12. **Jansen, E. F.,** *Food Sci. Technol.,* 1, 475, 1969.
13. **Sanders, I. O., Smith, A. R., and Hall, M. A.,** The measurement of ethylene binding and metabolism in plant tissue, *Planta,* 174, 97, 1989.
14. **Abeles, F. B.,** *Ethylene in Plant Biology,* Academic Press, New York, 1973.

15. **Jerie, P. H., Shaari, A. R., and Hall, M. A.,** Compartmentation of ethylene in developing cotyledons in *Phaseolus vulgaris* L., *Planta*, 144, 503, 1979.
16. **Beyer, E. M.,** $^{14}C_2H_4$: its incorporation and metabolism by pea seedlings under aseptic conditions, *Plant Physiol.*, 56, 273, 1975.
17. **Blomstrom, D. C. and Beyer, E. M.,** Plants metabolise ethylene to ethylene glycol, *Nature*, 283, 66, 1980.
18. **Jerie, P. H. and Hall, M. A.,** The identification of ethylene oxide as a major metabolite of ethylene in *Vicia faba* L., *Proc. R. Soc. London Ser. B*, 200, 87, 1978.
19. **Jakoby, W. B. and Fjellstedt, J. A.,** in *The Enzymes*, Boyer, P. D., Ed., Academic Press, New York, 1972, 199.
20. **Beyer, E. M. and Sundin, O.,** $^{14}C_2H_4$ metabolism in morning glory flowers, *Plant Physiol.*, 61, 896, 1978.
21. **Beyer, E. M.,** Effects of silver ion, carbon dioxide and oxygen on ethylene action and metabolism, *Plant Physiol.*, 63, 169, 1979a.
22. **Beyer, E. M.,** $^{14}C_2H_4$, its incorporation and oxidation to $^{14}CO_2$ by cut carnations, *Plant Physiol.*, 60, 203, 1977.
23. **Beyer, E. M.,** (^{14}C) ethylene metabolism during leaf abscission in cotton, *Plant Physiol.*, 64, 971, 1979b.
24. **Sisler, E. C.,** Measurement of ethylene binding in plant tissue, *Plant Physiol.*, 64, 538, 1979.
25. **Bengochea, T., Dodds, J. H., Evans, D. E., Jerie, P. H., Shaari, A. R., and Hall, M. A.,** Studies on ethylene binding by cell-free preparations from cotyledons of *Phaseolus vulgaris* L. I. Separation and characterisation, *Planta*, 148, 397, 1980a.
26. **Bengochea, T., Dodds, J. H., Evans, D. E., Jerie, P. H., Acaster, M. A., and Hall, M. A.,** Studies on ethylene binding by cell-free preparations from cotyledons of *Phaseolus vulgaris* L. II. Effects of structural analogues and of inhibitors, *Planta*, 148, 407, 1980b.
27. **Musa, S. K.,** Studies on the metabolism of ethylene by higher plants, Ph.D. thesis, University College of Wales, Aberystwyth, 1980.
28. **Smith, P. F., Venis, M. A., and Hall, M. A.,** Oxidation of ethylene by cotyledon extracts from *Vicia faba* L.: cofactor requirements and kinetics, *Plant Physiol.*, 52, 171, 1985.
29. **Beyer, E. M., Morgan, P. W., and Yang, S. F.,** Ethylene, in *Advanced Plant Physiology*, Wilkins, M. B., Ed., Pitman Books, London, 1980, 111.
30. **Colby, J., Stirling, D. I., and Dalton, H.,** The soluble methane monooxygenase of *Methylococcus capsulatus* (Bath). Its ability to oxygenate n-alkanes, n-alkenes, ethers and alicyclic, aromatic and heterocyclic compounds, *Biochem. J.*, 165, 395, 1977.
31. **Stirling, D. I., Colby, J., and Dalton, H.,** A comparison of the substrate and electron-donor specificities of the methane monooxygenase from three strains of methane-oxidizing bacteria, *Biochem. J.*, 177, 361, 1979.
32. **Beyer, E. M.,** Ethylene metabolism, in *Ethylene and Plant Development*, Roberts, J. A. and Tucker, G., Eds., Butterworths, London, 1985, 125.
33. **Hall, M. A., Evans, D. E., Smith, A. R., Taylor, J. E., and Al-Mutawa, M. M. A.,** Ethylene and Senescence, in *Growth Regulators in Plant Senescence*, Monograph 8, Jackson, M. B., Grout, B., and Mackenzie, I. A., Eds., British Plant Growth Regulator Group, 1982, 103.
34. **Dodds, J. H., Heslop-Harrison, J. S., and Hall, M. A.,** Metabolism of ethylene to ethylene oxide by cell-free preparations from *Vicia faba* L. cotyledons: effects of structural analogues and of inhibitors, *Plant Sci. Lett.*, 19, 175, 1980.
35. **Beyer, E. M.,** Rapid metabolism of propylene by pea seedlings, *Plant Physiol.*, 61, 893, 1978.
36. **Evans, D. E., Smith, A. R., Taylor, J. E., and Hall, M. A.,** Ethylene metabolism in *Pisum sativum* L.: kinetic parameters, the effects of propylene, silver and carbon dioxide and the comparison with other systems, *Plant Growth Regul.*, 2, 187, 1984.
37. **Sanders, I. O., Smith, A. R., and Hall, M. A.,** Ethylene metabolism and action, *Physiol. Plant.*, 66, 723, 1986.
38. **Abeles, F. B.,** Non-physiological binding of ethylene by plants, *Plant Physiol.*, 74, 535, 1984a.
39. **Abeles, F. B.,** A comparative study of ethylene oxidation in *Vicia faba* and *Mycobacterium paraffinicum*, *J. Plant Growth Regul.*, 3, 85, 1984b.
40. **Burg, S. P. and Burg, E. A.,** Molecular requirements for the biological activity of ethylene, *Plant Physiol.*, 42, 144, 1967.
41. **Sanders, I. O., Smith, A. R., and Hall, M. A.,** Ethylene metabolism in *Pisum sativum* L., *Planta*, 174, 104, 1989b.
42. **Sisler, E. C. and Yang, S. F.,** Anti-ethylene effects of cis-2-butene and cyclic olefins, *Phytochemistry*, 23, 2765, 1984.
43. **Konze, J. R. and Kwiatowski, G. H. K.,** Enzymic ethylene formation from 1-aminocyclopropane-1-carboxylic acid by manganese, a protein fraction and a cofactor of etiolated pea shoots, *Planta*, 151, 320, 1981.

44. **Sisler, E. C.,** Ethylene binding receptors—is there more than one?, in *Plant Growth Substances 1988*, Faris, R. P. and Rood, S. B., Eds., Springer-Verlag, New York, 1990, 192.
45. **Hall, M. A., Howarth, C. J., Robertson, D., Sanders, I. O., Smith, A. R., Smith, P. G., Starling, R. J., Tang, Zhao-Da, Thomas, C. J. R., and Williams, R. A. N.,** Ethylene binding proteins, in *Molecular Biology of Plant Growth Control*, Fox, J. E. and Jacobs, M., Eds., Alan R. Liss, New York, 1987, 335.
46. **Hall, M. A., Bell, M. H., Connern, C. P., Raskin, I., Robertson, D., Sanders, I. O., Smith, A. R., Turner, R., Williams, R. A. N., and Wood, C. K.,** Ethylene receptors, in *Molecular Aspects of Hormonal Regulation of Plant Development*, Kutacek, M., Elliot, M. O., and Machakova, I., Eds., SPB Academic Publishing, 1990, 233.
47. **Alwan, T. F.,** Ethylene metabolism in fruit tissue, Ph.D. thesis, University College of Wales, Aberystwyth, 1986.
48. **Siefermann, D. and Yamamoto, H. Y.,** Properties of NADPH and oxygen-dependent zeaxanthin epoxidation in isolated chloroplasts—a transmembrane model for the violaxanthin cycle, *Arch. Biochem. Biophys.*, 171, 70, 1975.
49. **Croteau, R. and Kolattukudy, P. E.,** Biosynthesis of hydroxyfatty acid polymers—enzymatic hydration of 18-hydroxy-cis-9,10-epoxystearic acid to *threo*-9,10,18-trihydroxystearic acid by a particulate preparation from apple *(Malus pumila)*, *Arch. Biochem. Biophys.*, 170, 73, 1975.
50. **Maynert, E. W., Foreman, R. L., and Watabe, T.,** Epoxides as obligatory intermediates in the metabolism of olefins to glycols, *J. Biol. Chem.*, 245, 5234, 1970.
51. **Filser, J. G. and Bolt, H. M.,** Exhalation of ethylene oxide by rats on exposure to ethylene, *Mutat. Res.*, 120, 57, 1983.
52. **Miller, R. and Guengerich, F. P.,** Oxidation of trichloroethylene by liver microsomal cytochrome P-450 evidence for chlorine migration in a transition state not involving trichloroethylene oxide, *Biochemistry*, 21, 1090, 1982.
53. **West, C. A.,** Hydroxylases, monooxygenases and cytochrome P-450, in *Biochemistry of Plants—A Comprehensive Treatise, Vol. 2*, Davies, D. D., Ed., Academic Press, New York, 1980, 317.
54. **Hall, M. A., Mutumba, G. C., Sanders, I. O., Smith, A. R., Starling, R. J., and Williams, R. A. N.,** Ethylene in seed physiology, in *Growth Regulators and Seeds*, Monograph 15, Pinfield, M. J. and Black, M., Eds., British Plant Growth Regulator Group, 1987, 53.
55. **Zeroni, M., Jerie, P. H., and Hall, M. A.,** Studies on the movement and distribution of ethylene in *Vicia faba* L., *Planta*, 134, 119, 1977.
56. **Sanders, I. O., Smith, A. R., and Hall, M. A.,** Ethylene binding in epicotyls of *Pisum sativum* L. cv "Alaska", *Planta*, in press.

Chapter 5

ETHYLENE-BINDING COMPONENTS IN PLANTS

Edward C. Sisler

TABLE OF CONTENTS

I. EARLY WORK

Neljubov[1] in 1901 reported the effects of ethylene on plants, and in 1904 Richards and MacDougal[2] reported that carbon monoxide also produced effects similar to ethylene. By 1913 it was known that acetylene, propylene, carbon monoxide, as well as ethylene gave the "triple response" in peas.[3]

Burg and Burg[4] were the first to make a comprehensive study of the molecular requirements for the biological activity of ethylene. Based on the growth of pea segments, several compounds were found to elicit an ethylene-like response. The approach used by Burg and Burg and their discussion and interpretation of results clearly recognized that a binding site should be present. It has been known since 1827 that olefins form complexes with metals, and based on the order of binding of olefins to silver, they postulated that ethylene was acting by binding to a metal. One important feature missed by Burg and Burg[4] was that some olefins bind to the site but rather than elicit a response, prevent it.

II. HOW DOES ETHYLENE ACT?

There are three ways that ethylene might act: (1) by serving as a cofactor in some reaction; (2) by being oxidized to some essential component and being incorporated into tissue; or (3) by binding to a receptor, providing some essential function, and then either diffusing away or being destroyed as is the case with other hormones. Each of these ways has been investigated and will be discussed. Beyer[5] and Abeles et al.,[6] using deuterium-labeled ethylene, investigated the possibility that ethylene acted as a cofactor in some reaction. They measured the possible exchange of deuterium on the deuterium-labeled ethylene in solution. No exchange of deuterium was detected, and this mode of action was thought to be unlikely. Beyer,[7] using specially purified ethylene, showed that some ethylene was metabolized by pea plants. Some of the ^{14}C from ethylene was converted to CO_2 and some was incorporated into tissue. Ethylene oxide was identified as the major intermediate. Using data obtained from the inhibition of ethylene oxidation by Ag^+ and CO_2, Beyer proposed a model based on ethylene oxidation. The oxidation data and the data for a physiological response did correlate well. A major objection was that the oxidation system did not saturate even at 140 μl/l of ethylene while the physiological response was saturated at 1 to 2 μl/l. Abeles[8] subsequently showed that CS_2 completely inhibits ethylene oxidation while not affecting ethylene action. These two findings make it difficult to ascribe more than a secondary role to ethylene oxidation. The oxidation of ethylene likely does involve the interaction of ethylene with a receptor or enzyme.[9-11] The oxidation of ethylene is the subject of another chapter in this book. The evidence at the present time favors the theory that ethylene binds to a receptor and brings about some essential change and then diffuses away unchanged. What function it provides is unknown.

III. CRITERIA FOR A PHYSIOLOGICAL RECEPTOR

An aqueous suspension of cuprous chloride will absorb ethylene well beyond the normal solubility of either.[12] Thus ethylene can bind without being bound to a receptor. Trewavas and Jones[13] list criteria needed for a component to be considered as a physiological receptor:

1. Binding is reversible
2. It is saturable and specific
3. The dissociation constant should be equal to or lower than the biologically active concentration
4. The number of sites is small

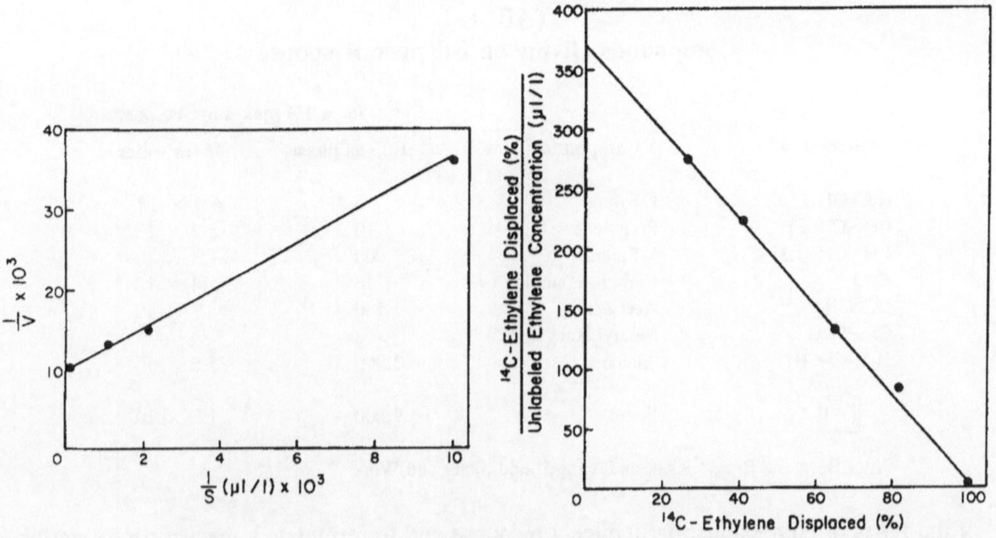

FIGURE 1. Lineweaver-Burk plot for ethylene action (tobacco leaf respiration) and Scatchard plot for ethylene binding in tobacco leaves; K_m (0.30 µl) and K_d (0.27 µl/l). (From Sisler, E. C., *Plant Physiol.*, 64, 538, 1979. With permission.)

5. Chemical modification leads to equivalent modification in binding and in eliciting a physiological response
6. Demonstration of biological function, which is a criterion for the future

At the present time there is evidence that more than one set of binding sites may be present,[14] but the first criterion, that the binding be reversible, seems to have been met. Although all binding sites appear to be reversible, there are considerable differences in the rates of association and dissociation. Some have a half-life of many hours; others have a half-life of a few minutes. All of the reported binding sites appear to be saturable and specific.

The criterion that the K_d of a binding site be close to or lower than the biological concentration in order to be considered as a physiological receptor has been shown to be the case in a number of instances, and the other criteria appear to be met (Figure 1). There are values for a one-half maximum physiological response available from a number of plants or plant parts. These range from 0.09 to 0.7 µl/l or perhaps higher, and the K_ds reported are within this range.

The criterion that the number of binding sites be low is met; however, the greatest number of binding sites has been found in tissue in which no physiological function for the site has been observed. Many of the tissues that do respond to ethylene have been shown to have very low amounts of binding sites.

IV. INTERACTION OF CHEMICALLY MODIFIED COMPONENTS AND OTHER COMPONENTS WITH BINDING SITES AND THE PHYSIOLOGICAL RECEPTOR

These fall into two categories: (1) those that interact with the site and give an ethylene response, and (2) those that interact with the site and do not give an ethylene response but compete with ethylene and can prevent the physiological response.

It was observed early that compounds other than ethylene gave a physiological response similar to ethylene (Table 1). Burg and Burg[4] attempted to make a systematic investigation

TABLE 1
Compounds Giving an Ethylene Response

Structure	Compound	Conc. for a 1/2 maximum response: μl/l gas phase	Conc. for a 1/2 maximum response: M (in water)
$H_2C{=}CH_2$	Ethylene	0.1	4.8×10^{-10}
$CH_3{-}CH{=}CH_2$	Propylene	10	5.2×10^{-8}
$CH_3{-}CH_2{-}CH{=}CH_2$	1-Butene	27,000	1.3×10^{-4}
$C{\equiv}O$	Carbon monoxide	270	5.4×10^{-8}
$HC{\equiv}CH$	Acetylene	280	1×10^{-6}
$CH_3{-}N{\equiv}C$	Methyl isocyanide	2	—
$H_2C{=}C{=}CH_2$	Allene	2,900	5.6×10^{-5}
	Furan	9,000	1.7×10^{-3}

After Burg and Burg,[4] Sisler and Yang,[64] and Sisler and Wood.[16]

of the types of compounds that induce a response and to formulate a mechanism of action. They showed that certain compounds gave similar kinetics when compared to ethylene in the pea growth test, and K_m values for a number of compounds were reported. Among these were ethylene, propylene, acetylene, allene, some substituted vinyl compounds, and carbon monoxide. Since that time several isocyanides have been reported to give an ethylene-like response,[15] and phosphorous trifluoride has been reported to be active.[15] Furan, a heterocyclic compound, is reported to give an ethylene response.[16]

Although Burg and Burg[4] tested some compounds which compete with ethylene for the binding site but do not give a response, they apparently never observed the effect. They tested *cis*-2-butene and 1,3-butadiene but they were simply reported as being inactive. Both of these compounds compete with ethylene and block the site. It is necessary to first have ethylene action to observe the blocking action of these compounds. The first report of an antiethylene effect of another vinyl compound was by Sisler and Pian,[17] in which 2,5-norbornadiene was shown to prevent ethylene action. Although it was not at first known whether the effect was direct, it was later shown that the compound interacts with the binding site[18,19] and gives competitive kinetics with ethylene in the physiological response.[20] Since the initial reports of the antiethylene effects of 2,5-norbornadiene, many other compounds have been found to elicit the response;[16,20,21] however, 2,5-norbornadiene remains one of the most effective (Table 2). Unfortunately, it has a very offensive odor, limiting its application.

V. COMPOUNDS INTERACTING WITH THE ETHYLENE RECEPTOR OR RESPONSE IN AN UNKNOWN MANNER

Burg and Burg[4] reported that CO_2 was a competitive inhibitor of ethylene action. Others have reported that the effect is probably not competitive, although the data frequently appear to be competitive data. The effect of CO_2 is complex,[22] with both stimulation of ethylene production as well as an inhibitory effect which does not appear to directly involve the binding site. How CO_2 causes an inhibition of ethylene action remains a mystery, yet in some instances it is one of the most practical inhibitors of ethylene action. Burg and Burg[4] also reported that oxygen was necessary for ethylene action at levels higher than those that would affect respiration.

Beyer[23] reported that Ag^+ ion specifically inhibits the action of ethylene. The effect probably is not competitive. Silver ion does inhibit ethylene binding both *in vitro*[19] and *in vivo*.[24] There is no conclusive experiment that shows that this is the way, or the only way,

TABLE 2
Inhibition of Ethylene Action by Various Compounds

Compound structure	Name	Plant material	K_i (gas)(μl/l)	K_i (water)(*M*)
	*trans*-Cyclooctene	Banana	0.78	2.5×10^{-8}
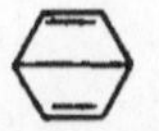	2,5-Norbornadiene	Pea Banana	170 55	6.4×10^{-6} 2.6×10^{-6}
	Cyclopentadiene	Banana	140	6.0×10^{-6}
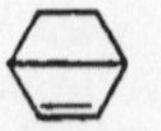	Norbornene	Pea	360	4.1×10^{-6}
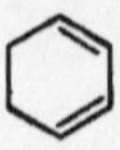	1,3-Cyclohexadiene	Pea	488	2.7×10^{-5}
	1,3-Cycloheptadiene	Pea	870	6.4×10^{-5}
	Cyclopentene	Pea	1,100	1.3×10^{-5}
	1,4-Cyclohexadiene	Pea	4,650	4.2×10^{-4}
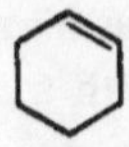	Cyclohexene	Pea	6,060	1.5×10^{-4}
	Cyclohexane	Pea	∞	—
	Benzene	Pea	∞	—
$HC{=}CH$ | | CH_3 CH_3	*cis*-Butene	Pea	7,100	3.1×10^{-4}
H H | | $H_2C{=}C{-}C{=}CH_2$	1,3-Butadiene	Banana	1,230	1.9×10^{-5}

After Sisler and Yang,[64] Sisler and Wood,[16] and Sisler et al.[65]

that it acts *in vivo* to block an ethylene response. Silver ion is a very potent inhibitor of sulfhydryl enzymes, and perhaps part of its action occurs in this manner. Yet silver remains a very powerful experimental tool to study ethylene responses and, despite the cost, has much practical value on such things as cut flowers.[25] A number of other compounds have been shown to inhibit binding; however, the mechanism is unknown (Table 3).

TABLE 3
Binding of ^{14}C-Ethylene to a Triton X-100 Extract of Mung Bean Sprouts

Compound	Concentration	Inhibition %
None	—	0
Ethylene	1,000 μl/l	100
Allyl alcohol	2,000 μl/l	100
2,5-Norbornadiene	2,000 μl/l	100
Ethylene oxide	10,000 μl/l	0
Carbon dioxide	10%	0
Potassium iodide	10^{-3} *M*	50
Thiourea	10^{-3} *M*	56
Diethyldithiocarbamate	10^{-3} *M*	54
o-Phenanthroline	10^{-3} *M*	66
Carbon monoxide	1,000 μl/l	65
NaN_2	10^{-2} *M*	0
HCN	2,000 μl/l	85
p-Hydroxymercuribenzoate	10^{-3} *M*	60
N-Ethylmaleimide	10^{-3} *M*	33
Urea	8 *M*	52
Mercaptoethanol	10^{-3} *M*	−13
Mercaptoethanol	6×10^{-1} *M*	100
Ethylenediaminetetraacetate	10^{-2} *M*	0
$AgNO_3$	10^{-4} *M*	95

After Sisler.[19]

VI. HOW DOES THE RECEPTOR WORK?

Ethylene, once bound, must do something to induce an ethylene response. Burg and Burg[4] proposed a model based on their experimental information that oxygen was necessary for ethylene to act. Their model proposed that ethylene formed a complex with a metal. Oxygen was also thought either to form a complex with the metal or to oxidize the metal so that ethylene could act.

Sisler[15] proposed a model based on a well-known phenomenon in inorganic chemistry known as the *trans* effect. This effect was first alluded to by Werner in 1893,[26] has developed over the years, but is still not completely understood. According to Chatt et al.[27] the *trans* effect, at least in some cases, is due to the ability of a coordinated ligand to accept by back-donation, electron density from the central metal; this is known as π-acceptance. According to this theory, ligands bind by donating electrons to a metal and then back-accept electrons into vacant orbitals, causing a change in the charge on the metal, which leads to a substitution of the ligand in the *trans* position. The *trans* effect series (more common ones) include:

$$H_2C{=}CH_2,\ C{\equiv}O,\ HC{\equiv}N > PR_3,\ H^- > S{=}C(NH_2)_2,\ CH_3^- > C_6H_5^-,\ NO_2^-,$$
$$I^-,\ SCN^- > Br^-,\ Cl^- > \text{Amines},\ NH_3,\ OH^-,\ H_2O$$

Not all of these are suited for biological systems, but this series (for platinum II) serves as a useful guide for types of compounds. The order of magnitude of ligand substitution from highest to lowest is 10^6.

Sisler[15] noted that all of the compounds that gave an ethylene response, i.e., ethylene, propylene, carbon monoxide, and probably cyanide, were high in the *trans* effect series. After testing isocyanides and phosphorous trifluoride, compounds which are also good π-acceptors, and finding them active in inducing an ethylene response, it was proposed that

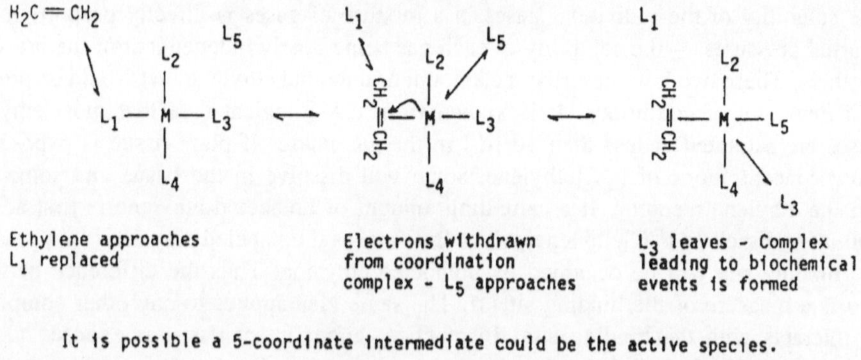

FIGURE 2. Model for possible mode of ethylene action. (From Sisler, E. C. and Goren, R., *What's New in Plant Physiol.*, 12, 37, 1981. With permission.)

π-acceptance was the character that caused the response. It was suggested that something similar to the *trans* effect was involved (Figure 2), and this might serve as a useful model for testing compounds as ethylene analogs or antagonists. Exactly how this mechanism would work in plants is not known, but some type of charge transfer is suggested, with the result that something is released or perhaps a conformational change takes place.

There is one report in the literature that propylene is a competitor of ethylene action.[28] How this would occur is unknown. Since propylene is much more rapidly oxidized by pea plants than ethylene, it is possible that products of propylene oxidation are giving the effect.

Vinyl compounds such as 2,5-norbornadiene and cyclopentadiene bind to the receptor but do not induce a response. These are highly strained compounds and π-acceptance probably is not sufficient to induce the response.

VII. ETHYLENE BINDING

Ethylene binding to plant material was shown by Sisler[29] and Sisler and Wylie[30] and about the same time Jerie et al.[31] reported measuring compartmentation of ethylene in plant tissue, which apparently also represents binding. Specific binding refers to binding of labeled ethylene which can be displaced by competition with unlabeled ethylene or with similarly acting compounds. Nonspecific binding as described by Abeles[32] does not represent displaceable ethylene binding and probably simply represents that the ethylene dissolved in the tissue. Unless samples are run in the presence and absence of unlabeled ethylene to determine the amount of binding by difference, or some other means is used to remove unbound ethylene, the amount bound cannot be distinguished from the amount of dissolved ethylene.

VIII. BASIS FOR MEASURING ETHYLENE BINDING

Gaseous compounds distribute themselves between the gaseous phase and liquid phase in a specific ratio depending on the gas, the partial pressure, the liquid, and the temperature. Thus, as Henry's law states, at a fixed temperature the weight of gas dissolved in a given quantity of liquid is proportional to its partial pressure. K = Pressure in mm of Hg/mol fraction in solution. The Henry's law constant for ethylene is given in the International Critical Tables for a number of temperatures in water.[33] Dalton's law states that in a mixture of gases, every gas exerts the same pressure as it would if it, alone, were confined to the same volume, and the total pressure is simply the sum of the partial pressures. Dalton showed that the solubility of the individual gases in a mixture of gases is directly proportional to their partial pressures — the solubility of each gas being nearly independent of the presence of the others. These two laws describe the amount of unbound ethylene that would be present under a given set of conditions. It is known from physiological data that most ethylene responses are saturated at less than 10 μl/l in the gas phase. If plant tissue is exposed to very low concentrations of [^{14}C]ethylene, some will dissolve in the tissue and some will bind to the ethylene receptor. If a saturating amount of unlabeled ethylene is first added, the amount of dissolved [^{14}C]ethylene will be the same as if unlabeled ethylene is not present, but the binding site will be occupied by unlabeled ethylene. Thus the difference between the two is a measure of the binding site(s). The same idea applies to any other compound which interacts with the binding site. In practice, either intact tissue is exposed to low concentrations of [^{14}C]ethylene in the presence and absence of unlabeled ethylene[29] or, for *in vitro* extracts, cellulose powder is added as a spacer to allow for easy exchange of gases.[34] Bengochea et al.[35] have exposed liquid samples to labeled ethylene and then blown air through the mixture for a short time (30 to 45 s) to remove the unbound ethylene, then determining the amount bound. This method was applied to samples in which the dissociation of ethylene was slow (many hours); however, it should be satisfactory even for samples having a relatively short dissociation time.

The distribution of ethylene between the gas phase and water phase is given by the Ostwald coefficient. The concentration in liquid/concentration in the gas phase is 0.23 at 0°, 0.1579 at 10°, 0.1412 at 15°, 0.1275 at 20°, 0.1162 at 25°, and 0.1068 at 30°. Thus, at 0° the liquid phase would contain 23% as much ethylene as the gas phase and at 25°, 11.62% as much.[36-38] Although more ethylene would be dissolved at cooler temperatures and more binding should occur, the background would be expected to be correspondingly higher.

IX. RESULTS OF BINDING *IN VIVO*

Binding of ethylene *in vivo* has been measured in tissue that responds to ethylene (leaves, sprouts, etc.) (Figures 3 and 4) and tissue known not to respond to ethylene (cotyledons, some seeds, etc.). The binding component(s) is widely distributed (Tables 4 and 5). In ethylene-responsive tissue, a rapid associative and dissociative component is found. The most rapid component dissociates with a half-life of 2 to 15 min (each tissue seems to be a little different). There appears to be another component with a half-life of 1 to 2 h in most tissues, and yet another component with a half-life of 40 to 50 h. Upon disruption of the tissue, the short half-life component (2 to 15 min) seems to disappear[14] while the longer dissociative components are stable and measurable (Figure 4). A fraction of previously bound ethylene rapidly dissociates when mung bean sprouts are blended.[14] This amount approximately corresponds to the short-time binding component.

In cotyledons of *Phaseolus vulgaris*[39] and mung bean seeds,[14] a long associative-dissociative half-life (20 to 50 h) appears to be present. Once bound, ethylene does not seem to be displaced by unlabeled ethylene from any of the binding components, nor does a vacuum increase the rate of diffusion from the site.

One of the ways an organism can regulate processes is to regulate the level of the receptor in tissues. Some experiments have been conducted to determine whether or not the receptor levels vary much. Experiments with tobacco leaves did not reveal much change over a period of 96 hours.[40] Similarly, not much change was observed in morning glory flowers, except a small decline did occur as the flowers began to fade.[41]

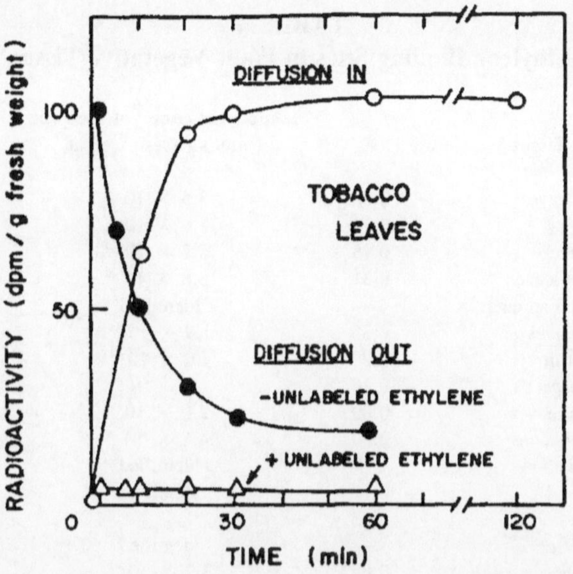

FIGURE 3. Association (○) and dissociation (●) of [^{14}C]ethylene in tobacco leaves. (From Goren, R. and Sisler, E. C., *Tob. Sci.*, 28, 110, 1984. With permission.)

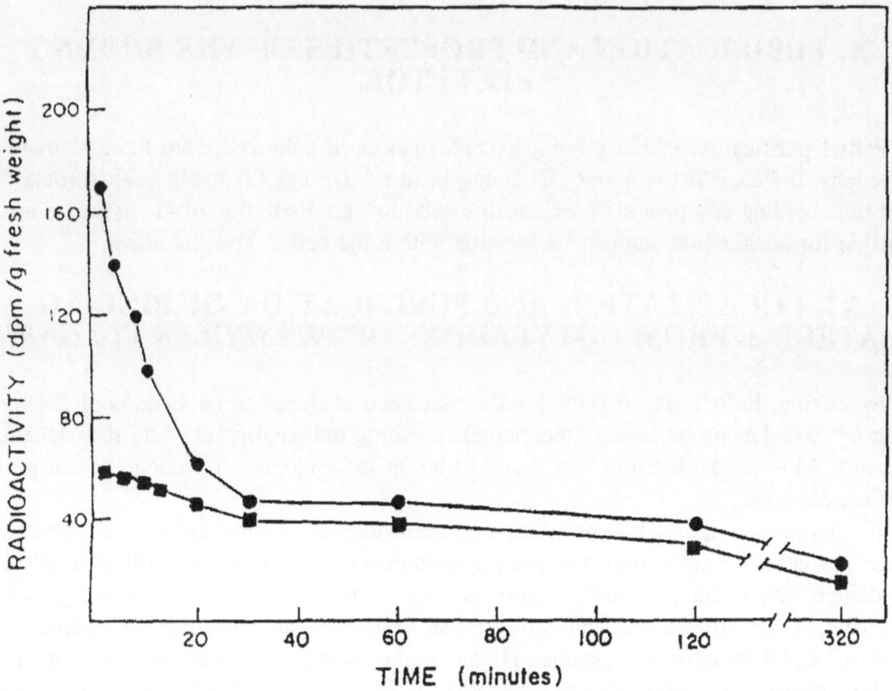

FIGURE 4. Diffusion of bound ethylene from (●) mung bean sprouts and (■) a mung extract from an equivalent amount of sprouts. (From Sisler, E. C., *Plant Growth Substances 1988*, Pharis, R. P. and Rood, S. B., Eds., Springer-Verlag, Berlin, 1990, 192. With permission.)

TABLE 4
Ethylene-Binding Sites in Plant Vegetative Tissue

Tissue	K_d	Apparent conc.[a] of receptor mol/kg fresh weight	Ref.
Tobacco leaves	0.30	3.5×10^{-9}	29
Bean leaves	0.14	2.0×10^{-9}	50
Citrus leaves	0.15	5.7×10^{-9}	50
Ligustrum leaves	0.31	6.8×10^{-9}	50
Mung bean sprouts	—	Identified[b]	—
Tomato leaves	0.30	1.9×10^{-9}	51
Tomato fruit	0.30	7.0×10^{-11}	51
Carnation petals	0.10	6.0×10^{-9}	25
Carnation leaves	0.09	2.0×10^{-9}	25
Bean cotyledons	0.30	4.8×10^{-9}	50
Bean roots	—	Identified[b]	—
Pea epicotyl	0.12 0.63	Identified	53
Wheat germ	—	Identified	—
Apple pulp	0.097	3.2×10^{-11}	41
Cucumber leaves	—	Identified	—
Mushroom sporophore	—	Not detected	—
Morning glory petals	0.10	3.8×10^{-9}	41

[a] Not corrected for the amount occupied by endogenous ethylene.
[b] The concentration was not determined.

X. PURIFICATION AND PROPERTIES OF THE BINDING RECEPTOR

Partial purification of the binding receptor has been achieved from three sources: (1) cotyledons of *Phaseolus vulgaris,* (2) mung bean seeds, and (3) mung bean sprouts. The short-time binding component, which seems to be lost on disruption of the tissue, is not yet available for purification studies. Its location within the cell is also unknown.

XI. LOCALIZATION AND PURIFICATION OF BINDING MATERIAL FROM COTYLEDONS OF *PHASEOLUS VULGARIS*

An attempt to localize ethylene binding has been made using cotyledons of *Phaseolus vulgaris*[31,42] and a combination of techniques, incuding radioautography and sucrose density gradients. Most of the binding was found to be on the endoplasmic reticulum and protein body membranes.

The greatest progress has been made on purification of binding sites in cotyledons. Bengochea et al.[35,39] solubilized the binding component from this source with Triton X-100, and demonstrated that proteolytic enzymes such as trypsin lowered the binding activity somewhat. They showed that the activity was heat-labile and that the pH optimum was between 7 and 9. In subsequent studies Thomas and co-workers[43,44] fractioned the component in a three-phase system and subjected it to other purification procedures such as ion exchange chromatography, Triton X-114 partitioning, and isoelectric focusing. The use of FPLC (fast protein liquid chromatography) anion exchange[45,46] and high pH fractionation have also been reported.

In a recent purification scheme, the binding component was extracted from cotyledons by blending with 8% sucrose and 0.2 *M* Tricine at pH 9.0. The fraction collected by centrifugation between 12,000 and 96,000 × *g* was collected and the binding component

TABLE 5
Ethylene Binding by Plant Seeds

Plant protein		Binding dpm/g	Binding dpm/g
Adzuki bean	*Vigna angularis*	18,000	80,000
Mung bean	*Vigna aureus*	16,300	67,650
Field pea	*Vigna sinensis*	6,870	30,100
Urd	*Phaseolus mungo*	6,000	
Cow pea	*Vigna sinensis*	5,400	23,700
Black bean	*Phaseolus vulgaris*	4,800	21,300
Black eye pea	*Vigna sinensis*	4,050	17,750
Pinto bean	*Phaseolus vulgaris*	3,700	16,700
Soybean	*Glycine max*	3,050	8,900
Bush bean	*Phaseolus vulgaris*	2,000	9,000
Fava bean	*Vicia faba*	600	2,423
Lentil	*Lens culinaris*	400	1,620
Garbanzo bean	*Cicer arietinun*	350	
Alfalfa	*Medicago sativa*	240	
Vetch	*Vicia sativa*	200	
Pea	*Pisum sativum*	47	195
Castor bean	*Ricinus communis*	83	
Cantaloupe	*Cucumis melo*	7,530	
Barley	*Hordeum vulgare*	3,780	39,380
Wheat	*Triticum aestivum*	1,030	7,370
Oat	*Avena sativa*	1,300	
Rye	*Secale cereale*	400	3,330
Watermelon	*Citrullus vulgaris*	540	
Sunflower	*Helianthus annuus*	380	1,490
Pumpkin	*Cucurbita pepo*	370	1,280
Tobacco	*Nicotiana tabacum*	320	
Coriander	*Coriandrum sativum*	330	
Buckwheat	*Fagopyrum esculentum*	290	2,470
Mustard	*Brassica*	240	4,020
Cotton	*Gossypium hirsutum*	230	480
Pop corn	*Zea mays*	180	
Millet	*Pennisetum*	140	1,260
Almond	*Prunus amygdalus*	0	
Flax	*Linum*	0	

From Sisler, E. C., in *Ethylene — Biochemical, Physiological and Applied Aspects,* Fuchs, Y. and Chalutz, E. Eds., Martinus Nijhoff/ Dr. W. Junk, The Hague, 1984, 45. With permission.

solubilized with 0.02 *M* Tris, 0.05 *M* NaCl, 20% Triton X-114, pH 7.0, using a detergent:protein ratio (mg:mg) of 4:1. Phase separation was accomplished by warming to 35°C and centrifuging at 1000 × *g*. The lower phase contained the binding component. Detergent was removed by Amberlite X D-2 resin beads. After centrifugation at 96,000 × *g* for 1 h, the pellet was activity-rich. The pellet was resuspended in 0.02 *M* Tris, 0.1% Triton X-100 at a pH of 8.9. This sample was applied to DEAE-Sepharose and eluted with 0.35 *M* NaCl in the same buffer at pH 8.9. Further purification was achieved by HPLC MONO Q chromatography, eluting with a 0 to 100% gradient of 0.35 *M* NaCL in Tris buffer at pH 8.9 (Table 6).

The component from *Phaseolus vulgaris* now has been purified to give a single band on two-dimensional gel electrophoresis and antibodies raised against it.[47] Also, polyacrylamide gel electrophoresis, attempts to develop antibodies, and a summary procedure are reported. The purified glycoprotein is reported to contain Cu, and the protein is now being sequenced.

TABLE 6
Purification and Recovery of Ethylene-Binding Activity from a Crude Homogenate of *Phaseolus vulgaris*

Stage	Specific activity (dpm/mg protein)	Purification	Recovery (%)
Crude homogenate	502		
96k × *g* Pellet	1,463	2.9	32
Triton X-114 detergent phase	3,289	6.6	25
96k × *g* pellet after detergent removal	8.52×10^4	170	10.6
DEAE-Sepharose anion exchange	3.19×10^5	635	3.6
FPLC Mono Q HR (5/5)	nd	nd	0.3

Note: Samples assayed for activity at each step (post-labeling); nd, not determined.

From Williams, R. A. N., Smith, A. R., and Hall, M. A., in *Plant Hormone Receptors,* Klambt, D., Ed., Springer-Verlag, Berlin, 1987, 303. With permission.

TABLE 7
Summary of the Properties of the Ethylene-Binding Site from *Phaseolus vulgaris*

	Membrane bound	Triton solubilized and partially purified protein
K_d Ethylene (*M*)	0.88×10^{-10}	5.5×10^{-10}
K_i Propylene (*M*)	5.6×10^{-7}	0.9×10^{-7}
K_i Acetylene (*M*)	1.03×10^{-5}	2.0×10^{-5}
K_i CO_2 (*M*)	1.15×10^{-3}	
k_1 (M^{-1} s^{-1})	2.97×10^3	0.042×10^3
k_{-1} (s^{-1})	1.4×10^{-6}	2.3×10^{-8} (calc)
pH Optimum	7.5—9.5	4—7
Sedimentation coefficient		2.2s
Stoke's radius (nm)		6.1
Frictional ratio		2.4
Molecular weight		52—60,000

From Hall, M. A. et al., in *Molecular Biology of Plant Growth Control,* Fox, J. E. and Jacobs, M., Eds., Alan R. Liss, New York, 1987, 335. With permission.

XII. SOME PROPERTIES OF THE BINDING COMPONENT FROM *PHASEOLUS VULGARIS*

Several properties of the ethylene-binding component has been reported[48] and these are summarized in Table 7. The binding of ethylene has an association rate of 2.97×10^3 M^{-1} s^{-1} and a dissociation rate of 1.4×10^{-6} s^{-1} at 25°C for the membrane-bound material. In a Triton X-100 extract the association rate is 4.2×10^1 M^{-1} s^{-1} and the dissociation rate is 2.3×10^{-8} s^{-1}. Many of the other properties were observed to change on solubilization and purification.

XIII. PURIFICATION OF THE BINDING COMPONENT FROM MUNG BEAN SPROUTS

Attempts were made to localize the binding receptor from tobacco leaves, but the amount was too low and the background too high to be practical. Mung bean sprouts, upon grinding,

TABLE 8
Purification of Ethylene-Binding Component from Mung Bean Sprouts

Fraction	Binding dpm/100 g	Binding dpm/g protein	Purification	Recovery %
Crude homogenate	102,700	6,848	1	100
pH 4.0 Precipitate	79,060	7,186	1	77
Triton X-100	91,500	183,000	27	89
CM-Sephadex	54,431[a]	1,088,620	159	53
Isoelectric focusing	48,000	2,800,000	409	47

[a] Approximately 20% of the binding component(s) does not adhere to CM-Sephadex and may represent a different component.

After Sisler[34] and Sisler and Wood.[67]

gave sufficient binding to provide a practical purification. Most of the binding was associated with the 12,000 to 100,000 × *g* fraction, indicating a membrane localization for the binding. Further localization has not been accomplished (Table 8).

Mung bean sprouts serve as a convenient source of ethylene-binding component. Sprouts are easily obtained by germinating mung bean seeds in the dark, and they also are usually available commercially. Sprouts are ground and extracted 3 times with buffer at pH 6.0 or above. After straining through cloth, the residue is discarded, the pH of the liquid adjusted to approximately 4.0, and centrifuged. The residue collected by centrifugation at pH 4.0 is adjusted to pH 7.0, extracted with 0.5% Triton X-100, readjusted to pH 4.0, and recentrifuged. This is repeated twice more. The resulting "soluble" extracts are combined and concentrated by making them 50% saturated in $(NH_4)_2SO_4$ and centrifuging. The floating material is collected and dialyzed. After dialysis, about 80% of the binding component will bind to CM-Sephadex in 10 m*M* potassium phosphate buffer at pH 5.5. The component is eluted with 200 m*M* potassium phosphate buffer at pH 7.0. After concentrating and dialyzing the material, it can be further purified by isoelectric focusing on pH 6 to 8 ampholite. Most of the binding component is found between pH 5.5 and 6.5. This gives about 400-fold purification. No attempt has been made to further purify this component by FPLC.

XIV. PURIFICATION OF THE BINDING COMPONENT FROM MUNG BEAN SEEDS

Mung bean seeds contain much more binding component on a weight basis than mung bean sprouts. This would appear to be a good source of material; however, little purification has been achieved from seeds. Mung bean seeds are soaked overnight at 5°C with distilled water. The hydrated beans are extracted successively with 5 mg, 2 mg, and 2 mg of Triton X-100 for each gram of beans, based on the original dry weight. Each time the bean material is centrifuged at 3000 × *g* for 5 min. Sufficient water is included to facilitate blending. The extracts are combined and additional Triton X-100 is added to make the extract 2% with respect to detergent. Without this additional Triton X-100 the mixture does not separate into two phases. After adjusting to 25% in NaCl and 10% in $(NH_4)_2SO_4$, the pH is adjusted to 7.5. The mixture will separate into two phases and most of the lower phase can be removed by siphoning. The upper phase is centrifuged and the floating material which contains most of the binding component is dialyzed. This material has been used for further purification studies; however, little purification has been achieved. After a 3-h exposure to [^{14}C]ethylene (0.02 μCi/l), seeds gave a binding value of 16,300 dpm/g seed with 67.6 dpm/mg protein, and after the extraction, 12,800 dpm/g with 145 dpm/mg protein.[49]

An acetone powder may be made from the hydrated beans in the usual way by blending with 5 volumes of acetone (−15°C), filtering, and drying. After soaking overnight, the binding component may be extracted and concentrated in the same way that it is obtained from seeds.

XV. BINDING PROPERTIES OF THE DIFFERENT BINDING COMPONENTS

Since the different binding components have widely different rates of association and dissociation, it might be expected that they would vary considerably in their other binding constants. A number of studies have been made on the binding of compounds which either elicit an ethylene-like response or prevent such a response. Although there is variation in the values obtained, they seem remarkably close for components differing so much in other properties. The K_d for ethylene, based on its solubility in water, *M*, and gas phase (μl/l), was observed to be

1. 4.6×10^{-10} *M* (0.09 μl/l) in apple pulp[41]
2. 4.8×10^{-10} *M* (0.1 μl/l) in morning glory flowers[41]
3. 2.0×10^{-9} *M* (0.27 μl/l) in tobacco leaves[29]
4. 1.1×10^{-9} *M* (0.15 μl/l) for *Citrus* leaves[50]
5. 2.3×10^{-9} *M* (0.31 μl/l) for *Ligustrum* leaves[50]
6. 1.0×10^{-9} *M* (~0.14 μl/l) for bean leaves[50]
7. 2.3×10^{-9} *M* (0.30 μl/l) for tomato leaves[51]
8. 4.8×10^{-10} *M* (0.10 μl/l) for carnation leaves[25]

The K_d for binding in a Triton X-100 extract of mung bean sprouts was 5.8×10^{-10} *M* (0.12 μl/l),[18] 5.8×10^{-10} *M* (0.12 μl/l) in mung bean seeds,[16] and 6.8×10^{-10} *M* (0.14 μl/l) in *Phaseolus vulgaris* cotyledons.[48] A high value of 3.1×10^{-8} (0.63 μl/l) was found by Sanders et al.[52] and Smith et al.[53] for a slowly dissociating component from pea after 200 h of binding, while the short-time binding component was 5.9×10^{-10} *M* (0.12 μl/l) after 1 h of binding. Although there is some variation in these values, most are well within the range of variation observed for most ethylene responses.[54]

Some K_m values for ethylene action are

1. 4.8×10^{-10} *M* (0.1 μl/l) for the pea growth test[16]
2. 1.9×10^{-9} *M* (0.30 μl/l) for tobacco leaf respiration[29]
3. 3.4×10^{-9} *M* (0.7 μl/l) for banana peel chlorophyll degradation[16]
4. 5.8×10^{-10} *M* (0.12 μl/l) for *Citrus* explant abscission[55]

Abeles[54] lists many gas phase K_m values obtained from the literature, and they range from 0.1 to 100 μl/l for higher plants, with most being 1 μl/l or less. All of the binding values are well within these limits.

The results for 2,5-norbornadiene for K_i values were

1. 6.4×10^{-6} *M* (170 μl/l) in pea[20]
2. 4.8×10^{-6} *M* (125 μl/l) in *Citrus*[55]
3. 8.6×10^{-6} *M* (225 μl/l) in tobacco leaves[16]
4. 8.6×10^{-5} *M* (2162 μl/l) in *Phaseolus vulgaris*[56]
5. 6.9×10^{-6} *M* (27 μl/l) in mung bean sprout extracts[18]
6. 5.9×10^{-6} *M* (25 μl/l) in mung bean seed extracts[16]

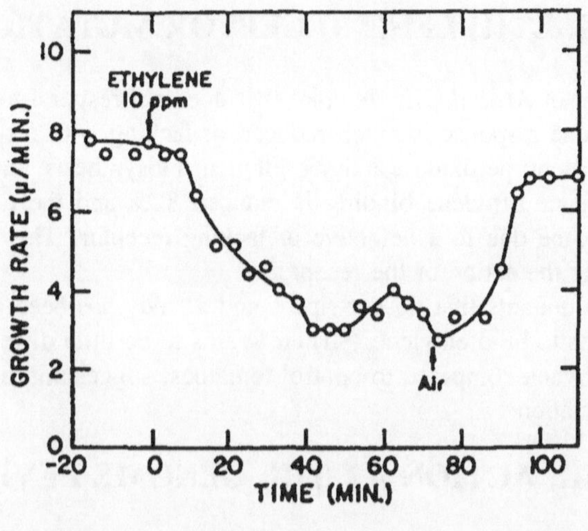

FIGURE 5. Time course of growth rate of intact pea following applications of ethylene at time 0, and the subsequent removal of the treatment 72 min later. (From Warner, H. L. and Leopold, A. C., *Biochem. Biophys. Res. Commun.*, 44, 989, 1971. With permission.)

The 50% displacement values are given by the equation, $K_d = C_{50}(1 + L^*/K_d^*)$, where L^* and K_d^* are the concentration and dissociation constant of labeled ethylene.[57]

XVI. INDUCTION OF A RESPONSE

When plants are exposed to ethylene, growth is inhibited in many plants. In rice and some other aquatic plants, growth is stimulated. These responses are thought to be short-time responses. When the ethylene is removed, the response ceases. Warner and Leopold[58] reported a response in pea to be within the range of 6 to 15 min (Figure 5).

Ethylene also controls the level of some enzymes not known to function in the growth and development of the plant itself, but which may serve in a protective role. Chitinase is an example. Abeles et al.[59] reported that as much as 4% of the soluble protein in bean leaves was chitinase after a 3-d exposure to ethylene. This represents a 50-fold increase in the level of chitinase, which is noted after 6 h of exposure to ethylene[59] and continues for at least 50 h. The endochitinase reported by Boller et al.[60] also requires about 6 h for induction. The synthesis of the enzyme depends on a continued exposure to ethylene. A withdrawal of the ethylene greatly reduces synthesis, and would not likely represent the action of the long-time binding component.

Another type of response induced by ethylene is the induction of ripening and abscission. In this type of response, enzymes are induced or their levels greatly increased. The time of exposure necessary to induce this type of response seems to be much longer. Treatment of *Citrus* explants with 2,5-norbornadiene 10 h after exposure to ethylene greatly retarded abscission, while little effect was observed after 21 h. The effect of a 6-h exposure to ethylene followed by 2,5-norbornadiene was not significantly different than that seen in a 2,5-norbornadiene control.[55] This suggests either a relatively slow process for the ethylene response (longer than 6 h) or else that the tissue is not sensitive to ethylene for several hours. This may be the case since 2,4-D can also suppress the response. It is not clear at the present time whether the same response is involved in abscission and ripening as in growth, but for a longer period of time in abscission and ripening, or whether a longer-time binding component is involved. Additional work is needed to clarify this point. Lincoln et al.[61] report an increase in mRNAs within 30 to 120 min in tomato after exposure to ethylene.

XVII. ETHYLENE RECEPTOR MUTATION

Recently a mutant of *Arabidopsis thaliana* that does not respond to ethylene has been reported.[62] The ethylene response is much reduced or lacking with respect to organ elongation, chlorophyll content, peroxidase activity, ethylene biosynthesis, and seed germination. The mutant is dominant. Ethylene binding is reduced 82% and the reduced response to ethylene is thought to be due to a defective or lacking receptor. This should be of great value in characterizing the action of the receptor.

There are tomato mutants that do not ripen, and a study has been made of the ability of *rin* and *nor* mutants to bind ethylene.[51] There seems to be little difference in binding in both leaves and fruit when compared to control tomatoes, suggesting that the receptor was not the site of the mutation.

XVIII. ACTION AT THE GENOME LEVEL

Attempts are being made to determine the mechanism of action of ethylene expression at the gene level. As examples, Broglie et al.[63] are conducting experiments with chitinase expression in tobacco, and the work by Bleecker et al.[62] on a mutant of *Arabidopsis* is promising. At the present time no evidence exists that ethylene acts directly at the genome level. Proof of whether or not it does may depend on development of satisfactory methods of measuring the short-time binding component *in vitro*. The longer-time binding components seem to be associated with membranes rather than the nucleus.

XIX. FUTURE WORK

Much work remains to be done. The question of whether or not there is more than one receptor needs to be answered. The location of the physiological receptor(s) remains a challenging problem. Once located, it needs to be determined how the receptor is able to control physiological events. Only when we can link the receptor directly to physiological events can we begin to understand the role of ethylene in plant growth and development. It is known that new proteins or altered protein levels are induced by ethylene, but the sequence of biochemical events that occurs at the receptor level is completely unknown. In addition, the exploitation of our knowledge that the receptor can be blocked and that mutants to ethylene are available may prove significant, both scientifically and economically.

REFERENCES

1. **Neljubov, D.,** Uber die horizontale Nutation der Stengel von *Pisium sativum* unde anderen Pflanzen, *Beih. Bot. Zentralbl.*, 10, 128, 1901.
2. **Richards, H. M. and MacDougal, D. T.,** The influence of carbon monoxide and other gases upon plants, *Bull. Torrey Bot. Club*, 3, 57, 1904.
3. **Knight, L. I. and Crocker, W.,** Toxicity of smoke, *Bot. Gaz.*, 55, 337, 1912.
4. **Burg, S. P. and Burg, E. A.,** Molecular requirements for the biological activity of ethylene, *Plant Physiol.*, 42, 144, 1967.
5. **Beyer, E., Jr.,** Mechanism of ethylene action. Biological activity of deuterated ethylene and evidence against exchange and *cis trans* isomerization, *Plant Physiol.*, 49, 672, 1972.
6. **Abeles, F. B., Ruth, J. M., Forrence, L. E., and Leather, G. R.,** Mechanism of hormone action. Use of deuterated ethylene to measure isotopic exchange with plant material and the biological effects of deuterated ethylene, *Plant Physiol.*, 49, 669, 1972.
7. **Beyer, E. M., Jr.,** Effect of silver ion, carbon dioxide, and oxygen on ethylene action and metabolism, *Plant Physiol.*, 63, 169, 1979.

8. **Abeles, F. B.,** A comparative study of ethylene oxidation in *Vicia faba* and *Mycobacterium paraffinicum*, *J. Plant Growth Regul.*, 3, 85, 1984.
9. **Evans, D. E., Smith, A. R., Taylor, J. E., and Hall, M. A.,** Ethylene metabolism in *Pisum sativum* L. Kinetic parameters, the effects of propylene, silver and carbon dioxide and comparison with other systems, *Plant Growth Regul.*, 2, 187, 1984.
10. **Smith, A. R. and Hall, M. A.,** Biosynthesis and metabolism of ethylene, in *The Biosynthesis and Metabolism of Hormones*, Crozier, A. and Hillman, J. R., Eds., Society for Experimental Biology Seminar 23, Cambridge University Press, New York, 1984, 201.
11. **Smith, A. R., Evans, D. E., Smith, P. G., and Hall, M. A.,** Ethylene metabolism in *Pisum sativum* L. and *Vicia faba* L., in *Ethylene and Plant Development*, Roberts, J. A. and Tucker, G. A., Eds., Butterworths, London, 1985, 139.
12. **Cotton, F. A. and Wilkinson, G.,** *Advanced Inorganic Chemistry*, 3rd ed., Interscience Publishers, New York, 1972, 1145 p.
13. **Trewavas, A. J. and Jones, A. M.,** Consequences of hormone-binding studies for plant growth substances, *What's New in Plant Physiol.*, 12, 5, 1981.
14. **Sisler, E. C.,** Ethylene binding receptors—is there more than one?, in *Plant Growth Substances 1988*, Pharis, R. P. and Rood, S. B., Eds., Springer-Verlag, Berlin, 1990, 192.
15. **Sisler, E. C.,** Ethylene activity of some π-acceptor compounds, *Tob. Sci.*, 21, 43, 1977.
16. **Sisler, E. C. and Wood, C.,** Competition of ethylene for unsaturated compounds for binding and action in plants, *Plant Growth Regul.*, 7, 181, 1988.
17. **Sisler, E. C. and Pian, A.,** Effect of ethylene and cyclic olefins on tobacco leaves, *Tob. Sci.*, 17, 68, 1973.
18. **Beggs, M. J. and Sisler, E. C.,** Binding of ethylene analogs and cyclic olefins to a Triton X-100 extract from plants: comparison with in vivo activities, *Plant Growth Regul.*, 4, 13, 1986.
19. **Sisler, E. C.,** Ethylene-binding properties of a Triton X-100 extract of mung bean sprouts, *J. Plant Growth Regul.*, 1, 211, 1982.
20. **Sisler, E. C. and Yang, S. F.,** Anti-ethylene effects of *cis*-2-butene and cyclic olefins, *Phytochemistry*, 23, 2765, 1984.
21. **Smith, A. R. and Hall, M. A.,** Ethylene binding, in *Ethylene and Plant Development*, Roberts, J. A. and Tucker, G. A., Eds., Butterworths, London, 1985, 101.
22. **Sisler, E. C. and Wood, C.,** Interaction of ethylene and CO_2, *Physiol. Plant.*, 73, 440, 1988.
23. **Beyer, E. M., Jr.,** A potent inhibitor of ethylene action in plants, *Plant Physiol.*, 58, 268, 1976.
24. **Goren, R., Mattoo, A. K., and Anderson, J. D.,** Ethylene binding during leaf development and senescence, *J. Plant Physiol.*, 117, 243, 1984.
25. **Sisler, E. C., Reid, M. S., and Yang, S. F.,** Effect of antagonists of ethylene action on binding of ethylene in cut carnations, *Plant Growth Regul.*, 4, 213, 1986.
26. **Werner, A.,** Beitrag zur Konstitution anorganisher Verbindungen, *Z. Anorg. Chem.*, 3, 267, 1893; English translation in *Classics in Coordination Chemistry*, Kauffman, G., Ed., Dover Press, New York, 1968, 9.
27. **Chatt, J., Duncanson, L. A., and Venanzi, L. M.,** Directing effects in inorganic substitution reactions, Part I, A hypothesis to explain the *trans* effect, *J. Chem. Soc.*, 4456, 1955.
28. **Dollwet, H. H. A. and Semens, R. E.,** Propylene—a competitor of ethylene action, *Plant Physiol.*, 56, 552, 1975.
29. **Sisler, E. C.,** Measurement of ethylene binding in plant tissue, *Plant Physiol.*, 64, 538, 1979.
30. **Sisler, E. C. and Wylie, P. A.,** In vivo measurement of binding to the ethylene binding site, *Plant Physiol.*, 61, S-131, 1978.
31. **Jerie, P. H., Shaari, A. R., and Hall, M. A.,** The compartmentation of ethylene in developing cotyledons of *Phaseolus vulgaris* L., *Planta*, 144, 503, 1979.
32. **Abeles, F. B.,** Nonphysiological binding of ethylene by plants, *Plant Physiol.*, 74, 525, 1984.
33. **Washburn, E. W., Ed.,** *International Critical Tables*, Vol. 3, McGraw-Hill, New York, 1928, 260.
34. **Sisler, E. C.,** Partial purification of an ethylene-binding component from plant tissue, *Plant Physiol.*, 66, 404, 1980.
35. **Bengochea, T., Acaster, M. A., Dodds, J. H., Evans, D. E., Jerie, P. H., and Hall, M. A.,** Studies on ethylene binding by cell free preparations of *Phaseolus vulgaris* L. II. Effects of structural analogs of ethylene and of inhibitors, *Planta*, 148, 407, 1980.
36. **Smith, A. R. and Hall, M. A.,** Mechanisms of ethylene action, *Plant Growth Regul.*, 2, 151, 1984.
37. **Wasik, S. P. and Tsang, W.,** Gas chromatographic determination of partition coefficients of some unsaturated hydrocarbons and their deuterated isomers in aqueous silver nitrate solutions, *J. Phys. Chem.*, 74, 2970, 1970.
38. **Wilhelm, E., Battino, R. and Wilcock, R. J.,** Low pressure solubility of gases in liquid water, *Chem. Rev.*, 77, 219, 1977.

39. **Bengochea, T., Dodds, J. H., Evans, D. E., Jerie, P. H., Niepel, B., Shaari, A. R., and Hall, M. A.**, Studies on ethylene binding by cell free preparations from cotyledons of *Phaseolus vulgaris* L. I. Separation and characterization, *Planta*, 148, 397, 1980.
40. **Goren, R. and Sisler, E. C.**, Ethylene binding: some parameters in excised tobacco leaves, *Tob. Sci.*, 28, 110, 1984.
41. **Blankenship, S. M. and Sisler, E. C.**, Ethylene binding changes in apple and morning glory during ripening and senescence, *J. Plant Growth Regul.*, 8, 37, 1989.
42. **Evans, D. E., Bengochea, T., Cairns, A. R., Dodds, J. H., and Hall, M. A.**, Studies on ethylene binding by cell-free preparations from cotyledons of *Phaseolus vulgaris* L. Subcellular localization, *Plant Cell Environ.*, 5, 101, 1982.
43. **Thomas, C. J. R., Smith, A. R., and Hall, M. A.**, The effect of solubilization on the character of an ethylene-binding site from *Phaseolus vulgaris* L. cotyledons, *Planta*, 160, 474, 1984.
44. **Thomas, C. J. R.**, Partial purification of an ethylene-binding site from *Phaseolus vulgaris* L. cotyledons, *Planta*, 164, 272, 1985.
45. **Hall, M. A., Smith, A. R., Thomas, C. J. R., and Howarth, C. J.**, Binding sites of ethylene, in *Ethylene—Biochemical, Physiological and Applied Aspects*, Fuchs, Y. and Chalutz, E., Eds., Martinus Nijhoff/Dr. W. Junk, The Hague, 1984, 55.
46. **Williams, R. A. N., Smith, A. R., and Hall, M. A.**, Characterization and purification of an ethylene binding component from developing cotyledons of *Phaseolus vulgaris* L., in *Plant Hormone Receptors*, Klambt, D., Ed., Springer-Verlag, Berlin, 1987, 303.
47. **Smith, A. R., Bell, M. H., Connem, C., Sandeis, I. O., Williams, R. A. N., Wood, C., and Hall, M. A.**, Ethylene binding sites in higher plants, *13th Int. Conf. Plant Growth Substances Abstr.*, S, 19, 1988.
48. **Hall, M. A., Howarth, C. J., Robertson, D., Sanders, I. O., Smith, A. R., Smith, P. G., Starling, R. J., Tang, Z.-D., Thomas, C. J. R., and Williams, R. A. N.**, Ethylene binding proteins in *Molecular Biology of Plant Growth Control*, Fox, J. E. and Jacobs, M., Eds., Alan R. Liss, New York, 1987, 335.
49. **Sisler, E. C.**, Purification of the ethylene-binding component from mung bean sprouts and seeds, in *Plant Hormone Receptors*, Klambt, D., Ed., Springer-Verlag, Berlin, 1987, 297.
50. **Goren, R. and Sisler, E. C.**, Ethylene-binding characteristics in *Phaseolus, Citrus,* and *Ligustrum* plants, *Plant Growth Regul.*, 4, 43, 1986.
51. **Sisler, E. C.**, Ethylene binding in normal, *rin* and *nor* mutant tomatoes, *J. Plant Growth Regul.*, 1, 219, 1982.
52. **Sanders, I. O., Robertson, D. R., Smith, A. R., and Hall, M. A.**, Ethylene binding and metabolism in *Pisum sativum* L. cv Alaska, in *Plant Hormone Receptors*, Klambt, D., Ed., Springer-Verlag, Berlin, 1987, 289.
53. **Smith, A. R., Robertson, D., Sanders, I. O., Williams, R. A. N., and Hall, M. A.**, Ethylene binding sites, in *Plant Hormone Receptors*, Klambt, D., Ed., Springer-Verlag, Berlin, 1987, 229.
54. **Abeles, F. B.**, *Ethylene in Plant Biology*, Academic press, New York, 1973, 302.
55. **Sisler, E. C., Goren, R., and Huberman, M.**, Effect of 2,5-norbornadiene on abscission and ethylene production in citrus leaf explants, *Physiol. Plant.*, 63, 114, 1985.
56. **Howarth, C. J., Smith, A. R., and Hall, M. A.**, Ethylene binding in *Phaseolus vulgaris* L. cotyledons, in *Ethylene in Plant Development*, Roberts, J. A. and Tucker, G. A., Eds., Butterworths, London, 1985, 117.
57. **Chen, Y. and Prusoff, W. H.**, Relationship between the inhibition constant K_i and the concentration of inhibitor which causes 50 percent inhibition (I_{50}) of an enzymatic reaction, *Biochem. Pharmacol.*, 22, 3099, 1973.
58. **Warner, H. L. and Leopold, A. C.**, Timing of growth regulation responses in peas, *Biochem. Biophys. Res. Commun.*, 44, 989, 1971.
59. **Abeles, F. B., Bosshart, R. P., Forrence, L. E., and Habig, W. H.**, Preparation and purification of glucanase and chitinase from bean leaves, *Plant Physiol.*, 47, 129, 1970.
60. **Boller, T., Gehri, A., Mauch, F., and Vogeli, U.**, Chitinase in bean leaves: induction by ethylene, purification, properties, and possible function, *Planta*, 157, 22, 1983.
61. **Lincoln, J. E., Cordes, S., Read, E., and Fischer, R. L.**, Regulation of gene expression by ethylene during *Lycopersicon esculentum* (tomato) fruit development, *Proc. Natl. Acad. Sci. U.S.A.*, 84, 2793, 1987.
62. **Bleecker, A. B., Estelle, M. A., Somerville, C., and Kende, H.**, Insensitivity to ethylene conferred by a dominate mutation in *Arabidopsis thaliana*, *Science*, 241, 1086, 1988.
63. **Broglie, K. E., Biddle, P., Cressman, R., and Broglie, R.**, Functional analysis of DNA sequences responsible for ethylene regulation of a bean chitinase gene in transgenic tobacco, *Plant Cell*, 1, 599, 1989.
64. **Sisler, E. C. and Yang, S. F.**, Ethylene, the gaseous plant hormone, *Bioscience*, 34, 234, 1984.

65. **Sisler, E. C., Blankenship, S., and Guest, M.,** Interaction of cyclooctenes and cyclooctadienes for ethylene binding and activity in plants, *Plant Growth Regul.*, 9, 157, 1990.
66. **Sisler, E. C.,** Distribution and properties of ethylene binding component from plant tissue, in *Ethylene—Biochemical, Physiological and Applied Aspects*, Fuchs, Y. and Chalutz, E., Eds., Martinus Nijhoff/Dr. W. Junk, The Hague, 1984, 45.
67. **Sisler, E. C. and Wood, C.,** Ethylene binding and evidence that binding in vivo and in vitro is to the physiological receptor, in *Plant Hormone Receptors*, Klambt, D., Ed., Springer-Verlag, Berlin, 1987, 239.
68. **Sisler, E. C. and Goren, R.,** Ethylene action—the basis for hormone action in plants, *What's New in Plant Physiol.*, 12, 37, 1981.

Chapter 6

ETHYLENE AND GENE EXPRESSION

Richard Broglie and Karen Broglie

TABLE OF CONTENTS

I. INTRODUCTION

Ethylene is perhaps the most extensively studied of all plant growth regulators. Many observations of the effects of this phytohormone on the growth and development of plants have been made (for reviews see References 1 to 3 and this volume) since its discovery in 1901.[4] Although much information is available on the role of ethylene in the regulation of such physiological processes as seed germination, seedling growth, fruit ripening, and plant senescence, and its function during conditions of environmental stress, little is known about the mechanism(s) by which these changes are brought about in plants. In recent years there has been increasing evidence to suggest that many of the responses of plants to ethylene involve changes in the pattern of gene expression.[5-8] Advances in recombinant DNA technology have allowed the isolation and characterization of a number of ethylene-regulated genes and studies are in progress to elucidate the mechanism of ethylene action at the level of gene expression. In this chapter, we will concentrate on recent advances in this area, with special emphasis on the role of ethylene in regulating gene expression during fruit ripening, plant senescence, and plant stress.

II. GENE EXPRESSION DURING TOMATO FRUIT RIPENING

During fruit ripening a number of physical and biochemical changes take place resulting in major changes in the appearance, texture, and taste of the fruit (for reviews see References 7, 9, and 10; and Chapter 13). In tomatoes, chloroplasts are converted into chromoplasts as they lose their starch, chlorophyll, and other photosynthetic pigments and accumulate β-carotene and lycopene. Structural changes in the cell wall lead to a progressive softening of the fruit, while sugars and organic acids accumulate in the vacuole. These changes are accompanied by the production of complex volatile compounds that affect flavor and aroma. Many of the physical and biochemical changes that take place during tomato fruit ripening are associated with changes in enzyme activities. For example, the activities of such enzymes as invertase, polygalacturonase, malic enzyme, phosphofructokinase, and ACC synthase increase substantially, while the activities of other enzymes such as citrate synthase and malic dehydrogenase decline. One of the first detectable signs of ripening is the dramatic rise in ethylene production associated with the "respiratory climacteric" and it is recognized that ethylene plays an essential role in the ripening process.[1-3] Exposure of unripe fruit to ethylene has been shown to accelerate the ripening process, while exposure of fruit to specific inhibitors of ethylene biosynthesis inhibit the ripening process.[3,11]

The first evidence for ethylene-associated changes in gene expression during fruit ripening was obtained when Rattanapone et al.[12] showed that the polypeptide profile of translation products was different depending upon whether mRNA from mature-green or red tomatoes was added to a cell-free protein synthesis system. Similar results were obtained subsequently by others studying either tomato[13-16] or avocado[17,18] fruit ripening or the effect of ethylene treatment on carrot roots[5] or soybean hypocotyl.[6] In order to study the changes in gene expression that accompany the increase in ethylene production associated with the ripening of climacteric fruits, cDNA libraries were constructed using mRNA extracted from ripe tomato[15,19-22] and avocado.[18,23] By using either differential or subtractive hybridization screening techniques or immunological detection methods, a number of ethylene- and/or ripening-induced cDNA clones have been identified. A list of clones whose protein products have been identified based on analysis of their nucleotide sequences is shown in Table 1. In addition, a number of cDNA clones complementary to mRNAs that encode proteins of unknown function have been studied.[15,27]

Lincoln et al.[15] have investigated the regulation of gene expression in unripe tomato fruit following exposure to exogenous ethylene. The expression of five different genes was

TABLE 1
Identification of Proteins Encoded by Cloned Ripening-Induced Genes

Clone	Source	Protein	Ref.
pAV5	Avocado	Cellulase	18
pAV363	Avocado	Cellulase	23
cDNA clone Fl	Tomato	Polygalacturonase	21
pTom6	Tomato	Polygalacturonase	19, 24
pPG16	Tomato	Polygalacturonase	21
pE41	Tomato	Polygalacturonase	15
pERI	Tomato	Protease inhibitor	15, 25
pPE1	Tomato	Pectinmethylesterase	26

determined by measuring mRNA levels at various stages of fruit development and after ethylene treatment using radioactively labeled cDNA clones as hybridization probes. During the course of this study evidence for a complex pattern of gene expression emerged. When early mature-green fruit were treated with exogenous ethylene a rapid accumulation of mRNAs corresponding to the E4, E8, E17 (proteinase inhibitor[25]), and J49 genes occurred.

In contrast, expression of the E41 gene, which encodes polygalacturonase, was not induced by this treatment, but occurred only in mature-green fruit concomitant with climacteric ethylene production. During the course of normal fruit ripening, mRNAs encoded by the E17 and J49 genes were found to accumulate prior to the onset of ethylene production, while mRNAs corresponding to the E4 and E8 genes accumulated approximately coincident with the onset of ethylene production. However, exposure to norbornadiene, a specific inhibitor of ethylene action, prevented the accumulation of mRNA encoded by each of these genes. Based on such experiments it was proposed that several mechanisms exist for the regulation of gene expression by ethylene during tomato fruit ripening. For example, expression of polygalacturonase is not only dependent upon ethylene production, but also on specific developmental cues since ethylene treatment of unripe fruit is not sufficient to induce polygalacturonase gene expression. In order to explain the differential timing of expression of E4, E8, E17, and J49 gene expression during fruit ripening, Lincoln et al.[15] have invoked the model of Trewavas,[28] which describes the interaction between a growth substance and a plant tissue. This model has been expressed as:

$$\text{growth substance} + \text{receptor} = \text{active complex} \rightarrow \rightarrow \text{biological response}$$

In this case, the growth substance is ethylene and the biological response is activation of gene expression. According to this model, activation of gene expression may result from increasing ethylene concentration, from a change in receptor concentration, or by a change in receptor activity. It has been proposed that the increased expression of the E4 and E8 genes results from an increase in ethylene concentration, while an increase in the capacity of the tissue to respond to low levels of ethylene is responsible for the early activation of E17 and J49 genes.[15] A similar pattern of gene expression has also been observed for the E4, E8, E17, and J49 genes when expression levels were determined by measuring changes in transcription rates in isolated nuclei,[29] rather than by measuring changes in steady-state mRNA levels.[15] The increases in transcription rates observed for the E4, E8, E17, and J49 genes were nearly as rapid as the increases observed for auxin-induced changes in gene transcription in soybean,[30] and provided further support for the hypothesis that hormonally induced changes in plant growth and development occur through rapid changes in the pattern of gene expression. Thus, in the case of ethylene-induced fruit ripening, the regulation of gene expression can occur at the level of gene transcription, either as a result of increased sensitivity to the phytohormone or an increase in ethylene concentration.

In addition to transcriptional control of gene expression, Lincoln et al.[15] have found that post-transcriptional processes, such as differential mRNA stability and/or rates of cytoplasmic transport, also play an important role in the modulation of gene expression by ethylene. In the case of E17 gene expression, an increase in the rate of transcription was observed prior to the detection of mRNA accumulation. Furthermore, exposure of unripe fruit to exogenous ethylene did not increase the transcription rate, but resulted in a sixfold increase in steady-state mRNA levels. Together, these experiments indicate that gene expression can be regulated by ethylene in unique ways by a combination of transcriptional and/or post-transcriptional regulatory processes.

In order to better understand the molecular mechanism(s) responsible for ethylene regulation of gene expression during fruit ripening, several ethylene-associated ripening-specific genes have been isolated and their expression studied in transgenic plants. The tomato polygalacturonase gene was found to be encoded by nine exons, spanning approximately 7 kb of genomic DNA.[31] RFLP mapping and genomic hybridization experiments revealed the presence of a single polygalacturonase locus containing a single gene.[9,31] These data suggest that the different isoforms of this enzyme observed previously[32] result from post-translational modifications of the primary polypeptide rather than from expression of different genes. In order to identify the regions of the polygalacturonase gene responsible for ripening-specific expression, a chimeric gene was constructed, which consisted of a 1.4-kb fragment of the polygalacturonase gene, derived from the region immediately upstream of the translation initiation site, fused to the reporter gene chloramphenicol acetyl transferase (CAT). Analysis of transgenic tomato plants containing this chimeric gene revealed that expression of the CAT enzyme was confined to ripening fruit; no activity could be detected in protein extracts from unripe fruit, leaves, stems, or roots, indicating that the 1.4-kb fragment of the polygalacturonase gene is sufficient to direct the ripening-specific expression of the reporter gene.[31] Further analysis of this promoter region by *in vitro* mutagenesis should allow a more definitive identification of the exact DNA sequences responsible for this regulation.

The expression of another ethylene-regulated, ripening-specific gene has also been studied in transgenic plants.[33] As noted previously, the expression of the tomato E8 gene is induced at the onset of ripening and also by treatment of unripe fruit with ethylene. The response of the E8 gene to ethylene is organ specific in that ethylene treatment of leaf tissue had little effect on gene expression. The E8 gene has been isolated and localized to a 4.4-kb fragment of tomato genomic DNA. The isolated gene was "tagged" with a small fragment of bacteriophage lambda DNA in order to follow its expression after reintroduction into transgenic tomato plants. Analysis of these plants showed that the expression of the tagged gene closely followed the expression of the endogenous E8 gene and additionally indicated that the 4.4-kb fragment contained the necessary sequence information for ripening-specific and ethylene-inducible expression. These regulatory sequences appeared to be situated in the 5′ upstream region since this fragment has been shown to direct the expression of the polygalacturonase coding region in transgenic *rin* fruit in response to treatment with ethylene or the ethylene analog, propylene.[34] Thus, the 5′ upstream region of both the polygalacturonase gene and the E8 gene plays an important role in mediating the developmental and ethylene-inducible expression of these genes.

One possible mechanism that can be invoked to explain the activation of gene expression by ethylene is analogous to the mechanism of glucocorticoid-induced gene expression in animals.[35] In this system, induction of gene expression is mediated by the binding of a hormone-receptor complex to regulatory sites either within or upstream of the transcribed region of inducible genes. To understand better the interaction of nuclear proteins with ethylene-responsive promoters, Deikman and Fischer[33] have characterized the interaction of the E8 promoter with nuclear proteins extracted from unripe and ripe tomato fruits using

DNA gel electrophoresis mobility retardation assays.[36] This technique utilizes labeled DNA fragments that are incubated with extracts of nuclear proteins and then resolved by gel electrophoresis; the interaction of proteins with specific DNA sequences is revealed by the formation of slowly migrating complexes. The specificity of the interaction can be studied by competition experiments using unlabeled DNA fragments. Through this analysis, a DNA-binding protein(s) associated with the expression of the E8 and E4 genes during fruit ripening has been identified. This protein factor interacts with several sites in the 5′ flanking region of the E8 gene and a single site in the E4 gene. It is present in significantly higher amounts in nuclear extracts from ripe fruit than in extracts from unripe fruit. This correlates well with the coordinated changes in E4 and E8 gene expression observed during fruit ripening. Comparison of the E4 and E8 binding sites revealed a high degree of conservation, both in nucleotide sequence and structural properties; however, the position of these sites differs significantly between the two genes.[37] The binding site for the E8 gene is located far upstream, about 1 kb from the transcriptional start site while, in the case of the E4 gene, binding is observed proximal to the TATA box, a conserved DNA motif responsible for both promoter activity and determination of the correct transcription initiation site.[38] The exact function of the binding of this factor and its role in regulating transcription of the E4 and E8 genes are unknown. The binding of the factor to upstream sequences in the E8 gene and the positive correlation between factor binding and transcription may indicate a role in mediating high-level expression in response to ethylene treatment. Alternatively, binding of this factor may play a role in the coordinate activation of E4 and E8 transcription at the onset of fruit ripening. Further experiments designed to test the significance of factor binding and a functional analysis of the DNA sequences involved in binding will be necessary to resolve these possibilities.

III. GENE EXPRESSION DURING PLANT SENESCENCE

The senescence of flower petals has been compared to the ripening of climacteric fruit, in that both tissues undergo senescence and exhibit an increase in respiratory activity and ethylene production.[39,40] The production of ethylene during petal senescence has been associated with a number of physiological and biochemical changes (for a review see Reference 40 and Chapter 12). These include the synthesis of hydrolytic enzymes, degradation of macromolecules, and the loss of cellular compartmentation. Treatment of flowers with inhibitors of ethylene biosynthesis[41] or ethylene action[42] has been shown to delay petal senescence, while exposure of preclimacteric petals to ethylene accelerates this process.[43] Analysis of total protein and mRNA profiles during the development of petal senescence indicates that this is an active process resulting in the appearance of new mRNA and protein species.[44,45] In addition, exposure of preclimacteric petals to ethylene results in the accumulation of several of these same mRNAs,[43] supporting the hypothesis that the changes in petal physiology observed following exposure to ethylene are due to rapid changes in gene expression.

In order to study specific changes in gene expression involved in petal senescence, Lawton et al.[46] have isolated several senescence-related cDNA clones from a cDNA library prepared with mRNA isolated from senescing carnation petals. Three cDNA clones have been characterized, one of which encodes a polypeptide of 81 kDa. This is consistent with results obtained from *in vitro* translation experiments,[43,44] which revealed the accumulation of several new mRNA species during petal senescence, including an mRNA that encodes an 81-kDa polypeptide. RNA hybridization experiments indicated that steady-state levels of mRNAs corresponding to these cDNA clones increase during flower petal senescence and after exposure of flowers to exogenous ethylene. However, the level of mRNA complementary to clone pSR5 was found to be present at low levels during the early stages of petal development and accumulated in mature petals prior to the onset of climacteric ethylene production. In contrast, mRNAs corresponding to clones pSR8 and pSR12 increased concomitant with the ethylene climacteric. The accumulation of these latter mRNAs was significantly reduced by the presence of ethylene biosynthesis inhibitors, whereas the accumulation of mRNA represented by pSR5 was only partially reduced in the presence of these inhibitors. These data suggest that several mechanisms exist for the regulation of gene expression during flower senescence and that expression of at least one class of mRNAs is strictly dependent on ethylene production. This is analogous to the situation described for the induction of ripening-specific mRNA synthesis in tomato.[15] Whether a similar mechanism exists for the regulation of gene expression during both fruit ripening and flower senescence remains to be determined; however, it is clear that the production of ethylene markedly influences the pattern of gene expression in both these tissues.

Ethylene has also been shown to play an important role in leaf abscission.[1-3] Early studies using inhibitors of RNA and protein synthesis demonstrated that this process involves the *de novo* synthesis of specific proteins.[47] Subsequently, changes in the profile of translatable mRNAs have been observed following ethylene treatment of abscission zone explants from *Phaseolus vulgaris.*[48] In this study, the levels of at least four translation products were found to increase upon ethylene treatment; three of these translation products with molecular weights of 32, 27, and 17 kDa were found to increase substantially, while a fourth translation product with a molecular weight of 42 kDa appeared only after ethylene treatment. Although the identity of these polypeptides has not been determined conclusively, the 32-kDa polypeptide may be an ethylene-induced chitinase.[48]

An important event in the abscission process is the secretion of cell wall hydrolases in the abscission layer (Chapter 11). In bean leaves, a strong correlation between the appearance of cellulase activity and the onset of abscission has been observed.[49] This increase in cellulase activity was shown to be due to the *de novo* synthesis of a basic, cell wall-associated form of the enzyme. A cDNA clone encoding abscission zone cellulase has been isolated from bean and used to study the hormonal and tissue-specific expression of the cellulase gene during leaf abscission.[50] Nucleotide sequence analysis of the bean cellulase cDNA clone revealed extensive amino acid and nucleotide sequence homology with the avocado cellulase cDNA clone.[10] Hybridization experiments have shown that the bean cellulase cDNA clone detects a 2-kb mRNA that encodes a polypeptide of 51 kDa. This polypeptide is immunoprecipitated with antibodies against bean abscission zone cellulase.

The cellulase cDNA clone has been used to show that the increase in cellulase activity following ethylene treatment of bean explants is accompanied by a concomitant accumulation of steady-state cellulase mRNA.[50] The induction of cellulase mRNA was greatest in abscission zones of these plants, but was also induced, though to a lesser extent, in stems and petioles. Removal of ethylene and addition of an inhibitor of ethylene action, norbornadiene, slows the increase in cellulase activity and leads to a reduction in the amount of cellulase mRNA associated with polysomes. Thus, ethylene is important for the accumulation of cellulase mRNA during leaf abscission; it is not known however, whether this is due to an increase in cellulase gene transcription or to an increase in mRNA stability. An interesting aspect of the abscission response is its inhibition by auxin treatment. Hybridization studies following auxin application to tissue explants indicate that this treatment inhibits the accumulation of cellulase mRNA.[50] This will be an interesting system in which to study the mechanism of the antagonistic role that these two phytohormones play in modulating cellulase gene expression during leaf abscission.

IV. GENE EXPRESSION DURING PLANT STRESS

In addition to functioning during normal physiological processes of plant growth and development, ethylene also plays a major role in the response of plants to conditions of environmental stress. Increased ethylene synthesis has been shown to be associated with oxygen deficiency and flooding,[51,52] wounding,[53] and pathogen attack.[54-56] In the case of bacterial or fungal infection, ethylene production has been correlated with the induction of several proteins that are thought to function in the defense response of the plant. Among these are enzymes involved in the production of phytoalexins,[57] hydroxyproline-rich glycoproteins,[56] and hydrolytic enzymes, such as β-1,3-glucanase[58,59] and chitinase.[59,60] cDNA clones that code for several of the enzymes involved in phytoalexin biosynthesis including L-phenylalanine ammonia lyase, 4-coumarate CoA ligase, and chalcone synthase have been used to study the effect of ethylene on the expression of these genes in carrot roots.[61] The results of RNA blot hybridization experiments indicate that ethylene treatment causes a dramatic increse in the steady-state levels of these mRNAs. In elicitor-treated plant cell cultures, the increase in expression of these genes has been correlated with an increase in the specific rate of transcription.[62,63] In the case of ethylene treatment, it is not known whether the increase in mRNA levels is also due to an increase in transcription rates.

In addition to enzymes involved in phytoalexin biosynthesis, other mRNAs encoding proteins involved in plant defense are also found to increase following ethylene treatment. These include hydroxyproline-rich glycoproteins,[61] β-1,3-glucanase,[59] and chitinase.[64] In the case of the hydroxyproline-rich proteins, ethylene treatment results in the accumulation of two mRNAs of 1.8 and 4.0 kb in length. In contrast, wounding causes an increase in the level of a 1.5-kb mRNA, which is reduced in abundance by ethylene treatment. Both the 1.8- and the 1.5-kb mRNAs are derived from the same gene, but differ in the length of their 5′ untranslated regions.[65] It has been proposed that each mRNA transcript has its own promoter,[65] one of which may be responsive to ethylene treatment and the other to wounding. Alternatively, it is possible that ethylene may regulate the differential processing of these two mRNA transcripts.

Another protein whose mRNA level has been shown to be influenced by the presence of ethylene is chitinase, an enzyme that catalyzes the hydrolysis of chitin, a β-1,4-linked homopolymer of *N*-acetyl-D-glucosamine and a ubiquitous component of many phytopathogenic fungi. In bean plants, chitinase is encoded by a small multigene family consisting of at least three members.[64] In healthy, uninfected plants, the level of this enzyme is low or undetectable. However, treatment with ethylene,[64] oligosaccharide elicitors,[66] or infection with fungal pathogens[67] results in an increase in chitinase mRNA levels and an increase in enzyme activity. Using cloned chitinase cDNAs as hybridization probes, the level of chitinase mRNA in bean and tomato seedlings was found to increase 50- to 75-fold following ethylene treatment (Figure 1). In order to investigate the molecular basis of ethylene regulation of chitinase gene expression and to understand better the role of ethylene in mediating the response to pathogen attack, we have isolated and characterized three chitinase genes from bean. Nucleotide sequence analysis of several chitinase cDNA clones indicated that at least two chitinase genes are expressed in response to ethylene treatment.[64] We have determined the complete nucleotide sequence of one of these genes, the CH5B gene, which resides on a 4.7-kb fragment of bean genomic DNA.[68] To define the DNA sequence elements involved in ethylene regulation of chitinase expression, we have introduced the CH5B gene into tobacco plants using Ti-plasmid derived vectors.[69] Figure 2 shows one of the vectors that was used to introduce the 4.7-kb fragment of bean DNA into tobacco plants. Analysis of the resulting transgenic plants containing this plasmid showed that expression of the bean chitinase gene is dependent upon the application of exogenous ethylene (Figure 3, lanes 1 to 8). In contrast, RNA from untreated or ethylene-treated control plants that contained only vector sequences, failed to hybridize to the bean chitinase cDNA probe (Figure 3, lanes 9 and 10). This indicates that the endogenous tobacco chitinase mRNAs are not detected under the stringent hybridization conditions used in this experiment. Similar results have been obtained using a binary Ti-plasmid vector to introduce the bean chitinase gene into tobacco.[68]

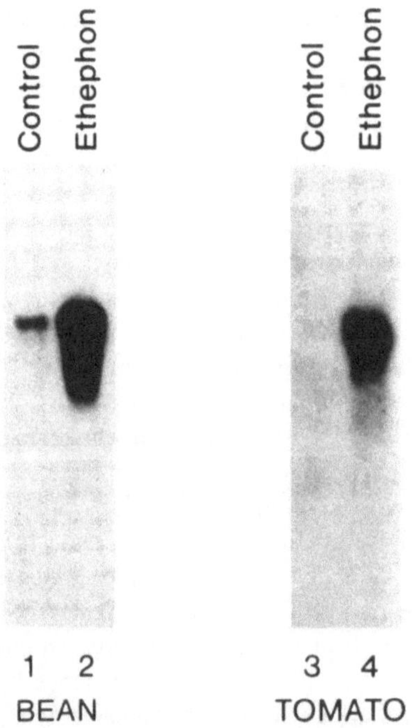

FIGURE 1. Induction of chitinase mRNA in ethephon-treated bean and tomato seedlings. Plants were treated with a solution of 1 mg/ml 2-chloroethylphosphonic acid (ethephon) for 36 h. Total RNA was isolated from control and treated plants, fractionated by agarose gel electrophoresis, and blotted onto nitrocellulose filters. Blots were hybridized with ^{32}P-labeled cDNA inserts from a bean (lanes 1 to 2) or tomato (lanes 3 to 4) chitinase cDNA clones.

By studying the expression of a chimeric gene composed of a fusion between 5′ flanking sequences of the CH5B gene and the coding region of β-glucuronidase in transgenic tobacco plants, we have found that DNA sequences situated upstream of the chitinase coding region are responsible for ethylene-regulated gene expression.

When the upstream sequences of two ethylene-regulated bean chitinase genes were compared, two regions of striking homology were observed: an upstream, AT-rich segment that shows 96% homology between the two genes and a region near the transcription start site of the CH5B gene that is exactly conserved with a region of the CH5A gene situated approximately 250 nucleotides further upstream. Deletion analysis of the CH5B promoter in stably transformed transgenic tobacco plants[68] and by transient gene expression in electroporated bean protoplasts[68a] have identified two promoter regions responsible for optimal chitinase gene expression: a quantitative region required for maximal gene expression and an ethylene-responsive region required for induction of gene expression by ethylene. A schematic summary of the results of these experiments is shown in Figure 4. In addition to identifying *cis*-acting DNA sequences, we have also begun to study the interaction of nuclear proteins with the chitinase promoter. A protein factor, which binds with high affinity to DNA sequences near the quantitative element, has been identified by gel retardation experiments.[68a] Competition experiments using promoter fragments of non-ethylene-regulated genes suggest that the binding of this factor to the chitinase promoter is specific. Additional experiments are necessary to determine the role of this protein in the regulation of chitinase gene expression.

Although it is clear from the above discussion that ethylene treatment of bean plants has a dramatic effect on chitinase gene expression, the role that this phytohormone plays in

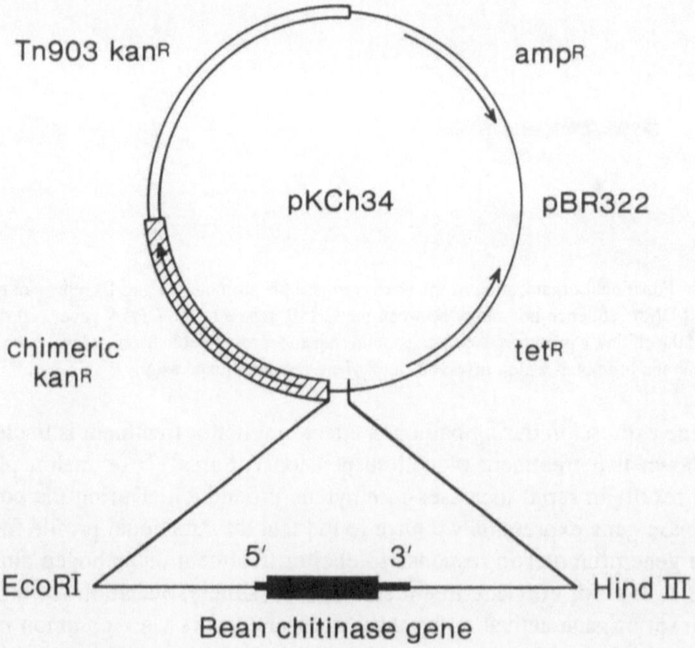

FIGURE 2. Diagrammatic representation of the intermediate vector, pKCh34. This vector was used in transformation experiments to introduce the bean chitinase gene into tobacco plants. The intermediate vector was transferred to *Agrobacterium tumefaciens* GV3850 using a tri-parental mating procedure. Co-integrate Ti-plasmids were formed in *Agrobacterium* by recombination with pBR322 sequences. The chimeric kan^R gene enabled selection of transformed plant cells on kanamycin, while the Tn903 kan^R gene was useful in selecting Ti plasmid co-integrates.

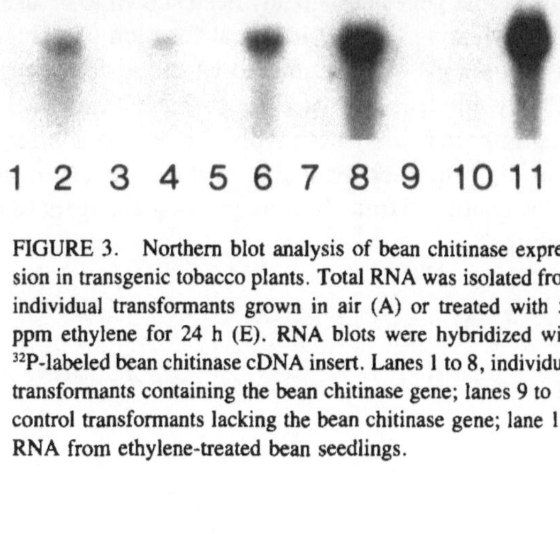

FIGURE 3. Northern blot analysis of bean chitinase expression in transgenic tobacco plants. Total RNA was isolated from individual transformants grown in air (A) or treated with 50 ppm ethylene for 24 h (E). RNA blots were hybridized with ^{32}P-labeled bean chitinase cDNA insert. Lanes 1 to 8, individual transformants containing the bean chitinase gene; lanes 9 to 10 control transformants lacking the bean chitinase gene; lane 11, RNA from ethylene-treated bean seedlings.

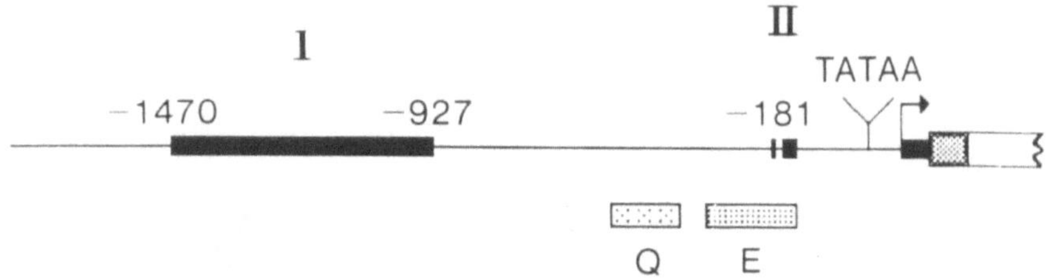

FIGURE 4. Functional organization of the bean chitinase 5B promoter; (I and II) represent regions of conserved DNA sequence homology between the CH5B gene and the CH5A gene, (Q) denotes the region of the chitinase promoter shown to contain sequences responsible for high level expression, (E) represents the promoter region involved in ethylene-regulated expression.

activation of gene expression during pathogen attack or elicitor treatment is unclear.[70] Other studies have shown that treatment of cell suspension cultures[71,72] or melon plants[56] with fungal elicitors results in rapid increases in ethylene production. During the course of our studies on chitinase gene expression we have found that the functional profile for activation of the chitinase gene promoter in response to elicitor treatment or pathogen attack is qualitatively similar to that for ethylene treatment. One possible explanation for this finding is that the mechanism of gene activation by these inducers occurs via a common intermediate in the signal transduction pathway.

V. SUMMARY

Advances in plant molecular biology research have contributed much toward our understanding of the molecular basis of hormonal control of gene expression in higher plants. An increasing number of ethylene-responsive genes have been isolated and characterized and analysis of their expression in transgenic plants have enabled the identification of important DNA sequence elements involved in regulating their expression. Additionally, *in vitro* DNA-protein binding experiments have allowed the identification of several transcription factors that interact with these sequences. Further characterization of these proteins and the genes that encode them will undoubtedly aid in unraveling the molecular events involved in activation of gene expression by the phytohormone ethylene. The availability of mutants blocked in their ability to respond to ethylene[73] will be invaluable for understanding the signal/transduction pathway as well as for the identification of the genes and proteins involved in this process. Similarly, the use of antisense RNA technology to specifically inhibit the expression of ethylene-inducible genes has already been shown to be useful as an alternative to genetic analysis for investigating the physiological function of polygalacturonase during fruit ripening.[74,75] This technology will continue to be useful for determining the function of other proteins encoded by ethylene-regulated genes. Finally, the isolation and characterization of genes encoding important enzymes involved in ethylene biosynthesis may lead to a better understanding of the processes regulating hormone levels in plants and offers the potential for improving the quality of fruits by manipulating endogenous ethylene production through genetic engineering.[76]

REFERENCES

1. **Abeles, F. B.,** *Ethylene in Plant Biology,* Academic Press, New York, 1973, 302.
2. **Lieberman, M.,** Biosynthesis and action of ethylene, *Annu. Rev. Plant Physiol.,* 30, 533, 1979.
3. **Yang, S. F. and Hoffmann, N. E.,** Ethylene biosynthesis and its regulation in higher plants, *Annu. Rev. Plant. Physiol.,* 35, 155, 1984.
4. **Neljubow, D.,** Ueber die horizontale nutation der stengel von *Pisum sativum* und einiger anderer, *Pflanzen Beih. Bot. Zentralbl.,* 10, 728, 1901.
5. **Christoffersen, R. E. and Laties, G. G.,** Ethylene regulation of gene expression in carrots, *Proc. Natl. Acad. Sci. U.S.A.,* 79, 4060, 1982.
6. **Zurfluh, L. L. and Guilfoyle, T. J.,** Auxin- and ethylene-induced changes in the population of translatable messenger RNA populations in basal sections and intact soybean hypocotyl, *Plant Physiol.,* 69, 338, 1982.
7. **Grierson, D.,** Gene expression in ripening tomato fruit, *Crit. Rev. Plant Sci.,* 3, 113, 1985.
8. **Nichols, S. E. and Laites, G. G.,** Ethylene-regulated gene transcription in carrot roots, *Plant Mol. Biol.,* 3, 393, 1984.
9. **Schuch, W., Bird, C. R., Ray, J., Smith, C. J. S., Watson, C. F., Morris, P. C., Gray, J. E., Arnold, C., Seymour, G. B., Tucker, G. A., and Grierson, D.,** Control and manipulation of gene expression during tomato fruit ripening, *Plant Mol. Biol.,* 13, 303, 1989.
10. **Tucker, M. L., Christofferson, R. E., Woll, L., and Laites, G. G.,** Induction of cellulase by ethylene in avocado fruit, in *Ethylene and Plant Development,* Roberts, J. A. and Tucker, G. A., Eds., Butterworths, London, 1985, 163.
11. **Hobson, G. E., Nichols, R., Davies, J. N., and Atkey, P. J.,** The inhibition of tomato fruit ripening by silver, *J. Plant Physiol.,* 116, 21, 1984.
12. **Rattanapone, N., Speirs, J., and Grierson, D.,** Evidence for changes in messenger RNA content related to tomato fruit ripening, *Phytochemistry,* 17, 1485, 1978.
13. **Grierson, D., Slater, A., Speirs, J., and Tucker, G. A.,** The appearance of polygalacturonase mRNA in tomatoes: one of a series of changes in gene expression during development and ripening, *Planta,* 163, 263, 1985.
14. **Biggs, M. S., Harriman, R. W., and Handa, A. K.,** Changes in gene expression during tomato fruit ripening, *Plant Physiol.,* 89, 117, 1986.
15. **Lincoln, J. E., Cordes, S., Read, E., and Fischer, R. L.,** Regulation of gene expression by ethylene during *Lycopersicon esculentum* (tomato) fruit development, *Proc. Natl. Acad. Sci. U.S.A.,* 84, 2793, 1987.
16. **Maunders, M. J., Holdsworth, M. J., Slater, A., Knapp, J. E., Bird, C. R., Schuch, W., and Grierson, D.,** Ethylene stimulates the accumulation of ripening-related mRNAs in tomatoes, *Plant Cell Environ.,* 10, 177, 1987.
17. **Christofferson, R. E., Warm, E., and Laites, G. G.,** Gene expression during fruit ripening in avocado, *Planta,* 155, 52, 1982.
18. **Christofferson, R. E., Tucker, M. L., and Laites, G. G.,** Cellulase gene expression in ripening avocado fruit: the accumulation of cellulase mRNA and protein as demonstrated by cDNA hybridization and immunodetection, *Plant Mol. Biol.,* 3, 385, 1984.
19. **Slater, A., Maunders, M. J., Edwards, K., Schuch, W., and Grierson, D.,** Isolation and characterization of cDNA clones for tomato polygalacturonase and other ripening-related proteins, *Plant Mol. Biol.,* 5, 137, 1985.
20. **Mansson, P.-E., Hsu, D., and Stalker, D.,** Characterization of fruit specific cDNAs from tomato, *Mol. Gen. Genet.,* 200, 356, 1985.
21. **Della Penna, D., Alexander, D. C., and Bennett, A. B.,** Molecular cloning of tomato fruit polygalacturonase: analysis of polygalacturonase mRNA levels during ripening, *Proc. Natl. Acad. Sci. U.S.A.,* 83, 6420, 1986.
22. **Sheehy, R. E., Pearson, J., Brady, C. J., and Hiatt, W. R.,** Molecular characterization of tomato fruit polygalacturonase, *Mol. Gen. Genet.,* 208, 30, 1987.
23. **Tucker, M. L., Durbin, M. L., Clegg, M. T., and Lewis, L. N.,** Avocado cellulase: nucleotide sequence of a putative full-length cDNA clone and evidence for a small gene family, *Plant Mol. Biol.,* 9, 197, 1987.
24. **Grierson, D., Tucker, G. A., Keen, J., Ray, J., Bird, C. R., and Schuch, W.,** Sequencing and identification of a cDNA clone for tomato polygalacturonase, *Nucl. Acids Res.,* 14, 8595, 1986.
25. **Margossian, L. J., Federman, A. D., Giovannoni, J. J., and Fischer, R. L.,** Ethylene-regulated expression of a tomato fruit ripening gene encoding a proteinase inhibitor I with a glutamic acid residue at the active site, *Proc. Natl. Acad. Sci. U.S.A.,* 85, 8012, 1988.
26. **Ray, J., Knapp, J. E., Grierson, D., Bird, C., and Schuch, W.,** Identification and sequence determination of a cDNA clone for tomato pectin esterase, *Eur. J. Biochem.,* 174, 119, 1988.
27. **Holdsworth, M. J., Schuch, W., and Grierson, D.,** Organization and expression of a wound/ripening related small multigene family from tomato, *Plant Mol. Biol.,* 11, 81, 1988.

28. **Trewavas, A. J.,** Growth substance sensitivity: the limiting factor in plant development, *Physiol. Plant.*, 55, 60, 1982.
29. **Lincoln, J. E. and Fischer, R. L.,** Diverse mechanisms for the regulation of ethylene-inducible gene expression, *Mol. Gen. Genet.*, 212, 71, 1988.
30. **Guilfoyle, T. J.,** Auxin-regulated gene expression in higher plants, *Crit. Rev. Plant Sci.*, 4, 247, 1986.
31. **Bird, C. R., Smith, C. J. S., Ray, J. A., Moureau, P., Beven, M. W., Bird, A. S., Hughes, S., Morris, P. C., Grierson, D., and Schuch, W.,** The tomato polygalacturonase gene and ripening specific expression in transgenic tomato plants, *Plant Mol. Biol.*, 11, 651, 1988.
32. **Tucker, G. A., Robertson, N. G., and Gierson, D.,** Changes in polygalacturonase isozymes during the "ripening" of normal and mutant tomato fruits, *Eur. J. Biochem.*, 112, 119, 1980.
33. **Deikman, J. and Fischer, R. L.,** Interaction of a DNA binding factor with the 5′-flanking region of an ethylene-responsive fruit ripening gene from tomato, *EMBO J.*, 7, 3315, 1988.
34. **Giovannoni, J. J., Della Penna, D., Bennett, A. B., and Fischer, R. L.,** Expression of a chimeric polygalacturonase gene in transgenic rin tomato fruits results in polyuronide degradation but not fruit softening, *Plant Cell*, 1, 53, 1989.
35. **Payyar, F., DeFranco, D., Firestone, G. L., Edgar, B., Wrange, O., Okret, S., Gustafsson, J., and Yamamoto, K. R.,** Sequence-specific binding of a glucocorticoid receptor to MTV DNA at sites within and upstream of the transcribed region, *Cell*, 35, 381, 1983.
36. **Singh, H., Sen, R., Baltimore, D., and Sharp, P. A.,** A nuclear factor that binds to a conserved sequence motif in transcriptional control elements of immunoglobulin genes, *Nature*, 319, 154, 1986.
37. **Cordes, S., Deikman, J., Margossian, L. J., and Fischer, R. L.,** Interaction of a developmentally regulated DNA-binding factor with sites flanking two different fruit-ripening genes from tomato, *Plant Cell*, 1, 1025, 1989.
38. **Shenk, T.,** Transcription control regions: nucleotide sequence requirements for initiation by RNA polymerase II and III, *Curr. Top. Microbiol. Immunol.*, 93, 25, 1981.
39. **Yang, S. F.,** Regulation of ethylene biosynthesis in relation to plant senescence, *Plant Growth Regul. Bull.*, 11, 9, 1983.
40. **Borochov, A. and Woodson, W. R.,** Physiology and biochemistry of flower petal senescence, *Hortic. Rev.*, 11, 15, 1989.
41. **Bufler, G., Mor, Y., Reid, M. S., and Yang, S. F.,** Changes in 1-aminocycloproprone-1-carboxylic acid content of cut carnation flowers in relation to their senescence, *Planta*, 150, 439, 1980.
42. **Wang, H. and Woodson, W. R.,** Reversible inhibition of ethylene action and interruption of petal senescence in carnation flowers by norbornadiene, *Plant Physiol.*, 89, 434, 1988.
43. **Woodson, W. R. and Lawton, K. A.,** Ethylene-induced gene expression in carnation petals. Relationship to autocatalytic ethylene production and senescence, *Plant Physiol.*, 87, 498, 1988.
44. **Woodson, W. R.,** Changes in protein and mRNA populations during the senescence of carnation petals, *Physiol. Plant.*, 71, 495, 1987.
45. **Woodson, W. R. and Handa, A. K.,** Changes in protein patterns and in vivo protein synthesis during presenescence and senescence of hibiscus petals, *J. Plant Physiol.*, 128, 67, 1987.
46. **Lawton, K. A., Huang, B., Goldsbrough, P. B., and Woodson, W. R.,** Molecular cloning and characterization of senescence-related genes from carnation flower petals, *Plant Physiol.*, 90, 690, 1989.
47. **Abeles, F. B., Leather, G. R., Ferrence, L. E., and Craker, L. E.,** Abscission: regulation of senescence, protein synthesis and enzyme secretion by ethylene, *HortScience*, 6, 371, 1971.
48. **Kelly, P., Trewavas, A. J., Lewis, L. N., Durbin, M. L., and Sexton, R.,** Translatable mRNA changes in ethylene induced abscission zones of *Phaseolus vulgaris* (Red Kidney), *Plant Cell Environ.*, 10, 11, 1987.
49. **Sexton, R. and Roberts, J. A.,** Cell biology of abscission, *Annu. Rev. Plant Physiol.*, 33, 133, 1982.
50. **Tucker, M. L., Sexton, R., de Campillo, E., and Lewis, L. N.,** Bean abscission cellulase. Characterization of a cDNA clone and regulation of gene expression by ethylene and auxin, *Plant Physiol.*, 88, 1257, 1988.
51. **Metraux, J. P. and Kende, H.,** The role of ethylene in the growth response of submerged deepwater rice, *Plant Physiol.*, 72, 441, 1983.
52. **Raskin, I. and Kende, H.,** Regulation of growth in stem sections of deepwater rice, *Planta*, 160, 66, 1984.
53. **Boller, T. and Kende, H.,** Regulation of wound ethylene synthesis in plants, *Nature*, 286, 259, 1980.
54. **Montalbini, P. and Elstner, E. F.,** Ethylene evolution by rust infected, detached bean leaves (*Phaseolus vulgaris* L.) susceptible and hypersensitive to Uromyces phaseoli (Pers.) Wint., *Planta*, 135, 301, 1977.
55. **Paradies, I., Hümme, H., Hoppe, H. H., Heitefuss, R., and Elstner, E. F.,** Induction of ethylene formation in bean (*Phaseolus vulgaris*) hypocotyl segments by preparations isolated from germ tube cell walls of Uromyces phaseoli, *Planta*, 146, 193, 1979.
56. **Roby, D., Toppan, A., and Esquerre-Tugaye, M.-T.,** Cell surfaces in plant microorganism interactions. V. Elicitors of fungal and plant origin trigger the synthesis of ethylene and of hydroxyproline-rich glycoproteins in plants, *Plant Physiol.*, 77, 700, 1985.

57. **Hahlbrook, K. and Grisebach, H.,** Enzymatic controls in the biosynthesis of lignin and flavonoids, *Annu. Rev. Plant Physiol.*, 30, 105, 1979.
58. **Abeles, F. B., Bonhart, R. P., Forrence, L. E., and Habig, W. H.,** Preparation and purification of glucanase and chitinase from bean leaves, *Plant Physiol.*, 47, 129, 1970.
59. **Felix, G. and Meins, F.,** Ethylene regulation of β-1,3-glucanase in tobacco, *Planta*, 172, 386, 1987.
60. **Boller, T., Gehri, A., Mauch, F., and Vögeli, U.,** Chitinase in bean leaves: induction by ethylene, purification, properties and possible function, *Planta*, 157, 22, 1983.
61. **Ecker, J. R. and Davis, R. W.,** Plant defense genes are regulated by ethylene, *Proc. Natl. Acad. Sci. U.S.A.*, 84, 5202, 1987.
62. **Chappell, J. and Hahlbrock, K.,** Transcription of plant defense genes in response to UV light or fungal elicitor, *Nature*, 311, 76, 1984.
63. **Lawton, M. A. and Lamb, C. J.,** Transcriptional activation of plant defense genes by fungal elicitor, wounding and infection, *Mol. Cell. Biol.*, 7, 335, 1987.
64. **Broglie, K. E., Gaynor, J. J., and Broglie, R. M.,** Ethylene-regulated gene expression: molecular cloning of the genes encoding an endochitinase from *Phaseolus vulgaris*, *Proc. Natl. Acad. Sci. U.S.A.*, 83, 6820, 1986.
65. **Chen, J. and Varner, J. E.,** An extracellular matrix protein in plants: characterization of a genomic clone for extensin, *EMBO J.*, 4, 2145, 1985.
66. **Hedrick, S. A., Bell, J. N., Boller, T., and Lamb, C. J.,** Chitinase cDNA cloning and mRNA induction by fungal elicitor, wounding and infection, *Plant Physiol.*, 86, 182, 1988.
67. **Roby, D. and Esquerre-Tugaye, M.-T.,** Induction of chitinases and of translatable mRNA for these enzymes in melon plants infected with Colletotrichum lagenarium, *Plant Sci.*, 52, 175, 1987.
68. **Broglie, K. E., Biddle, P., Cressman, R., and Broglie, R.,** Functional analysis of DNA sequences responsible for ethylene regulation of a bean chitinase gene in transgenic tobacco, *Plant Cell*, 1, 599, 1989.
68a. **Broglie, R. and Broglie, K. E.,** unpublished results.
69. **Horsch, R. B., Fry, J. E., Hoffmann, N. L., Eichholtz, D., Rogers, S. G., and Fraley, R. T.,** A simple and general method for transferring genes into plants, *Science*, 227, 1229, 1985.
70. **Mauch, F., Hadwiger, L. A., and Boller, T.,** Ethylene: symptom, not signal for the induction of chitinase and β-1,3-glucanase in pea pods by pathogens and elicitors, *Plant Physiol.*, 76, 607, 1984.
71. **Chappell, J. Hahlbrock, K., and Boller, T.,** Rapid induction of ethylene biosynthesis in cultured parsley by fungal elicitor and its relationship to the induction of phenylalanine ammonia-lyase, *Planta*, 161, 475, 1984.
72. **Tong, C. B., Labavitch, J. M., and Yang, S. F.,** The induction of ethylene production from pear cell cultures by cell wall fragments, *Plant Physiol.*, 81, 929, 1986.
73. **Bleeker, A. R., Estelle, M., Somerville, C., and Kende, H.,** Insensitivity to ethylene conferred by a dominant mutation in *Arabidopsis thaliana*, *Science*, 241, 1086, 1988.
74. **Smith, C. J. S., Watson, C. F., Ray,, J., Bird, C. R., Morris, P. C., Schuch, W., and Grierson, D.,** Antisense inhibition of polygalacturonase gene expression in tomatoes, *Nature*, 334, 724, 1988.
75. **Sheehy, R. E., Kramer, M., and Hiatt, W. R.,** Reduction of polygalacturonase activity in tomato fruit by antisense RNA, *Proc. Natl. Acad. Sci. U.S.A.*, 85, 8805, 1988.
76. **Sato, T. and Theologis, A.,** Cloning the mRNA encoding 1-aminocyclopropane-1-carboxylate synthase, the key enzyme for ethylene biosynthesis in plants, *Proc. Natl. Acad. Sci. U.S.A.*, 86, 6621, 1989.

Chapter 7

ETHYLENE INTERACTIONS WITH OTHER ENDOGENOUS GROWTH SUBSTANCES

Jeffrey C. Suttle

TABLE OF CONTENTS

I. INTRODUCTION

In common with other multicellular organisms, higher plants share the need to integrate growth and development in order to optimize their survival capabilities with respect to ever-changing environmental conditions. Although there may be other mechanisms, it is clear that this integration is achieved, at least in part, through the synthesis, transport, and recognition by the plant of a limited number of low molecular weight organic compounds: the endogenous plant hormones or phytohormones.

Historically, research has been directed towards the description of the five principal classes of phytohormones: auxins, gibberellins, cytokinins, abscisic acid, and ethylene (Figure 1). It has been argued that the limited number of endogenous regulators is insufficient to coordinate the complex and interlocking array of physiological processes that constitute plant development.[1-3] This argument is valid only if one considers each of these phytohormone classes in isolation and ignores the probability of other, yet uncharacterized, growth regulators. Further, if one considers the physiological ramifications of interhormone interactions as developmental determinants, then the spectrum of potential metabolic regulation becomes much broader.[3]

As illustrated in Figure 2, ethylene may interact with another hormonal system in a variety of ways that ultimately affect the biological activity of the second hormone. Ethylene may alter the relative rates of synthesis and metabolism of the hormone; thereby affecting the active hormone pool (Steps 1 and 2). Alternatively, ethylene may alter the availability or activity of hormone receptors by affecting the synthesis and/or inactivation (both reversible and irreversible) of the receptor molecule(s) itself (Steps 3 and 4). Finally, ethylene may affect the signal transduction mechanism that couple receptor occupancy to biochemical response (Steps 5 and 6). Not depicted in Figure 1, largely because the relevant information is not available, ethylene may alter the subcellular or tissue distributions of free hormone and receptor. The ability of ethylene to affect the metabolism of another hormone can no longer be viewed as simply an inactivation mechanism. As is the case in vertebrate organisms, it is becoming clear that metabolites of the principal plant hormones may have intrinsic biological activities and may influence developmental processes in a manner distinct from the parent hormone.

After reviewing much of the available literature, it is apparent that actual documentation of ethylene effects on other hormonal systems is relatively rare and biochemical descriptions even rarer. Much of the information is inferential in nature and must therefore be interpreted with caution. In this chapter, possible biochemical mechanisms through which ethylene may affect other hormone systems are discussed with the goal of stimulating research into this important aspect of plant hormone physiology.

II. AUXIN (INDOLE-3-ACETIC ACID)

A. PHYSIOLOGICAL INDICATIONS

Superficially, many effects of applied ethylene are opposite those elicited by auxin or mimic the effects of supraoptimal auxin levels.[4,5] Ethylene and auxin elicit opposing effects in the following assays: cell or tissue elongation, development of tropisms, organ abscission, and apical dominance. However, situations are known where ethylene and auxin appear to act in concert; two examples of this are epinastic growth and elongation of tissues of aquatic angiosperms.[6-8]

B. EFFECTS ON ENDOGENOUS LEVELS

There have been very few reports describing the effects of ethylene on endogenous indole-3-acetic acid (IAA) levels and most of these have employed bioassays to quantify

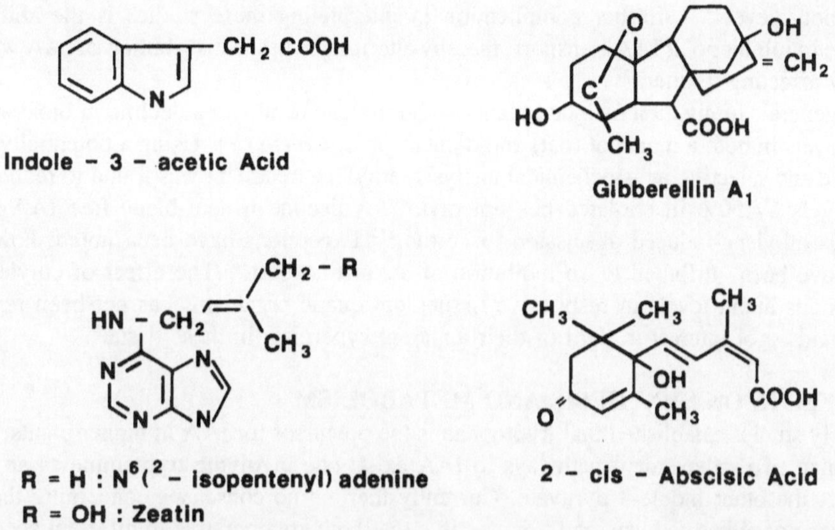

FIGURE 1. Chemical structures of representative compounds from each of the four principal plant hormone classes discussed in the text.

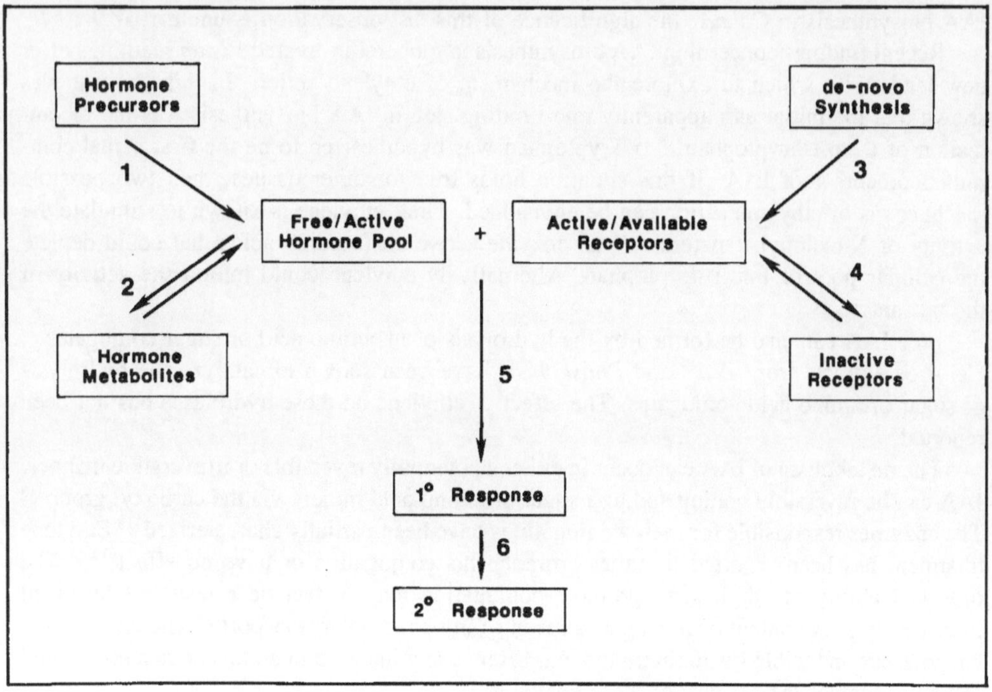

FIGURE 2. Possible biochemical mechanisms affecting plant hormone action. The active/available hormone pool can be affected by altered rates of hormone biosynthesis (Step 1) or metabolism (Step 2). Receptor levels can similarly be regulated by synthesis (Step 3) or inactivation (Step 4). The transduction mechanisms that lead from receptor occupation to biological responses (Steps 5, 6) can be selectively modified.

endogenous levels. A further complication in interpreting these studies is the ability of ethylene to inhibit polar IAA transport, thereby altering the tissue distribution of IAA without actually affecting its titer.

In general, fumigation of intact plants with ethylene results in a decline in bioassayable auxin levels in both a monocot (oat) and dicots (*Vicia, Coleus*).[9,10] Using a potentially more sensitive and selective physiochemical analysis, ethylene treatment was found to reduce free IAA levels by 50% in etiolated pea epicotyls.[11] A decline in leaf-blade free IAA levels preceeds ethylene-induced abscission in cotton.[11a] Exceptions have been noted; however, these have been attributed to an inhibition of auxin transport.[12] The effect of ethylene on endogenous auxin levels in responsive tissues of aquatic organisms has not been reported and would be of interest in light of their apparent synergism in these tissues.

C. EFFECTS ON SYNTHESIS AND METABOLISM

Early studies established that tryptophan is the precursor for IAA in higher plants. From tryptophan two independent pathways to IAA exist: one involving tryptamine as an intermediate, the other indole-3-pyruvate. Currently there is no consensus concerning the predominance of either pathway and it is possible that both are operative in different species.[13]

Fumigation of intact *Coleus* seedlings with ethylene results in a 50% reduction in the ability of cell-free extracts to convert tryptophan to IAA (as judged by bioassay of the incubation mixture). This reduction was accompanied by a reduced ability of cell-free extracts to decarboxylate tryptophan.[10,14] In the absence of knowledge concerning the route(s) of IAA biosynthesis in *Coleus*, the significance of this last observation is unclear.

Recent findings concerning IAA biosynthesis in gibberellin-treated *Pisum* seedlings offer new leads with which to explore the mechanism of ethylene action. In this study it was shown that the initial and apparently rate-limiting step in IAA biosynthesis was the racemization of L- to D-tryptophan.[15] D-Tryptophan was hypothesized to be the first actual committed precursor of IAA. If this situation holds true for other tissues, then two possible mechanisms of ethylene action can be envisioned. First, ethylene is known to stimulate the activity of *N*-malonyl transferase,[16] an enzyme active on D-amino acids that could deplete the cellular pool of free D-tryptophan. Alternatively ethylene could inhibit the activity of the racemase.

Free IAA can also be formed by the hydrolysis of its amino acid or sugar conjugates.[13] Cell-free extracts from *Zea*[17] and *Phaseolus*[13] have been shown to catalyze the hydrolysis of sugar or amino acid conjugates. The effect of ethylene on these hydrolases has not been reported.

The metabolism of IAA can occur in either a potentially reversible or irreversible manner. IAA can be reversibly conjugated to a sugar or amino acid moiety via the carboxyl group.[18] The enzymes responsible for these conjugations have been partially characterized.[18] Ethylene treatment has been reported to either promote this conjugation or have no effect.[19,20] The observed ability of ethylene to promote conjugation may in fact be a result of localized increases in IAA content occurring as a result of impaired polar transport.[21] The conjugating enzymes are inducible by auxin treatment.[22] Hence any increase in endogenous auxin would serve to increase the activity of these enzymes.

IAA can also be irreversibly oxidized to apparently inactive products.[18] This oxidation can be of two types: peroxidative or nondecarboxylative. An increase in peroxidase activity is a common effect of ethylene treatment and this results in an increase in IAA decarboxylation. However, the physiological relevance of this type of metabolism has been questioned. The nondecarboxylative pathway originally detected in *Zea* has now been detected in *Pinus sylvestris, Vicia faba, Oryza sativa, Ribes rubrum,* and *Brassica rapa.*[18] This oxidase system is somewhat unusual, requiring both molecular oxygen and a heat-stable, detergent-extractable cofactor (possibly a lipid peroxide).[23] The effect of ethylene on this newly characterized system has not been reported.

D. TRANSPORT

In contrast with other phytohormones, there exists a highly specific polar transport system for IAA.[24] The polar (typically basipetal) transport of IAA is thought to occur as a result of the asymmetric localization of phytotropin-sensitive IAA efflux proteins. These auxin transport proteins are located or active preferentially at the basal ends of cells in the transporting tissues. It has long been recognized[25,26] that ethylene treatment of intact seedlings results in an inhibition of this transport system often accompanied by growth perturbations (the triple response, geo- and photoinsensitivity, induction of organ abscission).

Studies,[27] using a tritiated phytotropin ligand (*N*-1-naphthylphthalamic acid) known to bind to a protein(s) intimately involved in polar IAA transport, have found that ethylene treatment of intact pea seedlings results in a loss of active phytotropin binding sites. Direct application of ethylene to the cell-free binding site preparation has no such effect. Thus, ethylene action is mediated by some, presently unknown, aspect of cellular metabolism.

E. ACTION

Excellent reviews of auxin action have been published.[28-30] Briefly, exogenous auxins such as IAA often elicit a rapid stimulation of proton secretion, membrane hyperpolarization, cell/tissue elongation, and altered gene expression (transcription). How these seemingly diverse tissue responses are related is still a matter of conjecture.

The mechanism(s) of auxin action have been extensively studied. At present no consensus concerning the molecular events underlying auxin action has emerged but some observations seem to be pertinent. Both soluble and particulate auxin binding proteins have been described[31-33] and the functional relevance of both types of binding proteins have been verified using *in vitro* transcription assays and immunological techniques. The sequence of events that follows receptor binding is only poorly understood at present. Altered cellular calcium fluxes have been proposed to mediate auxin action (i.e., act as second messengers).[30] This hypothesis has been weakened somewhat by the recent observation that artificial manipulation of intracellular calcium levels in auxin-responsive tissues has no effect on the transcription of auxin-regulated genes.[34]

Depending on the length of fumigation, treatment of intact etiolated seedlings with ethylene can inhibit or promote epicotyl or coleoptile elongation, suggesting an interaction with endogenous auxins.[35,36] The inhibitory response can be detected within 7 min of ethylene application.[37] The effect of simultaneous application of exogenous ethylene and auxin has not been examined. Ethylene and auxin act together in promoting the elongation of both *Oryza* coleoptiles and the petioles of certain aquatic dicots (*Nymphoides*).[7,8] Based on dose-response studies, it appears that ethylene treatment increases the sensitivity of rice coleoptiles to exogenous auxin. Further examination of these interactions is clearly warranted and may shed light on the mechanism of auxin action in these tissues.

An alteration of gene expression (i.e., modification of selected mRNA abundances) can be detected within 20 min of IAA application.[29,30] With respect to certain transcripts, this response is specific to IAA and other hormones (including ethylene) elicit no such response. Simultaneous application of ethylene, ethylene precursors, or ethylene synthesis inhibitors with the auxin 2,4-D in soybean hypocotyls had no effect on the levels of two auxin-induced mRNA species.[38] The effects of these treatments on the levels of auxin-regulated proteins were not studied. Given the ability of ethylene to affect many auxin responses, such studies may help to elucidate which of the auxin-induced (or repressed) proteins are relevant to particular responses.

III. GIBBERELLINS

A. PHYSIOLOGICAL INDICATIONS

The application of ethylene often results in gross symptomology that is opposite to that of gibberellin (GA) application. Thus, ethylene treatment generally results in a reduction in plant stature (height) while GAs promote internode elongation and thereby increase plant height.[39,40] When applied simultaneously, ethylene can antagonize the ability of GA to stimulate stem elongation.[41,42] This aspect of ethylene action underlies the agronomic use of ethylene-releasing compounds as antilodging agents in cereal and corn production. Inhibition of the synthesis or action of endogenous ethylene can enhance the activity of gibberellins.[42,43] Ethylene and GA are mutually antagonistic in other respects, including fruit senescence, hook opening of etiolated dicots, corolla expansion in *Ipomoea,* and sex expression in cucurbits.[44-47] Ethylene can inhibit or promote GA-induced α-amylase synthesis and secretion depending on the incubation time used. Conversely, ethylene and GA act together in promoting internode elongation of submerged aquatic angiosperms.[49,50] The application of ethylene to potato tubers prone to premature tuberization (rather than sprouting) resulted in the development of normal sprouts and an increase in biologically active GAs.[51]

B. EFFECTS ON ENDOGENOUS LEVELS

Most studies that examined endogenous GA levels have relied primarily on bioassay data. Recent evidence using GA biosynthesis/metabolism mutants have shown that, in certain instances, a single GA (i.e, GA_1 or GA_4) may be the biologically active hormone while all others are simply biological precursors or metabolites.[52] In contrast, most bioassays will respond to both GA_1 (or GA_4) and its metabolic precursors, thereby masking meaningful changes in hormonally relevant GA levels. Recently, it has been reported that application of ethylene at physiological levels to *Helianthus* seedlings results in a reduction in final hypocotyl length and a decrease in GA_1 levels (as measured by GC-MS).[53]

C. EFFECTS ON SYNTHESIS AND METABOLISM

The hypothesis that, in a given plant, only one of a metabolically related family of GAs is the active hormone forces one to consider the synthesis and metabolism together since they represent a metabolic sequence employing similar reactions (i.e., dehydrogenations and hydroxylations). These oxidations both lead to, and then away from, the active hormone species. The first committed step in GA biosynthesis is the conversion by microsomal enzymes of the diterpene geranylgeranyl pyrophosphate to ent-kaurene.[54] A series of microsomal oxidases (requiring NADPH; possibly P_{450} monoxygenases) convert ent-kaurene to GA_{12}-aldehyde. GA_{12}-aldehyde is metabolized by a series of soluble, iron-oxoglutarate-dependent hydroxylases to the various GAs present (including the bioactive species GA_1, GA_4). Further hydroxylations catalyzed by other soluble oxidases result in biologically inactive GAs (GA_8, GA_{34}, etc.). In addition, these inactive species may be conjugated with glucose as either *O*-glucosides or esters.[54]

Many of the enzymes catalyzing individual steps in GA biosynthesis and inactivation have been detected and characterized in cell-free preparations from several plants. This, coupled with the increasing availability of isotopically labeled GA precursors and intermediates, should allow for the exact determinations of any metabolic steps affected by ethylene treatment. This type of investigation has been attempted in *Helianthus* seedlings[53] but the results have been inconclusive.

Ethylene is a known suicide inhibitor of rat liver microsomal P_{450} monooxygenases.[55] Its effects on analogous plant enzymes have not been examined. Thus, any ethylene-induced reduction in GA levels could occur directly via inhibition of certain P_{450} enzymes or indirectly via either an alteration in gene expression/protein turnover or through the synthesis of endogenous modulators of these enzymes.

The ability of ethylene to antagonize the action of exogenous GA could be the result of enhanced metabolism. This possibility has not been examined.

D. ACTION

Current hypotheses concerning the mechanisms of GA action have been reviewed.[40,56,57] Application of GA to certain tissues can elicit measurable effects within 30 min (i.e., increase in pea elongation).[58] Most GA responses occur only after one or more hours. The relatively "slow" responses coupled with the superficial chemical similarities have prompted mechanistic analogies between GA action in plants and steroid action in vertebrates. Steroids have been shown to bind to cellular receptor molecules (cytoplasmic and/or nuclear) and these receptor-hormone complexes in conjunction with certain chromatin-associated proteins bind to select regulatory portions of responsive genes, thereby modifying their expression.[59]

GA binding proteins have been detected in soluble protein extracts from peas and cucumber.[60] Proteins in these extracts bind both GA and synthetic GA-like compounds in a manner consistent with the biological activities of these substances.[61,62] Recently it has been reported that isolated nuclei bind active GAs and in the presence of GA exhibit enhanced transcriptional activities.[63]

Cellular calcium has been shown to be important in GA action particularly in the stimulation of α-amylase activity in aleurone layers.[64] An alteration in polyamine metabolism has also been implicated in GA action.[65] At this point it is not at all clear whether either of these two agents act as intracellular mediators (i.e., 2nd messengers) of GA action or are simply required for proper cellular metabolism, hence GA action.[66]

In certain instances, ethylene application can significantly enhance or inhibit tissue responses to exogenous gibberellins. These interactions have been observed in both elongation and α-amylase assays.

In the case of α-amylase synthesis and secretion, ethylene has been reported to either inhibit or enhance GA-mediated synthesis and secretion of this enzyme.[67] This inhibition occurred only with extended treatment times (about 48 h) and was abolished with the inclusion of protease inhibitors.[48] An increase in external calcium was noted in ethylene-treated aleurone layers[68] and this may be relevant for two reasons. First, α-amylase is a calcium-containing metalloenzyme whose activity is strictly dependent on the presence of bound calcium. Second, calcium is required for both the intracellular processing and secretion of α-amylase.[64] Thus, alterations of intracellular calcium levels by ethylene could account for its action in this system.

Ethylene can interact with GAs in modifying stem or petiole elongation in either a positive or negative manner. Exogenous ethylene partially inhibits GA_3-mediated lettuce hypocotyl elongation.[41] From the limited data presented in these studies, it is not clear if ethylene is affecting the sensitivity of the tissues to GA_3 (i.e., altering the threshold concentration) or reducing the receptivity of the tissue. The application of GA to responsive tissues often elicits a small but reproducible increase in ethylene production.[42] Using inhibitors of ethylene synthesis and action, this endogenously produced ethylene was shown to antagonize GA-induced increase in epicotyl elongation.[42] From the data presented, endogenous ethylene reduces both the sensitivity and response capacity of the tissue.

In aquatic angiosperms such as *Oryza* and *Callitriche,* ethylene and GA act in a cooperative manner to enhance internode (stem) elongation.[49,50] In a detailed study with deepwater rice, ethylene enhanced both the sensitivity and response capacity of isolated rice internodes towards exogenous GA.[50] Other experiments have demonstrated a similar interaction between the endogenously produced hormones.[43]

In the absence of any accepted molecular explanation for GA action in elongating tissues, these observations are difficult to interpret. In cases where threshold sensitivity towards GA has been modified by ethylene, a case could be made for alterations in receptor affinity or decreased GA metabolism. With currently available methods, these hypotheses are testable.

IV. CYTOKININS

A. PHYSIOLOGICAL INDICATIONS

Exogenous cytokinins and ethylene have been shown to interact in a number of plant systems. Both cytokinins and ethylene stimulate the germination of certain dormant seeds, enhance femaleness in certain dioecious dicots, and inhibit primary root growth.[69-71] Exogenous ethylene acts antagonistically in the following cytokinin-induced responses: retardation of leaf and petal senescence, stimulation of cell division, and axillary bud outgrowth in pea seedlings and potato tubers.[5,72,73]

B. EFFECTS ON ENDOGENOUS LEVELS

There have been few studies that directly examined the effects of ethylene on endogenous cytokinin levels. Treatment of dormant *Spergula* seeds with ethylene resulted in an increase in cytokinin-ribotide activity prior to any observable effect on germination.[69] Many types of environmental stress such as temperature, dehydration, and flooding lead to a decline in cytokinin levels and an increase in ethylene production.[74] Such a relationship suggests an effect of ethylene on cytokinin levels. Likewise, leaf and petal senescence (both enhanced by ethylene) result in a decline in cytokinin levels.[75] The sprouting of potato buds (which is promoted by ethylene) is accompanied by a large increase in cytokinin levels.[76] These and other physiological correlations suggest that a more thorough study of the effects of ethylene on cytokinin levels is in order.

C. EFFECTS ON SYNTHESIS AND METABOLISM

The synthesis of biologically active cytokinins can occur as a result of *de novo* synthesis or by release from a conjugated form.[77] The cytokinin Δ^2-isopentenyladenine (IPA) is synthesized from 5′-adenosine monophosphate and Δ^2-isopentenylpyrophosphate initially forming the ribotide Δ^2-isopentenyladenosine-5′-monophosphate. After removal of the 5′-monophosphate, either the free base or the riboside can be stereospecifically hydroxylated on the isopentenyl side chain, forming zeatin or its riboside. In addition, the side chain of zeatin can be hydrogenated, forming dihydrozeatin. Depending on the bioassay system used, all of these cytokinins may be biologically active.

Further metabolism of cytokinins can occur by one or two routes: side chain cleavage or conjugation with either a sugar or amino acid moiety.[77] The oxidative cleavage of the side chain of IPA or zeatin results in the formation of adenine (or the respective ribosides). This cleavage is catalyzed by a soluble enzyme (termed cytokinin oxidase) and results in the irreversible loss of biological activity. Cytokinins can also be conjugated with sugars (typically glucose or ribose) in either the 9 or 7 position of the purine ring. The resultant *N*-glycosides are metabolically stable and may be biologically inactive end products. In contrast, zeatin can be *O*-glucosylated on the side chain and this appears to be a reversible step. The effect of ethylene on individual steps of cytokinin biosynthesis or metabolism has not been directly assessed. In spite of the paucity of data, modifications of cytokinin levels by ethylene appear likely and may partially explain some of the diverse responses to this hormone.

D. ACTION

Although the physiological effects of exogenous cytokinins are diverse and, in many instances, quite spectacular, little is known concerning the sequence of events that constitute the primary action of this class of hormones. As would be expected for a class of hormones eliciting dramatic alterations of plant development, cytokinins have been found to alter both the quantity and types of mRNA and proteins found in the treated tissues.[78,79] Typically, the studies have employed lengthy incubations (hours or days) in cytokinin solutions and, as such, do not shed much light on the early events.

Both soluble and particulate cytokinin-binding proteins have been described. One cytokinin-binding protein, present in fairly high amounts in wheat germ, has been studied in detail.[80] Arguing against its role as a receptor is the relatively low affinity for the native cytokinin zeatin, as opposed to synthetic cytokinins.[81] Recently, a soluble binding protein with good affinity for zeatin (Kd ~10 n*M*) has been detected in extracts prepared from barley leaves.[82]

Calcium ions (i.e., intracellular calcium) have been implicated as second messengers in certain cytokinin responses, including bud formation in the moss *Funaria* and betacyanin biosynthesis in *Amaranthus* seedlings.[83,84] In both cases cytokinins appear to act by increasing the level of intracellular calcium, and artificially raising internal calcium in the absence of the hormone can mimic the cytokinin response.

How these early (primary?) responses result in the more dramatic long-term cytokinin effects needs to be established. The ability of ethylene to affect any of these cytokinin-responsive systems remains to be tested. A possible candidate for such tests could be a leaf or petal senescence system. Exogenous ethylene and cytokinins act antagonistically, cytokinin-binding proteins can be detected, calcium ions can substitute for cytokinins. Thus, all the elements are present and functioning, and relationships can be experimentally assessed.

V. ABSCISIC ACID

A. PHYSIOLOGICAL INDICATIONS

Initially considered solely an inhibitory hormone, abscisic acid (ABA) is now known to both promote and inhibit various developmental processes.[85,86] In both instances ABA has been shown to elicit selective enhancement of gene expression.[87,88] Recent studies on the involvement of ABA in plant development have been facilitated by the availability of both chemical biosynthesis inhibitors and by ABA-deficient mutants.[89,90] Using these and more classical extraction and assay methods, ABA is now thought to participate in the following physiological processes: stomatal closure, inhibition of shoot elongation, promotion of root elongation, development of heterophylly in certain aquatic plants, maintenance and/or induction of quiescence in growing zones (embryos, buds), and development of stress tolerance.[86]

ABA is known to affect ethylene production in both a positive and negative fashion, depending on the tissue being studied. In addition, ethylene can affect some of the above-mentioned ABA-regulated processes. Like ABA, ethylene can inhibit shoot elongation, inhibit bud outgrowth, promote organ abscission, and affect stomatal aperature. An increase in ethylene production and ABA levels occurs following water stress. Thus, some type of physiological interaction between these two regulators is entirely possible, although largely untested.

B. EFFECTS ON BIOSYNTHESIS AND METABOLISM

ABA is a sesquiterpene that is ultimately derived from mevalonic acid. Two biosynthetic routes from mevalonate have been proposed.[86,91] In the first or so called direct route, ABA is formed directly from mevalonate in a manner similar to that found in ABA-producing fungi. In the indirect route, ABA is formed from the oxidative cleavage of a C_{40} (carotenoid) precursor. Both routes have experimental support and it has been suggested that both may operate in higher plants.

The catabolism of ABA is essentially an oxidative pathway ultimately resulting in conjugated metabolites. ABA is first hydroxylated by a microsomal P_{450} monooxygenase to phaseic acid. Phaseic acid in turn is reduced to 4′-dihydrophaseic acid which can then be converted into the 4′-*O*-glucoside. In addition, ABA can be directly linked with glucose, forming the glucose ester.

Treatment of detached citrus fruit with ethylene was found to result in the accumulation of both free and conjugated ABA in the peel tissue.[92] As both free and bound ABA were increased by ethylene treatment, it would appear that *de novo* ABA biosynthesis was stimulated.

However, the effect of ethylene on ABA biosynthesis has not been directly studied in either turgid or stressed plants. Application of ethylene to turgid seedlings results in a slight decrease in ABA levels.[93] Ethylene strongly promotes ABA metabolism in attached *Xanthium* leaves recovering from a transient episode of water stress.[93] From the data presented, it is not clear whether ethylene stimulated ABA oxidation, conjugation, or both. The ability to alter ABA metabolism is not trivial nor can it be viewed as solely an inactivation process. Phaseic acid is known to be biologically active in a number of assays, including inhibition of GA-induced α-amylase synthesis in barley aleurone layers.[94] It is possible (and experimentally verifiable) that the control of ABA oxidation to phaseic acid, coupled with unique biochemical targets for each, underlies some of the differential activity of ABA in various tissues.

C. ACTION

Plant responses to exogenous ABA can be quite rapid (<5 min: stomatal closure) or slow to develop (>30 min: alteration of gene expression). In certain systems these two types of responses exhibit differential sensitivities towards the enantiomers of ABA. This has led to the proposal that more than one binding site (receptor) for ABA exists.[95] The possibility of a distinct receptor for phaseic acid cannot be presently excluded. Using the intrinsic photolability of *cis*-ABA, three distinct binding proteins have been found in membranes of guard cell protoplasts.[96] To date, this work has not been confirmed and, others, using crude microsomal preparations isolated from barley aleurone, have found no evidence of saturable ABA binding. No reports concerning the existence (or lack thereof) of soluble binding sites have been published.

The effect of ABA on guard cell closure occurs as a result of rapid alterations of H^+/K^+ movement across the plasmalemma of guard cells.[97] In addition, it has been proposed that ABA acts as a calcium agonist affecting intracellular calcium levels.[98] Application of ABA can affect gene expression either by blocking the effects of another hormone (GA) or by eliciting a unique set of transcripts and proteins.[87] Little is known concerning the mechanism of ABA action in these situations.

Ethylene and ABA exert similar actions in a number of systems. Both inhibit extension growth in dicot seedlings, although only ethylene elicits the "triple response". Water stress-induced growth inhibition and accelerated organ abscission are accompanied by increases in both hormones; however, exact relationships between changes in either hormone and the physiological responses remain to be determined.

The ability of ABA to inhibit GA-induced α-amylase synthesis and secretion can be partially overcome by exogenous ethylene.[99] This antagonism was, at least partially, competitive and was attributed to an ability of ethylene to alleviate ABA-mediated inhibition of amylase mRNA translation. The transcription-inhibiting effect of ABA was apparently not affected by ethylene.

Recently, it has been shown that certain phenolic acids (i.e., *p*-coumaric acid) antagonize ABA action in a number of assays, including inhibition of α-amylase synthesis, induction of petiole abscission, growth inhibition, and stomatal closure.[100] Ethylene is known to affect the synthesis and metabolism of phenolic compounds. An exploration into the relationships between these three classes of endogenous regulators may shed light on some of these interactions.

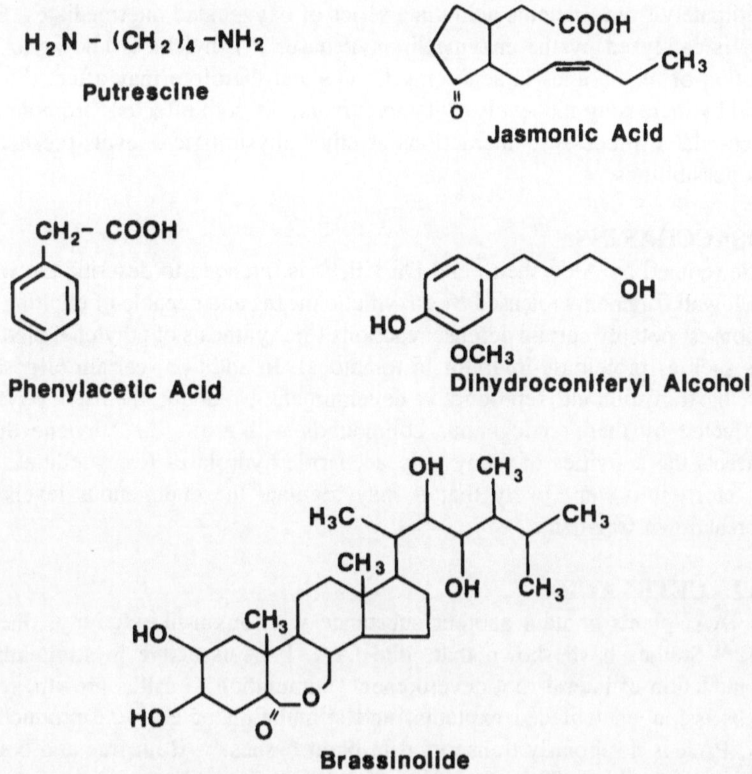

FIGURE 3. Chemical structures of selected compounds with demonstrated hormone-like activity in higher plants.

VI. MISCELLANEOUS BIOREGULATORS

In addition to the five principal classes of endogenous phytohormones, there exists an increasing number of incompletely understood endogenous substances capable of eliciting alterations in many aspects of plant development (Figure 3). The interactions between ethylene and any of these bioregulators are largely unknown. A partial listing of these diverse compounds is given below.

A. POLYAMINES

Polyamines are ubiquitous natural products found throughout the plant kingdom. Interest in polyamines as endogenous growth regulators was stimulated by the discovery of their ability to prevent or retard certain senescence-associated processes in plant tissue cultures and isolated leaves. Recent research has focused on their possible involvement as intracellular mediators (i.e., second messengers) for the principal plant hormones and in many instances their endogenous levels do fluctuate in accordance with this proposed physiological role.[101] Ethylene and polyamines are mutually inhibitory with respect to each other's biosynthesis and both share *S*-adenosyl-L-methionine as a biosynthetic intermediate. In spite of the activity in this area, little is known concerning the effect of ethylene on other aspects of polyamine physiology, such as compartmentation, binding, and action.

B. JASMONIC ACID

Both jasmonic acid and its methyl ester are widely distributed in higher plants.[102] Both compounds exert a strong senescence-accelerating effect in many bioassays. Jasmonic acid is derived ultimately from linolenic acid via a series of oxygenated intermediates. Entry into this pathway is catalyzed by the enzyme lipoxygenase.[103] Ethylene is known to stimulate the deterioration of membranes in senescing tissues and therefore may affect the levels of jasmonic acid by increasing the levels of its precursors. As both effectors promote a number of senescence-related processes, interactions at other physiological levels present distinct, but untested possibilities.

C. OLIGOSACCHARINS

This term, coined by Albersheim and Darvill,[104] is intended to describe the wide array of soluble cell wall fragments released by enzymatic means and capable of eliciting a variety of plant responses, notably certain defense reactions (the synthesis of ethylene, phytoalexins, and proteins such as proteinase-inhibitor in tomatoes). In addition, certain oligosaccharins can influence the transition into reproductive development. No doubt, the list of physiological processes affected by these endogenous compounds will grow. As ethylene directly or indirectly affects the activities of many polysaccharide hydrolases (i.e., cellulase, polygalacturonase, etc.), it is very likely that it may regulate the endogenous levels of these hydrolytic breakdown fragments.

D. PHENYLACETIC ACID

Besides IAA, plants contain another substance with auxin-like activity, phenylacetic acid (PAA).[105] Studies have shown that, like IAA, PAA is active in stimulating tissue elongation, induction of lateral root development, stimulation of callus growth, retardation of petiole abscission in debladed explants, and stimulation of ethylene production.[106,107] Unlike IAA, PAA is not polarly transported in plant tissues.[107] Both free and bound PAA are native to higher plants. The exact role of PAA as an endogenous regulator and its interactions with ethylene remain to be determined.

E. PHENOLIC COMPOUNDS

Widely regarded as growth inhibitors and nonspecific enzyme inhibitors, recent studies have demonstrated a more direct role of several classes of phenolic compounds in plant development. The phenolic acetosyringone and other related compounds have been found to be involved in the induction of the nodulation (nod) genes during the initiation and establishment of the symbiosis between legumes and *Rhizobium* bacteria.[108] The phenolic dihydroconiferol alcohol and certain derivatives can act as both IAA and GA synergists and may act as mediators of the cell division response to cytokinins.[109,110] Certain flavonoids such as quercitin compete with the phytotropin NPA for microsomal binding sites and may act as endogenous regulators of auxin transport.[111] In addition, phenolics can affect IAA metabolism and ABA action.

Ethylene is known to affect (generally stimulate) phenolic biosynthesis in many tissues. As such, ethylene may affect a wide variety of phenolic-sensitive developmental shifts depending on the species studied.

F. BRASSINOSTEROIDS

First detected (by its ability to induce a unique biological response in certain bioassays) in rape (*Brassica napus*) pollen, this class of compounds has now been detected in many species of higher plants.[112] Little is known about either the biosynthesis/metabolism or mechanisms of action of this class of plant bioregulators. Bioassay data indicate varying degrees of interactions between brassinosteroids and auxins and gibberellins. Little is known concerning the effects of the principal phytohormones on the synthesis, metabolism, or action of this poorly characterized class of potential endogenous regulators.

VII. PERSPECTIVES

To date, the available evidence indicates that the action of a given hormone in plant development is not only modified by, but directly depends on, the activity status of the remaining complement of hormones in the responding tissues. These interactions can manifest themselves in the forms of antagonism, complementation, or synergism.

In this chapter, experimental evidence has been provided that demonstrates or strongly indicates a variety of interactions between ethylene and other classes of plant hormones. For physiologists interested in these interactions, two immediate goals are evident; the first is to redefine the effects of ethylene on the levels of other hormones (or their precursors and metabolites), using the newer physiochemical or immunological detection methods, and subsequently to define, on a biochemical basis, how these effects occur.

In some cases, the experimental evaluation of ethylene effects on the receptor systems of other plant hormones is now within reach. Recent advances in the detection and partial purification of plant hormone receptors, including the molecular cloning of an auxin receptor in maize tissues,[113] will permit types of experimentation heretofore not possible.

A major uncertainty that hinders future progress in this area is our near-total ignorance of the biochemical events that follow and transduce receptor occupancy to the observed hormone response. However, it is encouraging that a resurgence of this type of research is occurring and has already produced significant results.[114]

In a different vein, it is highly likely that additional hormone-like substances remain undiscovered. The identification and characterization of the effects of these compounds will undoubtedly enhance our understanding of the hormonal basis of plant growth regulation.

In the concluding remarks of his 1979 review of ethylene physiology,[115] Morris Lieberman stated that the experimental evidence pertaining to the interactions between ethylene and other endogenous hormones was at the descriptive level. He went on to say " . . . information on mechanisms needs to be developed. Such findings should reveal much about the control and regulation of plants at the hormonal level . . . ". These sentiments still ring true today.

REFERENCES

1. **Canny, M. J.,** Ashby's Law and the pursuit of plant hormones, *Aust. J. Plant Physiol.*, 12, 1, 1985.
2. **Trewavas, A.,** Understanding the control of plant development and the role of growth substances, *Aust. J. Plant Physiol.*, 13, 447, 1986.
3. **Guern, J.,** Regulation from within: the hormone dilemma, *Ann. Bot.*, 60, 75, 1987.
4. **Burg, S. P. and Burg, E. A.,** The interaction between auxin and ethylene and its role in plant growth, *Proc. Natl. Acad. Sci. U.S.A.*, 55, 262, 1966.
5. **Burg, S. P.,** Ethylene in plant growth, *Proc. Natl. Acad. Sci. U.S.A.*, 70, 591, 1973.
6. **Reid, M. S., Mor, Y., and Kofranek, A. M.,** Epinasty of poinsettias—the role of auxin and ethylene, *Plant Physiol.*, 67, 950, 1981.
7. **Cookson, C. and Osborne, D. J.,** The stimulation of cell extension by ethylene and auxin in aquatic plants, *Planta*, 144, 39, 1978.
8. **Ishizawa, K. and Esashi, Y.,** Cooperation of ethylene and auxin in the growth regulation of rice coleoptile segments, *J. Exp. Bot.*, 34, 74, 1983.
9. **van der Laan, P. A.,** Effect of ethylene on growth hormone formation in *Avena* and *Vicia*, *Rec. Trav. Bot. Neerl.*, 31, 691, 1934.
10. **Ernest, L. C. and Valdovinos, J. G.,** Regulation of auxin levels in *Coleus blumei* by ethylene, *Plant Physiol.*, 48, 402, 1971.
11. **Lieberman, M. and Knegt, E.,** Influence of ethylene on indole-3-acetic acid concentration in etiolated pea epicotyl tissue, *Plant Physiol.*, 60, 475, 1977.

11a. **Suttle, J. C. and Hultstrand, J. F.**, Ethylene-induced leaf abscission in cotton seedlings. The physiological bases for age-dependent differences in sensitivity, *Plant Physiol.*, 95, 29, 1991.
12. **Burg, S. P., Apelbaum, A., Eisinger, W., and Kang, B. G.**, Physiology and mode of action of ethylene, *HortScience*, 6, 359, 1971.
13. **Cohen, J. D. and Bialek, K.**, The biosynthesis of indole-3-acetic acid in higher plants, in *The Biosynthesis and Metabolism of Plant Hormones*, Crozier, A. and Hillman, J. R., Eds., Cambridge University Press, Cambridge, 1984, 165.
14. **Valdovinos, J. G., Ernest, L. C., and Henry, E. W.**, Effect of ethylene and gibberellic acid on auxin synthesis in plant tissues, *Plant Physiol.*, 42, 1803, 1967.
15. **Law, D. M.**, Gibberellin-enhanced indole-3-acetic acid biosynthesis: D-tryptophan as the precursor of indole-3-acetic acid, *Physiol. Plant.*, 70, 626, 1987.
16. **Liu, Y., Hoffman, N. E., and Yang, S. F.**, Ethylene-promoted malonylation of 1-aminocyclopropane-1-carboxylic acid participates in autoinhibition of ethylene synthesis in grapefruit flavedo discs, *Planta*, 164, 565, 1985.
17. **Hall, P. J. and Bandurski, R. S.**, [^{3}H] Indole-3-acetyl-*myo*-inositol hydrolysis by extracts of *Zea mays* L. vegetative tissue, *Plant Physiol.*, 80, 374, 1986.
18. **Bandurski, R. S.**, Metabolism of indole-3-acetic acid, in *The Biosynthesis and Metabolism of Plant Hormones*, Crozier, A. and Hillman, J. R., Eds., Cambridge University Press, Cambridge, 1984, 183.
19. **Riov, J., Dror, N., and Goren, R.**, Effect of ethylene on [^{14}C]-indole-3-acetic acid metabolism in leaf tissues of woody plants, *Plant Physiol.*, 70, 1265, 1982.
20. **Goren, R., Bukovac, M. J., and Flore, J. A.**, Mechanism of indole-3-acetic acid conjugation: no induction by ethylene, *Plant Physiol.*, 53, 164, 1974.
21. **Beyer, E. M. and Morgan, P. W.**, Effect of ethylene on the uptake, distribution and metabolism of indole acetic acid-1-^{14}C and 2-^{14}C and naphthaleneacetic acid-1-^{14}C, *Plant Physiol.*, 46, 157, 1970.
22. **Venis, M. A.**, Auxin-induced conjugation systems in peas, *Plant Physiol.*, 49, 24, 1972.
23. **Reinecke, D. M. and Bandurski, R. S.**, Oxidation of indole-3-acetic acid to oxindole-3-acetic acid by an enzyme preparation from *Zea mays*, *Plant Physiol.*, 86, 868, 1988.
24. **Rubery, P. H.**, The mechanism of transmembrane auxin transport and its relation to the chemiosmotic hypothesis of the polar transport of auxin, in *Plant Growth Substances 1979*, Skoog, F., Ed., Springer-Verlag, Heidelberg, 1980, 50.
25. **Morgan, P. W. and Gausman, H. W.**, Effects of ethylene on auxin transport, *Plant Physiol.*, 41, 45, 1966.
26. **Burg, S. P. and Burg, E. A.**, Inhibition of polar auxin transport by ethylene, *Plant Physiol.*, 42, 1224, 1967.
27. **Suttle, J. C.**, Effect of ethylene treatment on polar IAA transport, net IAA uptake and specific binding of N-1-naphthylphthalamic acid in tissues and microsomes isolated from etiolated pea epicotyls, *Plant Physiol.*, 88, 795, 1988.
28. **Theologis, A.**, Rapid gene regulation by auxin, *Annu. Rev. Plant Physiol.*, 37, 407, 1986.
29. **Guilfoyle, T. J.**, Auxin-regulated gene expression in higher plants, *Crit. Rev. Plant Sci.*, 4, 247, 1986.
30. **Brummel, D. A. and Hall, J. L.**, Rapid cellular responses to auxin and the regulation of growth, *Plant Cell Environ.*, 10, 523, 1987.
31. **Batt, S., Wilkins, M. B., and Venis, M. A.**, Auxin binding to corn coleoptile membranes: kinetics and specificity, *Planta*, 130, 7, 1976.
32. **van der Linde, P. C. G., Bouman, L. H., and Libbenga, K. R.**, A soluble auxin-binding protein from cultured tobacco tissues stimulates RNA synthesis *in vitro*, *Planta*, 160, 102, 1984.
33. **Löbler, M., and Klämbt, D.**, Auxin binding protein from coleoptile membranes of corn (*Zea mays* L.). II. Localization of a putative auxin receptor, *J. Biol. Chem.*, 260, 9854, 1985.
34. **Reddy, A. S. N., Koshiba, T., Theologis, A., and Poovaiah, B. W.**, The effect of calcium antagonists on auxin-induced elongation and on the expression of two auxin-regulated genes in pea epicotyls, *Plant Cell Physiol.*, 29, 1165, 1988.
35. **van der Laan, P. A.**, Der einfluss van Aethylen auf die Wuchsstoffbildung bei *Avena* und *Vicia*, *Rec. L. Trav. Bot. Neerl.*, 31, 691, 1934.
36. **Michener, H. D.**, The action of ethylene on plant growth, *Am. J. Bot.*, 25, 711, 1938.
37. **Warner, H. L.**, The synthesis and mode of action of ethylene in etiolated pea seedlings, Ph.D. thesis, Purdue University, Lafayette, Indiana, 1970.
38. **Walker, J. C., Legocka, J., Edelman, L., and Key, J. L.**, An analysis of growth regulator interactions and gene expression during auxin-induced cell elongation using cloned complementary DNAs to auxin-responsive messenger RNAs, *Plant Physiol.*, 77, 847, 1985.
39. **Eisinger, W.**, Regulation of pea internode expansion by ethylene, *Annu. Rev. Plant Physiol.*, 34, 225, 1983.
40. **Jones, R. L.**, The role of gibberellins in plant cell elongation, *Crit. Rev. Plant Sci.*, 1, 23, 1983.
41. **Scott, P. C. and Leopold, A. C.**, Opposing effects of gibberellin and ethylene, *Plant Physiol.*, 42, 1021, 1967.

42. **Garcia-Martinez, J. L., Ben-Shalom, N., and Rappaport, L.**, Gibberellin-induced ethylene production and its effects on cowpea epicotyl elongation, *Plant Growth Regul.*, 2, 209, 1984.
43. **Ross, J. J. and Reid, J. B.**, Internode length in *Pisum*. The involvement of ethylene with the gibberellin-insensitive erectoides phenotype, *Physiol. Plant.*, 67, 673, 1986.
44. **Dostal, H. C. and Leopold, A. C.**, Gibberellins delay ripening of tomatoes, *Science*, 158, 1579, 1967.
45. **Kang, B. G. and Ray, P. M.**, Role of growth regulators in the bean hypocotyl hook opening response, *Planta*, 87, 193, 1969.
46. **Raab, M. M. and Koning, R. E.**, Interacting roles of gibberellin and ethylene in corolla expansion of *Ipomoea nil* (Convolvulaceae), *Am. J. Bot.*, 74, 921, 1987.
47. **Byers, R. E., Baker, L. R., Sell, H. M., Herner, R. C., and Dilley, D. R.**, Ethylene: a natural regulator of sex expression of *Cucumis melo* L., *Proc. Natl. Acad. Sci. U.S.A.*, 69, 717, 1972.
48. **Eastwell, K. C. and Spencer, M. S.**, Ethylene effects on amylase activity from isolated barley aleurone layers. Possible modification by proteolytic enzymes, *Plant Physiol.*, 70, 849, 1982.
49. **Musgrave, A., Jackson, M. B., and Ling, E.**, Callitriche stem elongation is controlled by ethylene and gibberellin, *Nature (London), New Biol.*, 238, 93, 1972.
50. **Raskin, I. and Kende, H.**, Role of gibberellin in the growth response of submerged deep water rice, *Plant Physiol.*, 76, 947, 1984.
51. **Dimalla, G. G. and van Staden, J.**, Effect of ethylene on the endogenous cytokinin and gibberellin levels in tuberizing potatoes, *Plant Physiol.*, 60, 218, 1977.
52. **Phinney, B. O., Freeling, M., Robertson, D. S., Spray, C. R., and Silverthorne, J.**, Dwarf mutants in maize—the gibberellin biosynthetic pathway and its molecular future, in *Plant Growth Substances 1985*, Bopp. M., Ed., Springer-Verlag, Heidelberg, 1986, 55.
53. **Pearce, D., Watson, K., Best, P. L., Reid, D. M., and Pharis, R. P.**, Ethylene-mediated regulation of gibberellin content and growth in *Helianthus annuus*, Proc. 13 Int. Conf. Plant Growth Substances, Calgary, Alberta, August, 1988, Abstr. 358.
54. **Graebe, J. E.**, Gibberellin biosynthesis and control, *Annu. Rev. Plant Physiol.*, 38, 419, 1987.
55. **Ortiz de Montellano, P. R. and Mico, B. A.**, Destruction of cytochrome P-450 by ethylene and other olefins, *Mol. Pharmacol.*, 18, 128, 1980.
56. **Stoddart, J. L.**, Gibberellin perception and its primary consequences, in *Plant Growth Substances 1982*, Wareing, P. F., Ed., Academic Press, London, 1982, 131.
57. **Akazawa, T. and Miyata, S.**, Biosynthesis and secretion of α-amylase and other hydrolases in germinating cereal seeds, *Essays Biochem.*, 18, 40, 1982.
58. **Warner, H. L. and Leopold, A. C.**, Timing of growth regulator responses in peas, *Biochem. Biophys. Res. Commun.*, 44, 989, 1971.
59. **Ringold, G. M.**, Steroid hormone regulation of gene expression, *Annu. Rev. Pharmacol. Toxicol.*, 25, 529, 1985.
60. **Srivastava, L. M.**, The gibberellin receptor, in *Plant Hormone Receptors*, Klämbt, D., Ed., Springer-Verlag, Heidelberg, 1987, 199.
61. **Yalpani, N. and Srivastava, L. M.**, Competition for *in-vitro* [^{3}H]gibberellin A_4 binding in cucumber by gibberellins and their derivatives, *Plant Physiol.*, 79, 963, 1985.
62. **Yalpani, N., Suttle, J. C., Hultstrand, J. F., and Rodaway, S. J.**, Competition for *in vitro* [^{3}H]gibberellin A_4 binding in cucumber by substituted phthalimids: comparison with *in vivo* gibberellin-like activity, *Plant Physiol.*, 91, 823, 1989.
63. **Sieciechowicz, K. A. and Srivastava, L. M.**, Effects of GA_4 and GA_4 binding protein(s) on gene expression in isolated nuclei, Proc. 13th Int. Conf. Plant Growth Substances, Calgary, Alberta, August, 1988, Abstr. 186.
64. **Jones, R. L. and Carbonell, J.**, Regulation of the synthesis of barley aleurone α-amylase by gibberellic acid and calcium ions, *Plant Physiol.*, 76, 213, 1984.
65. **Dai, Y., Kaur-Sawhney, R., and Galston, A. W.**, Promotion by gibberellic acid of polyamine biosynthesis in internodes of light-grown peas, *Plant Physiol.*, 69, 103, 1982.
66. **Smith, M. A., Davies, P. J., and Reid, J. B.**, Role of polyamines in gibberellin-induced internode growth in peas, *Plant Physiol.*, 78, 92, 1985.
67. **Eastwell, K. C. and Spencer, M. S.**, Effect of ethylene on the gibberellic acid-enhanced synthesis and release of amylase by isolated barley aleurone layers, *Plant Physiol.*, 69, 557, 1982.
68. **Eastwell, K. C. and Spencer, M. S.**, Modes of ethylene action in the release of amylase from barley aleurone layers, *Plant Physiol.*, 69, 563, 1982.
69. **van Staden, J., Olatoye, S. T., and Hall, M. A.**, Effect of light and ethylene upon cytokinin levels in seed of *Spergula arvensis*, *J. Exp. Bot.*, 24, 662, 1973.
70. **Chailakhyan, M. Kh.**, Genetic and hormonal regulation of growth, flowering, and sex expression in plants, *Am. J. Bot.*, 66, 717, 1979.
71. **Stenlid, G.**, Cytokinins as inhibitors of root growth, *Physiol. Plant.*, 56, 500, 1982.

72. **Mattoo, A. K. and Aharoni, N.**, Ethylene and plant senescence, in *Senescence and Aging in Plants*, Noodén, L. D. and Leopold, A. C., Eds., Academic Press, New York, 1988, 241.
73. **Blake, T. J., Reid, D. M., and Rood, S. B.**, Ethylene, indoleacetic acid and apical dominance in peas: a reappraisal, *Physiol. Plant.*, 59, 481, 1983.
74. **van Staden, J. and Davey, J. E.**, The synthesis, transport and metabolism of cytokinins, *Plant Cell Environ.*, 2, 93, 1979.
75. **Mayak, S., Halevy, A. H., and Katz, M.**, Correlative changes in phytohormones in relation to senescence processes in rose petals, *Physiol. Plant.*, 27, 1, 1972.
76. **Turnbull, C. G. N. and Hanke, D. E.**, The control of bud dormancy in potato tubers. Measurement of the seasonal pattern of changing concentrations of zeatin-cytokinins, *Planta*, 165, 366, 1985.
77. **McGaw, B. A., Scott, I. M., and Horgan, R.**, Cytokinin biosynthesis and metabolism, in *The Biosynthesis and Metabolism of Plant Hormones*, Crozier, A. and Hillman, J. R., Eds., Cambridge University Press, Cambridge, 1984, 105.
78. **Teyssendier de la Serve, B., Axelos, M., and Peaud-Lenoël, C.**, Cytokinins modulate the expression of genes encoding the protein of the light-harvesting chlorophyll a/b complex, *Plant Mol. Biol.*, 5, 155, 1985.
79. **Memelink, J., Hoge, J. H. C., and Schilperoot, R. A.**, Cytokinin stress changes the developmental regulation of several defence-related genes in tobacco, *EMBO J.*, 6, 3579, 1987.
80. **Brinegar, A. C. and Fox, J. E.**, Immunocytological localization of a wheat embryo cytokinin binding protein and its homology with proteins in other cereals, in *Plant Hormone Receptors*, Klämbt, D., Ed., Springer-Verlag, Berlin, 1987, 177.
81. **Polya, G. M. and Davis, A. W.**, Properties of a high-affinity cytokinin binding protein from wheat germ, *Planta*, 139, 139, 1978.
82. **Romanov, G. A., Taran, V. Ya., Chvojka, L., and Kulaeva, O. N.**, Receptor-like cytokinin-binding protein(s) from barley leaves, *J. Plant Growth Regul.*, 7, 1, 1988.
83. **Saunders, M. J. and Hepler, P. K.**, Calcium ionophore A23187 stimulates cytokinin-like mitosis in *Funaria*, *Science*, 217, 943, 1982.
84. **Elliott, D.**, Inhibition of cytokinin-regulated responses by calmodulin binding compounds, *Plant Physiol.*, 72, 215, 1983.
85. **Addicott, F. T., Ed.**, *Abscisic Acid*, Praeger, New York, 1983.
86. **Zeevaart, J. A. D. and Creelman, R. A.**, Metabolism and physiology of abscisic acid, *Annu. Rev. Plant Physiol.*, 39, 439, 1988.
87. **Lin, L. S. and Ho, T. H. D.**, Mode of action of abscisic acid in barley aleurone layers, *Plant Physiol.*, 82, 289, 1986.
88. **Williamson, J. D. and Quatrano, R. S.**, ABA-regulation of two classes of embryo-specific sequences in mature wheat embryos, *Plant Physiol.*, 86, 208, 1988.
89. **Gamble, P. E. and Mullet, J. E.**, Inhibition of carotenoid accumulation and abscisic acid biosynthesis in fluridone-treated dark-grown barley, *Eur. J. Biochem.*, 160, 117, 1986.
90. **Koorneef, M.**, Genetic aspects of abscisic acid, in *A Genetic Approach to Plant Physiology, Plant Gene Research*, Vol. 3, Blonstein, A. D. and King, P. J., Eds., Springer-Verlag, Berlin, 1986, 35.
91. **Neill, S. J., Horgan, R., and Walton, D. C.**, Biosynthesis of abscisic acid, in *The Biosynthesis and Metabolism of Plant Hormones*, Crozier, A. and Hillman, J. R., Eds., Cambridge University Press, Cambridge, 1984, 43.
92. **Goldschmidt, E. E., Goren, R., Even-Chen, Z., and Bittner, S.**, Increase in free and bound abscisic acid during natural and ethylene-induced senescence of citrus fruit peel, *Plant Physiol.*, 51, 879, 1973.
93. **Zeevaart, J. A. D.**, Metabolism of abscisic acid and its regulation in *Xanthium* leaves during and after water stress, *Plant Physiol.*, 71, 477, 1983.
94. **Walton, D. C.**, Structure-activity relationships of abscisic acid analogs and metabolites, in *Abscisic Acid*, Addicott, F. T., Ed., Praeger, New York, 1983, 113.
95. **Milborrow, B. V.**, A distinction between the fast and slow responses to abscisic acid, *Aust. J. Plant Physiol.*, 7, 749, 1980.
96. **Hornberg, C. and Weiler, E. W.**, High-affinity binding sites for abscisic acid on the plasmalemma of *Vicia faba* guard cells, *Nature*, 310, 321, 1984.
97. **Schauf, C. L. and Wilson, K. J.**, Effects of abscisic acid on K^+ channels in *Vicia faba* guard cell protoplasts, *Biochem. Biophys. Res. Commun.*, 145, 284, 1987.
98. **Owen, J. H.**, Role of abscisic acid in a Ca^{2+} second messenger system, *Physiol. Plant.*, 72, 637, 1988.
99. **Jacobson, J. V.**, Interaction between gibberellic acid, ethylene and abscisic acid in control of amylase synthesis in barley aleurone layers, *Plant Physiol.*, 51, 198, 1973.
100. **Laloraya, M. M., Nozzolillo, C., Purohit, S., and Stevenson, L.**, Reversal of abscisic acid-induced stomatal closure by *trans*-cinnamic and *p*-coumaric acid, *Plant Physiol.*, 81, 253, 1986.
101. **Galston, A. W.**, Polyamines as modulators of plant development, *BioScience*, 33, 382, 1983.
102. **Yamane, H., Takagi, H., Abe, H., Yokota, T., and Takahashi, N.**, Identification of jasmonic acid in three species of higher plants and its biological activities, *Plant Cell Physiol.*, 22, 689, 1981.

103. **Vick, B. A. and Zimmerman, D. C.,** The biosynthesis of jasmonic acid: a physiological role for plant lipoxygenase, *Biochem. Biophys. Res. Commun.*, 111, 470, 1983.
104. **Albersheim, P. and Darvill, A. G.,** Oligosaccharins, *Sci. Am.*, 253, 58, 1985.
105. **Wightman, F. and Lighty, D. L.,** Identification of phenylacetic acid as a natural auxin in the shoots of higher plants, *Physiol. Plant.*, 55, 17, 1982.
106. **Chamberlain, V. K. and Wain, R. L.,** Studies on plant growth-regulating substances. XXXIII. The influence of ring substituents on the growth-regulating activity of phenylacetic acid, *Ann. Appl. Biol.*, 69, 65, 1971.
107. **Suttle, J. C. and Mansager, E. R.,** The physiological significance of phenylacetic acid in abscising cotton cotyledons, *Plant Physiol.*, 81, 434, 1986.
108. **Redmond, J. W., Batley, M., Djordjevic, M. A., Innes, R. W., Kuewpel, P. L., and Rolfe, B. G.,** Flavones induce the expression of the nodulation genes in *Rhizobium*, *Nature*, 323, 632, 1986.
109. **Sakurai, N., Shibata, K., and Kamisaka, S.,** Stimulation of cucumber hypocotyl elongation by dihydroconiferyl alcohol. Interactions between dihydroconiferyl alcohol and auxin or gibberellin, *Plant Cell Physiol.*, 15, 709, 1974.
110. **Lee, T. S., Purse, J. G., Horgan, R., and Wareing, P. F.,** Dihydroconiferyl alcohol—a cell division factor from *Acer* species, *Planta*, 152, 571,1981.
111. **Jabobs, M. and Rubery, P. H.,** Naturally occurring auxin transport regulators, *Science*, 241, 346, 1988.
112. **Mandava, N. B.,** Plant growth-promoting brassinosteroids, *Annu. Rev. Plant Physiol.*, 39, 23, 1988.
113. **Inohara, N., Shimomura, S., Fukui, T., and Futae, M.,** Auxin-binding protein located in the endoplasmic reticulum of maize shoots: molecular cloning and complete primary structure, *Proc. Natl. Acad. Sci. U.S.A.*, 86, 3564, 1989.
114. **Boss, W. F. and Morre, J. D.,** *Second Messengers in Plant Growth and Development*, Alan R. Liss, New York, 1989.
115. **Lieberman, M.,** Biosynthesis and action of ethylene, *Annu. Rev. Plant Physiol.*, 30, 533, 1979.

Chapter 8

ETHYLENE AND SEED GERMINATION

Yohji Esashi

TABLE OF CONTENTS

I. INTRODUCTION

In nature, the germination of most seeds occurs in the semienclosed environment of the soil. Similarly, seeds of some aquatic plants, such as cattail and rice, will germinate in the restricted atmospheric conditions of submersion. Ethylene, which is emanated by seeds themselves, is a natural component of the soil atmosphere and, as a plant hormone, has aroused the interest of many investigators concerned with the control of seed germination (see reviews, References 1 to 5). Accumulation of ethylene in the ambient atmosphere is also accompanied by the accumulation of CO_2 and the consumption of O_2. Thus, the synthesis of ethylene in seeds and its action on seed dormancy and germination cannot be considered independently of either CO_2 or O_2. Other environmental factors, such as water, temperature, and light, as well as other plant hormones, such as gibberellins (Gib), cytokinins (Ckn), and abscisic acid (ABA), also have been shown to influence seed germination.

Historically, the importance of ethylene in the regulation of seed germination has been underestimated for several reasons. First, some seeds did not germinate in response to ethylene, while the germination of others was inhibited. Second, the maximum rate of emanation of ethylene from seeds occurred immediately after but not before germination. Third (the most crucial reason), aminoethoxyvinylglycine (AVG), a potent inhibitor of ethylene biosynthesis, had little influence on the germination of the ethylene-responsive seeds of lettuce, although ethylene emanation was clearly blocked. Thus, it was concluded[6,7] that the synthesis of ethylene in imbibing seeds was a result of growth rather than an initiator of the germination process. In this chapter, we will examine the action of ethylene in seed germination control.

II. WHAT IS SEED GERMINATION?

In many but not all species, ethylene appears to break seed dormancy and promote germination. Ethylene, like red (R) light[8] and Gib,[9] is ineffective on some seeds and acts negatively on others (i.e., ethylene inhibits the seed germination of *Plantago*[10] and *Potentilla*[11]). Therefore, the question is why these factors act differently in different plant species.

Seed germination occurs when the growing seed tissue pierces or breaks the seed coat, including the endosperm and/or pericarp. Therefore, the ease of seed germination depends upon the growth potential of the seed tissue and/or the mechanical resistance of the seed coat. Seeds display two basic types of dormancy: (1) primary or embryo dormancy due to the failure of seed tissues to grow, seed immaturity, or secondary dormancy due to the gradual loss of growth capacity of seed tissues during a period of water imbibition; (2) coat-imposed dormancy due to the mechanical restriction of the seed coat. It is possible that only the complete loss of the mechanical restraint of the seed envelope allows seed germination, with no need to alter the active growth potential of seed tissues. Thus, many researchers have tried to reduce the mechanical restraint of the seed coat by means of various germination-regulating factors, including ethylene.

The major component of cell walls in the seed coat of lettuce seeds is mannan.[12] However, unlike Ckn or Gib, ethylene does not exert any action on the production of mannanase.[13] Similar results were obtained with celery (*Apium graveolens* L.) seeds.[14] In lettuce seeds among the germination stimulators tested, only ethylene reduced the break strength of the seed coat.[15] However, this effect was observed only after germination had occurred. Thus, it was concluded that none of the germination promoters tested, including ethylene, functioned by lowering the mechanical resistance of the endosperm to embryonic growth. On the other hand, seed germination involves growth that occurs only when environmental conditions subsequent to germination are favorable for seedling establishment and growth. Moreover, changes in the mechanical resistance of the seed coat were barely detectable during imbibition of cocklebur seeds.[16] These data suggest that the diversity of ethylene action in regulating seed germination must be considered from the probable different modes of ethylene action on the growth of each seed tissue.

Ethylene is known to stimulate the germination and break the dormancy of many kinds of seeds, including peanut (*Arachis hypogaea*), lettuce (*Lactuca sativa*), clover (*Trifolium repens*), subterranean clover (*Trifolium subterraneum*), celery, sorrel (*Rumex acetosa*), lamb's-quarters (*Chenopodium album*), *Hypochoeris radicata, Matricaria recutita, Spergula arvensis,* Douglas fir (*Pseudotsuga menziesii*),[17] *Galeola septentrionalis,*[18] ragweed (*Ambrosia artemisiifolia*),[19] redroot pigweed (*Amaranthus retroflexus*),[20] sunflower (*Helianthus annuus*),[21] *Striga bermonthica,*[22] apple (*Malus domestica*),[23,24] curly dock (*Rumex crispus*),[25] and cocklebur (*Xanthium pennsylvanicum*).[26] Conversely, the germination of *Potentilla norvegica* (rough cinquefoil), responsive to R light, is almost completely prevented by ethylene at concentrations as low as 1 μl/l.[11] The sensitivity to ethylene inhibition was highest during and immediately after the irradiation period.[27] In tromban groundnut seeds, dormancy was effectively broken by the ethylene-releasing agent, ethephon (ETP), that often suppressed the germination of nondormant seeds.[28] These results demonstrate that the responsiveness of seeds to ethylene may change depending upon seeding conditions.

A cocklebur fruit contains dimorphic seeds exhibiting different germination behavior. Both the upper and lower seeds display primary dormancy for a certain period of time after harvest, but only the upper seeds remain dormant at moderate temperatures. This occurs as a result of their low growth capacity even after after-ripening (i.e., exhibiting seed coat-imposed dormancy). Thus, the upper seeds can enter secondary dormancy after long term incubation on water substratum at moderate temperatures. In cocklebur seeds, ethylene was effective in breaking primary dormancy,[26] seed coat-imposed dormancy,[26] and secondary dormancy,[29] and in accelerating the germination of nondormant lower seeds[26] at lower temperatures. However, ethylene severely inhibits germination when these seeds are presoaked for more than 8 h and at temperatures above 27°C.[30] The two opposing actions of ethylene in controlling seed germination may imply a threshold effect of ethylene such that it acts as a germination stimulator at higher concentrations, and an inhibitor at lower concentrations.

III. ETHYLENE PRODUCTION IN SEEDS

The fact that the removal of endogenously evolved ethylene from the ambient atmosphere by an ethylene absorbent significantly delays the germination of nondormant cocklebur,[26] lettuce,[31] and apple[23,24] seeds demonstrates that ethylene produced prior to germination functions as an endogenous germination stimulator. It has been established that ethylene evolved by imbibed seeds is produced by the pathway: methionine → *S*-adenosylmethionine → 1-aminocyclopropane-1-carboxylic acid (ACC) → ethylene.[7,32-34] Thus, incubation of seeds with ACC can promote germination in many species, including lettuce,[35] cocklebur,[33] and *Amaranthus caudatus*,[36,37] although it is less effective than ethylene. Moreover, the ACC content in seeds increases during water imbibition, and the addition of ACC greatly stimulates ethylene production.[7,33,35]

The patterns of ethylene production by seeds during imbibition differ markedly among seeds.[35,38-41] In all cases, the first peak of ethylene production coincides with the splitting of the seed coat due to radicle protrusion, suggesting that ethylene is produced in attendance with seed tissue growth. Nevertheless, ethylene production in many seeds is detectable long before any visual sign of germination is evident.[26,32,42] No ethylene production is detected in the dry state, suggesting that internal concentrations of ethylene in intact seeds are near the threshold required to induce germination. This possibility is supported by the following two facts. Lettuce seeds fail to germinate under hypobaric pressure unless exogenous ethylene is applied.[31] The germination of lettuce,[15,43] apple,[23] and cocklebur[33] seeds is significantly, although slightly, inhibited by AVG applied prior to germination.

In contrast, others using lettuce[15,31] and *A. caudatus*[36] seeds have reported that AVG drastically reduced ethylene production without affecting seed germination. A significant conversion of ACC to ethylene in thermodormant lettuce seeds occurs only after germination. These discrepancies have contributed to the confusion concerning the role of ethylene in seed germination.

Ethylene production in dormant peanut seeds is lower than that in nondormant ones,[44] being below that required for the induction of germination. Similar observations have been obtained with cocklebur seeds.[26,33] Ethylene production in both dormant and nondormant cocklebur seeds begins soon after the start of soaking and increases linearly until 12 h at 23°C. Thereafter the increase in ethylene production by nondormant seeds exceeds that of the dormant ones. After 24 h (prior to radicle protrusion) the former produces twice as much ethylene as the latter.

It has been reported that the state of secondary dormancy termed thermodormancy in lettuce results from a decline in ethylene production.[45,46] Similarly, cocklebur seeds do not produce sufficient ethylene to induce germination following a period of extended imbibition.[29] However, both the primary dormancy of peanut[44] and cocklebur,[26] and the secondary dormancy of lettuce[47,48] and cocklebur[29] seeds, were broken by exogenous ethylene. Further, addition of GA_3 alone was almost ineffective in breaking the secondary dormancy of lettuce[49,50] and cocklebur, unless combined with ethylene.[29] These results support an ethylene requirement for germination in some seed species. In wild oat[42] which exhibits no strict ethylene requirement for germination, there is little difference in ethylene production between dormant and nondormant seeds.

Ethylene production in plant organs may be significantly enhanced by CO_2 (see Chapters 1 and 2). Therefore in the soil, ethylene production in germinating seeds can be influenced by the CO_2 accumulated in the atmosphere surrounding the seeds. In fact, ethylene emanation in subterranean clover[51] and cocklebur[26,52,53] seeds was greater in the presence of CO_2. In cocklebur, endogenous CO_2 promoted release from primary dormancy and the germination of after-ripened nondormant seeds, independently of applied ethylene, at temperatures above 13°C.[26] Ethylene production was not enhanced by CO_2 at 13°C, but at 23°C the ethylene production in 1% CO_2 was enhanced about 10-fold in the dormant seeds and about 20-fold in the nondormant ones.[26]

IV. INTERACTIONS BETWEEN ETHYLENE AND OTHER NONVOLATILE GERMINATION-REGULATING FACTORS

A. LIGHT

Under certain temperature regimes, many species or varieties of seeds require R irradiation for germination. Both lettuce and upper cocklebur seeds respond to both ethylene and R light. Below 20°C, no variety of lettuce seeds requires R light to germinate, some require it around 25°C, and all require it at temperatures above 33°C.[54] In contrast, the upper seeds of cocklebur will not germinate without an exposure to R light even at the optimal temperature, while the lower seeds germinate in the dark even at suboptimal temperatures.[55] Many investigators using the lettuce variety 'Great lakes'[47,48,56] (nonphotoblastic at 25°C) have observed an interaction between ethylene and phytochrome. In contrast, in the cv. Grand Rapids[57-60] (positively photoblastic) ethylene and phytochrome act independently as in upper cocklebur seeds.[61]

The germination of thermodormant[47] or osmotically inhibited [56] 'Great Lakes' lettuce seeds is promoted by ethylene, applied during or after irradiation. The effect of ethylene increased with the duration of irradiation.[47] Thus, it was concluded that the effect of ethylene on seed germination is mediated by phytochrome.[56] The inhibition of germination in 'Great Lakes' lettuce seeds caused by osmoticum or ABA is partially reversed by ethylene only in the presence of light, suggesting that ethylene action depends absolutely upon light.[48] A similar interaction between ethylene and light occurs in *S. arvensis*[62,64] and *Melandrium noctiflorum*[63] seeds. Interestingly, the germination of the normally light-sensitive lettuce variety 'Mesa 659' was most effectively stimulated by endogenously evolved gases when irradiated.[65]

The germination of *P. norvegica* seeds is inhibited by ethylene.[27] Sensitivity to ethylene was greatest during and immediately after irradiation, as was the case with ethylene promotion in lettuce.[47] Delaying ethylene application until 24 to 30 h after irradiation was not inhibitory and this coincided with escape from phytochrome control. Ethylene inhibition was readily reversed upon the removal of gas and reirradiation of the seeds. Thus, ethylene mimics far-red (FR) light in reducing germination.

Other reports suggest that the germination-promoting effect of ethylene in 'Grand Rapids' lettuce seeds is independent of light.[57,59] In seeds with FR-induced dormancy, maximum germination occurred with R light plus ethylene. Apparently, R light does not stimulate seed germination through enhancement of ethylene production.[58] Thus, the primary requirement for germination of these secondarily dormant seeds was R light treatment; ethylene and GA_3 served to complement R light. This was also observed in celery seeds.[66]

The interaction of ethylene and light was studied in detail using positively photoblastic redroot pigweed seeds.[67] Ethylene and light demonstrated only an additive action, which indicated that they act independently. Thereafter, it was reported that ethylene was effective only when applied in light.[68]

CO_2 is required for phytochrome-induced germination of lettuce seeds.[69] As seen in Table 1, the germination-promoting effect of CO_2 in upper cocklebur seeds occurs only after R irradiation.[61] In contrast, the effectiveness of ethylene at 23°C was independent of pre-irradiation and the R light dependency of CO_2 disappeared completely in the presence of ethylene. Similarly, the germination-inhibiting effect of ethylene at 33°C required previous exposure to R light. Thus, the germination-regulating action of ethylene appears independent of phytochrome, whereas that of CO_2 is dependent upon Pfr (the active form of phytochrome).[61]

B. TEMPERATURE

The direction of ethylene action in *Potentilla*[27] and cocklebur[30] seeds depends upon temperature. The secondary dormancy of seeds is broken by ethylene[15,29,35,43,45,47,48,65,70-72] or ethylene-releasing chemicals.[50,73] The ability of ethylene to promote germination at non-optimal temperatures, in cocklebur,[26] lettuce,[43,57,60] and redroot pigweed,[68,74,75] indicates that ethylene can fully inhibit the establishment of secondary dormancy in these seeds. The induction of secondary dormancy in *Chenopodium bonushenricus* seeds was prevented by a combination of ETP, GA_3, and kinetin (KIN).[76] Exposure of imbibed sunflower seeds to

TABLE 1
Effects of 3% CO_2 and/or 3 μl/l Ethylene on the Germination of Upper Cocklebur Seeds Preirradiated for 5 min with R or 50 min with FR Light at Different Temperatures[61]

Temperature	CO_2 (3%)	Ethylene (3 μl/l)	Germination (%) Dark	R	FR
23°C	−	−	0	0	0
	−	+	35.5	32.0	36.4
	+	−	0	25.0	3.7
	+	+	98.2	92.5	100.0
33°C	−	−	51.6	90.3	53.3
	−	+	6.3	4.8	3.3
	+	−	75.9	100.0	77.5
	+	+	41.0	47.6	37.3

Note: Experiments were carried out using seeds presoaked in darkness for 1 week at 23°C (for 23°C Exp.) or for 3 weeks at 17°C (for 33°C Exp.) in order to remove any preexistent Pfr.

From Esashi, Y. et al., *J. Exp. Bot.*, 38, 702, 1987. With permission.

45°C for more than 48 h induced secondary dormancy, but seeds pretreated with ETP were less affected.[77] Preincubation at 45°C strongly reduced the capacity of imbibed seeds to convert ACC to ethylene at 20°C. Thus, the induction of secondary dormancy may be associated with reduced ability to convert ACC to ethylene.

The termination of primary dormancy of apple seeds by cold stratification is accompanied by an increase in ethylene production. However, ethylene is not required during the stratification period.[24] This is consistent with the suggestion that the delay of after-ripening of apple seeds within fruits stored at low temperature is not due to accumulated ethylene.[78] Nevertheless, the germination of apple seeds stratified under hypobaric conditions was reduced, suggesting an involvement of another volatile substance in the stratification process.[79]

It is natural that increasing temperature stimulates ethylene production in seeds by the activation of relevant enzymes.[26] However, prolonged incubation of seeds at elevated temperatures can reduce ethylene production by inducing secondary dormancy.[29,45] This is also associated with a temperature-dependent change in ethylene responsiveness of seeds. The sensitivity of cocklebur seeds to ethylene was lower at 33°C than at 23°C, and gradually decreased as preimbibition was extended.[29,80] Such a change in ethylene sensitivity during a soaking period is widespread, also occurring in lettuce,[47] *Spergula*,[64] and lamb's-quarters[81] seeds. The change in ethylene sensitivity in relation to temperature has been examined in redroot pigweed seeds.[67,75,82] The ethylene response threshold for germination was initially unaffected by temperature, but decreased by at least an order of magnitude during prolonged preincubation at constant temperature. In nature, buried seeds are exposed to daily and seasonal temperature fluctuations, and in redroot pigweed seeds this considerably enhances sensitivity to ethylene.[82] Similarly, the inhibition of R-dependent *Potentilla* seed germination by low levels of ethylene is partially overcome by alternating temperatures during imbibition.[27]

C. PLANT HORMONES

Since seed germination is a reflection of growth, plant growth regulators such as Gib and Ckn, can promote seed germination, while growth inhibitors (i.e., ABA) can inhibit germination. Ethylene can act synergistically or additively with Gib or Ckn to promote seed germination, whereas it antagonizes ABA. These interactions occur regardless of whether seeds are positively or negatively photoblastic. A combination of ethylene with Gib or Ckn may be required for the promotion of germination when the incubation temperature or the length of dark incubation are increased.

1. Gibberellins

The germination of lettuce seeds subjected to heat treatment, FR light,[57] extended dark storage,[58] high temperature soaking, water stress, or ABA,[48] was increased in the presence of both ethylene and Gib. The inhibition of germination of celery seeds by elevated temperature was reversed by a combination of ETP and GA_{4+7}.[73] In curly dock seeds,[25] secondary dormancy was partially broken by ethylene only if irradiated, but secondary dormancy (established under irradiation at high temperature) was broken only when GA_{4+7} was included.

Prior to germination, ethylene production is slightly promoted by Gib in lettuce[35,83] and peanut[84,85] seeds. The inhibition of germination of *A. caudatus* seeds by Tetcyclasis[86] or Paclobutrazol[87] (inhibitors of Gib synthesis) is reversed not only by GA_{4+7} but also by ETP or ACC. The effect of GA_{4+7} on germination disappears in the presence of AVG, and, conversely, AVG enhances the effect of Tetcyclasis. Thus, in these seeds Gib may act through ethylene production, or ethylene may sensitize the seeds to Gib. Secondarily dormant cocklebur seeds are responsive to GA_3 plus ethylene.[88] A similar phenomenon occurs in thermodormant lettuce seeds.[65]

2. Cytokinins

Ethylene production is stimulated by Ckn prior to germination in lettuce[35,83] and peanut[44,84,85] seeds. Thus, Ckn may promote germination by enhancing ethylene production.[44,84] In cocklebur, 6-benzylaminopurine (BA) promoted ethylene production only after germination.[88] Moreover, the promoting effect of BA in cocklebur seeds was reduced in the presence of an ethylene absorbent, and was further enhanced by ethylene or ETP, suggesting an independent action for both.[88,89] In lettuce seeds, KIN may alleviate salt stress-induced inhibition of germination by promoting both ACC synthesis and its conversion to ethylene.[90]

Thermodormant lettuce seeds that produced no ethylene regained this ability when subjected to various combinations of GA_3, Ckn, and CO_2. Seeds exposed simultaneously to all of these effectors exhibited maximal ethylene evolution prior to visible germination, and nearly complete germination resulted. As seen in Table 2,[43] AVG inhibits the actions of these agents that were completely reversed by exogenous ethylene. These results show that endogenous ethylene is essential for the relief of thermoinhibition of lettuce seed germination, and suggests that the insensitivity to AVG arises from a very low ethylene requirement.

An additive or synergistic interaction between KIN and ethylene or ETP is also common in seed germination.[25,50,89] The nature of the interaction between KIN and ethylene differs depending upon seeding condition: they act additively in breaking thermodormancy in 'Grand Rapids' lettuce seeds, but synergistically in reversing ABA inhibition. In *S. arvensis* seeds, ethylene increases the endogenous levels of Ckns.[62]

3. ABA

ABA inhibits the germination of many seeds and ethylene or ETP antagonizes ABA action in peanut,[44] lamb's-quarters,[81] cotton,[91] lettuce,[48,50] sunflower,[92] and *A. caudatus*.[37] Since ABA inhibits ethylene production during the pregermination period in peanut seeds, its action has been attributed to inhibition of ethylene production.[44,84] This possibility is supported by the fact that ABA inhibition of *Amaranthus* seed germination is reversed not only by ETP but also by ACC, and is augmented by AVG.[37] Germination in lettuce and cotton seeds was not completely reversed by ETP;[50,91] thus, ABA may act independently of the inhibition of ethylene production in these seeds.

TABLE 2
Effects of GA_3 + KIN, GA_3 + CO_2 or GA_3 + KIN + CO_2, Applied in the Presence or Absence of AVG and in Combination with Various Ethylene Concentrations on the Germination of Lettuce Seeds at 32°C

Treatment	Germination (%)		
	$- C_2H_4$	1 µl/l C_2H_4	10 µl/l C_2H_4
H_2O (control)	0	0.5	1.0
GA_3 + KIN	52.0	70.5	88.5
GA_3 + KIN + AVG	22.7	72.3	88.0
GA_3 + CO_2	55.6	71.0	92.5
GA_3 + CO_2 + AVG	9.6	58.0	91.5
GA_3 + KIN + CO_2	93.8	94.0	97.5
GA_3 + KIN + CO_2 + AVG	40.3	89.0	96.0

Note: GA_3, 0.5 m*M*; KIN, 0.05 m*M*; AVG, 1 m*M*; and CO_2, 10%.

From Saini, H. S., Consolacion, E. D., Bassi, P. K., and Spencer, M. S., *Plant Physiol.*, 81, 950, 1986. With permission.

D. WATER STRESS

Germination occurs by active water uptake by the seeds. Nevertheless, germination in the field proceeds under transient or permanent water stress. Ethylene or ETP induces partial germination of seeds under water stress in lettuce,[48,56,90] curly dock,[25] cocklebur,[93] *C. bonushenricus*,[76] redroot pigweed,[67,68,74] and *A. caudatus*.[36,37] In lettuce and curly dock seeds, ethylene acts in combination with either GA_3[25] or CO_2.[47,56] The involvement of endogenously evolved ethylene in KIN-enhanced germination of salt-stressed seeds has also been reported.[90]

In dormant redroot pigweed seeds, the ethylene-response threshold for germination decreased with decreasing water potentials.[67] The fungal toxin, cotylenin E, also counteracts water stress inhibition of lettuce seed germination.[90] Together with KIN, it increased pre-germination ethylene production and germination under saline condition; suggesting a need for ethylene production and action in the relief of salt stress.

E. NO_3

Germination in some seeds is stimulated by NO_3 ions. There is a close interrelation between NO_3 and ethylene action on seed germination.[94-98] The germination of common purslane (*Portulaca oleracea* L.) seeds in darkness was promoted by ethylene, and this effect was enhanced by NO_3 when applied simultaneously. The additive interaction disappeared when NO_3 or NO_2 was applied first. Thus, the rate of reduction of NO_3 to NO_2 may limit dark germination of common purslane seeds.[94]

The effect of ETP or ethylene on germination in relation to the availability of NO_3 was studied using different year-lots of lamb's-quarters seeds.[95] Both ETP and ethylene were more effective in the 1983 seeds than in the 1982 ones. A combination of ETP and NO_3 acted synergistically on 1982 seeds, but in 1983 seeds the combination was only marginally more effective than ETP (Figure 1). Interestingly, the 1982 seed contained an extremely low endogenous level of NO_3 as compared to the 1983 lot. Thus, high levels of either endogenous or applied NO_3 appear to enhance the responsiveness to ethylene. However, the reduced sensitivity of secondarily dormant lamb's-quarters seeds to ethylene could not be attributed to NO_3 loss from the seed during an imbibition period.[97] A similar additive interaction between ETP and NO_3 was also found in certain lots of wild oat seeds.[98]

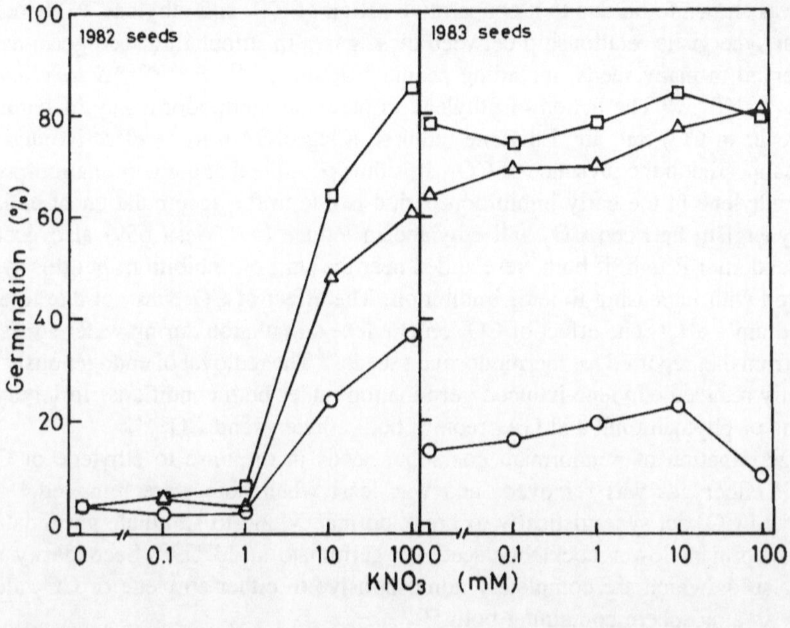

FIGURE 1. Effects of KNO_3 concentration on germination of lamb's-quarters seeds of different harvests in the presence of various concentrations of ETP: 0 (○), 10 (△), and 100 (□) mg/l. (From Saini, H. S., Bassi, P. K., and Spencer, M. S., *Plant Physiol.*, 77, 940, 1983. With permission.)

F. MISCELLANEA

Thiourea (TU) is a synthetic germination stimulant effective only at high concentrations. TU accelerates the germination-promoting effect of ethylene.[29,58,88] Conversely, the effectiveness of TU was reduced in the presence of an ethylene absorbent in cocklebur seeds, but it did not affect bulk ethylene production of seeds.[88,109] Triiodobenzoic acid and morphactin (both auxin-transport inhibitors) prevented lettuce seed germination and this was reversed by ETP.[99]

V. INTERACTIONS OF ETHYLENE AND OTHER GASES

A. CO_2

Semienclosed environments, such as soil, accumulate CO_2 as well as ethylene. In many instances CO_2 counteracts ethylene action (see Chapter 5). However, this antagonism does not appear to be true for germination.

1. Physiological Action

Conflicting results have been obtained concerning the effect of CO_2 on germination. Regardless of the variety, the germination inhibition in lettuce seeds induced by osmotic stress[56] or high temperature[47,60,65] was not reversed by CO_2 alone. Others have reported that CO_2 slightly stimulates lettuce seed germination when applied under hypoxia.[31] A lag period of 5 days is required for CO_2 to promote germination.[70] The germination of cocklebur seeds was delayed when CO_2 was removed.[26] CO_2 also increased the germination of nondormant lower cocklebur seeds under water stress in an ethylene-free atmosphere.[93] The effectiveness of CO_2 was less than that of ethylene. Similarly, removal of CO_2-inhibited germination of *S. arvensis* seed, but exogenous CO_2 had no effect.[64] In *Galeola septentrionalis* seeds, CO_2 was more effective than ethylene.[18]

The germination of subterranean clover seeds, previously reported to occur in response to CO_2, was shown to be due to a cooperative action of CO_2 and ethylene.[51] Thereafter, an additive or synergistic relationship between these gases in stimulating seed germination has been observed in many seeds, including peanut,[44] lettuce,[31,43,47,56,60,65,70] *S. arvensis*,[64] and cocklebur.[26,29,61,80,100] The action of ethylene in breaking thermodormancy of lettuce seeds did not occur in CO_2-free air. Ethylene required a lag of 24 h to be effective and this lag period disappeared in the presence of CO_2. It is thus possible that the germination-promoting action of ethylene in the early imbibition period is due to the accumulation of endogenous CO_2.[70] Synergism between CO_2 and ethylene in lettuce (cv. Mesa 659) also occurred in darkness and after R light if both were added near the start of imbibition, but this synergism disappeared with increasing time of imbibition. The effect of CO_2 was not due to a change in the medium's pH.[47] The effect of CO_2 on ethylene stimulation during water stress is quite different from that reported for thermodormant seeds.[56] The removal of endogenously evolved CO_2 greatly reduced ethylene-induced germination under both conditions. In lettuce seeds, the actions of phytochrome and GA_3 require both ethylene and CO_2.[65]

The germination of nondormant cocklebur seeds in response to ethylene or CO_2 was reduced if either gas was removed, and was least when both were removed.[26] Applied ethylene and CO_2 act synergistically to break dormancy and to stimulate germination, and enable nondormant lower cocklebur seeds to germinate at 13°C.[100] Secondarily dormant cocklebur seeds which are completely nonresponsive to either ethylene or CO_2 alone germinate in an atmosphere containing both.[29]

2. Ethylene Production

In cocklebur[26] and subterranean clover seeds,[51] the germination stimulation by CO_2 was greatly reduced by an ethylene absorbent, and CO_2 markedly increased ethylene production during the pregermination period. In nondormant cocklebur seeds, CO_2-enhanced ethylene production was not found at 13°C or under hypoxia.[26,100]

In contrast, ethylene production in thermoinhibited lettuce seeds was not enhanced by CO_2.[43] Treatments that counteracted this inhibition (GA_3 + CO_2 and GA_3 + CO_2 + KIN) resulted in approximately 10- to 40-fold increases in ethylene production, and the application of AVG reduced ethylene production to below control levels. These results imply that CO_2 enhancement of ethylene production during the pregermination period occurs only under conditions conducive to seed germination.

B. O_2

Oxygen tensions in soil are assumed to be lower than those in air, and to fluctuate depending upon temperature and rainfall. O_2 enrichment has been known to break dormancy and to promote germination in many seed species.

1. Physiological Action

The effect of ethylene removal on cocklebur seed germination is greater in hypoxic atmospheres,[26] and ethylene can act in anoxia.[102] Nondormant, upper cocklebur seeds are potentiated if subjected to anaerobic conditions for certain time periods after being presoaked under aerobic conditions (anaerobic induction).[101] This effect was significantly enhanced by ethylene added during the anoxic treatment.[102] This clearly indicates that ethylene is capable of acting on seed germination independently of O_2. Nevertheless, the germination-stimulating effect of ethylene in cocklebur seeds increased as O_2 tension is increased, reaching a maximum in pure O_2.[26,100] The secondary dormancy of these seeds is most effectively broken by ethylene when applied in an O_2-enriched atmosphere.[29] Similarly, release of ragweed seeds from dormancy is most rapid in ethylene plus O_2.[19]

2. Ethylene Production

The conversion of ACC to ethylene requires molecular O_2. In fact, ethylene production in cocklebur seeds increases with increasing O_2 tension from 0% to 20%, but is detectable in anoxic or hypoxic atmospheres.[103] Further O_2 enrichment inhibits ethylene production.

VI. CONTROL OF SEED TISSUE GROWTH BY ETHYLENE AND CO_2

A. GROWTH OF SEED TISSUES IN RESPONSE TO CO_2 AND/OR ETHYLENE

The force required to overcome the mechanical restraint imposed by the seed coat is generated by the growth of each seed tissue. In cocklebur seeds, both cotyledons and the embryonic axis produce the thrust. During germination of lower seeds, the embryonic axis supplies about 80% of the total thrust required, the remaining 20% is derived from the growth of cotyledons.[16] Each seed tissue has a different growth potential and responds differently to growth-regulating factors. Cocklebur seed germination occurs via different modes depending upon the treatment: normal germination in nondormant lower seeds begins by protrusion of the radicle tip through the seed coat, although a portion of the seed population exhibits atypical germination behavior.[88]

Primary dormancy of seeds results from insufficient growth potential in seed tissues. The inability of dormant axes and cotyledons of cocklebur seeds to grow disappears completely in the presence of ethylene. Regardless of dormancy status, trapping of endogenous ethylene reduces both axial and cotyledonary growth (although dormant seeds exhibit little or no overt ethylene production). Further, addition of ethylene strongly stimulated growth of both tissues, especially of the cotyledons. As a result, ethylene breaks primary dormancy in both the upper and lower cocklebur seeds and accelerates the release from the seed coat-imposed dormancy in upper ones.[104] Thus, ethylene-induced germination is atypical, occurring as a result of the preferential promotion of cotyledonary growth.[88,105]

Cucumber (*Cucumis sativus* L.) cotyledons produce ethylene in response to BA, but the BA-stimulated ethylene production can be separated from BA-stimulated growth through the use of KCl and $CaCl_2$.[106] Exogenous ethylene had little influence on the growth of cucumber cotyledons when KCl and $CaCl_2$ were added, and ethylene suppressed the BA-stimulated growth. Ethylene also inhibited KIN-induced growth of fenugreek (*Trigonella foenumgraecum* L.) cotyledons.[107]

Like ethylene, CO_2 is capable of promoting both axial and cotyledonary growth, the effect being greater in the axial tissue.[104] Thus, cotyledons isolated from nondormant seeds grow in response to ethylene, and respond only slightly to CO_2.[104] As a result, the germination pattern of cocklebur seeds was more normal in the presence of CO_2. However, expansion of excised fenugreek cotyledons was nearly equally promoted by ethylene or CO_2.[107] In cocklebur seeds, the growth-promoting effects of CO_2 were reduced when ethylene was absorbed and CO_2 enhanced the ethylene production in both tissues, especially the dormant ones.[104] The combined application of ethylene and CO_2 elicited the maximal growth of both tissues regardless of the dormancy status, suggesting that the maximal germination of cocklebur and lettuce seeds in response to ethylene + CO_2 is due to the maximum growth of both their axial and cotyledonary tissues.[104] Such a synergistic growth stimulation was not observed in fenugreek cotyledons.[107]

B. CHANGES IN SENSITIVITY TO ETHYLENE AND CO_2

Secondary dormancy in curly dock seeds may arise from the inability of the radicles to lower their water potential to a level which would enable them to grow by counteracting the surrounding seed structure.[25] The development of secondary dormancy is accompanied by decreased growth potentials in both axial and cotyledonary tissues during imbibition, coincident with a decline in ethylene production.[29,108] As mentioned in Section IV.B, the

TABLE 3
Effect of Preexposure of Axial Tissues to Ethylene on the Growth-Promoting Action of CO_2[80]

Pretreatment	Post-treatment: 3 μl/l C_2H_4	Post-treatment: 5% CO_2	Increase fresh weight
− C_2H_4	−	−	3.5 (%)
	−	+	4.4
	+	+	50.0
3 μl/l C_2H_4	−	−	3.6
	−	+	20.7
	+	+	54.1

Note: Axial segments were isolated from the secondarily dormant upper cocklebur seeds and exposed to CO_2 and/or ethylene for 16 h at 23°C after a 2 h pretreatment with ethylene.

(From Esashi, Y. and Ishizawa, K., *Biochemical and Physiological Aspects of Ethylene Production in Higher Plants,* Clijsters, H. et al., Kluwer Academic, 1989, 76. With permission.

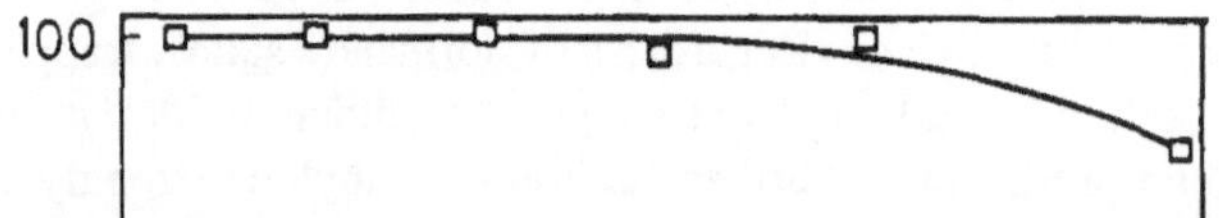

FIGURE 2. Effect of the duration of the presoaking period on the germination of upper cocklebur seeds in response to ethylene and/or CO_2 at 23°C:[80] (○) CO_2 and ethylene-free air; (△) 10 μl/l ethylene with NaOH; (▽) 5% CO_2 with Hg $(ClO_4)_2$; and (□) 10 μl/l ethylene plus 5% CO_2. (From Esashi, Y. et al., *Plant Physiol.*, 86, 39, 1988. With permission.)

germination-promoting effects of singly applied ethylene and CO_2, in general, decline with the development of secondary dormancy. When added early during imbibition, ethylene was more effective than CO_2 in promoting germination in cocklebur seeds. After 5 to 10 days of soaking, CO_2 became more effective than ethylene. Thereafter, it was less effective than ethylene (Figure 2).[80] This change corresponds roughly with that of sensitivity to CN, N_3,[109] anoxia, and chilling.[108] Furthermore, these changes parallel a change in the growth potential in the axial tissues of cocklebur seeds, which continues to rise during the 2 weeks of soaking,

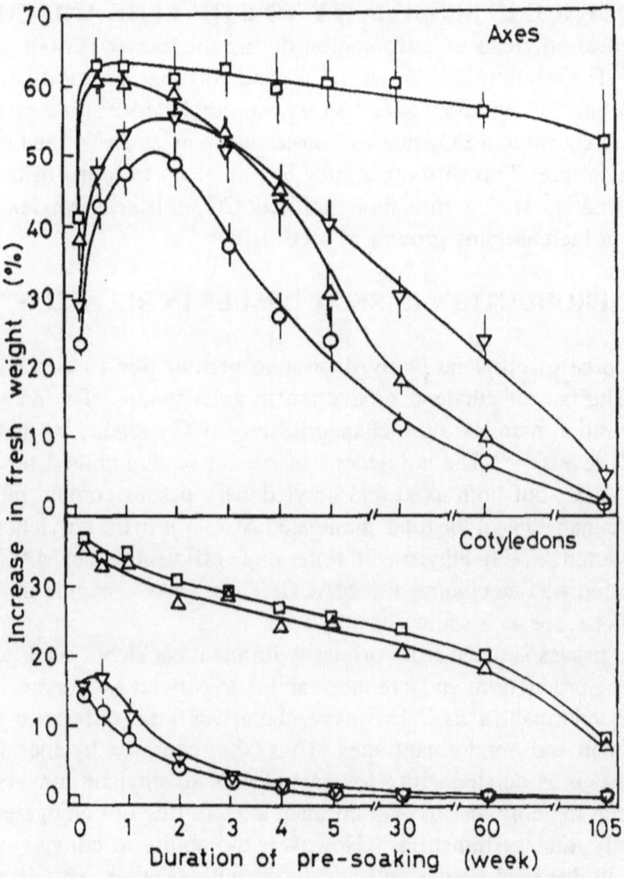

FIGURE 3. Effect of the duration of presoaking period and the growth responsiveness of the axial and cotyledonary tissues of upper cocklebur seeds to ethylene and CO_2 at 23°C.[80] Axial and cotyledonary tissues were fumigated for 17 and 44 h, respectively. (○) CO_2 and ethylene-free air; (△) 10 μl/l ethylene with NaOH; (▽) 5% CO_2 with $Hg(ClO_4)_2$; and (□) 10 μl/l ethylene plus 5% CO_2. (From Esashi, Y. et al., *Plant Physiol.*, 86, 39, 1988. With permission.)

and then begins to decline (Figure 3). That is, the growth of axial tissues, unlike the cotyledons, increases with increasing duration of soaking prior to isolation and reaches a maximum at 2 weeks, declining thereafter (Figure 3). These facts imply that secondary dormancy in cocklebur seeds actually develops after 2 weeks of imbibition at 23°C.[108]

The effects of ethylene and CO_2 are additive in the early periods of imbibition, becoming synergistic in later periods. The nonresponsiveness of upper cocklebur seeds to either of these gases was overcome in the presence of both.[80] Such changes in germination behavior were closely associated with the growth of the axial tissues, but not of the cotyledons (Figure 3). During the first 2 weeks, the promoting action of ethylene on axial growth was greater than that of CO_2, but after 5 weeks of soaking the reverse was true. On the other hand, the growth of cotyledons isolated from long term (>3 d) presoaked cocklebur was exclusively promoted by ethylene, and no interaction between ethylene and CO_2 was found. When secondary dormancy had been established before isolation, the effectiveness of ethylene and CO_2 was clearly manifested only in the presence of the other gas. These results may help to explain the conflicting findings that describe CO_2 and ethylene as being additive or synergistic in the same species.[31,47]

C. PHASE SEQUENCE IN SENSITIVITY TO ETHYLENE AND CO_2

Seeds respond to ethylene by germination during the earlier periods of water imbibition.[29,47,64,80,81,110] The relationships between CO_2 and ethylene sensitivity differ among seed species. In cocklebur[80,105] and *Spergula*[64] seeds, an ethylene-sensitive phase of seed germination occurs subsequent to a CO_2-sensitive phase; while in lettuce[65,70] and redroot pigweed[68] seeds, the reverse is true. This difference may be explained in terms of the dual nature of CO_2 action in promoting seed germination. Initially CO_2 enhances ethylene production[26,53] and subsequently it increases the growth of seed tissues.[104]

D. ETHYLENE PRODUCTION IN SEED TISSUES IN RELATION TO CO_2 AND O_2

The major source of ethylene evolved prior to germination in cocklebur seeds is the axial tissues.[104] The rate of ethylene production in axial tissues of nondormant cocklebur seeds during imbibition paralleled the change in free ACC content, but not that in MACC (*N*-malonyl ACC) levels.[32,33] The cotyledons of peanut seeds imbibed for 1 h contain the majority of seed ACC, but both axial and cotyledonary tissues contain more MACC than ACC.[7] Thus, more than 90% of the total conjugated ACC was in the cotyledons. Germinating peanut seeds converted ACC to ethylene 70 times more efficiently than MACC. The majority of ACC administered was metabolized to MACC. Thus, MACC occurs in peanut seeds but apparently does not serve as a source of ethylene.

Imbibed axial tissues isolated from primarily dormant cocklebur seeds are characterized by reduced ethylene production, and a reduced ability to convert exogenous ACC to ethylene as compared to nondormant axes.[33] However, there was little difference in ACC content between the dormant and nondormant axes. Thus, the reduced ethylene production in detached dormant axes is associated with a low rate of ACC to ethylene conversion. Significant differences between the dormant and nondormant axes in free or conjugated ACC contents were observed only after germination.[33] However, the ability to convert ACC to ethylene was not detected in dry seed tissues and developed with soaking. This is also the case for the cotyledonary tissues. Interestingly, the ACC content of the axial tissues increased gradually even in the case of dormant axes, that was accompanied by an increase in MACC accumulation.[33] This suggests that, even in dormant seed axes, ACC synthesis occurs and exceeds its conversion to ethylene.[33] Moreover, the ratio (1.8 to 2.0) of ACC to ethylene conversion rates between the dormant and nondormant cocklebur seeds was not affected by the concentration of ACC or by that of O_2, further supporting the view that the reduced ethylene production in dormant cocklebur seeds is largely due to their inability to convert ACC to ethylene.

Arguing against the view that ethylene production in germinating seeds is a result of growth of seed tissues are the results of Esashi et al.,[53] who found that the seed coat of cocklebur has no growth potential and includes a layer of endosperm that produces low levels of ethylene. Moreover, ethylene production in seed coat tissues was dependent on ACC and was increased by CO_2, without a marked increase in ACC content. The CO_2 enhancement of ethylene production in the seed coat is independent of the growth stimulation of seed tissues by CO_2. These results suggest that CO_2 enhances ethylene production in imbibed seeds by increasing both ACC supply and ACC to ethylene conversion.

As described previously,[102] the action of ethylene on germination was observed even in anoxia, where traces of ethylene were produced.[111,112] Axial and cotyledonary tissues isolated from dormant cocklebur seeds produced more ethylene under anoxic or hypoxic conditions relative to nondormant tissues.[103] In all seed tissues the optimal concentration of O_2 for ethylene production was 3%, and higher levels inhibited ethylene production.[103]

The enhanced ethylene production under hypoxia in the axes and cotyledons of dormant cocklebur seeds[103] may result from more active CO_2 evolution in primarily dormant vs.

nondormant seed tissues. Moreover, CO_2-enhanced ethylene production was completely blocked by AVG.[52] The concentration of CO_2 required to enhance ACC to ethylene conversion increased with increasing O_2 tensions, demonstrating that ACC to ethylene conversion may result from CO_2 action, which in turn is influenced by O_2. These facts imply that CO_2 acts on ethylene production by regulating certain more basic metabolic processes.

E. AXIAL AND COTYLEDONARY GROWTH IN RESPONSE TO GASES UNDER WATER STRESS

The inhibition of germination in cocklebur seeds by water stress exhibits two phases.[93] A steep decline in germination occurs after mild water stress, reflecting the growth response of axial tissues, while a less pronounced effect occurs with more severe water stress, reflecting the growth response of cotyledons. Oxygen enrichment improved both the axial and cotyledonary growth under water stress. Carbon dioxide primarily enhanced axial growth, while ethylene exclusively enhanced cotyledonary growth. When combined, these gases elicited greater growth of both axial and cotyledonary tissues, leading to germination under more severe water stress.

F. DIFFERENTIAL RESPONSES OF AXIAL TISSUES TO ETHYLENE AND CO_2

As mentioned in Section IV.B, the effect of ethylene on the germination of presoaked cocklebur seed depends on the incubation temperature.[30] Isolated axial tissues respond slightly to ethylene at 33°C.[104,113] Exogenous ethylene maximally stimulated elongation of bean (*Phaseolus vulgaris* L.) embryonic axes when applied during the first 12 h of dark incubation, but was inhibitory when application was delayed.[114,115] This shift in ethylene effectiveness resulted from a negative response by the axial zone to ethylene during incubation. The same shift was observed in cocklebur axes. Thus, ethylene displayed differential effects on axial growth in bean and cocklebur, depending upon the time of its exposure or the growth zone of the axis tested. Each zone of the axial tissue of cocklebur seed responds differently to R light, suggesting a physiological gradient within the axial tissue.[116] These findings may help to explain the nature of the reversal of the ethylene effects. As shown in Figure 4,[117] the growth-promoting action of ethylene in different zones of cocklebur seed axial tissues depended on temperature. At 23°C, ethylene stimulated growth in all zones, especially in the proximal portion of the axis corresponding to the future hypocotyl, while CO_2 acted only on the radicle tip zone. At 33°C, ethylene strongly suppressed the growth in the radicle tip zones, leading to an inhibition of seed germination. Similarly, Abeles[15] found different effects of ethylene on hypocotyl or root growth in germinating lettuce seeds. The inhibition of radicle tip growth in cocklebur axes at 33°C was enhanced by hypoxia and was alleviated by O_2 enrichment. Thus, the temperature-dependent shift of ethylene action is associated with a temperature-dependent responsiveness of the radicle tip zone to ethylene. These data further suggest that the synergistic or additive relationship in promotion of germination by CO_2 and ethylene occurs from the respective effects of CO_2 on zones A and B, with those of ethylene on zones B, C, and D.[117]

VII. VERIFICATION OF ETHYLENE PARTICIPATION IN SEED GERMINATION USING 2,5-NORBORNADIENE

Silver nitrate, an inhibitor of ethylene action, does not block ethylene-promoted germination of purslane seeds, although it reverses the typical ethylene inhibition of seedling growth that follows germination.[94] This suggests either that ethylene stimulates germination through a novel mechanism, or that the silver ion does not reach the proper site in the seeds. However, the germination of thermoinhibited lettuce seeds was prevented by inhibitors of

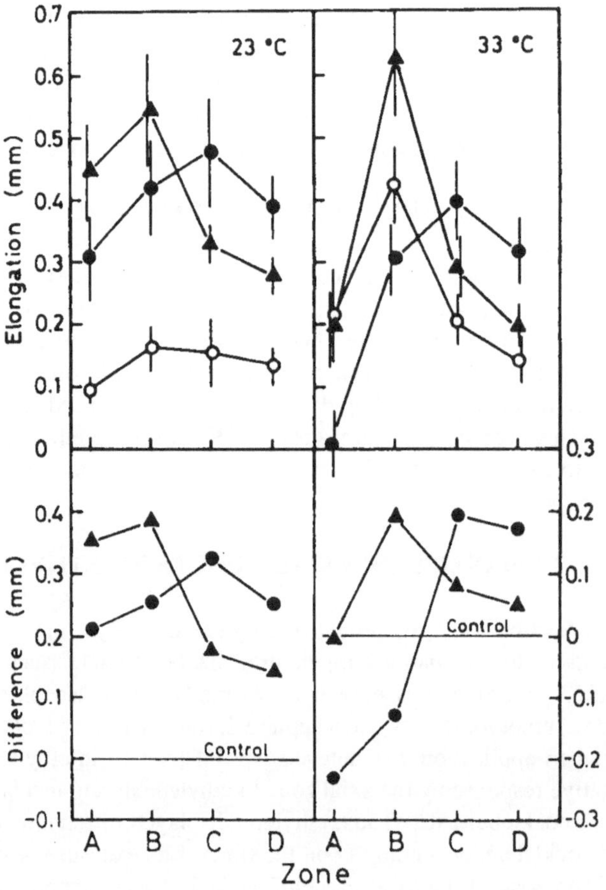

FIGURE 4. Growth profiles of each zone of isolated axes of upper cocklebur seeds in the presence of: 3 μl/l ethylene (●) or 3% CO_2 (▲) at 23°C for 19 h; or 33°C for 11.5 h (○), ethylene and CO_2 free control.[117] Axial segments were divided into 4 zones, A, B, C, and D from their tips by marking after being decoated. (From Esashi, Y. et al., *Plant Cell Physiol.*, 28, 163, 1987. With permission.)

ethylene action, such as silver thiosulfate[118] and 5-methyl-7-chloro-4-ethoxycarbonyl-methoxy-2,1,3-benzothiadiazole,[119] and their effects were reversed by the addition of ethylene.[15]

The germination of nondormant *A. caudatus* seeds is inhibited by 2,5-norbornadiene (NBD, a competitive inhibitor of ethylene action).[120] The inhibitory action is overcome by the addition of either ACC or ethylene and is enhanced by AVG. This suggests that endogenously produced ethylene is an indispensable factor for germination. Similar findings were obtained with lettuce seeds.[15,43] Khan and Huang[90] employed NBD to test the involvement of ethylene in the relief of salt stress by KIN. NBD inhibited the germination of salt-stressed lettuce seeds in the presence or absence of KIN in a dose-dependent manner (Figure 5). The addition of ACC or ethylene partially counteracted the effects of NBD. It was concluded that KIN may alleviate salt-stress inhibition of germination by promoting both ACC synthesis and its conversion to ethylene.

In cocklebur seeds, NBD strongly inhibited germination at 23°C and this effect was significantly reversed by exogenous ethylene, but at 33°C, a low concentration (3 μl/l) of ethylene enhanced the inhibitory NBD effect.[113] The germination inhibition by NBD was due to the inhibition of both axial and cotyledonary growth, and its inhibitory effect on cotyledonary growth was reversed by ethylene at both temperatures while the inhibition on

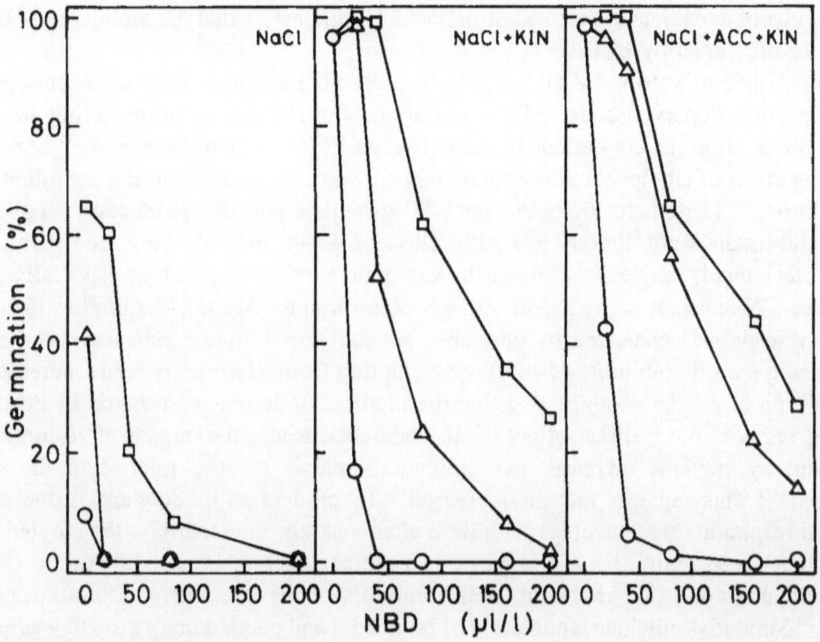

FIGURE 5. Inhibition by NBD of lettuce seed germination under salt stress induced by 0.1 *M* NaCl with or without 0.05 m*M* KIN and/or 10 m*M* ACC at 25°C, and its restoration by ethylene: (○) germination after 24 h NBD; (△) germination after 24 h NBD + 24 h air; (□) germination after 24 h + 24 h ethylene. (From Khan, A. A. and Huang, X. L., *Plant Physiol.*, 87, 847, 1988. With permission.)

axial growth was alleviated only at 23°C. The differential responses of cocklebur seeds to NBD reflected the different growth responses of the tip zones of axial tissues to NBD and ethylene. Maximum reversal of the inhibitory effects of NBD was elicited by simultaneous application of ethylene and CO_2. Both the germination-inhibiting effects of NBD in intact cocklebur seeds and isolated seed tissues were significantly counteracted not only by ethylene but also by endogenously evolved and exogenously applied CO_2 regardless of the incubation temperature, and suggested that NBD did not act specifically as a competitive inhibitor of ethylene action.[113]

VIII. MECHANISMS OF ACTION OF ETHYLENE IN STIMULATING SEED GERMINATION

A. RESPIRATION CONTROL

A rise in ATP content occurs in cocklebur and Indian rice grass seeds treated with ETP.[89] Regardless of the dormancy status, the ATP level in dry seeds is very low and undetectable on occasion. In imbibed seeds, ATP is produced via three processes: cytosolic glycolysis, the TCA cycle, and oxidative electron transfer in mitochondria. One of the steps of ATP production during glycolysis is the conversion of PEP to pyruvate that is catalyzed by pyruvate kinase in the presence of ADP. In a series of experiments using ethylene-responsive peanut seeds, the importance of this step for supplying ATP during the early imbibition stage of seed germination was demonstrated.[122-124] The addition of ETP during imbibition increased the ability of embryonic axes from dormant peanut seeds to accumulate ATP in the presence of PEP, but it had no effect on ATP accumulation in cotyledons.[122] It was demonstrated that ATP can be synthesized from PEP and AMP in cell-free extracts.[123,124] However, in this case, the activity of pyruvate kinase was not directly stimulated by ETP, although that of adenylate kinase was stimulated,[124] suggesting that the mobilization of AMP or ADP is stimulated by ETP.[123]

As described in Section IV.D, the growth-promoting action of ethylene occurs predominantly in cotyledonary tissues. ETP stimulation of ATP accumulation during imbibition was not observed in the cotyledons of peanut seeds.[122] In cocklebur seeds, the germination-promoting effect of ethylene was not found when it was applied only during the initial period of imbibition.[100] These facts imply that the ETP-stimulation of ATP production via glycolysis in the axial tissue is not directly related to ethylene action in controlling seed germination. On the other hand, ethylene is known to stimulate aerobic respiration, especially via the alternative CN-resistant pathway, in various plant organs. Aerobic respiration in imbibed cocklebur seeds was enhanced by ethylene, but the degree of the enhancement was very slight during early imbibition. Also, O_2 consumption from *S. arvensis* seeds increased 60 h after addition of ethylene, suggesting that the increase of aerobic respiration by ethylene in imbibing seeds is not a direct effect.[125] In cocklebur seeds, the degree of respiratory enhancement by ethylene increases as application time is delayed relative to the start of imbibition.[121] This suggests that the increased ATP production by ethylene is due to mitochondrial respiration that develops with time after soaking. Interestingly, the ethylene effect on cocklebur seed germination disappeared in the presence of benzohydroxamic (BHAM) or salicylhydroxamic (SHAM) acid, both inhibitors of the alternative respiration pathway (ARP).[126] Similarly, ethylene stimulation of both axial and cotyledonary growth was strongly inhibited by BHAM.[121] Moreover, ethylene-induced respiration was sensitive to BHAM and propylgallate (PG), another inhibitor of ARP, indicating an involvement of ARP in ethylene action in promoting seed germination.[72,127] This possibility was also suggested to occur in ragweed seeds.[19] However, ARP hardly participates in ATP production. In secondarily dormant cocklebur seeds, both O_2 and ethylene enhance not only the O_2 consumption via the ARP but also that via the cytochrome path (CRP), through which most ATP is produced.[72,127] The combination of O_2 and ethylene induces the maximal operation of both paths.

However, the respiration-stimulating effect of ethylene in cocklebur seeds decreased with increasing temperature, disappearing at temperatures above 33°C, and the ratio of the ARP flux to the CRP flux (V_{alt}/V_{cyt} ratio) was high at 33°C, where germination is suppressed by ethylene. At 33°C, the production of ATP was inhibited by ethylene.[72] The high V_{alt}/V_{cyt} ratio under ethylene in air at 33°C was reduced by application of O_2, that greatly increases O_2 consumption and overcomes the ethylene inhibition of germination.[72]

Carbon dioxide stimulates not only the germination of cocklebur seeds but also their respiration. The stimulation of respiration by CO_2 was greater at 33°C than at 23°C and CO_2, unlike ethylene, increased only the operation of CRP, thus reducing V_{alt}/V_{cyt} ratios regardless of temperature. Therefore, CO_2, unlike ethylene, increased ATP level in cocklebur axes prior to germination not only at 23°C but also at 33°C.[72] Thus, at 23°C where CO_2 and ethylene were additive or synergistic in controlling both axial and cotyledonary growth, they acted independently in stimulating seed respiration. On the other hand, an antagonism between CO_2 and ethylene appeared when the ARP flux rose. This temperature-dependent change in relationship between ethylene and CO_2 was also reflected by the ATP level, adenylate pool size, and energy charge (EC) of the axial tissues.[72] Thus, it is possible that the germination-regulating actions of ethylene, CO_2, and O_2, are mediated by the modification of aerobic respiratory systems.

Such a view is supported by the following facts. As described previously, the cotyledonary tissues, that are characterized by a slower growth rate than the axes, participate in germination at a later imbibition period. Delayed application of ethylene to cocklebur seeds promoted germination even at 33°C, although the germination was atypical. Moreover, ethylene inhibition of the high temperature-induced cocklebur seed germination was partially prevented by low concentrations of BHAM or PG. The restored germination was mostly abnormal (i.e., predominantly cotyledonary growth).[128] Cocklebur cotyledons are more responsive than the axes to ethylene and are characterized by lower EC and high ARP/CRP ratio. Thus, they could contribute to axial growth as a source of reduced carbon in addition to the formation of thrust by growth. In contrast, the axial tissue, especially the apical and subapical zones, depends more upon CO_2 for growth and requires a higher EC and low ARP/CRP ratios. As a result, the axial growth at higher temperature is slightly promoted by BHAM or PG.

Reducing equivalents (i.e., NADPH) necessary for sustaining anabolism during germination are produced by the pentose phosphate pathway (PPP). However, the breaking of seed dormancy in *S. arvensis* by ethylene did not involve a switch from the Embden-Mayerhof-Parnas pathway to PPP.[125] Therefore, in the presence of O_2, ethylene seems to stimulate ATP production via CRP activation and to increase carbon-skeleton supply via ARP activation. The former action of ethylene increases EC value, sustaining the growth of seed tissues, subapical zones, and cotyledons by increasing anabolism; while the latter action sustains the growth of apical zones of the axis by supplying carbon skeletons.

B. OSMOREGULATION

The respiration promotion in cocklebur seeds by ethylene occurred after 1 h exposure and disappeared shortly after ethylene removal.[121] However, ethylene required several hours to exhibit its germination-stimulating action.[100] In redroot pigweed seeds,[20] ethylene required at least 12 h and in Douglas fir seeds,[17] preexposure to pure ethylene containing no O_2 caused maximum germination in subsequent aerobic conditions. In upper cocklebur seeds, preexposure to ethylene in anoxia for 3 d induced a significant enhancement of germination.[102] These facts imply that ethylene acts to stimulate seed germination not only through the promotion of aerobic respiration but also through other, unknown, processes.

As previously described (Section VI.E), ethylene increased the growth potential of seed tissues, by lowering their water potential. Interestingly, preimbibition of seeds under water-stressed conditions markedly augments their germination potential (i.e., "seed priming").[129] The effectiveness of seed priming in cocklebur seeds is enhanced if ethylene is applied together with osmoticum during a preimbibition period.[130,131] The enhancing effects of ethylene on both the anaerobic induction and seed priming suggest that some ethylene actions, in contrast to those affecting respiration, can persist after its removal. Moreover, the respiration-stimulating effect of ethylene was reduced as the degree of water stress was increased, while its seed priming-enhancing effect was not affected by various types of respiration inhibitors. Thus, the action of ethylene in seed priming occurs independently of O_2 or aerobic respiration.[131]

In cocklebur seeds, the promoting effects of ethylene applied during both anoxic or osmotic pretreatment periods were cumulative. The subsequent growth of both the axial and cotyledonary tissues isolated from ethylene pretreated seeds was greater than that from untreated controls.[130,131] Table 4 shows that the ethylene treatment of cocklebur seeds during a priming period with a PEG solution increases the contents of osmotica in the cell saps obtained from the frozen and thawed seeds, thus leading to a rise in osmotic pressure (OSP). About 70% of OSP was contributed by the ethanol-soluble substances, while the remaining 30% came from EtOH-insoluble materials (perhaps soluble proteins). Ethylene augmented OSP in both fractions, increasing the contents of proteins, sugars, and especially amino acids in the cell sap (Table 4).[130] The ethylene-induced increases in OSP were found not only in the cell sap but also in water extracts of seed tissues, and its effect was much more pronounced in the cotyledonary tissue subjected to priming.[131] However, when cocklebur seed tissues were exposed to ethylene under water stress, they responded to ethylene by growth and a rise in OSP, without an increase in either soluble proteins or sugar. In all

TABLE 4
Effect of 30 μl/l Ethylene on the OSP and the Contents of Sugars, Amino Acids, and Soluble Proteins in the Cell Saps from Upper Cocklebur Seeds Incubated in a 200 mOsm/kg PEG Solution for 3 days[130]

Treatment	OSP (mOsm/kg)	Sugar (mmol)	Amino acid (mmol)	Protein (g/ml)
None	152	84.2	10.7	0.44
C_2H_4	277	132.9	26.6	0.78
% of Initial	182	158	249	177

cases, amino acids accumulated dramatically after a preexposure to ethylene. The accumulation of amino acids in seed tissues was also augmented when ethylene was applied during an anaerobic induction period.[130] In particular, the contents of glutamine, arginine, asparagine, γ-aminobutyrate, proline, alanine, and serine were increased by ethylene, and application of a mixture of these promoted both axial and cotyledonary growth in cocklebur seeds in ethylene- and CO_2-free air. Interestingly, the amino acid mixture increased the growth response of seed tissues to CO_2, and conversely, decreased that to ethylene. In particular, the growth response of the axial tissue to ethylene disappeared in the presence of the amino acid mixture.[130]

As described in Section VI.E, NO_3 is required for ethylene action in seed germination. The O_2 consumption in germinating *S. arvensis* seeds was greater in the presence of both ethylene and NO_2 than in ethylene alone,[125] suggesting that NO_3 does not function as an electron acceptor during seed germination. On the other hand, NH_3 gas has been reported to break the primary seed dormancy of grass weed species, such as ethylene-responsive wild oat.[132] Considering these facts, it is possible that NO_3 functions in ethylene action by being incorporated into amino acids. The resultant increase in the amino acid pool, produced independently of O_2, would act to raise OSP value in seed tissues, to sustain the formation of high molecular substances and, as a result, potentiate seed germination. Pretreatment of secondarily dormant cocklebur seeds with KCN or NaN_3 in the presence of O_2 restores the ethylene responsiveness which had been lost;[127] thus singly applied ethylene causes the full germination of secondarily dormant cocklebur seeds. Possibly, KCN and NaN_3 act by sparing the NADH necessary for the active operation of glutamate synthase (the formation of amino acids) and ARP (the supply of carbon-skeletons for amino acids), thereby permitting the germination-stimulating actions of ethylene.

IX. CONCLUSIONS

The concentration of ethylene in the soil where many seeds germinate is known to accumulate on occasion to physiologically effective levels. Thus, the restricted microenvironments of the soil provide conditions favorable for the regulation of germination in many seeds by accumulated gases. Such conditions are conducive to maintaining the internal concentration of endogenously produced ethylene at higher levels for longer periods in seeds.

From the data presented in this chapter, it is clear that seeds not only produce ethylene but respond to it as well. Thus, ethylene stimulates or inhibits the growth of seed tissues, thereby inducing the germination of some seeds or suppressing that of other seeds. The effects of ethylene on the various processes of cell metabolism (i.e., respiratory flux, amino acid accumulation, etc.) are intriguing and certainly warrant further investigation. However, at present, the mechanism of ethylene action in germination remains unsolved.

REFERENCES

1. **Abeles, F. B.,** *Ethylene in Plant Biology,* Academic Press, New York, 1972, chap. 6.
2. **Ketring, D. L.,** Ethylene and seed germination, in *The Physiology and Biochemistry of Seed Dormancy and Germination,* Khan, A. A., Ed., North-Holland, Amsterdam, 1977, 157.
3. **Hall, M. A., Acaster, M. A., Bengochea, T., Doddas, J. H., Evans, D. E., Jones, J. F., Jerie, P. H., Mutumba, G. C., Niepel, B., and Shaari, A. R.,** Ethylene and seeds, in *Plant Growth Substances 1979,* Skoog, F., Ed., Springer-Verlag, Berlin, 1980, 199.
4. **Bewley, J. D. and Black, M.,** *Physiology and Biochemistry of Seeds in Relation to Germination, 2,* Springer-Verlag, Berlin, 1982, chap. 3 and 4.
5. **Ross, J. D.,** Metabolic aspects of dormancy, in *Seed Physiology,* Vol. 2, Academic Press, New York, 1984, 45.
6. **De Greef, J. A. and De Proft, M.,** Case studies of ethylene release in higher and lower plant systems, in *Aspects and Prospects of Plant Growth Regulators,* Jeffcoat, B., Ed., British Plant Growth Regulator Group, Wantage, England, 1981, 9.
7. **Hoffman, N. E., Fu, J.-R., and Yang, S. F.,** Identification and metabolism of 1-(malonylamino)cyclopropane-1-carboxylic acid in germinating peanut seeds, *Plant Physiol.,* 71, 197, 1983.
8. **Hilton, J. R.,** An unusual effect of the far-red absorbing form of phytochrome: photoinhibition of seed germination in *Bromus sterilis* L., *Planta,* 155, 524, 1982.
9. **Okagami, N. and Kawai, M.,** Dormancy in *Dioscorea:* gibberellin-induced inhibition or promotion of *D. tokoro* and *D. tenuipes* in relation to light quality, *Plant Physiol.,* 60, 360, 1977.
10. **Olatoye, S. T. and Hall, M. A.,** Interaction of ethylene and light on dormant weed seeds, in *Seed Ecology,* Heydecher, W., Ed., Butterworths, London, 1972, 233.
11. **Taylorson, R. B.,** Response of weed seeds to ethylene and related hydrocarbons, *Weed Sci.,* 27, 7, 1979.
12. **Halmer, P., Bewley, J. D., and Thorpe, T. A.,** An enzyme to degrade lettuce endosperm cell walls. Appearance of mannanase following phytochrome- and gibberellin-induced germination, *Planta,* 130, 189, 1976.
13. **Halmer, P. and Bewley, J. D.,** Mannanase production by the lettuce endosperm: control by the embryo, *Planta,* 144, 333, 1979.
14. **Jacobsen, J. V., Pressman, E., and Pyliotis, N. A.,** Gibberellin-induced separation of cells in isolated endosperm of celery seeds, *Planta,* 129, 113, 1976.
15. **Abeles, F. B.,** Role of ethylene in *Lactuca sativa* cv 'Grand Rapids' seed germination, *Plant Physiol.,* 81, 780, 1986.
16. **Esashi, Y. and Leopold, A. C.,** Physical force in dormancy and germination of *Xanthium* seeds, *Plant Physiol.,* 43, 871, 1968.
17. **Borno, C. and Taylor, I. E. P.,** The effect of high concentrations of ethylene on seed germination of Douglas fir (Pseudotsuga menziesii (Mirb.) Franco), *Can. J. Forest Res.,* 5, 419, 1975.
18. **Nakamura, S. J., Uchida, T., and Hamada, M.,** Atmospheric condition controlling the seed germination of an achlorophyllous orchid, *Galeola septentrionalis, Bot. Mag. Tokyo,* 88, 103, 1975.
19. **Brennan, T., Willemsen, R., Rudd, T., and Frenkel, C.,** Interaction of oxygen and ethylene in the release of ragweed seeds from dormancy, *Bot. Gaz.,* 139, 46, 1978.
20. **Egley, G. H.,** Stimulation of common cocklebur (*Xanthium pennsylvanicum*) and redroot pigweed (*Amaranthus retroflexus*) seed germination by injections of ethylene into soil, *J. Weed Sci.,* 28, 510, 1980.
21. **Srivastava, A. K. and Dey, S. C.,** Physiology of seed dormancy in sunflower, *Acta Agron. Acad. Sci. Hung.,* 31, 70, 1982.
22. **Bebawi, F. F. and Eplee, R. E.,** Efficacy of ethylene as a germination stimulant of *Striga bermonthica* seed, *Weed Sci.,* 34, 694, 1986.
23. **Kepczyński, J., Rudnicki, R. M., and Khan, A. A.,** Ethylene requirement for germination of partly after-ripened apple embryo, *Physiol. Plant.,* 40, 292, 1977.
24. **Sińska, I. and Gladon, R. J.,** Ethylene and the removal of embryonal apple seed dormancy, *HortScience,* 19, 73, 1984.
25. **Samimy, C. and Khan, A. A.,** Secondary dormancy, growth-regulator effects, and embryo growth potential in curly dock (*Rumex crispus*) seeds, *Weed Sci.,* 31, 153, 1983.
26. **Katoh, H. and Esashi, Y.,** Dormancy and impotency of cocklebur seeds. I. CO_2, C_2H_4, O_2 and high temperature, *Plant Cell Physiol.,* 16, 687, 1975.
27. **Suzuki, S. and Taylorson, R. B.,** Ethylene inhibition of phytochrome-induced germination in *Potentilla norvegica* L. seeds, *Plant Physiol.,* 68, 1385, 1981.
28. **Joshi, R. K., Mishra, S. D., and Gaur, B. K.,** Release of dormancy in trombay groundnut seeds by plant growth regulators, *Acta Bot. Indica,* 6, 7, 1978.
29. **Esashi, Y., Okazaki, M., Yanai, N., and Hishinuma, K.,** Control of the germination of secondary dormant cocklebur seeds by various germination stimulants, *Plant Cell Physiol.,* 19, 1497, 1978.

30. **Esashi, Y., Saijoh, S., Ishida, S., Oota, H., and Ishizawa, K.,** Reversal of ethylene action on cocklebur seed germination in relation to duration of pre-treatment soaking and temperature, *Plant Cell Environ.*, 9, 121, 1986.
31. **Rudnicki, R. M., Braun, J. W., and Khan, A. A.,** Low pressure and ethylene seed germination, *Physiol. Plant.*, 43, 189, 1978.
32. **Satoh, S. and Esashi, Y.,** Ethylene production, 1-aminocyclopropane-1-carboxylic acid content and its conversion to ethylene in axial segments of dormant and nondormant cocklebur seeds, *Plant Cell Physiol.*, 24, 883, 1983.
33. **Satoh, S., Takeda, Y., and Esashi, Y.,** Dormancy and impotency of cocklebur seeds. IX. Changes in ACC-ethylene conversion activity and ACC content of dormant and nondormant seeds during soaking, *J. Exp. Bot.*, 35, 1515, 1984.
34. **Satoh, S. and Esashi, Y.,** Changes in ethylene production and in 1-aminocyclopropane-1-carboxylic acid (ACC) and malonyl-ACC contents of cocklebur seeds during their development, *Plant Cell Physiol.*, 25, 1277, 1984.
35. **Fu, J. R. and Yang, S. F.,** Release of heat pretreatment-induced dormancy in lettuce seeds by ethylene or cytokinin in relation to the production of ethylene and the synthesis of 1-aminocyclopropane-1-carboxylic acid during germination, *J. Plant Growth Regul.*, 2, 185, 1983.
36. **Kepczyński, J. and Karssen, C. M.,** Requirement for the action of endogenous ethylene during germination of non-dormant seeds of *Amaranthus caudatus*, *Physiol. Plant.*, 63, 49, 1985.
37. **Kepczyński, J.,** Inhibition of *Amaranthus caudatus* seed germination by polyethylene glycol-6000 and abscisic acid and its reversal by ethylene or 1-aminocyclopropane-1-carboxylic acid, *Physiol. Plant.*, 67, 588, 1986.
38. **Takayanagi, K. and Harrington, J. F.,** Enhancement of germination rate of aged seeds by ethylene, *Plant Physiol.*, 47, 521, 1971.
39. **Meheriuk, M. and Spencer, M.,** Ethylene production during germination of oat seeds and *Penicillium digitatum* spores, *Can. J. Bot.*, 42, 337, 1964.
40. **Ketring, D. L., Morgan, P. W., and Powell, R. D.,** Relation on ethylene production to germinability and growth of two oil seeds, cotton and peanut, in *Plant Growth Substances 1973*, Sumiki, Y., Ed., Hirokawa, Tokyo, 1974, 891.
41. **Spencer, M. and Olson, A.,** Ethylene production and lipid mobilization during germination of castor beans, *Nature*, 205, 699, 1965.
42. **Adkins, S. W. and Ross, J. D.,** Studies in wild oat seed dormancy. I. The role of ethylene in dormancy breaking and germination of wild oat seeds, *Plant Physiol.*, 67, 358, 1981.
43. **Saini, H. S., Consolacion, E. D., Bassi, P. K., and Spencer, M. S.,** Requirement for ethylene synthesis and action during relief of thermoinhibition of lettuce seed germination by combinations of gibberellic acid, kinetin, and carbon dioxide, *Plant Physiol.*, 81, 950, 1986.
44. **Ketring, D. L. and Morgan, P. W.,** Physiology of oil seeds. IV. Role of endogenous ethylene and inhibitory regulators during natural and induced after-ripening of dormant Virginia-type peanut seeds, *Plant Physiol.*, 50, 382, 1972.
45. **Burdett, A. N.,** Antagonistic effects of high and low temperature pretreatments on the germination and pre-germination ethylene synthesis of lettuce seeds, *Plant Physiol.*, 50, 201, 1972.
46. **Burdett, A. N.,** Ethylene synthesis in lettuce seeds: its physiological significance, *Plant Physiol.*, 50, 719, 1972.
47. **Negm, F. B., Smith, O. E., and Kumamoto, J.,** The role of phytochrome in an interaction with ethylene and carbon dioxide in overcoming lettuce seed thermodormancy, *Plant Physiol.*, 51, 1089, 1973.
48. **Dunlap, J. R. and Morgan, P. W.,** Reversal of induced dormancy in lettuce by ethylene, kinetin, and gibberellic acid, *Plant Physiol.*, 60, 222, 1977.
49. **Khan, A. A.,** Hormonal regulation of primary and secondary seed dormancy, *Isr. J. Bot.*, 29, 207, 1980/81.
50. **Rao, V. S., Sankhla, N., and Khan, A. A.,** Additive and synergistic effects of kinetin and ethylene on germination, thermodormancy, and polyribosome formation in lettuce seeds, *Plant Physiol.*, 56, 263, 1975.
51. **Esashi, Y. and Leopold, A. C.,** Dormancy regulation in subterranean clover seeds by ethylene, *Plant Physiol.*, 44, 1470, 1969.
52. **Esashi, Y., Saijoh, H., Sitoh, S., Ishida, S., and Satoh, S.,** CO_2 enhancement of aerobic and hypoxic C_2H_4 production in the tissues of cocklebur seeds at different dormant states, *Aust. J. Plant Physiol.*, 12, 59, 1985.
53. **Esashi, Y., Ooshima, Y., Abe, M., Kurota, A., and Satoh, S.,** CO_2-enhanced C_2H_4 production in tissues of imbibed cocklebur seeds, *Aust. J. Plant Physiol.*, 13, 417, 1986.
54. **Thompson, P. A.,** Geographical adaptation of seeds, in *Seed Ecology*, Heydecker, W., Eds., Butterworths, London, 1972, 31.
55. **Esashi, Y., Ishihara, N., Kuraishi, R., and Kodama, H.,** Light actions in the germination of cocklebur seeds. I. Differences in the light responses of the upper and lower seeds, *J. Exp. Bot.*, 34, 903, 1983.

56. **Negm, F. B. and Smith, O. E.,** Effects of ethylene and carbon dioxide on the germination of osmotically inhibited lettuce seeds, *Plant Physiol.*, 62, 473, 1978.
57. **Burdett, A. N. and Vidaver, W. E.,** Synergistic action of ethylene with gibberellin or red light in germinating lettuce seeds, *Plant Physiol.*, 48, 656, 1971.
58. **Speer, H. L., Hsiao, A. I., and Vidaver, W.,** Effects of germination-promoting substances given in conjunction with red light on the phytochrome-mediated germination of dormant lettuce seeds (*Lactuca sativa* L.), *Plant Physiol.*, 54, 852, 1974.
59. **Abeles, F. B. and Lonski, J.,** Stimulation of lettuce seed germination by ethylene, *Plant Physiol.*, 44, 277, 1969.
60. **Saini, H. S., Bassi, P. K., Consolacion, E. D., and Spencer, M. S.,** Interactions among plant hormones, carbon dioxide, and light in the relief of thermoinhibition of lettuce seed germination: studies in a flow-through gaseous system, *Can. J. Bot.*, 64, 2322, 1986.
61. **Esashi, Y., Hase, S., and Kojima, K.,** Light actions in the germination of cocklebur seeds. V. Effects of ethylene, carbon dioxide and oxygen on germination in relation to light, *J. Exp. Bot.*, 38, 702, 1987.
62. **Van Staden, J., Olatoye, S. T., and Hall, M. A.,** Effect of light and ethylene upon cytokinin levels in seeds of *Spergula arvensis*, *J. Exp. Bot.*, 24, 662, 1973.
63. **Hecker, M. and Bernhardt, D.,** Untersuchungen uber Protein- und RNA-Synthese in dormanten und uachgereiften Samen von *Melandrium noctiflorum*, *Biochem. Physiol. Pflanz.*, 169, 417, 1976.
64. **Jones, J. E. and Hall, M. A.,** Studies on the requirement for carbon dioxide and ethylene for germination of *Spergula arvensis* seeds, *Plant Sci. Lett.*, 16, 87, 1984.
65. **Keys, R. D., Smith, O. E., Kumamoto, J., and Lyon, J. L.,** Effect of gibberellic acid, kinetin, and ethylene plus carbon dioxide on the thermodormancy of lettuce seed (*Lactuca sativa* L. cv. Mesa 659), *Plant Physiol.*, 56, 826, 1975.
66. **Thomas, T. H., Palevitch, D., Biddington, N. L., and Austin, R. B.,** Growth regulators and phytochrome-mediated dormancy of celery seeds, *Physiol. Plant*, 35, 101, 1975.
67. **Schonbeck, M. W. and Egley, G. H.,** Effects of temperature, water potential, and light on germination responses of redroot pigweed seeds to ethylene, *Plant Physiol.*, 65, 1149, 1980.
68. **Schonbeck, M. W. and Egley, G. H.,** Phase-sequence of redroot pigweed germination responses to ethylene and other stimuli, *Plant Physiol.*, 68, 175, 1981.
69. **Bassi, P. K., Tregunna, E. B., and Purohit, A. N.,** Carbon dioxide requirements for phytochrome action in photoperiodism and seed germination, *Plant Physiol.*, 56, 335, 1975.
70. **Negm, F. B., Smith, O. E., and Kumamoto, J.,** Interaction of carbon dioxide and ethylene in overcoming thermodormancy of lettuce seeds, *Plant Physiol.*, 49, 869, 1972.
71. **Esashi, Y., Komatsu, H., Ushizawa, R., and Sakai, Y.,** Breaking of secondary dormancy in cocklebur seeds by cyanide and azide in combination with C_2H_4 and O_2 and their effects on cytochrome and alternative respiratory pathways, *Aust. J. Plant Physiol.*, 9, 97, 1982.
72. **Esashi, Y., Fuwa, N., Kurota, A., Oota, H., and Abe, M.,** Interrelation between ethylene and carbon dioxide in relation to respiration and adenylate content in the pre-germination period of cocklebur seeds, *Plant Cell Physiol.*, 28, 141, 1987.
73. **Palevitch, D. and Thomas, T. H.,** Thermodormancy release of celery seed by gibberellins, 6-benzylaminopurine, and ethephon applied in organic solvent to dry seeds, *J. Exp. Bot.*, 25, 981, 1974.
74. **Schonbeck, M. W. and Egley, G. H.,** Redroot pigweed (*Amaranthus retroflexus*) seed germination responses to afterripening, temperature, ethylene, and some other environmental factors, *J. Weed Sci.*, 28, 543, 1980.
75. **Schonbeck, M. W. and Egley, G. H.,** Changes in sensitivity of *Amaranthus retroflexus* L. seeds to ethylene during preincubation. I. Constant temperatures, *Plant Cell Environ.*, 4, 229, 1981.
76. **Khan, A. A. and Karssen, C. M.,** Induction of secondary dormancy in *Chenopodium bonushenricus* L. seeds by osmotic and high temperature treatments and its prevention by light and growth regulators, *Plant Physiol.*, 66, 175, 1980.
77. **Corbineau, F., Rudnicki, R. M., and Côme, D.,** Induction of secondary dormancy in sunflower seeds by high temperature. Possible involvement of ethylene biosynthesis, *Physiol. Plant.*, 73, 368, 1988.
78. **Kepczyński, J. and Rudnicki, R. M.,** Studies on ethylene in dormancy of seeds. II. The effects of endogenous ethylene on after-ripening of apple seeds in fruits stored at low temperature, *Fruit Sci. Rep.*, 3, 27, 1976.
79. **Kepczyński, J. and Rudnicki, R. M.,** Effect of exogenous ethylene on the after-ripening and germination of apple seeds, *Fruit Sci. Rep.*, 2, 25, 1975.
80. **Esashi, Y., Kawabe, K., Isuzugawa, K., and Ishizawa, K.,** Interrelation between carbon dioxide and ethylene on the stimulation of cocklebur seed germination, *Plant Physiol.*, 86, 39, 1988.
81. **Karssen, C. M.,** Two sites of hormonal action during germination of *Chenopodium album* seeds, *Physiol. Plant.*, 36, 264, 1976.
82. **Schonbeck, M. W. and Egley, G. H.,** Changes in sensitivity of *Amaranthus retroflexus* L. seeds to ethylene during preincubation. II. Effects of alternating temperature and burial in soil, *Plant Cell Environ.*, 4, 237, 1981.

83. **Stewart, E. R. and Freebairn, H. T.**, Ethylene, seed germination, and epinasty, *Plant Physiol.*, 44, 955, 1969.
84. **Ketring, D. L. and Morgan, P. W.**, Physiology of oil seeds. II. Dormancy release in Virginia-type peanut seeds by plant growth regulators, *Plant Physiol.*, 47, 488, 1971.
85. **Ketring, D. L. and Morgan, P. W.**, Physiology of oil seeds. I. Regulation of dormancy in Virginia-type peanut seeds, *Plant Physiol.*, 45, 268, 1970.
86. **Kepczyński, J.**, Ethylene-dependent action of gibberellin in seed germination of *Amaranthus caudatus*, *Physiol. Plant.*, 67, 584, 1986.
87. **Kepczyński, J., Kepczyńska, E., and Knypl, J. S.**, Effects of gibberellic acid, ethephon, and 1-aminocyclopropane-1-carboxylic acid on germination of *Amaranthus caudatus* seeds imbibed by paclobutrazol, *J. Plant Growth Regul.*, 7, 59, 1988.
88. **Esashi, Y., Hata, Y., and Katoh, H.**, Germination of cocklebur seeds: interactions between gibberellic acid, benzyladenine, thiourea, KNO_3 and gaseous factors, *Aust. J. Plant Physiol.*, 2, 569, 1975.
89. **Tao, K.-L., McDonald, M. B., and Khan, A. A.**, Synergistic and additive effects of kinetin and ethrel on the release of seed dormancy, *Life Sci.*, 15, 1925, 1975.
90. **Khan, A. A. and Huang, X-L.**, Synergistic enhancement of ethylene production and germination with kinetin and 1-aminocyclopropane-1-carboxylic acid in lettuce seeds exposed to salinity stress, *Plant Physiol.*, 87, 847, 1988.
91. **Halloin, J. M.**, Inhibition of cotton seed germination with abscisic acid and its reversal, *Plant Physiol.*, 57, 454, 1976.
92. **Kumar, M. U. and Sastry, K. S. K.**, Effect of exogenous application of growth regulators on germinating ability of developing sunflower seeds, *Indian J. Exp. Biol.*, 12, 543, 1974.
93. **Esashi, Y., Abe, Y., Ashino, H., Ishizawa, K., and Saitoh, K.**, Germination of cocklebur seeds and growth of their axial and cotyledonary tissues in response to C_2H_4, CO_2 and/or O_2 under water stress, *Plant Cell Environ.*, 12, 183, 1989.
94. **Egley, G. H.**, Ethylene, nitrate and nitrite interactions in the promotion of dark germination of common purslane seeds, *Ann. Bot.*, 53, 833, 1984.
95. **Saini, H. S., Bassi, P. K., and Spencer, M. S.**, Seed germination in *Chenopodium album* L.: relationships between nitrate and the effects of plant hormones, *Plant Physiol.*, 77, 940, 1983.
96. **Saini, H. S., Bassi, P. K., and Spencer, M. S.**, Seed germination in *Chenopodium album* L.: further evidence for the dependence of the effects of growth regulators on nitrate availability, *Plant Cell Environ.*, 8, 707, 1985.
97. **Saini, H. S., Bassi, P. K., and Spencer, M. S.**, Use of ethylene and nitrate to break seed dormancy of common lambsquarters *(Chenopodium album)*, *Weed Sci.*, 34, 502, 1986.
98. **Saini, H. S., Bassi, P. K., and Spencer, M. S.**, Interactions among ethephon, nitrate, and after-ripening in the release of dormancy of wild oat *(Avena fatua)* seed, *Weed Sci.*, 34, 43, 1985.
99. **Sankhla, N., Harsh, G. D., Vyas, O. P., Bohra, S. P., and Sankhla, D.**, Lettuce seed germination: counteraction of inhibition caused by morphactin and 2,3,5-triiodobenzoic acid by ethrel, *Isr. J. Bot.*, 22, 116, 1973.
100. **Katoh, H. and Esashi, Y.**, Dormancy and impotency of cocklebur seeds. II. Phase sequence in germination process, *Plant Cell Physiol.*, 16, 697, 1975.
101. **Esashi, Y., Kotaki, K., and Ohhara, Y.**, Induction of cocklebur seed germination by anaerobiosis: a question about the "inhibitor hypothesis" of seed dormancy, *Planta*, 129, 109, 1976.
102. **Esashi, Y., Okazaki, M., and Watanabe, K.**, The role of C_2H_4 in anaerobic induction of cocklebur seed germination, *Plant Cell Physiol.*, 17, 1151, 1976.
103. **Esashi, Y., Ohhara, Y., Kotaki, K., and Watanabe, K.**, Two C_2H_4-producing systems in cocklebur seeds, *Planta*, 129, 23, 1976.
104. **Esashi, Y. and Katoh, H.**, Dormancy and impotency of cocklebur seeds. III. CO_2- and C_2H_4-dependent growth of the embryonic axis and cotyledon segments, *Plant Cell Physiol.*, 16, 707, 1975.
105. **Esashi, Y., Katoh, H., and Leopold, A. C.**, Dormancy and impotency of cocklebur seeds. IV. Effects of gibberellic acid, benzyladenine, thiourea, and potassium nitrate on the growth of embryonic axis and cotyledon segments, *Plant Physiol.*, 59, 117, 1977.
106. **Green, J.**, The effect of potassium and calcium on cotyledon expansion and ethylene induced by cytokinins, *Physiol. Plant.*, 57, 57, 1983.
107. **Rijven, A. H. G. C.**, Ethylene and carbon dioxide in the growth regulation of isolated cotyledons of fenugreek (*Trigonella foenumgraecum* L.) in darkness, *Plant Sci. Lett.*, 2, 55, 1974.
108. **Esashi, Y., Kuraishi, R., Tanaka, N., and Satoh, S.**, Transition from primary dormancy to secondary dormancy in cocklebur seeds, *Plant Cell Environ.*, 6, 493, 1983.
109. **Esashi, Y., Ohhara, Y., Okazaki, M., and Hishinuma, K.**, Control of cocklebur seed germination by nitrogenous compounds: nitrate, nitrite, hydroxylamine, thiourea, azide and cyanide, *Plant Cell Physiol.*, 20, 349, 1979.
110. **Eplee, R. E.**, Ethylene: a witchweed seed germination stimulant, *Weed Sci.*, 23, 433, 1975.

111. **Esashi, Y., Watanabe, K., Ohhara, Y., Katoh, H.,** Dormancy and impotency of cocklebur seeds. V. Growth and ethylene production of axial segments in response to O_2, *Aust. J. Plant Physiol.*, 3, 701, 1976.
112. **Esashi, Y., Kotaki, K., and Ishidoya, O.,** Dormancy and impotency of cocklebur seeds. VI. Growth and C_2H_4 production of cotyledonary segments in response to O_2, *Aust. J. Plant Physiol.*, 3, 711, 1976.
113. **Ishizawa, K., Hoshina, M., Kawabe, K., and Esashi, Y.,** Effects of 2,5-norbornadiene on cocklebur seed germination and rice coleoptile elongation in response to CO_2 and C_2H_4, *Plant Growth Regul.*, 7, 45, 1988.
114. **Goto, N. and Esashi, Y.,** Stimulation by ethylene of axis and hypocotyl growth in bean and cocklebur seedlings, *Physiol. Plant.*, 31, 204, 1974.
115. **Goto, N. and Esashi, Y.,** Regulation of hypocotyl growth by ethylene, in *Proc. 8th Int. Conf. Plant Growth Substances,* Sumiki, Y., Ed., Hirokawa, Tokyo, 1974, 853.
116. **Esashi, Y., Saitoh, H., Saijoh, Y., Ishida, S., and Kodama, H.,** Light actions in the germination of cocklebur seeds. II. Possible origin of the diversity of light response in seed germination, *Plant Cell Physiol.*, 26, 361, 1985.
117. **Esashi, Y., Hase, S., Kojima, K., and Fuwa, N.,** Differential responsiveness in zonal growth of cocklebur seed axes to CO_2, C_2H_4, O_2 and respiration inhibitors, *Plant Cell Physiol.*, 28, 163, 1987.
118. **Beyer, E. M., Jr.,** A potent inhibitor of ethylene action in plants, *Plant Physiol.*, 58, 268, 1976.
119. **Van Daalen, J. J. and Daams, J.,** Substituted alkoxycarbonyl-methoxy-2,1,3-benzothiadizoles. A new group of synthetic plant growth regulators, *Naturwissenschaften,* 8, 395, 1970.
120. **Sisler, E. C. and Pian, A.,** Effect of ethylene and cyclic olefins on tobacco leaves, *Tob. Sci.*, 17, 68, 1973.
121. **Esashi, Y., Wakabayashi, S., Tsukada, Y., and Satoh, S.,** Possible involvement of the alternative respiration system in the ethylene-stimulated germination of cocklebur seeds, *Plant Physiol.*, 63, 1039, 1979.
122. **Perl, M.,** ATP accumulation in peanut seeds during seed-ripening and during the dormancy-breaking process, *J. Exp. Bot.*, 33, 449, 1982.
123. **Perl, M.,** The effect of ethylene and temperature on ATP accumulation from various substances in peanut *(Arachis hypogaea* L.) seeds, *J. Exp. Bot.*, 33, 456, 1982.
124. **Perl, M.,** ATP synthesis in cell-free extracts of peanut (*Arachis hypogaea* L.) seeds, *J. Exp. Bot.*, 33, 463, 1982.
125. **Jones, J. F. and Hall, M. A.,** The effect of ethylene on quantitative and qualitative aspects of respiration during the breaking of dormancy of *Spergula arvensis* L. seeds, *Ann. Bot.*, 48, 291, 1981.
126. **Schonbaum, G. R., Bonner, W. D., Jr., Strey, B. T., and Bahr, J. T.,** Specific inhibition of the cyanide-insensitive respiratory pathway in plant mitochondria by hydroxamic acids, *Plant Physiol.*, 47, 124, 1971.
127. **Esashi, Y., Komatsu, H., Ushizawa, R., and Sakai, Y.,** Breaking of secondary dormancy in cocklebur seeds by cyanide and azide in combination with ethylene and oxygen, and their effects on cytochrome and alternative respiratory pathways, *Aust. J. Plant Physiol.*, 9, 97, 1982.
128. **Esashi, Y., Kurota, A., Abe, M., and Fuwa, N.,** Further study on the involvement of cyanide-resistant respiration in the germination of cocklebur seeds, *Plant Cell Physiol.*, 28, 151, 1987.
129. **Hegarty, T. W.,** The physiology of seed hydration and dehydration, and the relation between water stress and the control of germination: a review, *Plant Cell Envrion.*, 1, 101, 1978.
130. **Esashi, Y. and Ishizawa, K.,** Oxygen-independent ethylene action in cocklebur seeds germination in relation to osmoregulation, in *Biochemical and Physiological Aspects of Ethylene Production in Lower and Higher Plants,* Clijsters, H., DeProft, M., Marcelle, R., and Van Poucke, M., Eds., Kluwer Academic Publishers, Dordrecht, 1989, 73.
131. **Esashi, Y., Matsuyama, S., Hoshina, M., Ashino, H., and Ishizawa, K.,** Mechanism of action of ethylene in promoting the germination of cocklebur seeds. I. Osmoregulation, *Aust. J. Plant Physiol.*, 17, 537, 1990.
132. **Cairns, A. L. P. and de Villiers, O. T.,** Breaking dormancy of *Avena fatua* L. seed by treatment with ammonia, *Weed Res.*, 26, 191, 1986.

Chapter 9

ETHYLENE IN ROOT GROWTH AND DEVELOPMENT

Michael B. Jackson

TABLE OF CONTENTS

I. INTRODUCTION

The evolution of roots approximately 400 million years ago opened up the biological colonization of the land. Today, roots remain responsible for extracting from soil most of the water and essential mineral elements utilized by autotrophic vascular plants, and through them, the inorganic nutrients needed by almost all of the terrestrial biosphere.[1] To effect these feats of extraction, the root system of a growing plant continually expands to exploit an ever increasing volume of soil, thereby escaping local depletions of soluble elements or of water. In so doing, roots come to constitute a large, if somewhat declining, proportion of total plant weight as the plant matures.[2] A week after germination, the root system of a barley plant (*Hordeum vulgare*) grown in nutrient solution may already weigh 470 mg (dry weight), comprise approximately 40% of the whole plant, be 50 m in total length and bear 2000 laterals.[3] A four-month-old plant of winter rye (*Secale cereale*) has been estimated by Dittmer[4] to possess over 13 million individual roots with a total length of 620 km, mostly as laterals at the quaternary level of branching. These roots present approximately 237 m^2 of potentially absorbing surface to the soil. If root hairs are included, this estimated surface area increases to 639 m^2. A myriad of individual apical meristems, growing zones and sites of new root or root hair formation are involved in achieving this impressive permeation of the soil. Together, they construct a three-dimensional root system with an internal and external morphology that is genus- and probably species-specific[5] and thus genetically regulated. Within these genetic constraints, root systems exhibit a remarkable degree of developmental plasticity[6] in the face of a complex and variable environment of solid, liquid, and gaseous components combined in various proportions, that teems with microorganisms.

The development of root systems is thus an elaborate and highly regulated process of impressive proportions that is responsive to environment and is central to the overall well being of plants and to ensuring heavy crop yields.[7] The development of roots is at least as sensitive to ethylene as that of other plant parts, and root tissues produce the gas in amounts that often reflect conditions at the root-soil interface. Furthermore, roots grow in a medium that can entrap ethylene, produce it, or oxidize it, depending on the circumstances. Ethylene may, therefore, be especially influential in fashioning the root system. This chapter examines the evidence for, and against, this point of view, with emphasis on morphology and anatomy. The approach taken is critical and analytical. The reader should thus be able to appreciate the reasoning behind the conclusions drawn, or readily construct an alternative interpretation of the evidence presented. Other chapters will utilize findings from research with roots that have increased our understanding of the biochemical or molecular biological basis of ethylene formation or action.

II. NEW ROOT FORMATION

A. ADVENTITIOUS ROOTING

The emergence of roots from tissue other than preexisting roots (i.e., adventitious rooting) is the culmination of a sequence of events, that commences with de-differentiation of particular cell clusters prior to organized cell divisions to form primordia, and completed by the outgrowth of these primordia to give visible roots. Spectacular promotion of rooting by supplying ethylene to whole plants (see Abeles,[8] page 129 for a photograph) was demonstrated in 15 species or cultivars over 50 years ago by Zimmerman and Hitchcock working at the Boyce Thompson Institute.[9] The same laboratory also reported that a range of naturally occurring and synthetic auxins stimulate both rooting and also ethylene evolution, as detected by bioassay.[10] These findings prompted the suggestion, made first by the Boyce Thompson workers, that endogenous ethylene is a natural regulator of root primordia formation or outgrowth and that the gas participates in auxin-promoted root formation.[11] These provocative ideas and their implications were largely ignored until the 1970s, partly because of prompt and largely unwarranted criticism of these ideas by Went and Thimann in their important book *Phytohormones* published in 1937.[12]

Unfortunately, research since about 1970 cannot be said to have dealt satisfactorily with the leading questions. These concern where applied ethylene acts in the catenary sequence leading to successful root production, whether endogenous ethylene functions as a regulator of rooting, and indeed if it has any part to play in root formation induced by either endogenous or externally applied auxin. The most detailed work has been with a convenient test rooting system[13] comprising the hypocotyl of light-grown mung beans (*Vigna radiata*), usually with the first leaves attached. DNA replication at primordial initiation sites in the hypocotyl begins within 12 h, the first cell divisions occur within 1 d and roots emerge from the hypocotyl within 5 to 7 d. Supplying ethylene gas can increase the number of roots formed in this system. The careful time-course study of Robbins et al.[14] shows that speed of primordia formation is unchanged by the gas but the number of primordia increases (from 19 to 45 per cutting in 6 $\mu l\ l^{-1}$ ethylene, given for 4 d). This suggests ethylene treatment can influence the earliest stages of initiation. Surprisingly high ethylene concentrations (15 to 30 $\mu l\ l^{-1}$) were needed for maximum effect. Results from using ethephon to enrich the system with ethylene have been contradictory, with stimulation,[15-17] suppression,[18] or no effect[19,20] on root number being reported by different laboratories. The uncertainty is increased by the absence of any marked promotion of rooting in mung bean cuttings given the ethylene precursor 1-aminocyclopropane-1-carboxylic acid (ACC).[21,22] Testing the involvement of endogenous ethylene by using inhibitors of ethylene biosynthesis (predominantly amino-ethoxyvinylglycine — AVG) has also failed to yield consistent or reliable results, probably because dose-response tests have not been made to establish the concentrations that are small enough to inhibit ethylene formation while minimizing nonspecific and possibly toxic effects. In several papers where AVG is reported to inhibit rooting,[17,18,23] extension of new roots was also severely repressed, while ACC treatments failed to reverse the inhibiting influence of AVG on root formation. This suggests AVG was acting in ways unconnected with its ability to suppress ethylene biosynthesis, a problem also encountered in research into somatic embryogenesis.[24] Little use has been made of inhibitors of ethylene action to examine ethylene-promoted rooting. Robbins et al.[14] found silver thiosulfate to be without effect on rooting of mung beans, while the volatile competitive inhibitor norbornadiene (NBD) suppresses rooting. Unfortunately, in this work the specificity of NBD action was not tested by attempting to reverse its effect with concomitant applications of ethylene.

Many experiments have explored the relationship between ethylene and the stimulation of rooting that results from auxin applications to mung beans. Since indole butyric acid (IBA) can promote rooting at concentrations that do not change the production of endogenous ethylene,[19] the necessary involvement of ethylene is in considerable doubt. Indeed, the additional ethylene formed when higher concentrations of auxin are given has been claimed to depress rooting.[18] Work with excised, dark-grown mung bean hypocotyls without leaves, and studies with more conventional light-grown cuttings have failed to reveal any obvious quantitative relationship between the contrasting abilities of different auxins to stimulate root formation and the extent to which they enhance ethylene production or ACC accumulation.[20,22] The inevitable conclusion must be that the role of endogenous ethylene in adventitious rooting in mung bean in the presence or absence of exogenous auxin still remains to be established.

The uncertainty concerning ethylene involvement is compounded by results with species other than mung bean. Submergence of hardwood cuttings, especially willow, in water, to encourage the accumulation of endogenous ethylene, or treatment of such cuttings with ethylene gas at high concentrations, or with ethephon, promote rooting by enhancing the outgrowth of preexisting primordia,[25] an effect also observed in whole plants of maize (*Zea mays*),[26] rice (*Oryza sativa*),[27] and marigold (*Tagetes erecta*).[28] In rice, the effect may require gibberellins, since inhibitors of gibberellin biosynthesis interfere with the efficacy of ethylene.[27] Linkins et al.[29] concluded that in auxin-treated *Phaseolus vulgaris* hypocotyls, ethylene was a prerequisite for primordial initiation rather than outgrowth because absorbing ethylene from the atmosphere surrounding plants enclosed *in vitro* prevented root formation. However, this finding must be set against their observation that ethylene applied alone to this plant does not promote rooting. Comprehensive studies by Geneve (cited by Hackett[30]), using ethylene treatments, inhibitors of ethylene action, or biosynthesis, led to a rejection of the hypothesis that ethylene is in any way causally involved in rooting of adult and juvenile ivy (*Hedera helix*) or in the promoting effects of auxin on root formation in juvenile ivy petioles. In contrast, a positive response to applied ethylene has been obtained with *Chrysanthemum* x *morifolium*,[31] while inhibition rather than promotion of rooting by ethylene application has been reported in sugar cane stems[32] and tomato (*Lycopersicon esculentum*) leaf discs maintained *in vitro*.[33] In the latter work, ethylene seemed to suppress rooting, since absorbing the extra ethylene evolved from discs treated with auxin, or applying silver ions as ethylene antagonists, enhanced root formation. Contrary results of Le Guyader[34] using thin cell layers of tobacco (*Nicotiana tabacum*) show that silver ions interfere with the switch from bud formation to rooting, thus implying a requirement for ethylene in the induction of roots. In cuttings of pea (*Pisum sativum*), small amounts of applied IAA (10 n*M*) inhibit rather than promote root initiation and outgrowth.[35] This rather surprising result was associated with increased ethylene formation, suggesting the gas inhibits rooting in this species. The conclusion was backed up by the finding that ACC, supplied in amounts that raised ethylene production rates to those obtained with 10 n*M* IAA, also inhibited root formation.[35]

Overall, this is a confused picture of conflicting results that can only begin to be resolved with the most careful experimentation in which (1) the amounts of ethylene supplied are accurately controlled in concentration, location, and duration, (2) inhibitors are used only after a thorough assessment of their specificity of action, and (3) responses are scored so as to discriminate between the various stages of root formation. One recurrent shortcoming in much published work is that absence of primordial outgrowth may have sometimes masked effects on initiation. Despite all the contradictory evidence, there remains a strong impression that in appropriate tissues, ethylene can exert a considerable promoting influence on root initiation and outgrowth. The need is to identify precisely the circumstances in which this activity takes place, the stages in root formation that ethylene can influence, and whether the effect is restricted to unnaturally large concentrations of ethylene rather than to the small amounts normally present in plant tissues. Mung bean cuttings may not be the most appropriate material to use and a return to the study of whole plants seems desirable in the first instance, since it is here that ethylene treatments have yielded their most spectacular effects.

B. LATERAL ROOTING

In a review by Torrey[36] of factors regulating lateral rooting, ethylene goes almost unmentioned, reflecting the paucity of experimental information. Radin and Loomis[37] found ethylene at 0.31 μl l^{-1} slightly suppresses lateral rooting in *in vitro* cultures of radish (*Raphanus sativus*), especially in the presence of 10^{-5} *M* IAA. However, the absence of laterals on main root axes of the reputedly ethylene-deficient tomato mutant 'diageotropica', and the resurgence of lateral formation if 0.7 μl l^{-1} ethylene is applied to whole plants, encouraged Zobel[38] to suggest that endogenous ethylene is important for inducing primordia in tomato. However, his evidence is less than straightforward. Subsequent studies have shown that the 'diageotropica' mutant is not especially deficient in ethylene; roots and shoots produce substantial mounts of the gas, both in the presence and absence of applied auxin, although the quantities from roots are approximately 50% smaller than normal.[39] A second difficulty is that applied ethylene does not stimulate lateral rooting through a direct effect on the roots. Instead, it seems to act indirectly through the shoots.[38] This indirect influence of ethylene on lateral roots was demonstrated by grafting a wild-type shoot on to diageotropic roots. This promoted lateral rooting on the mutant root axes,[39] implying that the relative amounts of a factor translocated from shoots to roots determines the extent of lateral initiation in this species. Ethylene given to the shoots of 'diageotropica' plants also stimulates the formation of lateral rooting, implying that the gas stimulates the production of this factor in aerial tissues. More work is needed to confirm and extend this interesting possibility.

Plants expressing hairy-root disease (i.e., the prolific production of lateral roots as a result of infection with the bacterium *Agrobacterium rhizogenes*) may offer opportunities for understanding the regulation of lateral rooting in a range of infectable species. Although T-DNA encoded auxin synthesis genes from the so called Ri plasmids present in the bacterium may be incorporated into the host genome, the extra roots may result more from an enhanced sensitivity to auxin rather than from additional auxin biosynthesis.[41] The consequences of this for endogenous ethylene formation and activity in lateral root initiation deserve examination. Overall, there is little reason to believe that lateral root formation is influenced markedly by ethylene.

III. ROOT EXTENSION

In this section, effects of applied ethylene are examined, together with evidence for the view that endogenous ethylene influences extension in ordinarily growing, i.e., unstressed roots. Approximately 75 years ago, Harvey and Rose[42] reported that exogenous ethylene at high concentrations (100 $\mu l\ l^{-1}$) suppresses root elongation and increases radial expansion. Likewise, the genetically based "overproduction" of endogenous ethylene in the *epi* mutant of tomato engenders a similarly stunted and swollen phenotype.[43] The ethylene-inhibited extension develops in association with ultrastructural changes such as cisternal stacking of rough endoplasmic reticulum, increased ribosome size, and increased cell wall thickness.[44] Minimum concentrations reported to decrease root length range from 0.001 $\mu l\ l^{-1}$ in rye (*Secale cereale*),[45] 0.04 $\mu l\ l^{-1}$ in excised radish roots,[37] 0.1 $\mu l\ l^{-1}$ in maize,[46] to over 1 $\mu l\ l^{-1}$ in rice.[47] The growth inhibition is usually reversible,[47,48] sometimes within a few minutes.[46,49] Species differ in the severity of the inhibition caused by supplying ethylene, with the growth of rice roots being notably less dampened at 1 $\mu l\ l^{-1}$ or more, than oats (*Avena sativa*), barley (*Hordeum vulgaris*), rye (*Secale cereale*), tomato (*Lycopersicon esculentum*), or white mustard (*Sinapis alba*).[47,50] Rice selections or cultivars also differ markedly between themselves in this regard. For example, Smith and Robertson[47] found 20 $\mu l\ l^{-1}$ ethylene inhibited rice cultivar 'Mas 2401' by 50%, but root extension of the cultivar 'OS6' was retarded by only half this amount. There may be some adaptive significance in this considerable resilience to ethylene in rice, since the crop is often grown in flooded conditions that favor ethylene accumulation.[51]

The question of whether the amounts of ethylene produced endogenously by unstressed roots are sufficient to restrain extension has been addressed by supplying roots with inhibitors of ethylene action or production. Acceleration of root extension by such treatments and their reversal with ACC or ethylene would constitute good evidence for believing that endogenous ethylene inhibits elongation. Few such experiments have been done well, since the inhibitors available can be toxic or have effects not connected with ethylene.[52] Consequently, in some published work it is uncertain if the inhibitors act with a useful level of specificity. Silver thiosulfate and 2,5-norbornadiene (NBD) *can* reverse the inhibiting effects of applied ethylene on root extension,[52,53] indicating nontoxic opposition of ethylene action is possible at appropriate doses. However, these inhibitors do not promote the extension of lettuce roots in the absence of added ethylene, and may even inhibit growth slightly, indicating that endogenous ethylene in unstressed roots (i.e., darkened roots growing in moist air) is unlikely to be retarding growth.[52] Similarly, maize roots in the dark and extending in moist air scrubbed free of ethylene do not elongate faster over 36 h in a concentration of NBD, shown by Moss et al.[53] to be capable of reversing the inhibitory influence of 5 μl l^{-1} ethylene. By contrast, Whalen[54] and Whalen and Feldman[46] have claimed that more than 100 μl l^{-1} NBD supplied for 24 h in a slow flow of moist air promotes maize root extension by 5%. However, the effect was apparently not statistically significant.[46]

Applying AVG at concentrations high enough to inhibit ethylene biosynthesis but without inducing toxicity symptoms has failed to give the expected promotion of extension in unstressed maize or lettuce roots,[52-54] again indicating that "basal" ethylene is unlikely to be retarding elongation. An exception to these findings is the work of Dubucq et al.,[56] who reported a 20% stimulation of extension in lentil roots by 10 μ*M* of an aminoethoxy-analog of rhizobitoxine, reputed to inhibit ethylene formation. However, this experiment was made in sealed tubes in which ethylene demonstrably accumulated to artificially high, growth-inhibiting, levels. The rhizobitoxine treatment presumably tempered this accumulation. But such an effect does not reliably inform us of the impact of the smaller amounts of ethylene ordinarily present in roots where gas exchange is unimpeded. Similar shortcomings are extant in the experiments of Mulkey et al.,[57-59] where AVG together with cobalt nitrate (an inhibitor of ACC oxidation to ethylene), promoted root extension in maize from approximately 1.5 mm to 2 mm h^{-1}. However, these roots were submerged in water and supplied with pure oxygen rather than air. Such conditions are likely to result in an artificially high level of ethylene, thus possibly invalidating the system for testing the role of "basal" ethylene. Further uncertainties arise from data of Mulkey et al. that show a growth stimulation by AVG + Co^{2+} being retained for 44 h, while ethylene accumulation merely halved in the first 4 h and reverted to control values by 44 h.[59] There is also a risk that the stimulation of extension by AVG may have been the result of promotion unconnected with its action as inhibitor of ethylene biosynthesis. As already mentioned, such an effect has been demonstrated in lettuce roots by Abeles and Wydoski.[52] This salutary finding emphasizes the importance of relying only on experiments in which thorough testing of the specificity of action of any inhibitors used is undertaken.

One obvious reason for the overall lack of convincing evidence that elongation in unstressed roots is stimulated by inhibitors of ethylene action or synthesis is that there is simply insufficient "basal" ethylene to be physiologically active. An alternative explanation is that these small concentrations of ethylene may *enhance* rather than restrain root extension. When roots of intact seedlings, are grown in well-aerated conditions (e.g., in moist, preferably flowing air to minimize artifactual accumulations of ethylene), small concentrations of ethylene, usually well below 1 μl l^{-1}, have been found to promote elongation in several species (Figure 1) including broad bean (*Vicia faba*),[60] rice,[47,50] tomato (*Lycopersicon esculentum*), especially the mutant form 'diageotropica',[50] maize,[103] and peanuts (*Arachis hypogea*).[61] Ethephon can also increase seedling root length in watermelon (*Citrillus lanatus*).[62] Konings and Jackson[50] failed to detect a significant promotion in the roots of white mustard (*Sinapis alba*), while observing differences in the extent of promotion between mutant and nonmutant tomatoes and rice. When the relative rates of ethylene production by these different plants were compared, the least ethylene was produced by roots that, in separate tests, responded the most positively to traces (0.02 μl l^{-1}) of exogenous ethylene and, conversely, that roots forming larger amounts of ethylene gave only a very weak (positive) response to traces of extra ethylene. This suggested that very small amounts of ethylene enhance root extension in a range of species in relation to the amount of endogenous ethylene present. In plants producing ethylene especially vigorously (e.g., white mustard), there may already be sufficient ethylene to completely saturate the capacity for a positive response, explaining why in these plants, small amounts of additional ethylene have little

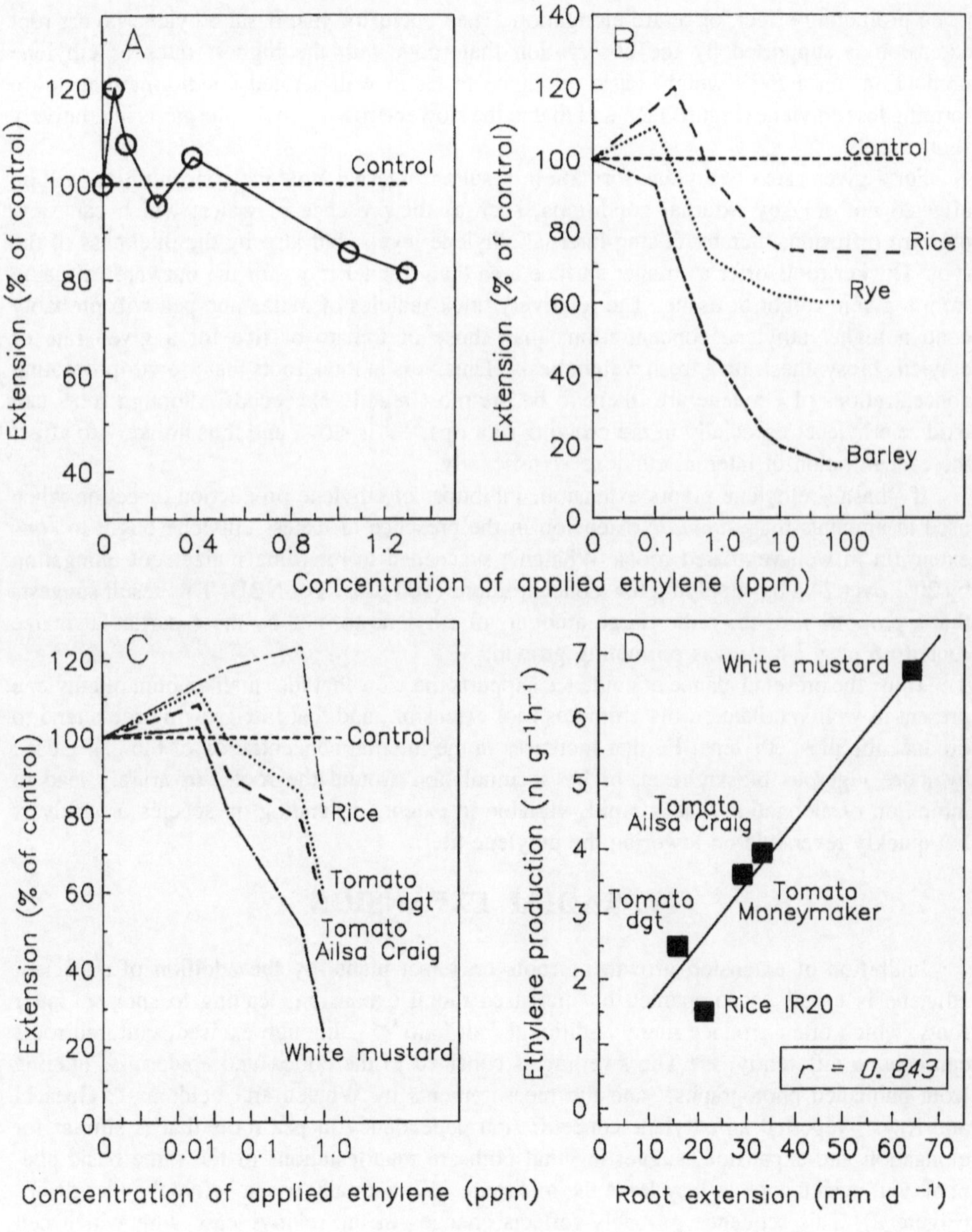

FIGURE 1. Four sets of results which indicate that the concentration of "basal" ethylene in well-aerated roots can lie on the ascending side of an optimal dose-response curve and thus favor root extension. (A) Extension by roots of broad bean after 48 h at different concentrations of ethylene. (Adapted from Kays et al.[60]) (B) Extension by roots of rice, cv. OS6; rye, cv. Ovari; and barley, cv. Maris Badger after 72 h in different concentrations of ethylene. (Adapted from Smith and Robertson.[47]) (C) Extension by roots of rice, cv. IR20; tomato, cv. Ailsa Craig; white mustard; and the tomato mutant 'diageotropica' (dgt) that form less ethylene than ordinary tomato roots after 24 h in different concentrations of ethylene. (Adapted from Konings and Jackson.[50]) (D) Relationship between rates of ethylene production and root extension in well-aerated conditions in five species or cultivars. (Adapted from Konings and Jackson.[50])

or no promoting effect, or retard elongation. The conclusion that basal ethylene favors root extension is supported by the observation that roots with the highest rates of ethylene production, on a fresh weight basis, elongate faster in well-aerated conditions than those forming less ethylene (Figure 1D), and that in the slower-growing roots, the gas is in relatively short supply.

For a given rate of ethylene formation, resultant internal ethylene concentration will be affected not only by external conditions, such as the presence of water, which can slow outward diffusion, thereby raising internal ethylene levels, but also by the thickness of the root. Thicker roots offer a smaller surface area than thinner roots for the outward diffusion from a given weight of tissue. The relatively thick radicles of maize and pea will probably contain higher ethylene concentrations than those of tomato or rice for a given rate of ethylene biosynthesis on a fresh weight basis. Thus, it is in thick roots that growth-promoting concentrations of ethylene are likely to be the most readily exceeded. Although roots can oxidize ethylene, especially in the growing root tips,[63] it is slow, and thus unlikely to affect the concentration of internal ethylene significantly.

If "basal" ethylene favors extension, inhibitors of ethylene production or action when used in amounts that *stimulate* extension in the presence of excess ethylene ought to *slow* extension in well-ventilated roots. Whalen[54] succeeded in retarding maize root elongation by 20% over 24 h by supplying low concentrations (100 $\mu l\ l^{-1}$) of NBD. This result suggests that a proportion of the rather large amounts of ethylene formed by these particular maize roots (6.6 $nl\ g^{-1}\ h^{-1}$) was promoting growth.

Thus, the present balance of evidence supports the view that the small amount of ethylene present in well-ventilated roots promotes root extension, and that fast-growing roots tend to produce the most ethylene. Further increases in the internal concentration of the gas caused by more vigorous biosynthesis, or by accumulation around the roots, invariably lead to inhibition of elongation that is rapid, variable in extent, depending on species or cultivar, but quickly reversible on lowering the ethylene titer.

IV. RADIAL EXPANSION

Inhibition of extension growth of roots on intact plants by the addition of sufficient ethylene is usually accompanied by increased radial expansion, leading to shorter, fatter roots, which often produce many additional root hairs[65-67] although excised, cultured roots can behave differently.[37,64] The swelling is confined to the cortex and epidermis, judging from published photographs[49] and the measurements by Whalen and Feldman.[46] Goeschl and Kays[61] reported an ethylene-concentration dependency in pea roots that is similar for elongation and expansion, suggesting that both are manifestations of the same basic phenomenon, even though as they leave the meristem, cells normally widen before they lengthen[68] (Figure 2). This sequence probably reflects changes in the relative ease with which cell walls extend laterally or longitudinally under turgor as they age. When 0.1 $\mu l\ l^{-1}$ ethylene is given for 8 d to inhibit barley root extension by 80%, these spatial and temporal relationships between the phases of cell widening and lengthening become changed (Figure 2). Although individual cells within the meristem are not affected,[46,68] indicating that the dimensions of these dividing cells are not susceptible to ethylene, the length of the meristem is shortened by about half[6] and, thus, cells stop dividing and begin expanding sooner and thus much closer to the root tip. According to Barlow,[69] cells in ethylene-treated roots that are dividing synthesize DNA at similar rates to the controls, and the duration of the mitotic cycle and the percentage of cells in mitosis at any one time are affected only slightly, except possibly for the first 4 h of exposure.[70] The results of Apelbaum and Burg[71] indicate more extensive suppression of mitotic activity.

The zone in which the cells elongate after leaving the meristem of barley roots is shortened

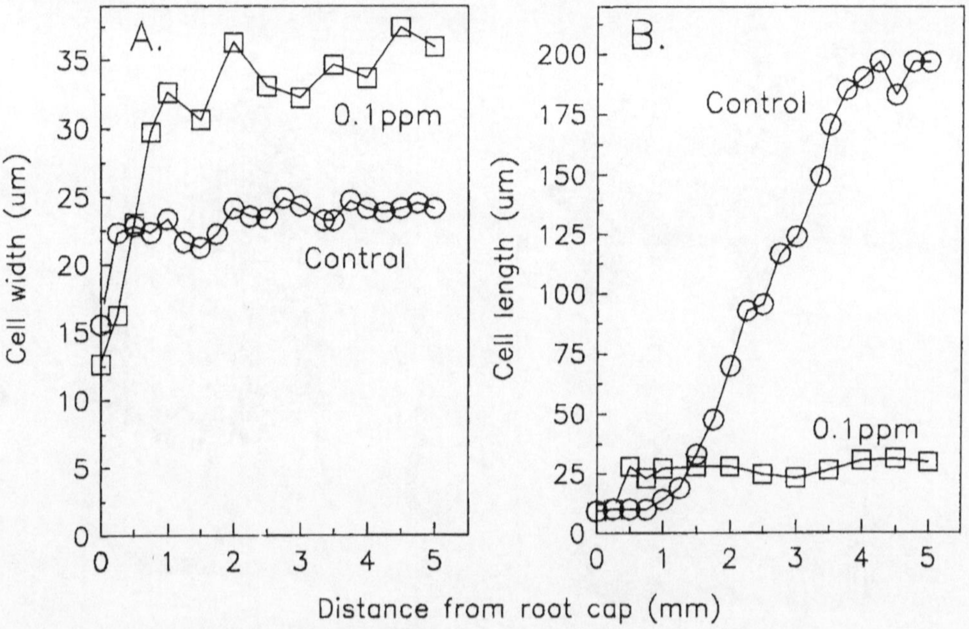

FIGURE 2. Effect of supplying 0.1 μl l^{-1} ethylene to root systems of *Hordeum vulgare* (barley) for 9 d on (A) cortical cell width and (B) cortical cell length in the apical 5 mm of main root axes. (From John[68])

by ethylene from 3750 μm to only 500 μm (Figure 2B). In contrast, the zone of cell widening is displaced away from the root tip and increases in overall length from the 250 μm typical of untreated roots to several millimeters (Figure 2A), to give, eventually, abnormally wide cells.[69] This work shows that ethylene inhibits root extension and promotes swelling by changing both the position and duration of rather distinct phases of latitudinal and longitudinal cell growth. The extra radial expansion induced by ethylene treatment does not, however, compensate entirely for the loss of cell length. Calculations of final cell volumes using the cell dimensions of John[68] indicate a 68% decrease after 8 d in 0.1 μl l^{-1} ethylene. This is probably the consequence of the loss of plastic extensibility that ethylene treatment also brings about in roots.[68] Studies of ethylene-treated and dark-grown pea or bean shoots suggest that ethylene favors lateral cell expansion by reorientating cellulose microfibrils.[69] Microfibril reorientation is probably linked causally to the pitch of microtubule helices that shroud the cell protoplast.[73,74] Since the orientation of microtubules in shoot cells alters from a latitudinal to a more longitudinal orientation within 30 min to 5 h of ethylene exposure,[74,75] it seems probable that similar events ordering the geometry of cellulose deposition also take place in ethylene-treated roots.

V. WHOLE ROOT SYSTEMS

Supplying sufficient ethylene to inhibit root extension and promote radial expansion of individual roots inevitably affects the appearance of the whole root system. In 20-d-old barley plants given 10 μl l^{-1} ethylene for 11 d in solution culture, the root system appears stunted, with seminal root axes, and first- and second-order laterals much shortened[66] (Figure 3). The roots are thicker, with curved, horizontally orientated tips and a dense mat of root hairs (Figure 3) but an unchanged number of laterals per unit length of seminal root axis.[66] Small decreases in dry weight gain of root systems given more than 1.0 μl l^{-1} have been reported in barley, maize, and tomato,[26,66,76] although Nakayama et al.,[77] using fog culture, reported *increases* of up to 41% in tomato and soybean roots, but at the expense of shoot

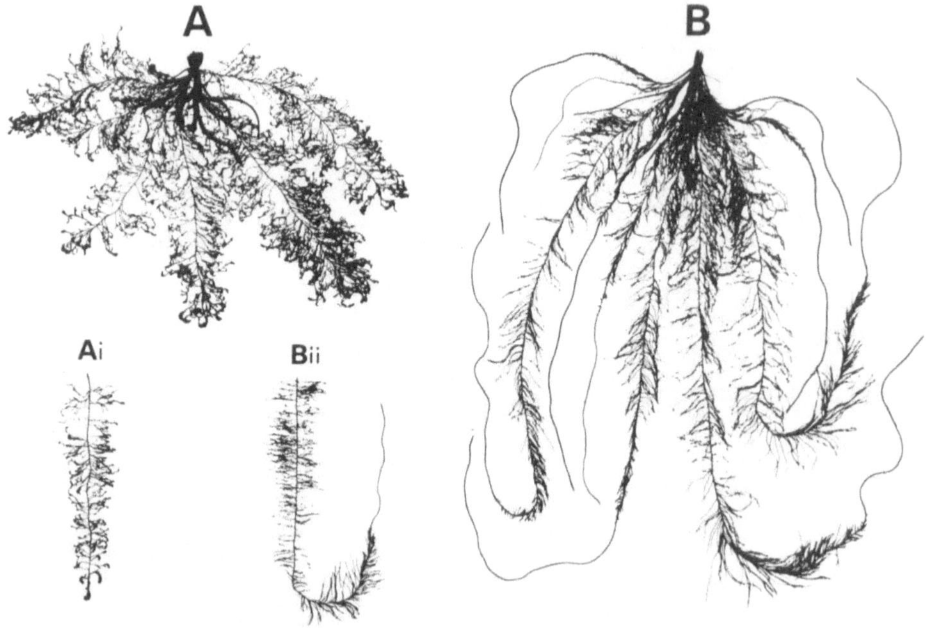

FIGURE 3. Morphology of whole root systems of barley (*Hordeum vulgare*) grown in nutrient solution aerated with 10 μl l^{-1} ethylene or with air. (A) Twenty-day-old root system after 11 d ethylene treatment; (B) As (A) but supplied with air. Inset (Ai): seminal root apex treated with ethylene for only 7 d; (Bii): as (Ai) but supplied with air. (Adapted from Crossett and Campbell.[66])

growth in the former species. In barley[66] but not in pea,[78] removal of applied ethylene after several days results in especially fast extension by those lateral roots originating on sections of main axis that extended while in the ethylene. This stimulation takes place at the expense of extension by the main axis, resulting in a shorter axis bearing unusually long laterals marking the region where axial extension was retarded by the preceding ethylene treatment.[66]

The function of root systems as absorbers of mineral nutrients is not much affected by ethylene. No change in concentration of the three major inorganic nutrients or in the rate of absorption of phosphorus, potassium, and strontium (as a tracer for calcium) were found in shoots or roots of barley plants grown in ethylene-enriched nutrient solution.[66] A report that 5 μl l^{-1} ethylene given for 7 to 14 d to maize roots in nutrient solution depresses phosphorus uptake and concentration in the foliage by 30%[26] suggests that nutrient absorption may be more susceptible to ethylene in maize than in barley.

VI. GROWTH RESPONSES TO APPLIED AUXIN

It is well known that applying auxin above a certain concentration (approximately 10^{-7} *M*) to roots enhances ethylene production within 30 min.[70] The auxin is thought to stimulate the action of preexisting enzymes associated with ethylene biosynthesis[79] and to increase synthesis of new enzymes, possibly including ACC synthase, that follows enhanced expression of the appropriate gene(s).[80] Acidification of the cytoplasm may be a prerequisite for the effect. Chadwick and Burg[65,82] were the first to argue that inhibition of extension and the swelling brought about by auxin treatment (10^{-7} to 10^{-4} *M*) to roots was a consequence of concomitant increases in ethylene. The principal points in their evidence are (1) the smallest concentration of auxin that inhibits extension is also the smallest to stimulate ethylene production; (2) IAA in the range 10^{-7} to 10^{-4} *M* inhibits extension in the absence but not in the presence of excess exogenous ethylene; (3) 10% carbon dioxide, a competitive antagonist of ethylene action, completely overcomes the inhibitory effects on root extension in 10^{-4} *M* to 10^{-7} *M* IAA. Carbon dioxide is not effective in the face of more concentrated IAA treatments[65] and these higher IAA concentrations are presumed by Chadwick and Burg to inhibit growth in ways unconnected with ethylene. Andreae et al.[83] (1968) have vigorously contested the claimed causal link between auxin-inhibited elongation and ethylene. In an excised pea-root culture system closely similar to that used by Chadwick and Burg, they succeeded in discriminating between effects of auxin and ethylene. For example, 2×10^{-5} *M* IAA was found to inhibit extension more quickly than 1000 $\mu l\ l^{-1}$ ethylene. Clearly, IAA has a rapid effect that is not attributable to ethylene.

The experiments by Andreae et al. and Chadwick and Burg probably overemphasized the effect of ethylene produced by roots, since the sealed culture vessels they used must have accumulated and retained the gas over approximately 18 h. In contrast, the effect of auxin per se may have been underestimated because of rapid metabolic depletion of the very limited total amount of IAA in the small volumes of medium employed. The notion that auxin retards root extension principally via ethylene has been further eroded by Rauser and Horton,[84] who found that the severity and reversibility of growth inhibition caused by small concentrations of IAA (5×10^{-5} *M*) cannot be equaled by ethylene, even at high concentrations (1000 $\mu l\ l^{-1}$). Similarly, Eliasson et al.[85] found IAA concentrations that stunted root extension in peas were too low to promote ethylene production. The findings of Dubucq et al.[56] and Scott and Norris[86] also cast doubt on the necessary involvement of ethylene in the inhibition of root extension caused by auxin application. The doubts are reinforced by measurements of hydrogen ion efflux that is thought to acidify cell walls of roots, rendering them more extensible. Amounts of IAA that retard extension also slow hydrogen ion efflux from maize[55,87] or *Lotus corniculatus* roots[41] and this may explain the inhibition of elongation. However, this is unlikely to be ascribed to ethylene since there is no evidence in the literature that applying ethylene inhibits hydrogen ion efflux. Furthermore, inhibitors of ethylene production (AVG and cobalt chloride) have failed to stimulate hydrogen ion efflux in the presence[41,87] or absence[58] of growth-inhibiting amounts of exogenous auxin. The question remains as to whether the extra ethylene induced by auxin is without any effect whatsoever on growth. This may be the situation if auxin blocks ethylene action, as it does in leaf abscission.

Overall, it appears that the involvement of ethylene in growth retardation caused by applied auxin has been over emphasized. Auxin obviously has the ability to inhibit root growth strongly and independently of ethylene, although any additional gas generated may add to the effect.

Auxin given to roots at concentrations too small to raise ethylene levels sometimes, but not always, stimulates rather than retards root extension. But such roots do, of course, still form some ethylene. The work of Mulkey et al.,[57] suggests this basal ethylene may oppose the growth-promoting action of very small amounts of auxin, similar to those occurring naturally in root tips.[88] Their key result is that the growth response to very low concentrations of applied IAA (10^{-9} *M* to 10^{-12} *M*) given by maize roots is changed from one of no effect to a substantial promotion of elongation if ethylene levels are diminished by supplying AVG and cobalt chloride in advance of the auxin. But, as already discussed, the roots used by these workers were probably artificially enriched with ethylene as a result of being underwater and aerated with pure oxygen. Thus, these results cannot reliably be used to deduce the role of "basal" ethylene in opposing auxin action.

VII. RESPONSES TO ENVIRONMENTAL FACTORS

A. LIGHT

Root extension can be inhibited by white light[89] at photon flux densities as low as 7.5 $\mu mol\ m^{-2}\ s^{-1}$. Short exposures for 1 s or more are effective,[90] and it is the root cap that senses the irradiation.[90,91] Light can also promote ethylene production by roots,[92] although it is unknown if this response too is mediated by the cap, or is as responsive as growth to short exposures of low irradiance. To investigate the possibility that the additional ethylene is responsible for the retardation of growth caused by light, Larqué-Saavedra et al.[93] exploited the ability of diiodo-4-hydroxybenzoic acid (DIHB) to block the growth-inhibiting effect of applied ethylene[93] and to slow ethylene production[94] in roots of cress (*Lepidium sativum*). They found DIHB overcame the inhibitory effect of white light on cress roots, indicating a role for ethylene in the light response. Unfortunately its effect was seen only when the roots were submerged deeply in unstirred water or supplied with methionine.[90] Both sets of conditions would encourage abnormally high concentrations of ethylene in the roots. DIHB failed to overcome the growth-inhibiting effect of light if the roots were covered by shallow films of water[93] where much of the additional ethylene generated in light would escape more readily to the air by diffusion. However, these results present a major difficulty by creating the need to explain how light inhibits the extension of roots under shallow water, where mediation by ethylene appears unlikely. This second mechanism may involve a nongaseous inhibitor of root extension produced by the cap (evidence reviewed by Jackson and Barlow[95]). Whatever this second mechanism may be, it can also be expected to operate in deeper water, where DIHB effectively counteracts it, in addition to overcoming ethylene-mediated growth inhibition. We must, therefore, conclude that light retards growth by more than one mechanism and that DIHB is not a specific inhibitor of ethylene production or action. Thus, DIHB has probably been of little help in reliably assessing the involvement of ethylene in the inhibition of root extension by light. More recent work with barley[96] and lettuce roots[52] has also raised further doubts concerning the specificity of DHIB action. A more dependable approach to the study of light-inhibited growth has been taken by Eliasson and Bollmark.[97] Using roots of pea seedlings, they show that when sufficient ACC is given to darkened roots to raise ethylene production to levels typical of those in the light, extension is slowed to rates that closely resemble those of illuminated roots. Furthermore, tip curvature and plagiotropism that is typical of illuminated roots of pea is reproduced in the dark by the ACC, while inhibitors of ethylene production or action such as amino-oxyacetic acid (AOA), cobalt chloride, or silver ions, enhance elongation of roots in the light to rates approaching those in the dark. Thus, ethylene seems to play some part in light-inhibited growth, but more detailed work is needed to test if it is involved in cap-mediated responses to very small amounts of irradiance, and to explore the relationship between this and the nongaseous inhibitor that has also been implicated in the retardation of root extension by light.

B. GRAVITY

Most roots elongate downwards, although not necessarily vertically. Gravity is the most powerful guiding influence and is sensed in statocyte cells at the core of the root cap. The cap may also be responsible for producing one or more growth inhibitors that are transported across the quiescent center and apical meristem to the elongating zone.[95] If the root is displaced from its preferred equilibrium position (the liminal angle), the gravity-stimulated cap is thought to effect an asymmetric distribution of inhibitor towards the lower side of the elongating zone, resulting in downward curvature. The chemical identity of the inhibitor(s) involved is controversial,[95] but auxin, linked with Ca^{2+}, are strong candidates.[98] Ethylene has also been implicated in the regulation of normal gravitropic curvature because the slower extension on the lower side of the root could result from the activity of additional ethylene produced in response to enrichment with auxin from the cap. Chadwick and Burg[65] initiated and supported this hypothesis by showing that applying 10% CO_2 to pea roots placed horizontally in moist air, a treatment previously shown to oppose the growth-inhibiting action of applied ethylene, slows the rate of curvature development over the first 4 h. The result implies that gravicurvature is in some measure a result of ethylene-retarded growth on the lower side of the root that can be overcome by the CO_2. However, an alternative explanation is that since the CO_2-treated roots do eventually attain a similar angle to control roots after a delay of 4 to 6 h, the CO_2 may have acted simply to slow root elongation which, *a priori,* must result in slower curvature. Indeed, Tackett and Pearson[99] have shown clearly that CO_2 can inhibit root extension. In the air-grown roots of pea used by Chadwick and Burg[65] to study gravitropism, endogenous ethylene may well have been poised at low, growth-promoting concentrations[100] and CO_2 could have interfered with such activity at the presumptive ethylene receptor sites. There is also a question over the specificity of action of CO_2 on root growth, since according to Curtis[107] it can reduce the effects of malformin in ways that are probably unconnected with ethylene.

Chadwick and Burg[65] were able to prevent gravicurvature almost completely by applying excess ethylene (100 $\mu l\ l^{-1}$). They interpreted this as being a consequence of overriding the internal gradient of endogenous ethylene between the upper and lower sides of the root. However, it is much more likely to have been simply the result of a severely inhibited rate of root extension. This latter interpretation is entirely borne out by the data of Hensel and Iversen[102] showing that inhibition of gravicurvature by 0.2 $\mu l\ l^{-1}$ ethylene given to cress roots could be accounted for by a retardation of longitudinal extension. The view that ethylene is inextricably implicated in normal root gravicurvature is eroded further by experiments with horizontally positioned maize root segments growing in moist air. Modest concentrations of ethylene (0.1 to 0.8 $\mu l\ l^{-1}$) that should be sufficient to destroy small internal gradients of endogenous ethylene within such roots, were found by Bucher and Pilet[103] to influence curvature only slightly, in association with minimal changes in the rate of extension. In conclusion, it seems unlikely that subtle gradients in ethylene across the root can explain root gravicurvature, otherwise these ethylene exposures should have eliminated the gravity response. Furthermore, direct evidence of gradients in ethylene production across horizontally positioned roots is lacking (cf. gravitropic grass nodes, Wright et al.[104]) and overall, gravistimulated roots do not produce more ethylene than vertical roots.[92] It is also uncertain that amounts of endogenous IAA in roots are in any case sufficient to influence ethylene production (see the above account of growth responses to applied auxin).

Although evidence unambiguously implicating *endogenous* ethylene in gravitropic curvature is minimal, there are many reports of applied ethylene interfering with normal gravitropism. As already discussed, the simplest way this can happen is through an inhibition of the cell extension that underpins curvature. But, the direction of this remaining growth often reorientates from the near-vertical towards the horizontal (plagiogravitropism), giving characteristically curved tips[66,78,103,105] (Figure 3 inset) that may even turn upwards and extend against the direction of gravity.[61,78] In some instances, spectacular coiling growth has been seen[9,42,48,67,106] (Figure 4), that can be prevented by applying the inhibitors of ethylene action, NBD and silver nitrate.[48,107] Obviously, the usual sensing of position within the field of gravity, or responses to it, are either over-ridden or modified. But where ethylene acts in the sequence of events between sensing gravity and its transduction into differential growth, or if the root cap plays a part, are unknown. It is notable that roots of the tomato mutant 'diageotropica', which are ageotropic,[48] and possess modified patterns of ethylene response and production,[39,109,110] do not form helical spirals in ethylene, as can roots of ordinary tomatoes.[48] Roots subjected to flooding of the soil or submergence, have been reported to grow upwards rather than downwards.[110-112] This may enable the roots to reach a source of oxygen before they asphyxiate. Since such roots are likely to be enriched with ethylene because of restrictions on the rate of diffusive loss imposed by water, the upward growth may be mediated by the entrapped gas. The possible involvement of ethylene in this important adaptive response merits further scrutiny.

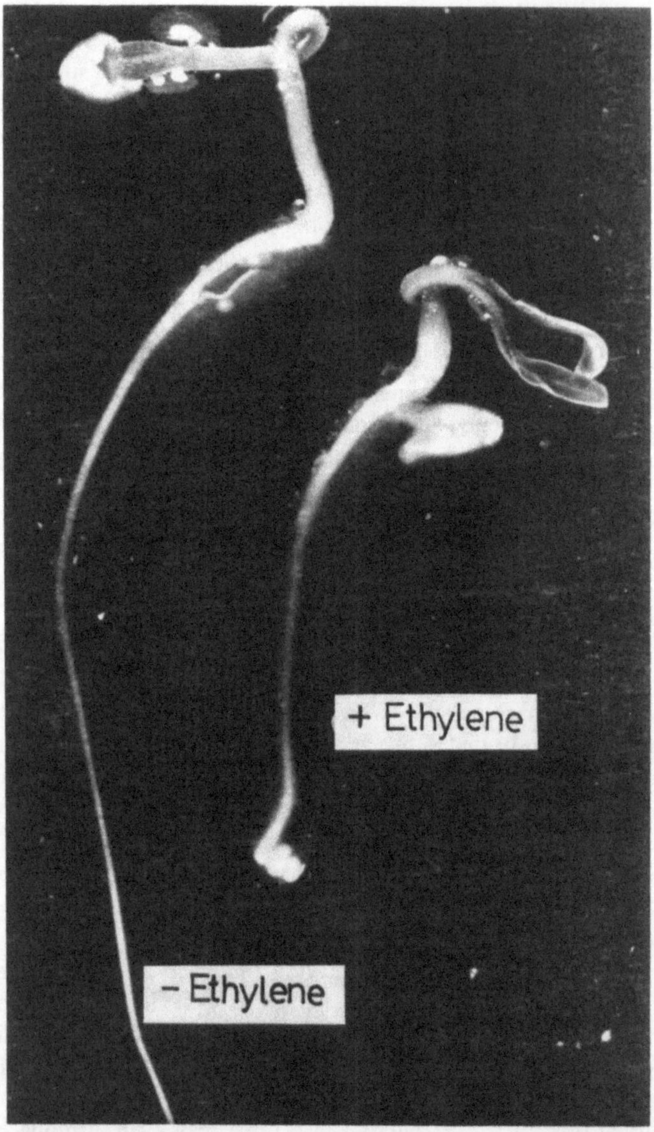

FIGURE 4. Root coiling in seedlings of tomato (*Lycopersicon esculentum*) after 72 h treatment with 10 μl l^{-1} ethylene. (From Woods, S. L., Roberts, J. A., and Taylor, I. B., *Plant Growth Regul.*, 2, 217, 1984. With permission.)

C. MECHANICAL IMPEDANCE AND PHYSICAL WOUNDING

1. Mechanical Impedance

All soil offers some resistance to extending roots, although quantifying how much resistance they experience is difficult. The rigid "penetrometer" probes commonly used to make measurements of soil strength cannot follow the tortuous path of least resistance typically adopted by a root, and thus they overestimate the pressures involved.[113] However, only very small mechanical pressures (approximately 20 kPa) would be needed to inhibit root extension and promote lateral swelling.[114] These are well below the maximum axial pressures of several hundred kPa that roots are capable of exerting.[115] The changed morphology of mechanically impeded roots (see photographs by Atwell[116] and Moss et al.[53]), that includes a more transverse orientation of cellulose microfibrils in cell walls,[117] is strongly reminiscent of ethylene-treated roots. The swelling behind the meristem may be an adaptation, helping to increase the maximum sustainable load, while tending to decrease the actual load on the root apex by opening up the soil ahead of the tip.

In support of the idea that ethylene mediates the effects of small mechanical pressures, Kays et al.[60] reported that when root tips of *Vicia faba* reach a physical obstruction, and are also constrained laterally to keep the tip pressed against the barrier, ethylene production increases within 1 h and is then maintained for a further 20 to 30 h. If, however, roots are allowed to become deflected and to extend across the surface of the impeding barrier, ethylene production is not enhanced[60,118] and may even decrease, while root extension remains substantially unchanged.[118] Although the specificity or effectiveness of DIHB as an inhibitor of ethylene production or action is questionable (see sections on root extension and light), there are several reports that this compound can stimulate extension by mechanically impeded roots[119-123] although in absolute terms the extent of the effect has usually been small. These results suggest that ethylene plays a part in inhibiting elongation under mechanical stress. Unfortunately, recent work with maize seedlings, using AVG and norbornadiene under well-aerated through-flow conditions,[53] has failed to confirm such a role for ethylene. In this study, low, nontoxic concentrations of AVG, that demonstrably reduced ethylene production in mechanically impeded roots to rates below even those of nonimpeded roots, fail to ameliorate the effects of mechanical impedance caused by roots extending into a bed of 3.0-mm diameter glass spheres. NBD at a concentration that totally overcomes the influence of 5 $\mu l\ l^{-1}$ ethylene also fails to reduce the impact of mechanical stress on root growth. Furthermore, there is a delay of several hours between the roots first penetrating between the glass spheres and the increase in ethylene production.[53] This suggests that the extra ethylene is produced as a consequence of intense mechanical pressures caused by increased lateral expansion against the surrounding glass spheres rather than a result of the apical contact at the root cap.[53] According to Goss and Russell[124] it is the cap that should be the site for sensitive detection of axial pressure. Thus, the involvement of endogenous ethylene as an integral component of root responses to small external pressure seems unlikely. However, it is clear that pressure can increase ethylene formation, but where along the root this is most sensitively perceived, and whether the minimum pressures known to retard elongation are also sufficient to stimulate ethylene production remain to be determined. The finding of Whalen[118] that extension by maize roots is increased rather than suppressed shortly after they first touch an obstacle, shows there is still much to learn about the subtleties of impedance effects and the possible involvement of ethylene.

2. Physical Wounding

In common with other plant parts, roots increase their rate of ethylene production if tissue is crushed or cut.[96,125] The impact of the additional gas on root extension has not been examined, although it may well contribute to a slowing of extension and gravicurvature that follows wounding of the cortex.[126] Rost and colleagues[70,127] find cell division is temporarily slowed by excision of the terminal 1 cm of pea roots, resulting in a reduction in the percentage of cells in mitosis over approximately the first 4 h. Transverse cuts into intact root have the same effect. Ethylene treatment (50 $\mu l\ l^{-1}$) of nonwounded roots can also temporarily lower the mitotic index.[127] Since AVG and silver ions can eliminate the wounding effect on growth, which may, however, be recovered by giving ethylene,[70] it seems probable that short-term suppression of cell division by wounding is mediated by ethylene.

D. IMPEDED AERATION AND OXYGEN SHORTAGE

Excessively wet soil is a common cause of poor root aeration that enriches the root with ethylene in several ways. Most obviously, enrichment occurs as a result of water around the root slowing ethylene loss by radial diffusion, leading to an accumulation of the gas.[50]

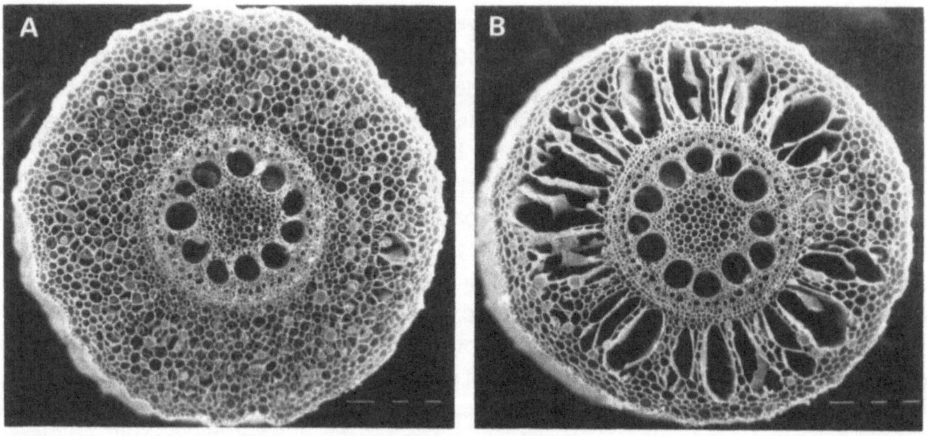

FIGURE 5. The effect of supplying ethylene to well-aerated roots of maize (*Zea mays*) for 7 d on aerenchyma formation in the cortex. Scanning electron micrographs of 5-d-old tissue from (A) controls given no additional ethylene and (B) roots supplied with 5 μl l^{-1} ethylene. (From Jackson, M. B., *Growth Regulators in Root Development, Monogr. 10,* Jackson, M. B. and Stead, A. D., Eds., British Plant Growth Regulator Group, Wantage, 1983, 103. With permission.)

Similarly, ethylene produced by soil microorganisms is trapped within the soil and enters into solution.[128] The water also restricts the entry of atmospheric oxygen into the soil and roots to replace that used in respiration.[129] Any resulting decrease in oxygen concentrations stimulate ethylene production in roots of *Zea mays* and *Hordeum vulgare*[96,130,131] provided a small amount of oxygen remains. Most of this ethylene is produced in the apical 10 mm of root where the extra ethylene production is mirrored by increased amounts of ACC.[132] A third source of increased ethylene may be invoked during prolonged soil waterlogging, which can encourage strictly anaerobic microorganisms that generate ethylene[133] while suppressing microbial oxidation of the gas.[134] The impact on roots by ethylene formed under the extremely reducing environments will usually be minimal, since without oxygen there can be little growth for ethylene to influence. However, even in anaerobic soil, the roots of some species gain limited access to atmospheric oxygen that moves *internally* by diffusion and convective flow. In these plants, internal oxygen transport is facilitated by the development of large interconnected gas spaces (aerenchyma).[130] Such roots thus retain the capacity to respond to anaerobically generated soil ethylene. In *Zea mays* at least, the formation of aerenchyma is itself thought to be a response to extra ethylene.[136] Roots extending into moist air develop little aerenchyma, while some slight promotion is seen in well-aerated nutrient solution.[137] However, if aeration of the solution is stopped and oxygen concentrations drift down to below 10%, a highly porous, aerenchymatous cortex (Figure 5) is formed within 7 d,[137,138] in association with at least a doubling of vacuum-extractable ethylene and substantially increased ethylene production.[131,138] This additional ethylene is implicated in aerenchyma development because ethylene treatment of well-aerated roots stimulates aerenchyma formation, thus mimicking poor aeration.[138] Further, silver ions[137,139] and AVG[131] successfully reverse aerenchyma development under conditions of partial oxygen shortage. The effect of these inhibitors seems reasonably specific, since no inhibition of root extension takes place at the concentration of silver nitrate or AVG used, and the suppressing effect of AVG on aerenchyma is almost totally reversible by simultaneously supplying ACC.[131] Cobalt chloride and amino-oxyacetic acid can also inhibit aerenchyma formation in poorly aerated maize roots,[137] but the specificity of action of these inhibitors of ethylene biosynthesis is less uncertain in this case. Nitrate deficiency can promote extensive aerenchyma even in well-aerated roots of maize.[140] The explanation is thought to be a much-enhanced sensitivity to substantially unchanged amounts of endogenous ethylene.[141]

Rupture of the plasmalemma is one of the first signs of physical degeneration in cells that collapse later to create the gas-filled lacunae, which constitute maize aerenchyma.[142] Nothing is known of the biochemistry and molecular biological events that underpin these events. Roots of rice are also notably highly aerenchymatous, but in contrast to maize, cell collapse commences with degeneration of the wall rather than the tonoplast[143] and is *not* strongly regulated by ethylene.[144]

E. NODULATION AND NITROGEN FIXATION

A chance observation that nodulation of cultured pea roots inoculated with *Rhizobium* bacteria was inhibited if culture vessels were tightly sealed[145] led Grobbelaar et al.[146] to test the influence of ethylene. They found that absorbing accumulated olefins with mercuric perchlorate increased nodulation from 0 to 42 nodules per cultured root system, while 0.4 $\mu l\ l^{-1}$ ethylene supplied to the roots in a through-flow system reduced the number of nodules from 35 to only 4 per root system over 18 d. Furthermore, when ethylene treatment (10 $\mu l\ l^{-1}$) was delayed until after nodulation, the ability of the nodules to incorporate ^{15}N was suppressed by 90%.[146] These startling findings have been repeated in essence by Drennan and Norton[148] and Goodlass and Smith[78] using *Pisum sativum* and *Trifolium repens.* That ethylene inhibits N-fixation by existing nodules has important implications for the acetylene reduction assay so widely used to quantify N-fixation. The validity of this convenient measure of nitrogenase activity is obviously at risk from the copious production of ethylene generated by acetylene reduction that is the measure of N-fixing potential.

It is possible that the well-known depressing effect of nitrate fertilizer on the extent of nodulation by *Rhizobium* may involve ethylene, since in *Medicago sativa,* ethylene production increases with the concentration of nitrate.[149] A more obscure role for ethylene in nodulation has been established by Zaat et al.[150] in *Vicia sativum* roots infected with one particular strain of *Rhizobium* that causes stunting, swelling, copious root hair development, and abnormal nodulation that is restricted to emerging lateral roots. These symptoms of aberrant morphology induced by the bacteria, or by extracts from bacterial cultures, can be reversed with AVG, suggesting that high rates of endogenous ethylene production arising from the bacterial colonization may explain the phenomenon. Unfortunately, direct measurement of high rates of ethylene formation induced by the bacterium was not actually made by Zaat et al. Furthermore, the specificity of the AVG was not tested by attempting to reverse the effect of the inhibitor with exogenous ACC or ethylene. However, the ability of AVG to stimulate extension of infected roots does suggest that the concentrations chosen were sufficiently low to be nontoxic.

Overall, these findings indicate an important but strangely neglected effect of ethylene on nodulation and N-fixation, which holds significance for practical farming, for assaying of nitrogenase activity, and for the study of nodulation by roots infected with *Rhizobium.* One goal for the future will be to find where ethylene acts in the sequence from first infection of root hairs, through the initiation of nodule meristems, the cell expansion that finally forms the visible nodule, and to the chemistry and molecular biology of N-fixation. Since changes in the porosity of nodules in response to environment play an important part in ensuring optimal oxygen tensions for the nitrogen-fixing bacteroids,[151] the possible involvement of ethylene in regulating this "nodular aerenchyma" in legumes also deserves examination.

VIII. CONCLUDING REMARKS

There is certainly no shortage of fascinating and varied responses that can be secured by exposing roots to ethylene. Thus, roots may be induced or their induction repressed, elongation can be speeded up or stunted in association with distinct swelling, while a normally dependable sense of direction becomes corrupted. Add to this, a tendency for ethylene-treated roots to proliferate root hairs, to hollow out their cortices, to decline to cohabit with *Rhizobium* bacteriods, to inhibit atmospheric nitrogen fixation within preexisting nodules, then the potency of ethylene to alter the course of root-system development and functioning become immediately apparent. With many of these responses being mirrored in the behavior of roots exposed to different kinds of environmental stress or to treatments with other hormones, interest is heightened by the possibility that the actions of ethylene may explain these phenomena. It is in demonstrating convincingly that naturally produced ethylene does indeed operate in this way and perhaps also plays a part in the growth of unstressed roots, that much evidence remains rudimentary and less than convincing. This state of affairs exists despite the considerable advantages conferred by the availability of gas chromatographs possessing a sensitivity to ethylene rivalling that of the plants themselves, the possibility of dosing plants or plant organs with precisely controlled amounts of the hormone, and the existence of inhibitors of ethylene action or biosynthesis that are potentially invaluable for testing not only the essentiality of ethylene for a given process but also for assessing the physiological impact of given rates of ethylene formation. The lack of appropriate testing of the specificity of action of inhibitors is probably the most serious shortcoming in much recent work and may well have led to conflicting or inappropriate conclusions. However, with the advent of more sensitive detectors for ethylene (e.g., the photoacoustic laser detector[152]) and an increased awareness of the need for much greater care and precision in the use of inhibitors and the measurement of the phenomenon under study, the prospect for clarifying the involvement of ethylene in the everyday life of roots still appears to be good. Such work will need to go hand in hand with molecular and biochemical studies of ethylene action if the latter are not to risk ending in futility when, too late, it is realized that some favorite response to ethylene has little to do with what really goes on in roots outside the plant science laboratory.

REFERENCES

1. **Epstein, E.**, The role of roots in the chemical economy of life on earth, *BioScience*, 27, 783, 1977.
2. **Wilson, J. B.**, A review of evidence on the control of shoot:root ratio; in relation to models, *Ann. Bot.*, 61, 433, 1988.
3. **Sen, D. N.**, Root system and root ecology, in *Environment and Root Behaviour*, Sen., D. N., Ed., Geobios, Jodhpur, India, 1980, 1.
4. **Dittmer, H. J.**, A quantitative study of the roots and root hairs of a winter rye plant *(Secale cereale)*, *Am. J. Bot.*, 24, 417, 1937.
5. **Kutschera, L.**, *Wurzelatlas mitteleuropäischer Ackerunkräuter und Kulturpflanzen*, DLG-Verlags-GmbH, Frankfurt-am-Main, 1960.
6. **Barlow, P. W.**, Anatomical controls of root growth, in *Roots and the Soil Environment, Aspects of Applied Biology*, 22, Association of Applied Biology, Wellesbourne, 1989, 57.
7. **Gales, K.**, Yield variation of wheat and barley in Britain in relation to crop growth and soil conditions—a review, *J. Sci. Food Agric.*, 34, 1085, 1983.
8. **Abeles, F. B.**, *Ethylene in Plant Biology*, Academic Press, New York, 1973, 128.
9. **Zimmerman, P. W. and Hitchcock, A. E.**, Initiation and stimulation of adventitious roots caused by unsaturated gases, *Contrib. Boyce Thompson Inst. Plant Res.*, 5, 351, 1933.
10. **Zimmerman, P. W. and Wilcoxon, F.**, Several chemical growth substances which cause initiation of roots and other responses in plants, *Contrib. Boyce Thompson Inst. Plant Res.*, 7, 209, 1935.
11. **Crocker, W., Hitchcock, A. E., and Zimmermann, P. W.**, Similarities in the effects of ethylene and the plant auxins, *Contrib. Boyce Thompson Inst. Plant Res.*, 7, 231, 1935.
12. **Went, F. W. and Thimann, K. V.**, *Phytohormones*, Macmillan, New York, 1937, 181.
13. **Hess, C. E.**, Internal and external factors regulating root initiation, in *Root Growth*, Whittington, W. J., Ed., Butterworths, London, 1969, 42.
14. **Robbins, J. A., Reid, M. S., Paul, J. L., and Rost, T. L.**, The effect of ethylene on adventitious root formation in mung bean (*Vigna radiata*) cuttings, *J. Plant Growth Regul.*, 4, 147, 1985.

15. **Krishnamorthy, H. N.,** Promotion of rooting in mung bean hypocotyl cuttings with ethrel, *Plant Cell Physiol.*, 11, 979, 1970.
16. **Roy, B. N., Basu, R. N., and Bose, T. K.,** Interaction of auxins with growth-retarding, -inhibiting and ethylene-producing chemicals in rootings of cuttings, *Plant Cell Physiol.*, 13, 1123, 1972.
17. **Robbins, J. A., Kays, S. J., and Diss, M. A.,** Enhanced rooting of wounded mung beans cuttings by wounding and ethephon, *J. Am. Soc. Hortic. Sci.*, 108, 325, 1983.
18. **Geneve, R. L. and Heuser, C. W.,.** The relationship between ethephon and auxin in adventitious root initiation in cuttings of *Vigna radiata* (L.) R. Wilcz., *J. Am. Soc. Hortic. Sci.*, 330, 1983.
19. **Mudge, K. W. and Swanson, B. T.,** Effect of ethephon, indole butyric acid and treatment solution pH on rooting and on ethylene levels within mung bean cuttings, *Plant Physiol.*, 61, 271, 1978.
20. **Batten, D. J. and Mullins, M. G.,** Ethylene and adventitious root formation in hypocotyl segments of etiolated mung bean (*Vigna radiata* (L.) Wilczek) seedlings, *Planta*, 138, 193, 1978.
21. **Jusaitis, M.,** Rooting of intact mung bean hypocotyls stimulated by auxin, ACC and low temperature, *HortScience*, 21, 1024, 1986a.
22. **Riov, J. and Yang, S. F.,** Ethylene and auxin-ethylene interaction in adventitious root formation in mung bean (*Vigna radiata*) cuttings, *Plant Growth Regul.*, 8, 131, 1989.
23. **Jusaitis, M.,** Rooting response of mung bean cuttings to 1-aminocyclopropane-1-carboxylic acid, *Sci. Hortic.*, 29, 77, 1986.
24. **Meijer, E. G. M.,** Developmental aspects of ethylene biosynthesis during somatic embryogenesis in tissue cultures of *Medicago sativa*, *J. Exp. Bot.*, 40, 479, 1989.
25. **Kawase, M.,** Causes of centrifugal root promotion. II. Ethylene, *HortScience*, 6, 282, 1971.
26. **Jackson, M. B., Drew, M. C., and Giffard, S. C.,** Effects of applying ethylene to the root system of *Zea mays* on growth and nutrient concentration in relation to flooding tolerance, *Physiol. Plant.*, 52, 23, 1981.
27. **Suge, H.,** Ethylene and gibberellin. Regulation of internodal elongation and nodal root development in floating rice, *Plant Cell Physiol.*, 26, 607, 1985.
28. **Mullins, M. G.,** Auxin and ethylene in adventitious root formation in *Phaseolus aureus* (Roxb.), in *Plant Growth Substances, 1970*, Carr, D. J., Ed., Springer-Verlag, Berlin, 1972, 526.
29. **Linkins, A. E., Lewis, L. N., and Palmer, R. L.,** Hormonally induced changes in the stem and petiole anatomy and cellulase enzyme patterns in *Phaseolus vulgaris* L., *Plant Physiol.*, 52, 544, 1973.
30. **Hackett, W. P.,** Donor plant maturation and adventitious root formation, in *Adventitious Root Formation in Cuttings*, Davis, T. D., Haissig, B. E., and Sankhla, N., Eds., Dioscorides Press, Portland, OR, 1988, 11.
31. **Soffer, H., Mayak, S., Burger, D. W., and Reid, M. S.,** The role of ethylene in the inhibition of rooting under low oxygen tensions, *Plant Physiol.*, 89, 165, 1989.
32. **Byther, R. S. and Moore, P. G.,** Inhibition of sugar cane rooting by *Ceratocystis paradoxa*, *Can. J. Bot.*, 52, 761, 1974.
33. **Coleman, W. K., Huxter, T. J., Reid, D. M., and Thorpe, T. A.,** Ethylene as an endogenous inhibitor of root regeneration in tomato leaf discs cultured *in vitro*, *Physiol. Plant.*, 48, 519, 1980.
34. **Le Guyader, H.,** Éthylène et différenciation de racines par des cultures de couches cellulaires minces de tabac, *C.R. Acad. Sci., Paris, t. 305, Série III*, 591, 1987.
35. **Nordström, A.-C. and Eliasson, L.,** Regulation of root formation by auxin ethylene interaction in pea *Pisum sativum* cultivar Weilbulls-Marma stem cuttings, *Physiol. Plant.*, 61, 298, 1984.
36. **Torrey, J. G.,** Endogenous and exogenous influences on the regulation of lateral root formation, in *New Root Formation in Plants and Cuttings*, Jackson, M. B., Ed., Martinus Nijhoff, Dordrecht, 1986, 31.
37. **Radin, J. W. and Loomis, R. S.,** Ethylene and carbon dioxide in the growth and development of cultured radish roots, *Plant Physiol.*, 44, 1584, 1969.
38. **Zobel, R. W.,** The genetics of root development, in *The Development and Function of Roots*, Torrey, J. G. and Clarkson, D. T., Eds., Academic Press, New York, 1976, 261.
39. **Jackson, M. B.,** Is the diageotropic tomato ethylene-deficient?, *Physiol. Plant.*, 46, 347, 1979.
40. **Zobel, R. W.,** Rhizogenetics (root genetics) of vegetable crops, *HortScience*, 21, 956, 1980.
41. **Shen, W. H., Petit, A., Guern, J., and Tempé, J.,** Hairy roots are more sensitive to auxin than normal roots, *Proc. Natl. Acad. Sci. U.S.A.*, 85, 3417, 1988.
42. **Harvey, E. M. and Rose, R. C.,** The effects of illuminating gas on root systems, *Bot. Gaz.*, 60, 27, 1915.
43. **Fujino, D. W., Burger, D. W., Yang, S. F., and Bradford, K. J.,** Characterization of an ethylene overproducing mutant of tomato (*Lycosperison esculentum* Mill. cultivar VFN8), *Plant Physiol.*, 88, 774, 1988.
44. **Freytag, A. H., Berlin, J. D., and Linden, J. C.,** Ethylene-induced fine structure alterations in cotton and sugar beet radicle cells, *Plant Physiol.*, 60, 140, 1977.
45. **Plhàk, F.,** Oxidative and phosphorylative metabolism of rye roots during the inhibition of elongation growth by ethylene, in *Structure and Function of Primary Root Tissues*, Kolek, J., Ed., Veda, Bratislava, 1974, 285.

46. **Whalen, C. M. and Feldman, L. J.**, The effect of ethylene on root growth of *Zea mays* seedlings, *Can. J. Bot.*, 66, 719, 1988.
47. **Smith, K. A. and Robertson, P. D.**, Effect of ethylene on the root extension of cereals, *Nature*, 234, 148, 1971.
48. **Woods, S. L., Roberts, J. A., and Taylor, I. B.**, Ethylene-induced root coiling in tomato seedlings, *Plant Growth Regul.*, 2, 217, 1984.
49. **Jackson, M. B.**, Regulation of root growth and morphology by ethylene and other externally applied growth substances, in *Growth Regulators in Root Development, Monogr. 10*, Jackson, M. B. and Stead, A. D., Eds., British Plant Growth Regulator Group, Wantage, 1983, 103.
50. **Konings, H. and Jackson, M. B.**, A relationship between rates of ethylene production by roots and the promoting and inhibiting affects of exogenous ethylene and water on root elongation, *Z. Pflanzenphysiol.*, 92, 385, 1979.
51. **Jackson, M. B. and Pearce, D. M. E.**, Hormones and morphological adaptation to aeration stress in rice, in *Plant Life under Oxygen Deprivation Ecology, Physiology and Biochemistry*, Jackson, M. B., Davies, D. D., and Lambers, H., Eds., S.P.B. Academic, The Hague, 1990, 47.
52. **Abeles, F. B. and Wydoski, S. G.**, Inhibition of ethylene synthesis and action: a comparison of their activities in a lettuce root growth model system, *J. Am. Soc. Hortic. Sci.*, 112, 133, 1987.
53. **Moss, G. I., Hall, K. C., and Jackson, M. B.**, Ethylene and the responses of roots of maize (*Zea mays* L.) to physical impedance, *New Phytol.*, 109, 303, 1988.
54. **Whalen, M. C.**, The role of ethylene in root development of *Zea mays*, Ph.D. thesis, University of California, Berkeley, 1986.
55. **Lüthen, H., Hilgendorf, B., and Böttger, M.**, Effect of auxin on growth, proton secretion and transmembrane electron transfer in intact maize roots, in *Structural and Functional Aspects of Transport in Roots*, Loughman, B. C., Gasparikovà, O., and Kolek, J., Eds., Kluwer Academic, Dordrecht, 1989, 63.
56. **Dubucq, M., Hofinger, M., and Gaspar, T.**, Auxin-controlled root growth and ethylene production, *Plant Cell Environ.*, 1, 151, 1978.
57. **Mulkey, T. J., Kuzmanoff, K. M., and Evans, M. L.**, Promotion of growth and shift in the auxin dose/response relationship in maize roots treated with ethylene biosynthesis inhibitors aminoethoxyvinylglycine and cobalt, *Plant Sci. Lett.*, 25, 43, 1982.
58. **Mulkey, T. J., Kuzmanoff, K. M., and Evans, M. L.**, Promotion of growth and hydrogen ion efflux by auxin in roots of maize *Zea mays* cultivar hydrid WF-9-X-38 pre-treated with ethylene biosynthesis inhibitors, *Plant Physiol.*, 70, 186, 1982.
59. **Mulkey, T. J., Evans, M. L., and Kuzmanoff, K. M.**, The kinetics of abscisic acid action on root growth and gravitropism, *Planta*, 157, 150, 1983.
60. **Kays, S. J., Nicklow, C. W., and Simons, D. H.**, Ethylene in relation to the response of roots to physical impedance, *Plant and Soil*, 40, 565, 1974.
61. **Goeschl, J. D. and Kays, S. J.**, Concentration dependences of some effects of ethylene on etiolated pea, peanut, bean and cotton seedlings, *Plant Physiol.*, 55, 670, 1975.
62. **Wilson, G. T.**, Morphological variations in water melon *Citrullus lanatus* induced by Alar and Ethrel, *HortScience*, 6, 277, 1971.
63. **Abeles, F. B.**, A comparative study of ethylene oxidation in *Vicia faba* and *Myrobacterium paraffinicum*, *J. Plant Growth Regul.*, 3, 85, 1984.
64. **Svensson, S. B.**, A comparative study of the changes in root growth induced by coumarin, auxin, ethylene, kinetin and gibberellic acid, *Physiol. Plant.*, 26, 115, 1972.
65. **Chadwick, A. V. and Burg, S. P.**, An explanation of the inhibition of root growth caused by indole-3-acetic acid, *Plant Physiol.*, 42, 415, 1967.
66. **Crossett, R. N. and Campbell, D. J.**, The effects of ethylene in the root environment upon the development of barley, *Plant and Soil*, 42, 453, 1975.
67. **Gold, A. H. and Faccioli, G.**, Comparative effect of curly top infection and ethylene treatment on tomatoes and other plants, *Phytopathol. Mediterr.*, 11, 145, 1972.
68. **John, A.**, The effect of ethylene on the growth of roots of *Hordeum vulgare* and *Oryza sativa*, Ph.D. thesis, University College of Wales, Aberystwyth, 1976.
69. **Barlow, P. W.**, The effect of ethylene on root meristems of *Pisum sativum* and *Zea mays*, *Planta*, 131, 235, 1976.
70. **Rost, T. L., Jones, T., and Robbins, J. A.**, The role of ethylene in the control of cell division in cultured pea root tips—a mechanism to explain the excision effect, *Protoplasma*, 130, 68, 1986.
71. **Apelbaum, A. and Burg, S. P.**, Effect of ethylene on cell division and deoxyribonucleic acid synthesis in *Pisum sativum*, *Plant Physiol.*, 50, 117, 1972.
72. **Apelbaum, A. and Burg, S. P.**, Altered cell microfibrillar orientation in ethylene-treated *Pisum sativum* stems, *Plant Physiol.*, 48, 648, 1971.
73. **Steen, D. A. and Chadwick, A. J.**, Ethylene effects in pea stem tissue, evidence of microtubule mediation, *Plant Physiol.*, 67, 460, 1981.

74. **Roberts, I. N., Lloyd, C. W., and Roberts, K.,** Ethylene-induced microtubule re-orientations: mediation by helical rays, *Planta,* 164, 439, 1985.
75. **Lang, J. M., Eisinger, W. R., and Green, P. B.,** Effects of ethylene on the orientation of microtubules and cellulose microfibrils of pea epicotyl cells with polylamellate cell walls, *Protoplasma,* 110, 5, 1982.
76. **Jackson, M. B. and Campbell, D. J.,** Effects of benzyladenine and gibberellic acid on the responses of tomato plants to anaerobic root environments and to ethylene, *New Phytol.,* 82, 331, 1979.
77. **Nakayama, M., Shimura, K., and Oota, Y.,** Physiological action of ethylene in crop plants. 4. Effect of ethylene application to roots on the growth of tomato and soyabean plants, *Proc. Crop Sci. Soc. Jpn.,* 42, 493, 1973.
78. **Goodlass, G. and Smith, K. A.,** Effects of ethylene on root extension and nodulation of pea (*Pisum sativum* L.) and white clover (*Trifolium repens* L.), *Plant Soil,* 51, 387, 1979.
79. **Steen, D. A. and Chadwick, A. V.,** Effects of cycloheximide on indoleacetic acid-induced ethylene production in pea root tips, *Plant Physiol.,* 52, 171, 1973.
80. **Imaseki, H., Nakajima, N., and Todaka, I.,** Biosynthesis of ethylene and its regulation in plants, in *Biomechanisms Regulating Growth and Development,* Steffens, G. L. and Rumsey, T. S., Eds., Kluwer Academic, Dordrecht, 1988, 205.
81. **Zoglaüer, K., Dembny, H., and Goring, H.,** Inhibition of IAA-induced ethylene production and proline accumulation in wheat coleoptiles by fusicoccin, *Biochem. Physiol. Pflanz.,* 182, 23, 1987.
82. **Chadwick, A. V. and Burg, S. P.,** Regulation of root growth by auxin-ethylene interaction, *Plant Physiol.,* 45, 192, 1970.
83. **Andreae, W. A., Venis, M. A., Jursic, F., and Dumas, T.,** Does ethylene mediate root growth inhibition by indole-3-acetic acid?, *Plant Physiol.,* 43, 1375, 1968.
84. **Rauser, W. and Horton, R. F.,** Rapid effects of IAA and ethylene on the growth of intact pea roots, *Plant Physiol.,* 55, 443, 1975.
85. **Eliasson, L., Bertell, G., and Bolander, E.,** Role of ethylene in relation of root growth, in *6th Congr. Fed. Eur. Soc. Plant Physiol., Abstr. Vol., Abstr.* 7-16, 1988.
86. **Scott, P. C. and Norris, L. A.,** Separation of effects of auxin and ethylene in pea roots, *Nature,* 227, 1366, 1970.
87. **Evans, M. L., Mulkey, T. J., and Vesper, M. J.,** Auxin action on proton influx in corn roots and its correlation with growth, *Planta,* 148, 510, 1980.
88. **Pilet, P.-E.,** Control of root growth by endogenous IAA and ABA, in *Growth Regulators in Root Development Monogr. No. 10,* Jackson, M. B. and Stead, A. D., Eds., British Plant Growth Regulator Group, Wantage, 1983, 15.
89. **MacDonald, I. R. and Gordon, D. C.,** The regulation of root growth in cress seedlings by light and gravity, *J. Exp. Bot.,* 29, 1051, 1978.
90. **Wilkins, H., Larqué-Saavedra A., and Wain, R. L.,** Studies on plant growth regulating substances. XXXVII. The effects of light and hormone inhibitors on plant root growth, *Ann. App. Biol.,* 78, 169, 1974.
91. **Pilet, P.-E. and Ney, D.,** Rapid, localized light effect on root growth in maize, *Planta,* 144, 109, 1978.
92. **Bucher, D. and Pilet, P. E.,** Ethylene production in growing and gravireacting maize and pea root segments, *Plant Sci. Lett.,* 22, 7, 1981.
93. **Larqué-Saavedra, A., Wilkins, H., and Wain, R. L.,** Promotion of cress root elongation in white light by 3,5-diiodo-4-hydroxy-benzoic acid, *Planta,* 126, 269, 1975.
94. **Robert, M. L., Taylor, H. F., and Wain, R. L.,** Ethylene production by cress roots and excised cress root segments and its inhibition by 3,5-diiodo-4-hydroxybenzoic acid, *Planta,* 126, 273, 1975.
95. **Jackson, M. B. and Barlow, P. W.,** Root geotropism and the rôle of growth regulators from the cap, *Plant Cell Environ.,* 4, 107, 1981.
96. **Jackson, M. B., Dobson, C. M., Herman, B., and Merryweather, A.,** Modification of 3,5-diiodo-4-hydroxybenzoic acid (DIHB) activity and situmulation of ethylene production by small concentrations of oxygen in the root environment, *Plant Growth Regul.,* 2, 251, 1984.
97. **Eliasson, L. and Bollmark, M.,** Ethylene as a possible mediator of light induced inhibition of root growth, *Physiol. Plant.,* 72, 605, 1988.
98. **Pickard, B. G.,** Roles of hormones, protons and calcium in geotropism, in *Encyclopedia of Plant Physiology, New Series, Vol. II, Hormonal Regulation of Development III, Environmental Factors,* Pharis, R. P. and Reid, D. M., Eds., Springer-Verlag, Berlin, 1985, 193.
99. **Tackett, J. L. and Pearson, R. W.,** Effect of carbon dioxide on cotton seedling root penetration of compacted soil cores, *Soil Sci. Soc. Am. Proc.,* 28, 741, 1964.
100. **Konings, H. and Jackson, M. B.,** Production of ethylene, and the promoting and inhibiting effects of applied ethylene on root elongation, in various species, in Agricultural Research Council Letcombe Laboratory Annual Report, 1974/1975, Agricultural Research Council, London, 23.
101. **Curtis, R. W.,** Mediation of a plant response to malformin by ethylene, *Plant Physiol.,* 42, 76, 1968.
102. **Hensel, W. and Iversen, T.-H.,** Ethylene production during clinostat rotation and effect on root geotropism, *Z. Pflanzenphysiol.,* 97, 343, 1980.

103. **Bucher, D. and Pilet, P.-E.,** Ethylene effects on growing and gravireacting maize root segments, *Physiol. Plant.*, 55, 1, 1982.
104. **Wright, M., Mousdale, D. M. A., and Osborne, D. J.,** Evidence for a gravity-regulated level of endogenous auxin controlling cell elongation and ethylene production during geotropic bending in grass nodes, *Biochem. Physiol. Pflanz.*, 172, 581, 1978.
105. **Borgström, G.,** Theoretical suggestions regarding the ethylene responses of plants and observations on the influence of apple-emanations, *K. Fysiogr. Saellsk. Lund. Foerh.*, 9, 135, 1939.
106. **Vyas, S. P., Bohra, S. P., and Sankhla, N.,** Antagonism between morphactin and ethylene in root-coiling of *Ipomoea pentaphylla*, *Z. Pflanzenphysiol.*, 69, 185, 1973.
107. **Curtis, R. W.,** Antiethylene properties of $AgNO_3$ and 2,5-norbornadiene in light and dark in *Vigna radiata*, *J. Plant Growth Regul.*, 6, 41, 1987.
108. **Zobel, R. W.,** Some physiological characteristics of the ethylene-requiring tomato mutant diageotropica, *Plant Physiol.*, 52, 385, 1973.
109. **Kelly, M. O. and Bradford, K. J.,** Insensitivity of the *Diageotropica* mutant to auxin, *Plant Physiol.*, 82, 713, 1986.
110. **Bond, G.,** Some features of root growth in nodulated plants of *Myrica gale* L., *Ann. Bot.*, 16, 467, 1952.
111. **Sena Gomes, A. R. and Kozlowski, T. T.,** Responses of *Melaleuca quinquenervia* seedlings to flooding, *Physiol. Plant.*, 49, 473, 1980.
112. **Ellmore, G. S.,** Root dimorphism in *Ludwigia peploides* (Onagraceae) structure and gas content of mature roots, *Am. J. Bot.*, 68, 557, 1981.
113. **Whiteley, G. M., Utomo, W. H., and Dexter, A. R.,** A comparison of penetrometer pressures and the pressures exerted by roots, *Plant Soil*, 61, 351, 1981.
114. **Goss, M. J.,** Effects of mechanical impedance on root growth in barley (*Hordeum vulgare* L.). I. Effects on elongation and branching of seminal roots, *J. Exp. Bot.*, 28, 96, 1977.
115. **Misra, R. K., Dexter, A. R., and Alston, A. M.,** Maximum axial and radial growth pressures of plant roots, *Plant Soil*, 95, 315, 1980.
116. **Atwell, B. J.,** Physiological responses of lupin roots to soil compaction, in *Structural and Functional Aspects of Transport in Roots*, Loughman, B. C., Gasparikova, U., and Kolek, J., Eds., Kluwer Academic, Dordrecht, 1989, 251.
117. **Veen, B. W.,** The influence of mechanical impedance on the growth of maize roots, *Plant Soil*, 66, 101, 1982.
118. **Whalen, M. C.,** The effect of mechanical impedance on ethylene production by maize roots, *Can. J. Bot.*, 66, 2139, 1988.
119. **Wilkins, S. M., Wilkins, H., and Wain, R. L.,** Chemical treatment of soil alleviates effects of soil compaction on pea seedling growth, *Nature*, 259, 392, 1976.
120. **Wilkins, H., Wilkins, S. M., and Wain, R. L.,** Studies on plant growth-regulating substances. XLVIII. The effects of 3,5-diiodo-4-hydroxybenzoic acid on seedling growth in compact and loose soils, *Ann. Appl. Biol.*, 87, 415, 1977.
121. **Wilkins, S. H., Wilkins, H., and Wain, R. L.,** Studies on plant growth regulating substances. LI. The effects of 3,5-dibromo- and 3,5-dichloro-4-hydroxybenzoic acids on seedling growth in compact and loose soils, *Ann. Appl. Biol.*, 89, 265, 1978.
122. **Goss, M. J., Dexter, A. R., and Evans, M.,** Mechanics of root elongation and the effects of 3,5-diiodo-4-hydroxybenzoic acid (DHIB), *Plant Soil*, 99, 211, 1987.
123. **Saini, G. R.,** Root elongation and plant growth in a basal till compact soil treated with 3,5-diiodo-4-hydroxybenzoic acid and gibberellic acid, *Agron. J.*, 71, 1067, 1979.
124. **Goss, M. J. and Russell, R. S.,** Effects of mechanical impedance on root growth in barley (*Hordeum vulgare* L.). III. Observations on the mechanism of response, *J. Exp. Bot.*, 31, 577, 1980.
125. **Schierle, J., Rohwer, F., and Bopp, M.,** Distribution of ethylene synthesis along the etiolated pea shoot and its regulation by ethylene, *J. Plant Physiol.*, 134, 331, 1989.
126. **Bjorkman, T. and Cleland, R. E.,** The role of the epidermis and cortex in gravitropic curvature of maize roots, *Planta*, 176, 513, 1988.
127. **Rost, T. and Sammut, M.,** Regulation of cell division in pea root tips after wounding: a possible role for ethylene, *Protoplasma*, 111, 1, 1982.
128. **Smith, K. A. and Russell, R. S.,** Occurrence of ethylene, and its significance, in anaerobic soil, *Nature*, 222, 769, 1969.
129. **Jackson, M. B.,** Aeration in the nutrient film technique of glasshouse crop production and the importance of oxygen, ethylene and carbon dioxide, *Acta Hortic.*, 98, 61, 1980.
130. **Jackson, M. B.,** Ethylene as a growth promoting hormone under flood conditions, in *Plant Growth Substances 1982*, Wareing, P. F., Ed., Academic Press, London, 1982, 291.
131. **Jackson, M. B., Fenning, T. M., Drew, M. C., and Saker, L. R.,** Stimulation of ethylene production and gas-space (aerenchyma) formation in adventitious roots of *Zea mays* L. by small partial pressures of oxygen, *Planta*, 165, 486, 1985.

132. **Atwell, B. J., Drew, M. C., and Jackson, M. B.,** The influence of oxygen deficiency on ethylene synthesis, 1-aminocyclopropane-1-carboxylic acid levels and aerenchyma formation in roots of *Zea mays*, *Physiol. Plant.*, 72, 15, 1988.
133. **Smith, K. A. and Restall, S. W. F.,** The occurrence of ethylene in anaerobic soil, *J. Soil Sci.*, 22, 430, 1971.
134. **Cornforth, I. S.,** The persistence of ethylene in aerobic soils, *Plant Soil*, 42, 85, 1975.
135. **Armstrong, W.,** Aeration in higher plants, in *Advances in Botanical Research*, Woolhouse, H. W., Ed., Academic Press, London, 1979, 226.
136. **Jackson, M. B.,** Regulation of aerenchyma formation in roots and shoots by oxygen and ethylene, in *Cell Separation in Plants. Physiology, Biochemistry and Molecular Biology*, Osborne, D. J. and Jackson, M. B., Eds., Springer-Verlag, Berlin, 1989, 263.
137. **Konings, H.,** Ethylene-promoted formation of aerenchyma in seedling roots of *Zea mays* L. under aerated and non-aerated conditions, *Physiol. Plant.*, 54, 119, 1982.
138. **Drew, M. C., Jackson, M. B., and Giffard, S.,** Ethylene-promoted adventitious rooting and development of cortical air spaces (aerenchyma) in roots may be adaptive responses to flooding in *Zea mays* L., *Planta*, 147, 83, 1979.
139. **Drew, M. C., Jackson, M. B., Giffard, S. C., and Campbell, R.,** Inhibition by silver ions of gas space (aerenchyma) formation in adventitious roots of *Zea mays* L. subjected to exogenous ethylene or to oxygen deficiency, *Planta*, 153, 217, 1981.
140. **Konings, H. and Verschuren, G.,** Formation of aerenchyma in roots of *Zea mays* in aerated solutions and its relation to nutrient supply, *Physiol. Plant.*, 49, 265, 1980.
141. **Drew, M. C., Chuan-Jie, H., and Morgan, P. W.,** Ethylene synthesis and sensitivity in the formation of aerenchyma in response to deficiencies of N and P in roots of *Zea mays*, in *Biochemical and Physiological Aspects of Ethylene Production in Lower and Higher Plants*, Clijsters, H., de Proft, M., Marcelle, R., and van Poucke, M., Eds., Kluwer Academic, The Hague, 1989, 323.
142. **Campbell, R. and Drew, M. C.,** Electron microscopy of gas space (aerenchyma) formation in adventitious roots of *Zea mays* L. subjected to oxygen shortage, *Planta*, 157, 350, 1983.
143. **Webb, J. and Jackson, M. B.,** Transmission and cryo-scanning electron microscopy study of the formation of aerenchyma (cortical gas-filled space) in adventitious roots of rice, *J. Exp. Bot.*, 37, 832, 1986.
144. **Jackson, M. B., Fenning, T. M., and Jenkins, W.,** Aerenchyma (gas-space) formation in adventitious roots of rice (*Oryza sativa* L.) is not controlled by ethylene or small partial pressures of oxygen, *J. Exp. Bot.*, 36, 1566, 1985.
145. **Small, J. G., Hough, M. C., Clarke, B., and Grobbelaar, N.,** The effect of temperature on nodulation of whole plants and isolated roots of *Phaseolus vulgaris* L., *S. Afr. J. Sci.*, 64, 218, 1968.
146. **Grobbelaar, N., Clark, B., and Hough, M. C.,** The inhibition of root nodulation by ethylene, *Agroplantae*, 2, 81, 1970.
147. **Grobbelaar, N., Clarke, B., and Hough, M. C.,** The nodulation and nitrogen fixation of isolated roots of *Phaseolus vulgaris* L. III. The effect of carbon dioxide and ethylene, in *Biological Nitrogen Fixation in Natural and Agricultural Habitats, Plant and Soil*, Special Vol., Lie, T. A. and Mulder, E. G., Eds., Martinus Nijhoff, The Hague, 1971, 223.
148. **Drennan, D. S. H. and Norton, C.,** The effect of ethrel on nodulation of *Pisum sativum*, *Plant Soil*, 36, 53, 1972.
149. **Ligero, F., Lluch, C., and Olivares, J.,** Evolution of ethylene from roots and nodulation rate of *Alfalfa medicago-sativa* L. plants inoculated with *Rhizobium-meliloti* as affected by the presence of nitrate, *J. Plant Physiol.*, 129, 461, 1987.
150. **Zaat, S. A. J., Van Brussel, A. A. N., Tak, T., Lugtenberg, B. J. J., and Kijne, J. W.,** The ethylene inhibitor aminoethoxyvinylglycine restores normal nodulation by *Rhizobium leguminosarum* biovar vicia on *Vicia sativum nigra* by suppressing the 'thick and short roots' phenotype, *Planta*, 177, 141, 1989.
151. **Sprent, J. I.,** Nitrogen fixation, in *Advanced Plant Physiology*, Wilkins, M. B., Ed., Pitman, London, 1984, 249.
152. **Woltering, E. J., Haren, F., and Boerrichtere, H. A. M.,** Use of a laser driven photoacoustic detection system for measurements of ethylene production in *Cymbidium* flowers, *Plant Physiol.*, 88, 506, 1988.

Chapter 10

ETHYLENE IN VEGETATIVE GROWTH

Ilya Raskin

TABLE OF CONTENTS

I. INTRODUCTION

A large number of observations on the effects of ethylene on vegetative growth have been made since the discovery of the hormonal effects of this gas.[1] The role of ethylene in regulating such physiological processes as dormancy, germination, root and leaf development, and abscission are discussed elsewhere in this book. This chapter will concentrate on the effects of ethylene on the vegetative growth of dry land and semiaquatic plants with the emphasis on the cellular mechanisms of ethylene action. The recent work dedicated to understanding ethylene action at the molecular level and the future approaches to ethylene research will also be discussed.

II. DRY LAND PLANTS

The inhibitory effects of ethylene on vegetative shoot growth were extensively reviewed by Abeles.[2] In short, ethylene inhibits shoot elongation in most terrestrial dicotyledonous plants and, with few exceptions,[3,4] in monocotyledonous plants. Ethylene causes swelling of the elongating regions of shoots, roots, and leaf petioles by inhibiting longitudinal growth (for review see References 2, and 5 to 8). Ethylene-induced closure of the hook region of monocot and dicot seedlings is also based on the ability of ethylene to selectively inhibit cell growth on the inner part of the hook.[2] In some plants, the inhibitory effects of ethylene on shoot growth have been attributed to the reduction in auxin levels[9] possibly via ethylene-induced acceleration of auxin metabolism.[10] Ethylene also inhibits IAA transport in plant systems where ethylene inhibits elongation.[11-13] Understanding the mechanism by which ethylene modifies cell growth is an important step in learning about the molecular mode of action of this hormone as well as more general questions concerning plant development and organogenesis. Therefore, in recent years investigators have been concentrating on the intracellular processes triggered by ethylene rather than on the phenomenological studies of ethylene effects on the whole plant.

At the cellular level, ethylene-induced swelling of etiolated tissue is caused by the inhibition of cell division and alteration of cellular growth in the elongating region in which lateral cell expansion induced by ethylene is accomplished at the expense of longitudinal extension.[14-19] The direction of the cell expansion is thought to be determined by the direction of the cellulose microfibril deposition in the cell wall.[20-22] Electron and polarized-light microscopy has demonstrated that the transverse orientation of the cellulose microfibrils occurs during longitudinal elongation, while longitudinal microfibril deposition produces lateral expansion. The ethylene effect on microfibril deposition in laterally expanding organs has confirmed that ethylene-induced swelling is accompanied by a change in cellulose microfibril orientation from transverse to longitudinal.[8,23,24] The orientation of the microfibrils in the cell wall is hypothesized to be determined by the orientation of the microtubules.[25,26] The study of ethylene-treated tissues has supported this view by demonstrating that ethylene treatment leads to a net reorientation of the microtubules in the longitudinal direction.[27-29] In addition to the physical structure, the chemical composition of the cell wall of terrestrial plants is also modified by ethylene. The amount of hydroxyproline in the cell wall increases following treatment with ethylene.[30,31] Since cell well elongation may be at least partially controlled by the amount of cross-linking between hydroxyproline residues in the wall proteins and cell wall polysaccharides,[32,33] increases in wall hydroxyproline may result in the cessation of cell elongation observed as a result of ethylene treatment. However, other studies have indicated that ethylene may affect free proline pool sizes rather than hydroxyproline itself.[34]

III. WET LAND AND AQUATIC PLANTS

In contrast to dry land plants, ethylene commonly enhances growth in semiaquatic and aquatic plants. For instance, Musgrave et al.[35] were first to observe that the growth of nonsubmerged *Callitriche platycaepa* in an ethylene atmosphere was stimulated to the same extent as that of the fully submerged plants. Therefore, it was proposed that growth of submerged *Callitriche platycaepa* was enhanced because of the accumulation of ethylene in the tissue. The ethylene-induced elongation in *Callitriche* was also dependent on the presence of endogenous gibberellins. Ethylene was also implicated in the promotion of growth in such semiaquatic plants as *Ranunculus sceleratus,*[36,37] *Sagittaria pygmaea,*[38] *Potamogeton distinctus,*[38] and in the semiaquatic fern *Regnellidium diphyllum.*[39-42]

Of all semiaquatic plants, the growth-promoting effect of ethylene has been most extensively studied in rice. Low concentrations of ethylene increase the elongation of the rice mesocotyl and coleoptile.[43-46] Auxin-induced elongation of rice coleoptile segments is promoted by ethylene, indicating that ethylene effects may be partially mediated by auxin.[47] This possibility is partially downgraded by the fact that ethylene may still exert its stimulatory effects on rice coleoptiles even when auxin levels are very low.[48] In comparison to ethylene inhibition of auxin transport, which occurs in terrestrial plants,[11-13] ethylene promotes auxin transport in rice mesocotyls.[49] In addition to ethylene, high CO_2 and low oxygen also contribute to the enhancement of mesocotyl and coleoptile elongation in rice seedlings.[50] In contrast, the leaf growth in rice is inhibited by high CO_2 and low O_2. The additive effect of ethylene, CO_2, and low oxygen allows rice seedlings to emerge from shallow water on waterlogged soils.

The growth-promoting effects of ethylene were also studied in deepwater rice varieties. These varieties are grown in the low-lying areas of Southeast Asia which are flooded every year with up to 5 m of water during the rainy season. The survival of deepwater rice depends on the ability of its internodes to elongate up to 25 cm a day, which is usually fast enough to keep a part of the foliage above the water. It was shown that the signal for the enhanced internodal growth in deepwater rice is the reduced level of oxygen in the submerged internodes.[51,52] Low oxygen tensions stimulate ethylene production by increasing the activity of ACC-synthase.[53] Ethylene accumulation in the submerged internodes is facilitated by the slow rate of ethylene diffusion in water. Ethylene promotes internodal growth by increasing the responsiveness of the tissue to endogenous gibberellin.[54] The growth response is based on enhanced cell division in the intercalary meristem and on an increase in the final length of internodal cells.[55] Subsequent to ethylene treatment, the lag time for the acceleration of internodal elongation is 60 min.[56] The ultrastructural studies of the rapidly elongating regions of rice internodes showed the stimulation of cell wall synthesis at the outer epidermal wall and suppression of secondary wall formation in the inner internodal tissue.[57] Since the outer epidermal wall is the site where the wall loosening required for growth takes place, it is possible that ethylene exerts some of its growth-promoting effects in rice by increasing the outer wall plasticity, thus permitting organ expansion. The elongation rates in the internodes that reach the water surface decrease rapidly because of the faster diffusion of ethylene into the atmosphere and lower rates of ethylene biosynthesis under atmospheric oxygen tensions.

IV. MORPHOGENIC RESPONSES

Poor aeration conditions imposed by partial flooding or growth in the waterlogged soils also leads to the accumulation of ethylene in the tissues of many mono- and dicotyledon species not adapted to wet lands. The accumulation of ethylene, in turn, causes the death and separation of cortical cells, resulting in the formation of connected intercellular lacunae that facilitate gas movement in the submerged tissues.[58] At the same time, increases in the ethylene concentration in waterlogged plants stimulates epinastic curvature of tomato petioles.[59,60] The increase in the ethylene content in the shoots of waterlogged tomato plants could be at least partially explained by the export of the immediate ethylene precursor ACC from the anoxic roots.[61] Rapid ethylene accumulation in waterlogged *Phaseolus vulgaris* was also shown to contribute to leaf wilting.[62]

Application of ethylene stimulates xylem formation in pine.[63] Ethylene also promotes xylem-cell differentiation in explants of Jerusalem artichoke (*Helianthus tuberosus*) via a regulatory mechanism that may be regulated by protein phosphorylation.[64] In lettuce pith explants, ethylene was suggested to play a role in controlling lignification during xylogenesis by inducing wall-bound peroxidase activity.[65] Xylem formation also increased following the application of such ethylene precursors as ACC and methionine.[66]

V. GEOTROPISM AND APICAL DOMINANCE

One of the earliest mentioned effects of ethylene on the etiolated seedlings of dicotyledon plants is the loss of geotropism,[1,67] reviewed in Reference 2. However, all the evidence linking endogenous ethylene to the regulation of the geotropic response is indirect. For example, in the digeotropic mutant of tomato, isolated as a result of spontaneous mutation, the shoots elongate in the horizontal plane. Exogenous ethylene causes prostrate shoots of the digeotropic tomato to grow upward.[68,69] On withdrawing the exogenous ethylene, the digeotropic growth habit is resumed.[59] It was also noted that in several plant species ethylene evolution was higher in plants placed on their side[70] or rotated in the clinostat — a device that rotates the plant around a horizontal axis.[71,72] Asymmetrical distribution of auxin in gravitropically stimulated shoots results in asymmetric ethylene production.[73] Also antagonists of ethylene action, such as Ag^+ and CO_2 and inhibitors of ethylene synthesis such as AVG and Co^+ can significantly delay the gravitropic response.[74] It is difficult to prove that the gravity-imposed difference in ethylene production is the cause and not the result of geotropism[75] and that the effect of inhibitors is specific enough to suggest ethylene involvement.

The role of ethylene in the control of apical dominance was reviewed by Yeang and Hillman in 1984.[76] Application of exogenous ethylene to the apical region of shoots is conducive to lateral bud outgrowth, whereas application directly to the lateral buds is inhibitory for the later stages of bud development. Recent studies have presented evidence that, at least in *Pharbitis nil,* the release of the lateral buds from apical dominance, as a result of shoot inversion or other mechanical perturbations, is mediated by the ethylene-induced restriction of main shoot growth.[77-79] This indicates that the ethylene effect on the lateral buds is rather indirect and resembles the effects of decapitation. A similar explanation may apply to *Avena sativa,* where ethylene is capable of promoting tiller release from apical dominance.[80] Therefore, it is still not known whether endogenous ethylene is involved in the regulation of geotropism and apical dominance.

The question of which processes in plant vegetative growth are actually regulated by endogenous ethylene is much more difficult to answer than the question of how vegetative growth is affected by the application of exogenous ethylene. Although a lot of phenomenological data exists that pertains to the second question, we know far less about the regulatory functions of endogenous ethylene. The discovery that certain cyclic olefins, most notably 2,5-norbornadiene,[81,82] are specific and reversible inhibitors of ethylene action provided tools for examining the effects of endogenous ethylene on vegetative growth. For example, the addition of low concentrations of 2,5-norbornadiene to etiolated pea seedlings resulted in slight acceleration of longitudinal growth over the air control.[82] 2,5-Norbornadiene-induced acceleration of elongation was also observed in etiolated pigweed (*Amaranthus retroflexus*) seedlings.[83] These findings indicate that endogenously produced ethylene may optimize shoot

TABLE 1
Responses to Ethylene in Wild-Type and Ethylene-Insensitive *etr Arabidopsis*

Parameter	Ethylene response (percent of air control)	
	Wild type	*etr* Mutant
Organ elongation		
Root	17	111
Hypocotyl	36	104
Chlorophyll content		
Leaf	35	97
Stem	110	94
Peroxidase activity		
Leaf	303	102
Stem	421	92
Ethylene biosynthesis		
Leaf	15	96
Seed germination	664	100

From Bleecker, A. B., Estelle, M. A., Somerville, C., and Kende, H., *Science*, 241, 1089, 1989. With permission.

strength and emergence from the soil by increasing shoot thickness at the expense of longitudinal growth. In contrast, in rice coleoptile and internodes, where ethylene was shown to promote growth (see above), the application of 2,5-norbornadiene inhibited elongation, unless exogenous ethylene was also added.[84,85]

VI. A LOOK INTO THE FUTURE

The recent development of *Arabidopsis thaliana* mutants deficient in their ability to respond to ethylene has provided us with the best tools yet to study the regulatory effect of endogenous ethylene on plant growth. The study of these mutants, using the tools of molecular biology, should permit the identification of genes and proteins that are involved in ethylene signal transduction in plants. The most detailed account of ethylene-insensitive mutants obtained from a M2 population of ethylmethylsulfonate (EMS)-mutagenized seed came from Bleecker et al.[86] The mutant line that had a dominant mutation at a locus designated *etr*, lacked several responses to ethylene that are present in the wild-type plant. In contrast to the wild type, ethylene did not inhibit hypocotyl elongation in etiolated *etr* seedlings, and did not enhance germination, peroxidase activity, and leaf senescence (Table 1). The *etr* mutant also lacked the feedback suppression of ethylene biosynthesis by ethylene. In our laboratory,[12] ethylene-insensitive *Arabidopsis* mutants were isolated after screening 500,000 seedlings grown from M2 EMS-mutagenized seeds. The mutant seedlings, in which hypocotyl elongation and/or hook closure was not inhibited by ethylene, were selected in darkness in a flow-through chamber through which air containing 5 μl/l ethylene was passed at a constant rate. Genetic crosses were performed with 6 mutants. Two were dominant ethylene-insensitive (*eti*) mutations, whereas the other four were recessive. The four recessive mutations belonged to two complementation groups. Figure 1 shows the effects of different ethylene concentrations on hypocotyl length and hook angle in *eti* 3, which is a dominant phenotype, and *eti* 5 and *eti* 10, which are recessive phenotypes.[87] In *etr* 3, ethylene had very little effect on hypocotyl length whereas the hook angle was significantly increased. In contrast, ethylene significantly inhibited growth in *eti* 10 and did not affect the hook angle. In *eti* 5, ethylene had no effect on elongation or hook opening. In both *etr*[86] and *eti* 5,

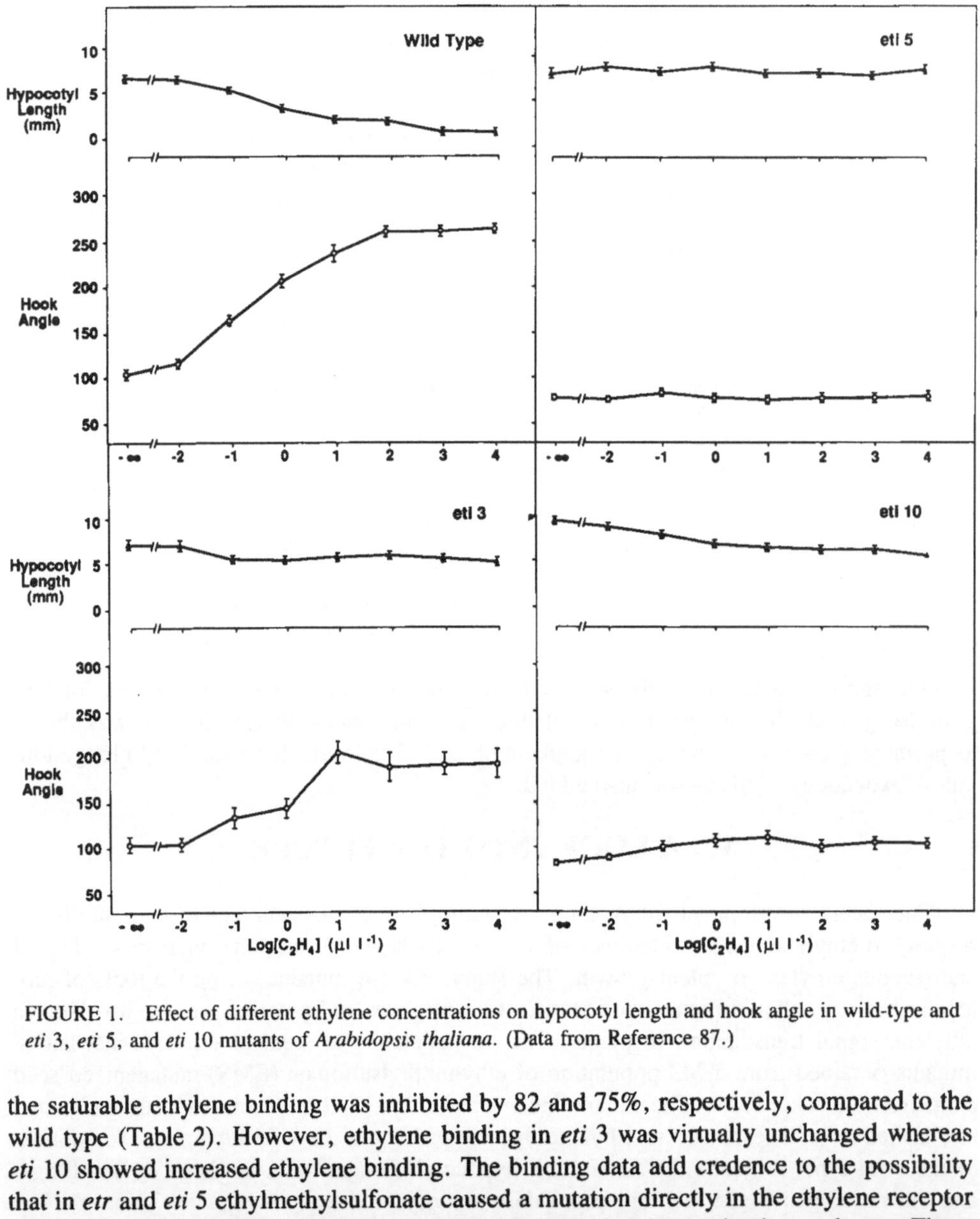

FIGURE 1. Effect of different ethylene concentrations on hypocotyl length and hook angle in wild-type and *eti* 3, *eti* 5, and *eti* 10 mutants of *Arabidopsis thaliana.* (Data from Reference 87.)

the saturable ethylene binding was inhibited by 82 and 75%, respectively, compared to the wild type (Table 2). However, ethylene binding in *eti* 3 was virtually unchanged whereas *eti* 10 showed increased ethylene binding. The binding data add credence to the possibility that in *etr* and *eti* 5 ethylmethylsulfonate caused a mutation directly in the ethylene receptor gene rather than in other components of the ethylene signal transduction pathway. These results suggest that cloning the ethylene receptor gene is possible. Also, cloning genes from the mutants such as *eti* 3 and *eti* 10, that could be impaired in later steps in the ethylene signal transduction pathway, should help to unravel the molecular sequence of ethylene effects on vegetative growth. Most of the tools necessary to accomplish such tasks are already available and are being constantly improved. Improvements in the *Arabidopsis* genetic[88] and RFLP[89] maps, coupled with better methods for chromosome walking, make mapping and cloning the genes responsible for a certain phenotype easier. Also new techniques for direct tagging of the genes of interest will soon be available. T-DNA of *Agrobacterium tumefaciens* was recently used as an insertional mutagen in order to tag a dwarf gene in *Arabidopsis.*[90] Similarly, the technique for the transposon tagging of genes in plants is being developed.[91,92] These rapid developments in biotechnology should further accelerate

TABLE 2
***In vivo* Saturable Binding of [^{14}C]Ethylene in *etr*[a] and *eti Arabidopsis*[b]**

Mutant	Saturable binding (dpm/10 g) ± S.D.	% of the Wild-type
etr 78	59 ± 41	18
eti 5	72 ± 19	25
eti 3	230 ± 31	81
eti 10	442 ± 36	154

[a] Data from Reference 86.
[b] Data from Reference 87.

the process of learning about the molecular mechanisms that transmit and regulate ethylene effects on the vegetative growth. Understanding the effects of ethylene on the genomic level will, in turn, allow the manipulation of the plant processes that are regulated by ethylene.

REFERENCES

1. **Neljubow, D.,** Uber die horizontale Nutation der Stangel von *Pisum sativum* und einiger anderen Pflanzen, *Beih. Bot. Centralbl.*, 10, 128, 1901.
2. **Abeles, F. B.,** *Ethylene in Plant Biology,* Academic Press, New York, 1973.
3. **Suge, H.,** Stimulation of oat and rice mesocotyl growth by ethylene, *Plant Cell Physiol.*, 12, 831, 1971.
4. **Cornforth, I. S. and Stevens, R. J.,** Ethylene and the germination and early growth of barley, *Plant Soil,* 38, 581, 1973.
5. **Burg, S. P.,** Ethylene and plant growth, *Proc. Natl. Acad. Sci. U.S.A.*, 70, 591, 1973.
6. **Osborne, D. J.,** Ethylene and target cells in the growth of plants, *Sci. Prog. Oxford,* 64, 51, 1977.
7. **Lieberman, M.,** Biosynthesis and action of ethylene, *Annu. Rev. Plant Physiol.*, 30, 533, 1979.
8. **Eisinger, W.,** Regulation of pea internode expansion by ethylene, *Annu. Rev. Plant Physiol.*, 34, 225, 1983.
9. **Lieberman, M. and Knegt, E.,** Influence of ethylene on indole-3-acetic acid concentration in etiolated pea epicotyl tissue, *Plant Physiol.*, 60, 475, 1977.
10. **Riov, J., Dror, N., and Goren, R.,** Effect of ethylene on [^{14}C]indole-3-acetic acid metabolism in leaf tissues of woody plants, *Plant Physiol.*, 70, 1265, 1982.
11. **Morgan, P. W. and Gausman, H. W.,** Effects of ethylene on auxin transport, *Plant Physiol.*, 41, 45, 1966.
12. **Beyer, E. M. and Morgan, P. W.,** Effect of ethylene on the uptake, distribution, and metabolism of indoleacetic acid-1-^{14}C and -2-^{14}C and naphthaleneacetic acid-1-^{14}C, *Plant Physiol.*, 46, 157, 1970.
13. **Malloch, K. R. and Osborne, D. J.,** Auxin and ethylene control of growth in seedlings of *Zea mays* L. and *Avena sativa* L., *J. Exp. Bot.*, 27, 992, 1976.
14. **Marinos, N. G.,** Some responses of Avena coleoptile to ethylene, *J. Exp. Bot.*, 11, 227, 1960.
15. **Chadwick, A. V. and Burg, S. P.,** Regulation of root growth by auxin-ethylene interaction, *Plant Physiol.*, 45, 192, 1970.
16. **Apelbaum, A. and Burg, S. P.,** Effects of ethylene and 2,4-dichlorophenoxyacetic acid on cellular expansion in *Pisum sativum, Plant Physiol.*, 50, 125, 1972.
17. **Apelbaum, A. and Burg, S. P.,** Effects of ethylene on cell division and deoxyribonucleic acid synthesis in *Pisum sativum, Plant Physiol.*, 50, 117, 1972.
18. **Nee, M., Chiu, L., and Eisinger, W.,** Induction of swelling in the internode tissue by ethylene, *Plant Physiol.*, 62, 902, 1978.
19. **Taiz, L., Rayle, D. L., and Eisinger, W.,** Ethylene-induced lateral expansion in etiolated pea stems, *Plant Physiol.*, 73, 413, 1983.
20. **Green, P. B.,** Organogenesis—a biophysical view, *Annu. Rev. Plant Physiol.*, 31, 51, 1980.
21. **Richmond, P. A., Metraux, J.-P., and Taiz, L.,** Cell expansion patterns and directionality of wall mechanical properties in Nitella, *Plant Physiol.*, 65, 211, 1980.

22. **Takeda, K. and Shibota, H.,** Changes in microfibril arrangement on the inner surface on the epidermal cell walls in the epicotyl of *Vignia angularis* Ohwi et Ohashi during cell growth, *Planta,* 151, 385, 1981.
23. **Burg, S. P. and Burg, E. A.,** Auxin stimulated ethylene formation: its relationship to auxin inhibited growth, root geotropism and other plant processes, in *Biochemistry and Physiology of Plant Growth Substances,* Wightman, F. and Setterfield, G., Eds., Range, Ottawa, 1967, 1275.
24. **Apelbaum, A. and Burg, S. P.,** Altered cell microfibrillar orientation in ethylene-treated *Pisum sativum* stems, *Plant Physiol.,* 48, 648, 1974.
25. **Newcomb, E. H.,** Plant microtubules, *Annu. Rev. Plant Physiol.,* 20, 253, 1969.
26. **Lloyd, C. W.,** Toward a dynamic helical model for the influence of microtubules on wall patterns in plants, *Int. Rev. Cytol.,* 86, 1, 1986.
27. **Lang, J. M., Eisinger, W. R., and Green, P. B.,** Effect of ethylene on the orientation of microtubules and cellulose microfibrils of pea epicotyl cells with polylamellate cell walls, *Protoplasma,* 110, 5, 1962.
28. **Steen, D. A. and Chadwick, A. V.,** Ethylene effect in pea stem tissue, evidence of microtubule mediation, *Plant Physiol.,* 67, 460, 1981.
29. **Roberts, I. N., Lloyd, C. W., and Roberts, K.,** Ethylene-induced microtubule reorientations: mediation by helical arrays, *Planta,* 164, 439, 1985.
30. **Ridge, I. and Osborne, D. J.,** Role of peroxidase when hydroxyproline-rich protein in plant cell walls is increased by ethylene, *Nature New Biol.,* 229, 205, 1971.
31. **Savada, D. and Chrispeels, M. J.,** Hydroxyproline-rich cell wall protein: role in the cessation of elongation in excised pea epicotyls, *Dev. Biol.,* 30, 49, 1973.
32. **Lamport, D. T. A.,** The protein component of primary cell walls, *Adv. Bot. Res.,* 2, 151, 1965.
33. **Savada, D., Walker, F., and Chrispeels, M. J.,** Hydroxyproline-rich cell wall protein (extensin) biosynthesis and accumulation in growing pea epicotyls, *Dev. Biol.,* 30, 42, 1973.
34. **Nee, M., Chiu, L., and Eisinger, W.,** Induction of swelling in pea internode tissue by ethylene, *Plant Physiol.,* 62, 902, 1978.
35. **Musgrave, A., Jackson, M. B., and Ling, E.,** *Callitriche* stem elongation is controlled by ethylene and gibberellin, *Nature New Biol.,* 238, 93, 1972.
36. **Musgrave, A. and Walters, J.,** Ethylene-stimulated growth and auxin transport in *Ranunculus sceleratus,* *New Phytol.,* 72, 783, 1973.
37. **Samarakoon, A. B. and Horton, R. F.,** Petiole growth in *Ranunculus sceleratus* L.: ethylene synthesis and submergence, *Ann. Bot.,* 54, 263, 1984.
38. **Suge, H. and Kusanagi, T.,** Ethylene and carbon dioxide: regulation of growth in two perennial aquatic plants, arrowhead and pondweed, *Plant Cell Physiol.,* 16, 65, 1975.
39. **Musgrave, A. and Walters, J.,** Ethylene and buoyancy control of the semi-aquatic fern *Regnellidium diphyllum, Planta,* 121, 51, 1974.
40. **Cookson, C. and Osborne, D. J.,** The stimulation of cell extension by ethylene and auxin in aquatic plants, *Planta,* 144, 39, 1978.
41. **Cookson, C. and Osborne, D. J.,** The effect of ethylene and auxin on cell wall extensibility of the semi-aquatic fern *Regnellidium diphyllum, Planta,* 146, 303, 1979.
42. **Walters, J. and Osborne, D. J.,** Ethylene and auxin-induced cell growth in relation to auxin transport and metabolism and ethylene production in the semi-aquatic fern *Regnellidium diphyllum, Planta,* 146, 309, 1979.
43. **Ku, H. S., Suge, H., Rappaport, L., and Pratt, H. K.,** Stimulation of rice coleoptile growth by ethylene, *Planta,* 90, 333, 1970.
44. **Imaseki, H., Pjon, C. J., and Furuya, M.,** Phytochrome action in *Oriza sativa L.* IV. Red and far red reversible effect on the production of ethylene in excised coleoptiles, *Plant Physiol.,* 48, 241, 1971.
45. **Suge, H.,** Stimulation of oat and rice mesocotyl growth by ethylene, *Plant Cell Physiol.,* 12, 831, 1971.
46. **Miller, J. H. and Miller, P. M.,** Ethylene and the responses to light of rice seedlings, *Physiol. Plant.,* 30, 206, 1974.
47. **Imaseki, H. and Pjon, C.-J.,** The effect of ethylene on auxin-induced growth of excised rice coleoptile segments, *Plant Cell Physiol.,* 11, 827, 1970.
48. **Katsura, N. and Suge, H.,** Does ethylene induce elongation of the rice coleoptile through auxin?, *Plant Cell Physiol.,* 20, 1147, 1979.
49. **Suge, H.,** Stimulation of oat and rice mesocotyl growth by ethylene, *Plant Cell Physiol.,* 12, 831, 1971.
50. **Raskin, I. and Kende, H.,** Regulation of growth in rice seedlings, *J. Plant Growth Regul.,* 2, 193, 1983.
51. **Metraux, J.-P. and Kende, H.,** The role of ethylene in the growth response of submerged deep water rice, *Plant Physiol.,* 72, 441, 1983.
52. **Raskin, I. and Kende, H.,** Regulation of growth in stem sections of deep-water rice, *Planta,* 160, 66, 1984.
53. **Cohen, E. and Kende, H.,** In vivo 1-aminocyclopropane-1-carboxylate synthase activity in internodes of deepwater rice, *Plant Physiol.,* 84, 282, 1987.

54. **Raskin, I. and Kende, H.,** Role of gibberellin in the growth response of submerged deep water rice, *Plant Physiol.*, 76, 947, 1984.
55. **Bleecker, A. B., Schuette, J. L., and Kende, H.,** Anatomical analysis of growth and developmental patterns in the internode of deepwater rice, *Planta*, 169, 490, 1986.
56. **Rose-John, S. and Kende, H.,** Short-term growth response of deep-water rice to submergence and ethylene, *Plant Sci.*, 38, 129, 1985.
57. **Kutscera, U. and Kende, H.,** Particles associated with the outer epidermal wall in the internodes of deepwater rice, *Ann. Bot.*, 63, 385, 1989.
58. **Jackson, M. B.,** Ethylene and responses of plants to soil waterlogging and submergence, *Annu. Rev. Plant Physiol.*, 36, 145, 1985.
59. **Jackson, M. B. and Campbell, D. J.,** Movement of ethylene from roots to shoots, a factor in the responses of tomato plants to waterlogged soil conditions, *New Phytol.*, 74, 397, 1975.
60. **Bradford, K. J. and Dilley, D. R.,** Effects of root anaerobiosis on ethylene production, epinasty and growth of tomato plants, *Plant Physiol.*, 61, 506, 1978.
61. **Bradford, K. J. and Yang, S. F.,** Xylem transport of 1-aminocyclopropane-1-carboxylic acid, an ethylene precursor, in waterlogged tomato plants, *Plant Physiol.*, 65, 322, 1980.
62. **van Schravendijk, W. and van Andel, O. M.,** The role of ethylene during flooding of *Phaseolus vulgaris*, *Physiol. Plant.*, 66, 257, 1986.
63. **Barker, J. E.,** Growth and wood properties of *Pinus radiata* in relation to applied ethylene, *N. Z. J. For. Sci.*, 9, 15, 1979.
64. **Koritsas, V.M.,** Effect of ethylene and ethylene precursors on protein phosphorylation and xylogenesis in tuber explants of *Helianthus tuberosus* (L.), *J. Exp. Bot.*, 39, 375, 1988.
65. **Miller, A. R., Crawford, D. L., and Roberts, L. W.,** Lignification and xylogenesis in *Lactuca* pith explants cultured *in vitro* in the presence of auxin and cytokinin: a role for endogenous ethylene, *J. Exp. Bot.*, 36, 110, 1985.
66. **Miller, A. R. and Roberts, L. W.,** Ethylene biosynthesis and xylogenesis in *Lactuca* pith explants cultured *in vitro* in the presence of auxin and cytokinin: the effect of ethylene precursors and inhibitors, *J. Exp. Bot.*, 35, 691, 1984.
67. **Knight, L. I., Rose, C. R., and Crocker, W.,** Effect of various gases and vapors upon etiolated seedlings of the sweet pea, *Science*, 31, 635, 1910.
68. **Zobel, R. W.,** Some physiological characteristics of the ethylene requiring tomato mutant digeotropica, *Plant Physiol.*, 52, 385, 1973.
69. **Zobel, R. W.,** Control of morphogenesis in the ethylene requiring tomato mutant digeotropica, *Can. J. Bot.*, 52, 735, 1974.
70. **Wright, M., Mousdale, D. M. A., and Osborne, D. J.,** Evidence for gravity-regulated level of endogenous auxin controlling cell elongation and ethylene production during geotropic bending of grass nodes, *Biochem. Physiol. Pflanz.*, 172, 581, 1978.
71. **Hensel, W. and Iversen, T.-H.,** Ethylene production during clinostat rotation and effect on root geotropism, *Z. Pflanzenphysiol. Bd.*, 97, 343, 1980.
72. **Salisbury, F. B. and Wheeler, R. M.,** Interpreting plant responses to clinostating. I. Mechanical stresses and ethylene, *Plant Physiol.*, 67, 677, 1981.
73. **Abeles, F. B. and Rubenstein, R.,** Regulation of ethylene evolution and leaf abscission by auxin, *Plant Physiol.*, 39, 963, 1964.
74. **Salisbury, F. B. and Wheeler, R. M.,** Gravitropism in higher plant shoots. I. A role of ethylene, *Plant Physiol.*, 67, 686, 1981.
75. **Firn, R. D. and Digby, J.,** The establishment of tropic curvatures in plants, *Annu. Rev. Plant Physiol.*, 31, 131, 1980.
76. **Yeang, H. Y. and Hillman, J. R.,** Ethylene and apical dominance, *Physiol. Plant.*, 60, 275, 1984.
77. **Prasad, T. K. and Cline, M. G.,** Mechanical perturbation-induced ethylene releases apical dominance in *Pharbitis nil* by restricting shoot growth, *Plant Sci.*, 41, 217, 1984.
78. **Prasad, T. K. and Cline, M. G.,** Shoot inversion-induced ethylene in *Pharbitis nil* induces the release of apical dominance by restricting shoot elongation, *Plant Sci.*, 38, 163, 1985.
79. **Prasad, T. K. and Cline, M. G.,** The control of apical dominance: localization of the growth region in the *Pharbitis nil* shoot, *J. Plant Physiol.*, 125, 185, 1986.
80. **Harrison, M. A. and Kaufman, P. B.,** Does ethylene play a role in the release of lateral buds (tillers) from apical dominance in oats?, *Plant Physiol.*, 70, 811, 1982.
81. **Sisler, E. C. and Plain, A.,** Effect of ethylene and cyclic olefins on tobacco leaves, *Tob. Sci.*, 17, 68, 1973.
82. **Sisler, E. C. and Yang, S. F.,** Anti-ethylene effects of cis-2-butene and cyclic olefins, *Phytochemistry*, 23, 2765, 1984.
83. **Raskin, I. and Beyer, E. M., Jr.,** Role of ethylene metabolism in *Amaranthus retroflexus*, *Plant Physiol.*, 90, 1, 1989.

84. **Ishizawa, K., Hoshina, M., Kawabe, K., and Esashi, Y.**, Effects of 2,5-norbornadiene on cocklebur seed germination and rice coleoptile elongation in response to CO_2 and C_2H_4, *J. Plant Growth Regul.*, 7, 45, 1988.
85. **Bleecker, A. B., Rose-John, S., and Kende, H.**, An evaluation of 2,5-norbornadiene as a reversible inhibitor of ethylene action in deepwater rice, *Plant Physiol.*, 84, 395, 1987.
86. **Bleecker, A. B., Estelle, M. A., Somerville, C., and Kende, H.**, Insensitivity to ethylene is conferred by a dominant mutation in *Arabidopsis thaliana*, *Science*, 241, 1089, 1989.
87. **Smith, A. R., Harpham, N. V., Sanders, I. O., Hall, M. A., and Raskin, I.**, unpublished data, 1989.
88. **Koornneef, M., Van Eden, J., Hanhart, C. J., Stam, P., Braaksma, F. J., and Feenstra, W. J.**, Linkage map of *Arabidopsis thaliana*, *J. Hered.*, 74, 265, 1983.
89. **Chang, C., Bowman, J. L., DeJohn, A. W., Lander, E. S., and Meyerowitz, E. M.,**. Restriction fragment length polymorphism linkage map for *Arabidopsis thaliana*, *Proc. Natl. Acad. Sci. U.S.A.*, 85, 6856, 1988.
90. **Feldmann, K. A., Marks, M. D., Christianson, M. L., and Quatrano, R. S.**, A dwarf mutant of *Arabidopsis* generated by T-DNA insertion mutagenesis, *Science*, 243, 1315, 1989.
91. **Federoff, N. V., Furtek, B. D., and Nelson, O. E.**, Cloning of the *bronze* locus in maize by a simple and generalizable procedure using the transposable controlling element *Activator (Ac)*, *Proc. Natl. Acad. Sci. U.S.A.*, 81, 3825, 1984.
92. **Cone, K. C., Burr, F. A., and Burr, B.**, Molecular analysis of the maize anthocyanin regulatory locus *C1*, *Proc. Natl. Acad. Sci. U.S.A.*, 83, 9631, 1986.

Chapter 11

ETHYLENE IN LEAF ONTOGENY AND ABSCISSION

Daphne J. Osborne

TABLE OF CONTENTS

I. INTRODUCTION

A. HIGHER PLANTS

Leaves come in all shapes, forms, and sizes and all of them produce ethylene.[1] In higher plants these flattened multilayered photosynthetic structures of palisade and mesophyll are traversed by networks of vascular bundles and enclosed between nonphotosynthetic epidermal cells with thickened cuticles perforated by arrays of complex stomatal apertures. The typical dicotyledonous leaf exhibits dorsiventral asymmetry with respect to its stem but lateral symmetry with respect to the centrally positioned petiole. The petiole base adjoins the stem and subtends a lateral bud. During the course of its development a leaf differentiates many types of cells, each with a specific recognition and response to chemical (hormonal) and physical signals. It is this diversity of cell potential that provides for the shape and form that the leaf eventually acquires and which is then a characteristic of the species. A particularly important cell type to differentiate is the abscission zone cell, for it is by the separation of a plate of these cells, one from another, under the influence of endogenously produced ethylene, that the shedding of the leaf from the stem is achieved, so terminating the functional life of the organ.

Dorsiventral asymmetry is generally less evident in the leaves of monocotyledons and the form is commonly straplike with parallel venation as in the Gramineae and most Liliaceae. The differentiation of leaf abscission cells in these groups is rare. Instead, at senescence, the blade dies back to the junction of the leaf blade with the stem and an active cell separation, with resulting leaf shedding, does not occur. In some species of the Palmaceae *(Elaeis guineensis)*, bases of dried-up leaves may then remain attached to the stem for many years. Nonetheless, some barrier exists between the cells of the senescing leaf and those of the adjoining stem, for the stem tissue retains its fully functional and nonsenescing state quite unimpaired, despite the immediate proximity of the senescing cells of the leaf base.

B. LOWER PLANTS

All lower plants tested so far produce ethylene (see Section VI.B) but in bryophytes and pteridophytes, shedding of leaves (i.e., foliose structures of the gametophytic or sporophytic generation) is unknown, though other cell separation events, such as spore shedding and abscission of gemmae, are normal occurrences. A feature of the gametophytes of many ferns, liverworts, and mosses is the development of a thallus which is mostly only one cell thick and with an apparently limited number of distinguishable cell types. Vascular connections as we know them in xylem elements of higher plants are not differentiated and growth derives from the divisions of a single apical cell. Defined cell lineages are therefore absent and mechanical support resides in a thickening of the walls in certain specialized regions of the simple thallose development. Of particular interest for this review is the recently demonstrated operation of an alternative pathway for the biosynthesis of ethylene by the leafy structures of lower plants. The significance and implications of this are discussed in more detail later in this chapter.

II. LEAF ONTOGENY

A. GENERAL EFFECTS OF ETHYLENE

1. Higher Plants

In most plants, ethylene is an effective inhibitor of leaf expansion, but there are exceptions. Concentrations of applied ethylene as low as 0.25 nl l^{-1} will reduce growth in all cell types in the dicotyledons, for example *Cucumis*,[1] *Pisum sativum*, and *Phaseolus vulgaris*.[2] That modifications of endogenous ethylene production can also regulate leaf growth *in vivo* is indicated by the transient enhancement of expansion that follows a red light pulse.

This enhancement is attributed to an associated phytochrome-mediated red light repression of ethylene production as in etiolated pea seedlings.[3] A promotion of leaf expansion has also been observed in the presence of Co^{2+}. Since the latter ion represses the enzymatic conversion of 1-aminocyclopropane-1-carboxylic acid (ACC) to ethylene then here too, greater leaf expansion seems attributable to a reduced level of ethylene production.[4]

In the Gramineae (*Poa pratensis* and *Avena sativa*) total leaf cell number is also found to be reduced by ethylene, reflecting inhibitory effects on cell division in the continuously functioning meristem at the base of the leaf.[5] The effect of ethylene upon cell enlargement, however, depends on the state of cell maturity. In young leaves, expansion is reduced by ethylene but, interestingly, in mature leaves enlargement can be enhanced. This indicates that some temporal and stage-specific factors must operate to determine the response of different ages of leaf cells to ethylene.

2. Lower Plants

Little is known of the role of ethylene in leaf (or foliose) development in lower plants, although there is good evidence that the elongation of cells in the filamentous stage of gametophyte development in ferns (for example, in *Onoclea sensibilis*) can be enhanced by ethylene concentrations as low as 0.01 to 0.1 $\mu l\ l^{-1}$, though cell division is inhibited as in higher plants.[6] The differentiation of mega- rather than microsporangia on the reproductive leaves (sporophylls) of *Selaginella* plants can be determined by exposure to ethylene.[7] In this respect the response resembles that of the ethylene enhancement of pistillate flower formation in higher plants.[8,9]

All lower plants so far tested produce ethylene at rates similar to those of higher plants (about 0.1 to 5.0 $nl\ g^{-1}$ F. wt h^{-1}) but despite the several parallels with the leaves of higher plants, evidence is increasing that the ethylene involvement with growth and differentiation has certain characteristics that are unique to the primitive groups (see the following sections).

3. Semiaquatic Plants

Many higher and lower genera have evolved as amphibious species and these have a special significance in considering the part that ethylene plays in the growth of leaves.[10] Plants that colonize marshy and shallow-water habitats have the capacity to differentiate organs with structures and functions that suit them to either the aqueous or the aerial environment. Most, however, differentiate selected organs that are specifically adapted to the water (the roots and leaves) or to the air (the flowering parts). Frequently, the aerial leaf forms develop initially upon wholly submerged parts and exhibit not only polymorphy of leaves and shoots but the precise differentiation of a biochemically distinct cell type in the petiole. This cell type is characterized by a capacity for overall enlargement and an enhanced rate of extension growth in the presence of added ethylene.[10] Submergence of the shoot — the condition that results in a build up of the endogenous concentrations of ethylene within the tissues — evokes a similar growth response. There is no evidence however, that ethylene plays a direct role in the regulation of leaf form, but it is very clear that ethylene-enhanced cell extension in petioles or stems is a characteristic of all genera of semiaquatic plants investigated so far. The differentiation of these ethylene-responsive cells (called Type 3 cells)[11] that exhibit overall enhanced cell enlargement with ethylene, can be seen as critical for the growth adjustment of stems and petioles in the maintenance of the leaf blades at their preferred position with respect to the water surface (see Section IV.A). In the foliose lower plant (*Riella helicophylla*) that grows primarily submerged, the presence of ethylene-responsive Type 3 cells has been demonstrated in the central column region of the leafy thallus of the gemmalings.[12,13] When exposed to ethylene therefore, the final shape and form of these thalli is changed from ovate to one that is considerably more elongate than normal.

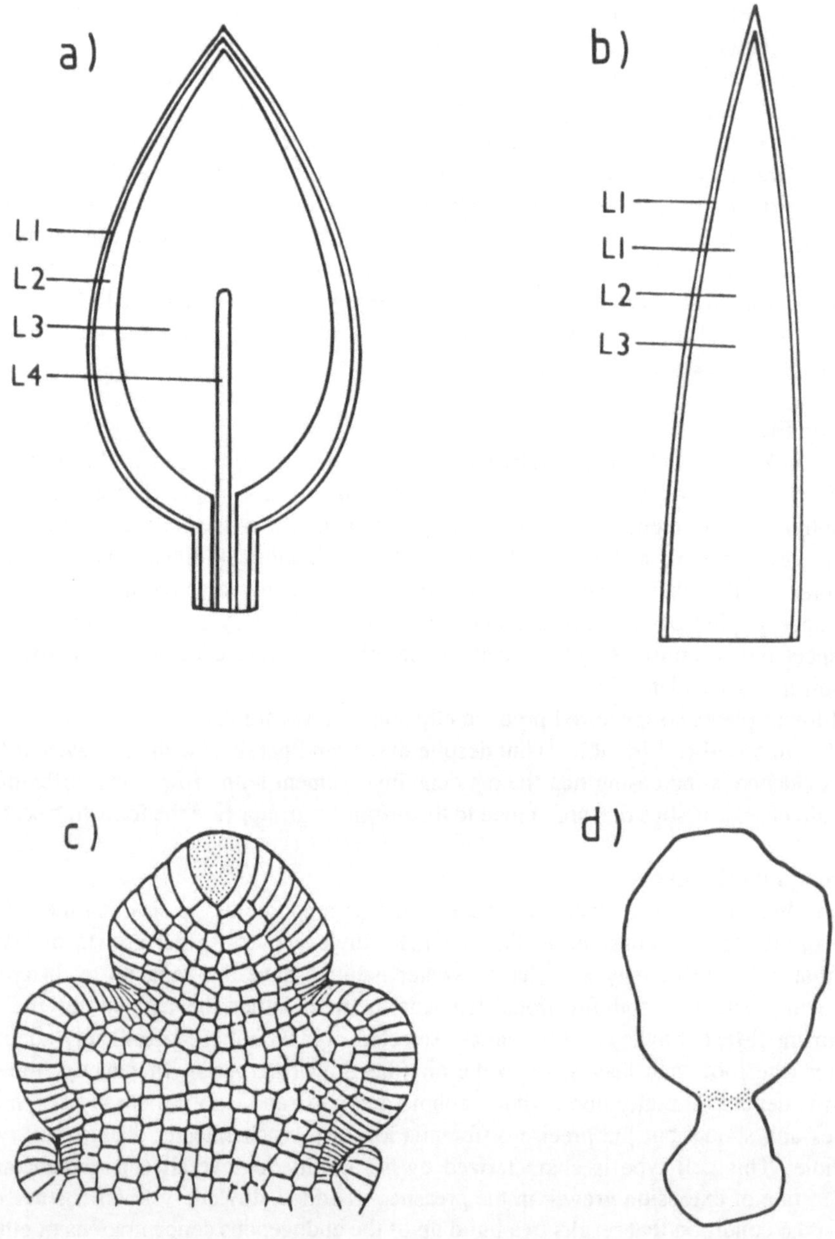

FIGURE 1. Primordia and cell lineages in the foliose development of higher and lower plants: (a,b) multilineage patterns from multicellular primordia of dicotyledon and monocotyledon leaves,[16] L1 (epidermis), L1, L2 and L3 (mesophyll, palisade), L4 (central core); (c) single primordium, single lineage in leaves of lower plants[17] (bryophytes, pteridophytes); (d) single lineage in intercalary meristem of unistratose thallus of lower plant (gemmaling, *Riella helicophylla*[12]).

B. PRIMORDIAL DEVELOPMENT

Leaves and foliose organs of plants are generated either from primordial tissue of several cell layers derived from the apical meristem as in higher plants, or from single apical cell initials as in the more primitive orders (Figure 1). This distinction now appears important for following the regulatory roles of ethylene in plant development.

Although newly initiated leaf primordia in angiosperms (di- or monocotyledons) possess a multicellular distal meristem, intercalary divisions take place and by the time a primordium is some 80 to 200 μm long, new marginal and submarginal growth zones are established along the lateral flanks. The development of these lateral regions of expansion and the attenuation of the distal meristem marks the onset of lamina formation in dorsiventrally flattened leaves (Figure 1a). Early growth of the primordium is predominantly by cell division which may reach 20 or more cycles before leaf unfolding. However, the major period of cell division (>90%) still occurs during lamina expansion, when cells may increase in volume some 15- to 20-fold.[14,15] In the monocotyledons with straplike leaves,[16] the lateral margins have limited growth and instead a broad meristem is established at the junction of the leaf base and the stem. This contributes to the continuous production of cells so that those of the leaf tip are usually already fully expanded while divisions are still continuing at the base (Figure 1b).

In the leaves of ferns, a single apical cell generally founds the cell population and the shape and form of the pinnules and pinnae is dictated by the relative growth of the cells cleaved from that initial.[17] The whole blade is therefore derived from a single source rather than from multiple cell layers as in higher plants (Figure 1c).

Single apical primordial cells are also typical of the Bryophytes and those at the apex of the leafy liverwort *Plagiochila artica* may be subject to some regulation by ethylene. In these plants two of the three cell faces normally give rise to the two rows of leafy outgrowths, but if $AgNO_3$ is added (theoretically to block ethylene binding and, thus, ethylene action) then all three faces of the apical cell produce leafy outgrowths. It is suggested from this that ethylene suppresses divisions at one cell face.[18] Additions of 1-aminoethoxyvinylglycine (AVG) or 1-aminooxyacetic acid (AOA), which in higher plants will block the production of ethylene from methionine via ACC, were also found to permit the development of the three-rowed phenotype variant. However, ethylene appears not to be synthesized via the methionine-ACC pathway in these lower plants (see Section VI.B), so the proposed roles for AVG and AOA in primordial regulation should be reevaluated.

In gemmalings of certain liverworts (e.g., *Riella helicophylla*) an intercalary meristem is established in the tiny thallus prior to its shedding from the parent plant (Figure 1d). These thalli are one cell thick so all the cells derive from a single lineage, as in the adults with single apical cells.

III. CELL DIFFERENTIATION

A. THE LEAF BLADE

Differentiation of cell types commences very soon after formation of the primordium. In angiosperms the primary vein procambium develops in the acropetal direction when the primordial initial is no more than 50 to 100 μm in length, and secondary veins are formed by the time the primordium reaches 250 μm. Increase in thickness during lamina expansion is due primarily to an increase in palisade cell height, the cells being taller when they expand at high light intensities.[19] The expansion process itself appears to be dependent upon a light-promoted cell wall acidification (proton secretion) and a resulting enhanced wall extensibility.[20] Bearing in mind how cell shape and size can be modified by the relative levels of auxin and ethylene to which a tissue is exposed,[21] and the importance of cell commitment in determining signal recognition and response,[22] it becomes increasingly necessary to understand the performance of individual cell types to imposed chemical and physical cues. It is then the collective interaction of the cells that frames the final expression of pattern attained in the complex organization of a leaf.

Determination of the leaf primordium as a committed leaf structure occurs very early in apical development. For example if an immature leaf of *Helianthus annuus* or the fern *Osmunda cinnamomea* is excised when as little as 1 mm in length and then maintained on an agar culture medium of sugars and salts, a small but mature leaf is eventually produced, demonstrating an autonomous control of the completion of pattern development. However, with even smaller primordia, a regeneration of whole plants takes place, indicating that the ability to develop as a leaf, without further signals from the rest of the plant, is not determined at the first cell division of the leaf initial. Rather, determination is acquired at some subsequent but still very early stage of primordial development.[23,24]

In angiosperms, leaf meristems are multicellular and cell lineage analysis of genetically identifiable cell clones in chimeric plants has shown that the epidermis (including the chloroplast-containing guard cells of the stomatal complex) are derived from the L1 layer of the apex, whereas the internal leaf structures — vascular tissue, palisade, and mesophyll cells — are derived primarily from the L2 and L3 layers, L1 may contribute in maize. In tobacco an L4 layer is also recognizable in the central midrib (Figure 1a).[16] Of special interest in the present context is the demonstration in maize leaves of the interaction between neighboring cells derived from different mixed clonal lineages (Figure 1b). For example, L2 cells of certain mutants can reduce the phenotypic response (in this case, pigmentation) in adjacent nonmutant L2 cells.[25] Such cell interactions are of importance in the consideration of factors controlling the differentiation of abscission zones in the leaf.

B. THE ABSCISSION ZONE

1. Derivation

Analysis of cell layer interactions during the development of flower pedicel abscission zones in chimeric tomato plants has demonstrated the importance of signal perception and response between the different cell lineages of epidermis (L1) and the internal tissues (L2 or L3) in determining the site of zone cell differentiation.[26] In *Lycopersicon esculentum* the abscission zone cells develop midway along the pedicel and in *L. pennellii* the abscission zone forms at the base of the pedicel. Interspecific chimeras between the two plants have shown that the *position* of the zone is determined by the genetic status of L3 cells. The L1 and L2 tissue responds by forming normal abscission cells in the position on the pedicel dictated by the genotype of the internal L3 tissue. In a chimera of *L. esculentum,* with *L. esculentum,* mutant *jointless* (*jointless* lacks an abscission zone in the pedicel), in which the L2 and L3 layers were wild type, abscission zones were formed in the usual midpedicel position, but in a chimeric plant with the L2 and L3 cells derived from the mutant *jointless* lineages, no zone cells were formed at all.[26] Both the differentiation, and the position of differentiation, of zone cells are therefore determined by the inner cell lineages.

Chimeras of this type are valuable tools in differentiation studies and the results indicate that neither L1- nor L2-derived cells can develop autonomously but instead, respond to transmitted signals from internal cell lineages.

2. Timing

Once an abscission zone has been formed, the potential of these cells to separate makes the organ they subtend vulnerable to premature shedding whenever a critical level of ethylene is experienced. The formation of the abscission zone can therefore occur relatively late in leaf primordial differentiation. It is not surprising also that such cells appear to be differentiated only *after* the major period of ethylene production by the apex has subsided. Examples of a precise timing of the differentiation of abscission zone cells during a leaf ontogeny can be found in the embryos of certain seeds and in the differentiation of winter buds in temperate deciduous shrubs and trees. There is now evidence too, that the differentiation of zone cells is under hormonal control, with the ethylene produced from the apical tissues playing a determining role in directing the position at which the abscission zone is formed (see Section III.C).

Possibly, abscission has been longest studied, and is best understood, in the bean, *Phaseolus vulgaris,* in which two abscission zones are differentiated in the primary leaf, one where the petiole joins the stem and another (the distal abscission zone) where the pulvinus immediately below the blade joins the petiole.[27,28] In mature seeds of *Phaseolus*

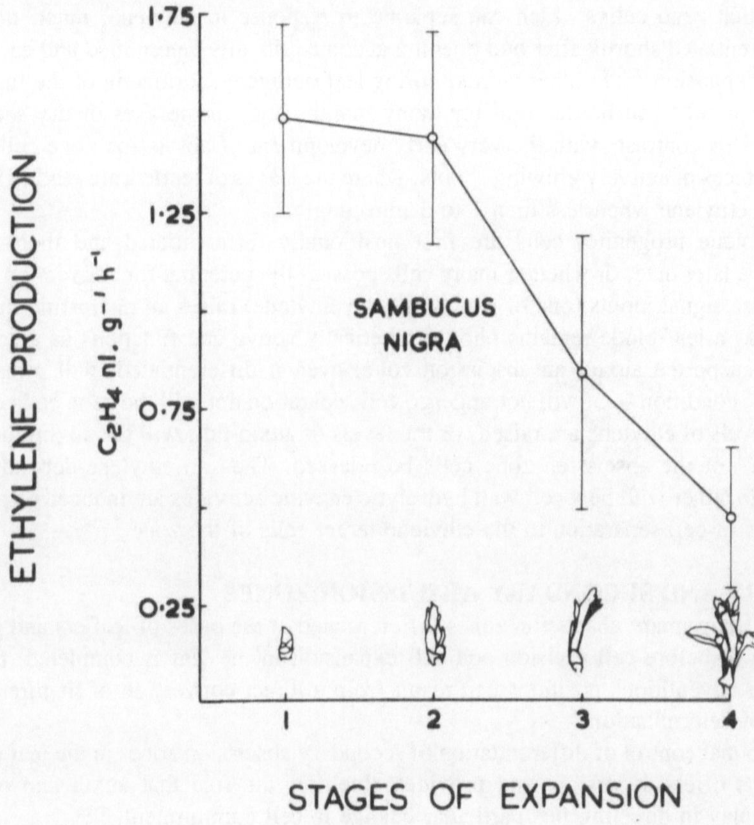

FIGURE 2. Ethylene production by winter buds of *Sambucus nigra* at early bud break (stage 1) to the release of expanding leaves (stage 4). (Data from Roberts, J. A. and Osborne, D. J., unpublished.)

vulgaris the embryo axis already possesses the two primary leaves but they are, apparently, without abscission zones. On the imbibition of the seed and for the first 4 days of germination at 24°C neither an attached nor an excised axis can be induced to shed at blade or petiole junctions in even the highest concentration of ethylene (1,000 μl 1^{-1}). However, after day 5, even low concentrations of ethylene (1 to 10 μl 1^{-1}) will induce cell separation at the normal zone positions.[28a] Furthermore, only after day 5 will the walls of the zone region of ethylene-treated axes react positively to phloroglucinol in concentrated hydrochloric acid (indicating pentoses), which is an early chemical marker for the initiation of cell wall separation in the abscission zones of *Phaseolus* leaves. It would appear therefore that the zone cells are either not differentiated, or if they are differentiated, they are not responsive to the ethylene signal, before the 5th day of germination. High levels of ethylene are generated by germinating seeds, both from the active expansive growth of the axis and from the pressures of the testa and soil as the shoot emerges. Functional abscission zones would normally be induced to separate at such concentrations. Their suppression until a late stage of germination is therefore an important aspect of temporal development in the leaf.

A somewhat similar situation may exist in winter buds when after breaking dormancy, the newly emerging leaves very rarely abscind, and cannot be induced to do so in 1000 μl 1^{-1} ethylene. During the period of bud break, very high levels of ethylene are produced, primarily by the bud scales (Figure 2), and these levels are sufficient to initiate the separation of leaf abscission zones in a normally extending summer shoot. It is concluded that functional abscission zone cells are specifically *not* differentiated in primordial leaves that are destined to become enclosed within winter buds. Experiments with *Sambucus nigra* have shown that zone cells, which can separate in response to ethylene, must, however, become differentiated shortly after bud opening at some critically determined and early stage of rapid leaf expansion.[28b] In other words, during leaf ontogeny, formation of the functional abscission zone cells can be deferred for many months and, in the axes of dry seeds, for many years. This contrasts with the very early development of abscission zone cells in the leaves from apices of actively growing shoots, where the leaves or leaflets are readily induced to abscind in ethylene when less than 1 to 3 mm long.

Whether zone progenitor cells are first positionally differentiated and then become committed at a later date, or whether many cells possess the potential for conversion to zone cells by precise signal inputs (one of which may be ethylene) raises an interesting question.

As long as a leaf blade remains photosynthetically active and functions as a source of basipetally transported auxin, an abscission zone, even if differentiated, will remain in a "suppressed" condition — it will not undergo cell separation nor will the zone cells expand. Only if the levels of ethylene are raised, or the levels of auxin fall, will the auxin-controlled "suppression" of the abscission zone cells be released. Then an ethylene-dependent cell enlargement together with new cell wall hydrolytic enzyme activities are induced as preludes to the process of cell separation in the ethylene target cells of the zone.[27,28]

C. PRIMARY AND SECONDARY ABSCISSION ZONES

Whereas the primary abscission zones differentiated at the bases of leaflets and petioles are formed well before cell division and cell expansion in the leaf is completed, those in secondary, or adventitious positions can result from a direct conversion of mature cells to a new state of determination.

The hormonal control of differentiation of secondary abscission zones in the leaf petioles of *P. vulgaris* offers an insight and provides clues for the role that auxin and ethylene together can play in directing this particular change in cell commitment.[28,29]

When segments containing the distal pulvinus, the distal abscission zone, and a part of the petiole (i.e., an explant) are excised from a primary leaf and maintained at a controlled high humidity in air, the pulvinus undergoes a premature senescence. As a result, the production of ethylene increases as the tissue becomes progressively senescent. It is clear that pulvinar senescence is primarily due to auxin deprivation through a termination of auxin transport from the blade, for it can be delayed or prevented in explants by external applications of an auxin to the cut pulvinus surface. In concert with the rising ethylene production and the lowering of auxin fluxes across the zone, the critical threshold levels of ethylene to which the zone cells respond are eventually reached and the abscission program of cell separation is set in train.[27-29]

The speed at which separation is initiated and completed is determined by the levels of auxin repressor and ethylene inducer to which the committed zone cells are exposed. In air, separation of explants is usually complete in 48 hours. If additional ethylene is supplied in the ambient air the minimum time for the completion of separation can be reduced to 32 to 36 hours, but if an auxin is supplied to the pulvinar end of the explant from the start of the experiment, sufficiently high applications (a 2 μl drop of 1×10^{-3} *M* IAA) can permanently repress senescence of the pulvinus and also block abscission. That the zone target cells respond to the signals of both repressor and inducer is seen from the results of combined auxin and ethylene treatments — high levels of ethylene will always overcome auxin applications.[30,31] Furthermore, provided the ethylene-induced abscission process has not proceeded too far and actual cell separation has not occurred, the process can be reversed by an auxin application or by the removal of ethylene. Then the so called "breakstrength" (the force required to separate the tissue at the abscission zone) again increases. Fruit loosening and retightening has frequently been observed with incomplete chemically induced shedding of citrus under field conditions and this reversibility of wall adhesion has been directly related to falling levels of internal ethylene within the fruit.[32]

TABLE 1
The Positional Induction of Secondary Abscission Zones in Petiole Stumps of Abscinded Explants of *Phaseolus vulgaris* in the Presence of Ethylene (10 μl l^{-1}) and Different Auxin Gradients

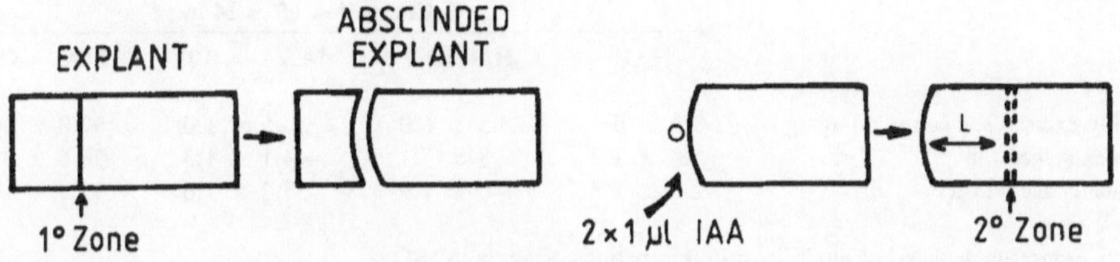

IAA Treatment	Distance L between primary and secondary zones (mm)
H_2O	1.3 ± 0.25
1 × 10^{-5} *M*	1.5 ± 0.25
1 × 10^{-4} *M*	3.0 ± 0.52
1 × 10^{-3} *M*	4.9 ± 0.62

Note: Auxin applied as aqueous drops to the exposed surface of separated zone cells.

From Osborne, D. J., *Hormone Action in Plant Development,* Hoad, G. V., Lenton, J. R., Jackson, M. B., and Atkin, R. K., Eds., Butterworths, London, 1987, 265. With permission.

In explants of *P. vulgaris,* once the pulvinus has abscinded, the remaining petiole part with the zone cells firmly attached can develop secondary abscission zones if maintained in low levels of ethylene and supplied with auxin to the distal end — i.e., to the surface of the zone cells that have just separated. Under these conditions, certain cells of the cortex — but only those of the cortical parenchyma — change from a cortical cell competence to that of zone cell competence and exhibit all the characteristics of normally differentiated primary abscission zone cells.[28,29] The position at which these events occur is controlled, in part, by the apex-to-base gradient of auxin. The distance from the surface of the primary abscission zone to the site of formation of the secondary abscission zone is determined by the concentration of IAA supplied. The higher the concentration of IAA, the more distant the formation of the zone (Table 1). If the explants are maintained in air or in ethylene-depleted air, the incidence of secondary zones is rare and the formation of zone-type cells with the potential to separate is diffuse. It appears therefore, that ethylene too plays an important role in precisely defining where cells will or will not become the newly committed zone cells. Furthermore, it is evident that cells of L1 lineages are not subject to the same controls as those of L2 or L3 for the L1 epidermal cells at an abscission zone do not enlarge, but instead are forced apart by the pressures of the expanding L2-derived cells below them. Cells of the pith tissue have not been observed to be part of secondary abscission zone formation and it would seem that in *P. vulgaris,* a differentiation sequence to abscission zone target cell competence can pass from cortical cell type to zone cell type but not via pith cell type as an intermediate.[28]

The absence of an abscission zone in the fruit pedicel of tomato mutant *Lycopersicon esculentum,* var Ailsa Craig *jointless,*[80a] and in the leaves and fruits of another nonabscinding mutant *Lupinus angustifolius,* ABS−[80b] is of special interest, for in neither of these examples is there evidence of an anatomical differentiation of zone cells and even when exposed to very high levels of ethylene, cell separation cannot be induced. It must be concluded therefore, that given the apparent normality of these plants in other observed respects, the signal sequence that regulates the differentiation of functional ethylene-responsive zone cells can be determined by a single gene mutation.

TABLE 2
The Elongation of Subapical Segments Excised from Petioles of the Semiaquatic Fern *Regnellidium diphyllum*, Illustrating the Dependence of Auxin for Expression of Ethylene-Induced Cell Expansion

	% Elongation after 24 h			
	H_2O	C_2H_4 6 μl l^{-1}	IAA 1 × 10^{-4}	IAA + C_2H_4
Freshly excised segments, no aging	8.0 ± 0	23.5 ± 6.0	55.5 ± 1.0	92.0 ± 10.0
Segments aged 6 h	8.0 ± 2.2	22.8 ± 7.1	44.1 ± 1.3	78.6 ± 3.3
Segments aged 23 h	2.8 ± 1.8	8.0 ± 1.9	38.0 ± 7.6	77.7 ± 10.9

Note: Segments depleted of auxin by different periods of aging in H_2O.

IV. ETHYLENE AND LEAF TARGET CELLS

A. CELL GROWTH

With respect to cell enlargement, and the regulation by auxin and ethylene, three general cell types can be distinguished in plants. The one that has been most studied occurs in elongating parts of immature dicotyledonous shoots and the coleoptiles and mesocotyls of monocotyledons. These cells enlarge or elongate in response to auxin but they do not do so in response to ethylene. Rather, ethylene regulates the *direction* of the auxin-controlled cell expansion, not the *extent* of expansion. As a result, a lateral growth, rather than an elongation growth, of the cell occurs and the final shape tends to an isodiametric rather than an elongate form. As far as has been determined, most cells of a leaf blade conform to this type — Type 1.[11]

Cells of the abscission zone behave in a quite opposite way. Auxin *suppresses* expansion but ethylene can *induce* it, and in many plants, zone cell expansion is a visible prelude to the separation event. Cells that respond in this way are called Type 2.[11,28]

The third type, Type 3, is common to the petioles and stems of semiaquatic plants.[11] They elongate upon shoot submergence and exhibit large increases in overall extension in either auxin, *or* ethylene. Significantly, this ethylene-induced enlargement has an absolute requirement for the presence of adequate levels of endogenous (or applied) auxin to permit the enlargement to occur.[33]

The requirement for auxin has been convincingly demonstrated in the semiaquatic plants *Regnellidium diphyllum* and *Nymphoides peltata*. If the leaf blade is removed, the petiole will no longer elongate when it is submerged or exposed to ethylene but replacement of the blade with auxin restores elongation in ethylene. For example, freshly excised segments of petiole of *Regnellidium* show a little elongation with ethylene, a large induced growth elongation with auxin, and an even greater elongation with both together (Table 2). However, excised segments that have been aged for more than 6 h in water (i.e., 23 h) to deplete their free endogenous auxin, show practically no elongation response to ethylene but their growth can be restored if auxin is added; their response to ethylene in the presence of auxin is then more than additive. This has been called the "supergrowth" effect of auxin with ethylene (Table 2).

In the petiole of *N. peltata* ethylene has been shown to enhance cell number as well as cell size, and the effect of ethylene or submergence upon petiole growth depends upon the stage of leaf ontogeny. In young leaves, before full expansion of the lamina, both cell division and cell enlargement in the petiole are increased by ethylene,[34] but in mature leaves, cell elongation is the major response.[35] The ecological impact of submergence adjustment and the differentiation of Type 3 cells can be followed in the distribution of three species of *Ranunculus* that colonize the ridge and furrow grassland that has resulted from the medieval practice of strip farming in the U.K.[36] Surveys show that *R. repens* is preferentially located in the furrows that are subject to winter flooding, *R. acris* predominates in intermediate positions midway up the ridge, while *R. bulbosus* occupies only the well-drained land along the tops of the ridges. A study of seedlings of these three species has shown that elongation growth of the petioles of excised leaves of *R. repens* is enhanced by 10 $\mu l\ l^{-1}$ ethylene or by submergence, those of *R. acris* show an attenuated response, while *R. bulbosus* shows no enhanced elongation attributable to ethylene.[10] These responses are less readily observed in the young leaves of mature plants and they are, therefore, an interesting example of physiological change during ontogenetic development.

The differentiation of cells that enlarge or elongate in response to ethylene in what are ostensibly land plants may be much more common than is currently known. The distribution of *Rumex* species on river banks can also be related to the ability of the plants to tolerate flooding by elongation of stems or petioles,[37] and the nine species of semiaquatic *Polygonum* studied with respect to a cline of phenotypic plasticity in response to submergence raises the question whether these too have differing abilities to differentiate Type 3 cells.[38]

Although Type 3 cells are common to semiaquatics, some are found in limited clusters in land plants. It is to groups of these cells, positionally differentiated along the upper adaxial flanks of the bases of the leaf petioles that we may attribute the auxin *or* ethylene induction of epinastic responses by the leaves of plants such as tomato and sunflower.[11] The curvatures result from either an auxin- or ethylene-induced cell expansion of the Type 3 cells, which in turn causes the downward bending of the whole petiole base. Such responses are typical of certain plants when the soil is flooded. Under the conditions of partial anoxia that then prevail, conversion of ACC to ethylene in the roots is restricted. Epinasty of the shoots then results from the transport of ACC from the roots and its conversion to ethylene in the leafy aerial parts.[39] In plants such as the tomato, which possess Type 3 cells on their petioles, the ethylene-induced epinasty can be blocked by a pretreatment of the region with $CoCl_2$ (Co^{2+}), which inhibits the conversion of ACC to ethylene, or by $AgNO_3$ (Ag^+) which is presumed to bind to, and inactivate, some part of the perception site for the ethylene response[81] (see Chapter 5). As in the semiaquatic plants, removal of the endogenous source of auxin prevents the elongation response of the Type 3 cells to ethylene. It has been shown that removal of the blade portion of the leafy bract in *Poinsettia* plants will prevent epinasty of the petioles held under mechanical restraint,[40] however when lanolin containing IAA is applied to the cut end to replace the blade, epinastic curvature is fully restored. Spraying with silver thiosulfate (which blocks many ethylene responses) will also abolish epinasty in plants exposed to ethylene if the thiosulfate treatment is given 18 hours before the ethylene, but it does not block restraint-induced epinasty when applied just prior to the restraint.

Epinasty of the leaf blade itself may be under a similar control to that in the petiole. A study of the interaction of auxin and ethylene in the curvature responses of the blades of *P. vulgaris* var Pinto has shown that the upward (hyponastic) bending is initiated within 15 min and can readily be measured within 2 to 4 h when 1 m*M* auxin (IAA) is applied to either ventral or dorsal surfaces and is the result of cell growth which is greater on the lower than the upper side.[41] Within 24 h of an auxin treatment, the leaf returns to its normal planar position, and the return has been attributed to an auxin-induced ethylene production that commences 1 h after the auxin application. Ethylene alone does not induce the hyponasty but results indicate that the return to a planar orientation *is* ethylene dependent. For example, return does not commence until at least 2 h after the auxin-induced hyponasty and after auxin-induced ethylene production is well established (i.e., at 6 h). Also inhibitors of ethylene biosynthesis (AVG) or competitors of ethylene action (CO_2) maintain hyponastic bending and block the epinastic return to planar orientation. That auxin and ethylene act in concert is not in doubt and although other explanations have been put forward, the simplest would be that an auxin and ethylene-induced growth of Type 3 cells differentiated in upper and lower sides of the leaf blade would determine the extent of nastic curvature attained in the different auxin and ethylene concentrations to which the leaf becomes exposed.[41]

Studies of mutant plants with unusual nastic responses help to resolve the intriguing possibilities of Type 3 cells in the leaves of land plants. The tomato mutant *diageotropica* has dark green leaves that are hyponastic at the edges. Exposure of these plants to a concentration of 5 nl l^{-1} ethylene will, however, normalize the growth of new young leaves but has a lesser effect upon older leaves.[42] *Diageotropica* has, in addition, a low epinastic response to applied auxin compared with normal plants, 100 μM IAA being required to produce curvatures that can be elicited by 10 μM in the parent isogenic plants. With respect to hyponasty of the leaf in these plants, one must conclude that the perception/response of auxin- and ethylene-responsive Type 3 cells is modified by the single gene point mutation.[43]

The *Epi* mutant of tomato is another single gene mutant that has a special interest for leaf growth for it expresses many of the phenotypic symptoms that would normally result from an exposure of the parent (VFN8) to ethylene. These include severe leaf epinasty and swelling of the cortical cells of the stem and petiole. Although these effects are associated with an elevated production (about × 2) of ACC and ethylene, attempts to suppress the ethylene overproduction by inhibitors of ethylene biosynthesis (AVG or Co^{2+}) or to compete the sites of ethylene action (by silver thiosulfate or norbornadiene) fail to normalize the mutant phenotype. Assays have shown that the unusual growth habit cannot be attributed to abnormal levels of free endogenous auxin compared with the parent VFN8,[44] and exposure of the mutant to ethylene exaggerates the epinastic and swelling responses. The target cells for epinastic growth in the *Epi* plants appear therefore to have acquired a quantitatively different perception/response from the parent in respect of auxin and ethylene signals. The mutant *diageotropica* may be another example of a modification of Type 3 cells, for in these plants, auxin-induced ethylene production is lower than the parent, and a 10-fold higher concentration of IAA is needed to elicit the same angle of petiole curvature.

From the indications we have available today, it seems likely that many more modifications of auxin- and ethylene-responsive cell types may exist in the leaves and petioles of land plants than we have presently discovered. The differences in response of the *Ranunculus* species that colonize the ridge and furrow referred to in Section IV.A could represent the extremes of the range. Lenticels, that occur along the leaf petioles of many species of trees and shrubs may be a further example of a distinct cell type. Certainly the enlargement of lenticel cells can be induced by ethylene.[1] The use of single gene mutants and a greater knowledge of the interactions between ethylene and different cell lineages will doubtless advance our understanding of leaf cell growth and its control.

B. IDENTIFICATION OF ETHYLENE-RESPONSIVE TARGET CELLS

Whereas the identification of highly differentiated cells such as vascular elements, statocytes, guard cells, or trichomes requires only a simple microscopic assessment, the identification of cells with specific biochemical characteristics or with specific competence to respond to hormonal signals is not so easily achieved. Immunological recognition of cell-specific proteins, glycoproteins, or oligosaccharide sequences has become a powerful tool for following differentiation and for mapping cells expressing type-specific marker molecules. In the leaves of *Vicia faba,* for example, the stomatal guard cells can not only be recognized microscopically but their protoplasts are also detectable immunologically by the expression of a guard cell-specific plasmamembrane protein that distinguishes them from all other cells of the leaf.[45] In contrast, cells of abscission zones before the initiation of separation are not usually recognizable structurally from their nonseparating neighboring cells. In certain species, however, the size of the zone cells may be somewhat smaller than those of the neighboring cells. Extraction and fractionation of zone cell tissue excised from the rachis of nonseparating leaves of *Sambucus nigra* has shown however that several polypeptides (110, 95, 43, and 28 kDa) are preferentially expressed in zone cells compared with nonzone cells.[31] When such protein extracts are competed with rabbit antiserum IgG produced in response to antigenic extracts of nonzone leaf tissue, low-abundancy proteins (or antigenic determinants) unique to uninduced zone cells are enriched. Fractionation of these enriched zone protein preparations on sodium dodecyl sulfate (SDS) gels followed by immunodiffusion and Western blotting confirms the presence of unique zone cell peptides of approximately 32 and 34 kDa.[46]

With similar competition techniques, peptides unique to zone cells have also been identified in the petioles of *P. vulgaris* leaves. Segments of SDS gels containing fractionated proteins from abscission zones have been competed through affinity columns of anti-pulvinus + petiole IgG coupled to sepharose and the eluates then used to raise further antisera. Using immunogold procedures and electron microscopy, the antiserum to one of these bands (30 to 45 kDa) was found to contain antibodies that preferentially recognize a component of the middle lamella and primary walls of zone cells compared with nonzone cells.[47] By direct tissue blotting onto nitrocellulose paper,[48] a monoclonal antibody, raised against abscission cells of *Sambucus* leaves and which possesses the N-linked oligosaccharide sequence that recognizes the epitope Manα3 (Manα6) (Xylβ2) Manβ4GlcNAcβ4 (Fucα3) GlcNAc,[49] has been shown to give locally high binding to regions that correspond to abscission zone cells. This can be demonstrated *before* an induction of cell separation has been initiated.[47]

It seems clear that *in vivo* the abscission process in the leaf is induced by biochemical events that are initiated in the leaf blade. Ethylene produced from senescing or stressed tissue adjacent to the abscission zone plays a critical part. In *P. vulgaris* the tissue adjacent to the zone (i.e., the pulvinus) produces high levels of ACC and ethylene at senescence,[11] and experiments with explants have shown that regulation of pulvinus senescence can control the onset of separation events.[27,28] It may be that the presence in freshly excised pulvinar tissue of a 92-kDa membrane polypeptide that shows an auxin-enhanced and Ca^{2+}-dependent phosphorylation, is an example of a molecular marker for the nonsenescent state; it may also indicate a potential for high ethylene production when senescence occurs.[50] The differentiation of cell types with specific molecular markers that denote their specific physiological and biochemical function is a general feature of differentiation. It is evident that in the leaf, at least one of these cell types, the abscission zone cell, can now be recognized in this way as a specific class of ethylene-responsive target cell at both the biochemical and ultrastructural levels.

C. THE COMMITTED ABSCISSION ZONE CELL

Antibodies have been used successfully (again in *P. vulgaris*) to demonstrate the zone cell-specific production of an isozyme of β-1,4-glucanhydrolase (an endocellulase) after the induction of zone cell separation by ethylene.[27,28] This 9.5-pI isozyme is distinct from the 4.5-pI constitutive enzyme of the leaf and is synthesized *de novo* in response to ethylene. Secretion of this, and presumably other enzymes also, is associated with cell to cell dissociation in the precise location of the differentiated zone cells — other cells adjacent to the zone cells do not separate. This indicates that not only are the zone cells the sole cell type to secrete the specific pI isozyme, but the walls of zone cells are the only ones that separate under the influence of their secreted enzymes. The differences in the distribution of immunogold labeling of the cell walls of zone cells compared with nonzone cells in bean are the only evidence to date of a specific molecular identity for the middle lamella and primary wall of a differentiated zone cell compared with its neighbor.[47] That leaf abscission zone cells are examples of a committed differentiated state with specific gene expressions suppressed by auxin and induced by ethylene has been further demonstrated in *P. vulgaris* in the following way.[31]

In explants in which the zone cells have been induced to produce β-1,4-glucanhydrolase and the pulvinus has abscinded, the intact but enlarged zone cells remain attached by their proximal ends to the petiole cells below them. At this stage the zone cells (and only the zone cells) are active secretory cells as the continued production of hydrolase enzymes and the dilated, vesicle-associated ultrastructure of their dictyosomes demonstrates. However, if the inducer (ethylene) is removed or depleted by inclusion of an ethylene adsorbent, or the repressor (an auxin, 2,4-D or IAA) is supplied in microliter drops to the surface of the zone cells, within 24 hours production of the glucanhydrolase falls and the membranes of the dictyosome stacks become compressed and no longer associate with vesicles.[31] Nonetheless, the auxin-treated cells remain fully competent to respond to a second period of induction by ethylene. However, while still under the influence of the applied auxin they require a higher concentration of ethylene ($\times$ 50) than that of the original ethylene induction to evoke a renewed production of the glucanhydrolase and to show a renewed activation of the dictyosome stacks. None of these changes are observed in adjacent nonzone cells. Measurements of the enzyme activation and the numbers and membrane widths of the dictyosome stacks are given in detail elsewhere,[31] but this example serves to illustrate how cell-specific responses can be elicited by ethylene in leaf cells and how these responses can be switched "on" or "off" by the levels of ethylene to which they are exposed.

V. ETHYLENE PRODUCTION BY LEAVES

A. DURING DEVELOPMENT

Most rapidly dividing and expanding tissues show relatively high rates of ethylene production, and this may well reflect the high endogenous metabolic turnover of young cells. Leaves are no exception and usual rates for expanding leaves range from 0.1 to 0.4 nl g^{-1} F. wt h^{-1} falling to <0.1 in fully expanded annual leaves; levels as low as 0.01 μl g^{-1} F. wt h^{-1} are reported for the 2nd-year mature leaves of the evergreen *Euonymus japonica*.[51] With the onset of senescence, ethylene production levels can rise again dramatically and in deciduous leaves (*Prunus serrulata*) rates as high as 16.0 nl g^{-1} F. wt h^{-1} occur during the yellowing associated with degeneration of the chloroplasts.[52] Generally herbaceous leaves produce less than this at senescence, levels seldom rising above 5 to 6 nl g^{-1} F. wt h^{-1}. Some herbaceous leaves, particularly those of grasses, show little rise in ethylene production at senescence and this may be the result of leaf dehydration that also occurs during the yellowing stages. It has been shown that ethylene production from the glumes, lemma, and palea of the fruits of *Avena fatua* declines with desiccation at maturation of the grain.[53] The coordinated and membrane-associated conversion of ACC to ethylene (the ethylene-forming enzyme activity, EFE) by these leafy organs is presumed to be disrupted by loss of membrane integrity as the cells become progressively dehydrated. The high levels of production maintained by deciduous leaves as they senesce and also by the pulvinus of the bean leaf (up to 6 nl g^{-1} F. wt h^{-1}) may indeed be attributable to their maintenance of a high moisture content throughout the period of yellowing.[52] Indeed, the yellow leaves of many deciduous and herbaceous plants will readily convert ACC to ethylene, suggesting that EFE activity still functions.[52a] Further study to distinguish desiccation-controlled from senescence-regulated ethylene production and enzymic from nonenzymic conversion of ACC is much needed for a fuller understanding of this aspect of ethylene in leaf development.

B. STRESS AND HORMONAL FACTORS

Almost any perturbation to the cells of higher plants leads to an increase in their production of ethylene — the so called "wound ethylene" response. Shaking a box of seedlings, stroking a leaf surface, chilling, mild desiccation, pathogenic infections, or gaseous pollutants such as ozone or sulfur dioxide all lead to increases of different duration and all are probably attributable to disturbances of membrane organization and the passage of cellular components between different membrane-bound domains. Again, leaves are no exception. The increases seem to be linked to the availability of ACC and the activity of ACC synthase and the ethylene-forming enzyme complex (EFE). Because the synthesis of ACC synthase can be regulated by the level of auxin in a tissue (the addition of auxin enhancing transcription of the gene and its translation),[54,55] it is of special interest to consider the relationship between auxin and ethylene production. In young leaves, where auxin levels are high, ethylene production is also high. It is not clear however, that auxin levels alone determine the rate of flux of the C3 to C4 carbons of methionine to ethylene, for cytokinins also stimulate ethylene formation and cytokinins too are high in young leaves.[56] (See Chapters 7 and 14.) Furthermore, auxin levels can be extremely low in senescing leaves when ethylene production is highest and the addition of auxin at this stage does not enhance ethylene production in species so far reported.[52]

It now seems that the enzymes of the biosynthesis pathway from methionine through ACC to ethylene are distributed in different subcellular compartments. One of these, ACC synthase, can be cytosolic,[57] but EFE appears to be located at the cytosol-vacuolar interface of the tonoplast membrane.[58,59] Whereas the conversion of ACC to malonyl-ACC (MACC) seems to be localized in the cytosol, the product readily crosses the tonoplast membrane and accumulates within the vacuole.[60] At different stages of differentiation, cell expansion, and senescence, the permeability of these membranes to metabolites of all kinds must change and their response to endogenous hormonal regulators will be modified in consequence. Not all cells will behave in the same way and not all species will conform to a common rule. But whereas there is abundant evidence that a change in the rate of ethylene production is one of the most rapid biosynthetic responses of a green and functional leaf to either internal or external signals, a senescent leaf may no longer respond to the signals of wounding or to the addition of hormones by an increase in ethylene release. It has been proposed therefore that membrane-confined cofactors play a critical and limiting role in the ethylene formation pathway *in vivo* and that the availability of these cofactors to the sites of enzyme-regulated reactions is most likely to be determined by the effectiveness of transmembrane function.[52] Changes in the ionic environment within a cell due to perturbation or "wounding" are sufficient to alter membrane permeabilities. In young leaves, where active membrane synthesis occurs, membrane organization can be rapidly restored so that cofactor efflux is strictly limited in time and rises in ethylene production are of short duration. However, at senescence, as membrane turnover decreases and membrane integrity becomes irreversibly impaired, cells become permanently leaky. An unrestricted release of cofactors may play an important part in determining the long-term and uncontrolled rise in ethylene production that is typical of many senescent conditions.[51,52] Although we do not know which cofactors in leaves might operate in this way to enhance ethylene biosynthesis *in vivo* some may well be phenolic in nature. Other compounds can function as ethylene biosynthesis inhibitors. Salicylic acid for example, suppresses the EFE-mediated conversion of ACC to ethylene and its lipid solubility and effectiveness only at pH values below 5.0 make it of special interest.[61] The ubiquity and sequestration of such substances in plant cells would accord them a role as endogenous regulators of ethylene formation.

C. DIFFERENT CELL TYPES

It has long been known that the expansion growth of epidermal tissue from stems and coleoptiles determines the extent to which the internal cortical and vascular elements can elongate or expand.[62] It is also clear that epidermal cells derive from a different cell lineage (L1) from the meristem than those of the internal tissues (L2 and L3). Furthermore, epidermal cells frequently show different growth responses to hormones from those of their neighbors.[62] This is perhaps not surprising since the absence of chloroplasts from all but the guard cells provides an immediate distinction between epidermis and the adjoining internal tissues. The discovery that epidermal strips from mung bean hypocotyls do not respond to IAA by an enhanced production of ethylene, and are also unable to convert ACC to ethylene, was, however, the first evidence that epidermal cells might not possess the normal EFE activity.[63] The cells responsible for ethylene production in the mung bean hypocotyl were, it seemed, confined to the cortex and pith.

We have investigated this question in leaves, using the pea mutant Argenteum, whose epidermis only loosely adheres to the internal palisade and mesophyll. These silvery leaves can be readily stripped for epidermal material. We have confirmed that in this variety of *Pisum sativum,* the epidermis produces normal levels of ethylene, but like the epidermis of mung bean hypocotyls, the production is not enhanced by ACC. Whereas palisade and mesophyll cells of all ages will actively convert applied ACC to ethylene, epidermis of these leaves appears to lack the EFE capacity. Such data point to the existence of an alternative ethylene production pathway in epidermal cells of this genetic line.

An interesting aspect of this discovery derives from results of experiments designed to test whether the leaf cuticle or the stomata is the pathway for release of the enhanced ethylene production induced by exposure to ozone. Ozone-induced release was found only from the stomatal side in the hypostomatous leaves of *Eucalyptus globulus* and *Hedera helix,* with the nonstomatal side failing to show any ozone-induced increase at all.[64] Perhaps these epidermal cells lack a stress-induced (wound) ethylene synthesis. In this respect, therefore, they may possess an ethylene biosynthesis control similar to that reported for a number of lower plants.[10,11,13] (See Section VI.B.)

VI. ETHYLENE BIOSYNTHESIS

A. HIGHER PLANTS

The ethylene biosynthetic pathway from methionine via *S*-adenosylmethionine and ACC to ethylene is well established for higher plants (see Chapter 1), but we still possess little information on the relative ability of different cell types to execute the pathway, or on the limitations imposed on the pathway during ontogeny and development. Such studies are not easy to carry out for excision of tissues and the production of protoplasts invariably activates the formation of ACC, either enzymically via ACC synthase or possibly, in some circumstances, by conversion from the inactive malonyl-ACC. But we do now know a whole range of plants that do not show these wound responses although they are equally subject to the regulation of growth and development by ethylene. These are the lower plants. It would seem that their leafy parts all possess the capacity to produce ethylene but resemble the epidermal tissue of certain higher plants in their lack of an ACC-stimulated pathway.[65]

B. LOWER PLANTS

The semiaquatic fern *Regnellidium diphyllum* provided the first evidence for a plant with leaves and petioles possessing an alternative pathway for the biosynthesis of ethylene.[10] First, the production of ethylene was found not to be reduced by additions of analogs of rhizobitoxine which, in other plants, block the pyridoxal-dependent enzymic conversion of *S*-adenosylmethionine to ACC.[66] The leaves also exhibited no IAA-induced ethylene production,[33] no wound ethylene production,[33] no senescence-associated ethylene production, and failed to convert ACC to ethylene.[11] ACC did not substitute for ethylene in the Type 3 cell elongation response of the petioles,[11] and rhizobitoxine did not reduce the ethylene-dependent elongation of petioles upon submergence.[66] When ^{14}C-methionine or ^{14}C-ACC was supplied to the leaves, no ^{14}C-labeled ethylene was produced.[82]

Studies with another semiaquatic lower plant, the liverwort *Riella helicophylla,* have also shown that ACC will not substitute for ethylene in the enhancement of cell growth in the central column region of the thallus.[13] Indeed, concentrations of ACC between 5×10^{-4}

TABLE 3
Distribution of EFE Activity in the Expanding Leaves of a Range of Lower and Higher Plants[65]

	C_2H_4 production nl g^{-1} F. wt h^{-1} ACC concentration (*M*)			
	0	5×10^{-5}	1×10^{-4}	1×10^{-3}
Liverworts				
Marchantia polymorpha (HSI)[a]	1.5	—	—	1.8
	0.8	0.8	—	—
Riella helicophylla[a]	1.4	1.0	1.2	—
Mosses				
Sphagnum sp.	0.2	—	0.2	—
	2.4	—	—	2.9
Polytrichum commune	1.4	1.0	—	—
	1.5	—	—	1.1
Ferns				
Matteuccia struthiopteris	3.6	—	—	3.8
Azolla caroliniana	0.7	0.8	—	—
Asplenium scolopendrium	0.5	—	—	0.6
Regnellidium diphyllum	1.2	1.1	1.3	1.3
Gnetaceceae				
Gnetum gnemon	0.7	5.2	—	—
Ephedraceae				
Ephedra equisetina	1.7	10.3	—	—
Gingkoaceae				
Gingko biloba	1.9	9.4	—	—
Cycads				
Zamia floridana	1.3	11.5	—	—
Cycas revoluta	1.0	9.7	—	—
Gymnosperms				
Larix europea	4.6	15.0	—	—
Pinus sylvestris	4.8	9.8		
Angiosperms				
Phaseolus vulgaris	0.3	6.0	12.0	33.2

Note: Ethylene production over 3 to 6 h incubation period following surface treatment with aqueous solutions. Values underlined are significant enhancements above their controls.

[a] Axenic cultures

and 5×10^{-6} *M* will inhibit auxin-induced elongation of these cells. In this plant too, there is no IAA-enhanced ethylene production, no wound-induced ethylene production, and no conversion of ACC to ethylene.[13]

Absence of the EFE conversion pathway is not restricted to those lower plants that colonize semiaquatic habitats, for it has now been established that neither the gametophytes of liverworts or mosses, nor the sporophytes of all the pteridophytes so far tested, have any significant capacity to convert ACC to ethylene (Table 3).[65] However, the Gnetales, cycads, gymnosperms, and angiosperms all do so (Table 3) and enhanced levels of ethylene can usually be measured within 30 min of the addition of ACC to leafy tissue.[65]

It is tempting to speculate that the biosynthesis pathway for ethylene, first developed by the lower land plants, may have persisted in the L1 epidermal tissues of a restricted number of higher plants — although it has not persisted in the epidermis of the bulb scales of *Allium cepa,* for example.[65a] The methionine, ACC route could well have arisen at a later date in response to changes in environmental pressure. In considering leaf ontogeny, the derivation of epidermal tissue from L1 lineages in higher plants may relate most closely to the tissues of liverworts and mosses, which all arise from a single apical meristemic cell, while genera possessing multicellular meristems with multiple cell lineages (the Gnetales, cycads, gymnosperms, and angiosperms) acquired the methionine-ACC pathway to ethylene with their new internal cell lineages. On further investigation, we may find that in some of the pteridophytes, cells are present in the multicellular apex of the sporophytes that can function as a source of lineages with some EFE capacity.

There has been a report in the literature that leaves of the fern, *Matteuccia* will convert ACC to ethylene if they are treated with high concentrations (1×10^{-3} *M*) and then enclosed (1 g in 25 ml) for a long period (24 h).[67] They do not, however, do so in the more usual short term assays at lower concentrations. Levels of 1×10^{-3} *M* can be toxic to some plant tissues and prolonged exposure to such concentrations may well lead to a nonenzymic breakdown of ACC to ethylene. In *Riella,* for example, a rise in ethylene formation will occur after 2 d in 1×10^{-3} *M* ACC as the fronds become chlorotic and start to fragment.

VII. THE ROLES OF ETHYLENE

We now have antibodies to ACC-synthase,[55,68,69] and methods for obtaining membrane vesicles from leaf material[57] that retain stereospecificity of EFE activity (i.e., the vesicles select and convert the same stereoisomer of the ACC analog 1-amino-2-ethylcyclopropane-1-carboxylic acid to 1-butene as intact tissues *in vivo*).[70] Isolation and cloning of one of the ACC-synthase genes has now been achieved.[71] There is good reason, therefore, to expect exciting new information on the regulation and the pathway of ethylene biosynthesis in plants, which will open further opportunities for studying the capacities of specific cell types in leaf and other tissues for the production and perception of, and response to, ethylene.

The mechanism by which ethylene accomplishes signal transmission is as yet little understood, but the unique possession by plants of a finely tuned biosynthesis pathway for a gaseous hormone must command our study. That ethylene from one fruit can ripen neighboring fruit and that ripening in certain tropical trees is highly synchronized is very well established.[1] In this connection, a potentially important communication route has been reported to result from wounding of plant tissues. When 7 to 10% of the leaves of young poplar trees were mechanically damaged, as in attacks by herbivores, the levels and synthesis of phenolic components rose 100% in 52 h, not only in the damaged trees, but also in those that were nearby and undamaged.[72] Whether this response might have protected neighboring trees is unclear, but ethylene produced from the damaged foliage was invoked as the transmitted signal.

Perhaps more readily understood is the role that ethylene can play in regulating the transport of other growth regulatory chemicals within the plant from one part to another. Polar auxin transport in segments of petioles of *P. vulgaris* leaves is reduced by 25% within 6 h and by 50% after 8 h of exposure to 500 $\mu l\ l^{-1}$ ethylene,[73] and the inhibition is dose dependent up to 100 $\mu l\ l^{-1}$ over 18 h in segments of etiolated pea epicotyls.[74] Of special significance however is the 60% decline detected in the density of binding sites to the auxin transport inhibitor, *N*-1-naphthylphthalamic acid, on microsomes prepared from the ethylene-treated epicotyls.[74]

Considerable effort has been extended in the search for binding sites in plant cells which show a high affinity for ethylene and which could be part of the physiological transduction of the hormone signal (see Chapter 5). Binding of radioactive ethylene to a cellular component, as distinct from dissolved ethylene, was first shown over 10 years ago for tobacco leaves.[75] Protein fractions now isolated from cotyledon or shoot tissues, suggest that the perception and transduction of ethylene in plants could indeed be initiated by binding to a specific protein component of the cell.[76,77,78] However, the number of different proteins that might function as receptors, their location within the cell, and their diversity from one cell type to another remains unknown.

Although ethylene is a ubiquitous environmental gas, the plant must perceive the molecule in solution in aqueous or lipid phases. Because of their large surface to volume ratio, leaves are the sites of major ethylene loss by diffusion to the air. The levels of ethylene retained by different cell types will depend upon the solubility coefficients and diffusion pathways between the cell wall and the cell contents. In addition, the membrane composition of the organelles, the total lipid content, and the organization of cytoplasmic compartments will differ from one tissue to another.[79]

During ontogeny, physiological responses of an organ to ethylene must change and these must in part be controlled at the cell level by developmental differences in membrane compositions accompanying changes in gene expression and cell determination. In the leafy structures of both higher and lower plants, cell lineage and the differentiation of specific responsive cell classes could provide unique specifications for ethylene recognition and for the ensuing mechanisms of hormone signal transduction.

ACKNOWLEDGMENTS

I wish to thank Mrs. Ann Rogers, Mrs. Susan Jackson, and Mrs. Janice Henderson for their valuable assistance with the preparation of this manuscript.

REFERENCES

1. **Abeles, F. B.,** *Ethylene in Plant Biology,* Academic Press, New York, 1973.
2. **Kays, S. J. and Pallas, J. E.,** Inhibition of photosynthesis by ethylene, *Nature,* 285, 51, 1980.
3. **Goeschl, J. D., Pratt, H. K., and Bonner, B. A.,** An effect of light on the production of ethylene and the growth of the plumular portion of etiolated pea seedlings, *Plant Physiol.,* 42, 1077, 1967.
4. **Lau, O. L. and Yang, S. F.,** Inhibition of ethylene production by cobaltous ion, *Plant Physiol.,* 58, 114, 1976.
5. **Van Andel, O. M. and Verkerke, D. R.,** Stimulation and inhibition by ethylene of stem leaf growth of some Gramineae at different stages of development, *J. Exp. Bot.,* 29, 639, 1978.
6. **Miller, P. M., Sweet, H. C., and Miller, J. H.,** Growth regulation by ethylene in fern gametophytes. 1. Effects on protonemal and rhizoidal growth and interaction with auxin, *Am. J. Bot.,* 57, 212, 1970.
7. **Brooks, K. E.,** Reproductive biology of *Selaginella.* 1. Determination of megasporangia by 2-chloroethylphosphonic acid, an ethylene releasing compound, *Plant Physiol.,* 51, 718, 1973.
8. **Rudich, J., Haley, A. H., and Kedar, N.,** Increase in femaleness of three cucurbits by treatment with Ethrel, an ethylene releasing compound, *Planta,* 86, 69, 1969.
9. **Mohan-Ram, H. Y. and Jaiswal, V. S.,** Induction of female flowers on male plants of *Cannabis sativa* by 2-chloroethanephosphonic acid, *Experientia,* 26, 214, 1970.
10. **Osborne, D. J.,** Ethylene and plants of aquatic and semi-aquatic environments: a review, *Plant Growth Regul.,* 2, 167, 1984.
11. **Osborne, D. J.,** The ethylene regulation of cell growth in specific target tissues of plants, in *Plant Growth Substances 1982,* Wareing, P. F., Ed., Academic Press, London, 1982, 279.
12. **Stange, L. and Osborne, D. J.,** Cell specificity in auxin — and ethylene-induced 'supergrowth' in *Riella helicophylla, Planta,* 175, 341, 1988.
13. **Stange, L. and Osborne, D. J.,** Contrary effects of ethylene and ACC on cell growth in the liverwort *Riella helicophylla,* in *Biochemical and Physiological Aspects of Ethylene Production in Lower and Higher Plants,* Clijsters, H., de Proft, M., Marcelle, R., and van Poucke, M., Eds., Kluwer Academic Publishers, Dordrecht, The Netherlands, 1989, 341.
14. **Dale, J. E. and Milthorpe, F. L.,** General features of the production and growth of leaves, in *The Growth and Function of Leaves,* Dale, J. E. and Milthorpe, F. L., Eds., Cambridge University Press, Cambridge, 1983, 151.
15. **Goodwin, P. B. and Erwee, M. G.,** Hormonal influences in leaf growth, in *The Growth and Function of Leaves,* Dale, J. E. and Milthorpe, F. L., Eds., Cambridge University Press, Cambridge, 1983, 207.

16. **Poethig, S.**, Cellular parameters of leaf morphogenesis in maize and tobacco, in *Contemporary Problems in Plant Anatomy*, White, R. A. and Dickison, W. C., Eds., Academic Press, New York, 1984, 235.
17. **Hagemann, W.**, Morphological aspects of leaf development in ferns and angiosperms, in *Contemporary Problems in Plant Anatomy*, White, R. A. and Dickison, W. C., Eds., Academic Press, New York, 1984, 301.
18. **Basile, D. V. and Basile, M. R.**, Desuppression of leaf primordia of *Plagiochila artica* (Hepaticae) by ethylene antagonists, *Science*, 220, 1051, 1983.
19. **Dengler, N. G.**, Comparative histological basis of sun and shade leaf dimorphism in *Helianthus annuus*, *Can. J. Bot.*, 58, 717, 1980.
20. **Van Volkenburgh, E., Cleland, R. E., and Schmidt, M. G.**, The mechanism of light-stimulated leaf cell expansion, in *Control of Leaf Growth*, Baker, N. R., Davies, W. J., and Ong, C. K., Eds., Cambridge University Press, Cambridge, 1985, 223.
21. **Osborne, D. J.**, Control of cell shape and cell size by the dual regulation of auxin and ethylene, in *Perspectives in Experimental Biology*, Vol. 2, Sunderland, N., Ed., Pergamon Press, Oxford, 1976, 89.
22. **Osborne, D. J.**, Concepts of target cells in plant differentiation, *Cell Differ.*, 14, 161, 1984.
23. **Steeves, T. A. and Sussex, I. M.**, Studies on the development of excised leaves in sterile culture, *Am. J. Bot.*, 44, 665, 1957.
24. **Steeves, T. A. and Sussex, I. M.**, *Patterns in Plant Development*, Prentice-Hall, Englewood Cliffs, NJ, 1972.
25. **Langdale, J. A., Lane, B., Freeling, M., and Nelson, T.**, Cell lineage analysis of maize bundle sheath and mesophyll cells, *Dev. Biol.*, 133, 128, 1989.
26. **Szymkowiak, E. J. and Sussex, I. M.**, Chimeric analysis of cell layer interactions during development of the flower pedicel abscission zone, in *Cell Separation in Plants: Physiology, Biochemistry and Molecular Biology*, Osborne, D. J. and Jackson, M. B., Eds., Springer-Verlag, Heidelberg, 1989, 363.
27. **Sexton, R. and Roberts, J. A.**, Cell biology of abscission, *Annu. Rev. Plant Physiol.*, 33, 133, 1982.
28. **Osborne, D. J.**, Abscission, *Crit. Rev. Plant Sci.*, 8, 103, 1989.
28a. **Thompson, D. S.**, unpublished data.
28b. **Roberts, J. A. and Osborne, D. J.**, unpublished data.
29. **Osborne, D. J. and McManus, M. T.**, Flexibility and commitment in plant cells during development, *Curr. Top. Dev. Biol.*, 20, 383, 1986.
30. **Osborne, D. J. and Sargent, J. A.**, The positional differentiation of abscission zones during the development of leaves of *Sambucus nigra* and the response of the cells to auxin and ethylene, *Planta*, 132, 197, 1976.
31. **Osborne, D. J., McManus, M. T., and Webb, J.**, Target cells for ethylene action, in *Ethylene and Plant Development*, Roberts, J. A. and Tucker, G. A., Eds., Butterworths, London, 1985, 197.
32. **Holm, R. E. and Wilson, W. C.**, Ethylene and fruit loosening from combinations of citrus abscission chemicals, *J. Am. Soc. Hortic. Sci.*, 102, 576, 1977.
33. **Walters, J. and Osborne, D. J.**, Ethylene and auxin-induced cell growth in relation to auxin transport and metabolism and ethylene production in the semi-aquatic plant, *Regnellidium diphyllum*, *Planta*, 146, 309, 1979.
34. **Ridge, I. and Amarasinghe, I.**, Ethylene and growth control in the fringed waterlily *(Nymphoides peltata)*: stimulation of cell division and interaction with buoyant tension in petioles, *Plant Growth Regul.*, 2, 235, 1984.
35. **Ridge, I. and Osborne, D. J.**, Wall extensibility, wall pH and tissue osmolality: significance for auxin and ethylene-enhanced petiole growth in semi-aquatic plants, *Plant Cell Environ.*, 12, 383, 1989.
36. **Sarukhan, J. and Harper, J. L.**, Studies on plant demography: *Ranunculus repens* L., *R. bulbosus* L. and *R. acris* L., *J. Ecol.*, 61, 675, 1973.
37. **Voesenek, L. A. C. J. and Blom, C. W. P. M.**, Growth response of *Rumex* species in relation to submergence and ethylene, *Plant Cell Environ.*, 12, 425, 1989.
38. **Mitchell, R. S.**, Submergence experiments on nine species of semi-aquatic *Polygonum*, *Am. J. Bot.*, 63, 1158, 1976.
39. **Bradford, K. J. and Yang, S. F.**, Xylem transport of 1-aminocyclopropane-1-carboxylic acid, an ethylene precursor, in waterlogged tomato plants, *Plant Physiol.*, 65, 322, 1980.
40. **Reid, M. S., Mor, Y., and Kofranek, A. M.**, Epinasty of *Poinsettias* — the role of auxin and ethylene, *Plant Physiol.*, 67, 950, 1981.
41. **Hayes, A. B.**, The interaction of auxin and ethylene in the maintenance of leaf blade form in *Phaseolus vulgaris* L. var pinto, *Am. J. Bot.*, 68, 733, 1981.
42. **Zobel, R. W.**, Some physiological characteristics of the ethylene-requiring tomato mutant *diageotropica*, *Plant Physiol.*, 52, 385, 1973.
43. **Ursin, V. M. and Bradford, K. J.**, Auxin and ethylene regulation of petiole epinasty in two developmental mutants of tomato, *diageotropica* and *epinastic*, *Plant Physiol.*, 90, 1341, 1989.

44. **Fujino, D. W., Nissen, S. J., Jones, A. D., Burger, D. W., and Bradford, K. J.,** Quantification of indole-3-acetic acid in dark grown seedlings of *diageotropica* and *epinastic* mutants of tomato (*Lycopersicon esculentum* Mill.), *Plant Physiol.*, 88, 780, 1988.
45. **Key, G. and Weiler, E. W.,** Monoclonal antibodies identify common and differentiation-specific antigens in the plasmamembrane of guard cells of *Vicia faba* L., *Planta*, 176, 472, 1988.
46. **McManus, M. T. and Osborne, D. J.,** Identification of polypeptides specific to rachis abscission zone cells of *Sambucus nigra*. I. Immunological investigation, *Physiol. Plant.*, 79, 471, 1990.
47. **McManus, M. T. and Osborne, D. J.,** Identification of leaf abscission zones as a specific class of target cells for ethylene, in *Cell Separation in Plants, Physiology, Biochemistry and Molecular Biology*, Osborne, D. J. and Jackson, M. B., Eds., Springer-Verlag, Heidelberg, 1989, 201.
48. **Spruce, J., Mayer, A. M., and Osborne, D. J.,** A simple histochemical method for locating enzymes in plant tissue using nitrocellulose blotting, *Phytochemistry*, 11, 2901, 1987.
49. **McManus, M. T., McKeating, J., Secher, D. S., Osborne, D. J., Ashford, D., Dwek, R. A., and Rademacher, T. W.,** Identification of a monoclonal antibody to abscission tissue that recognises xylose/fucose containing N-linked oligosaccharides from higher plants, *Planta*, 175, 506, 1988.
50. **Poovaiah, B. W., Friedmann, M., Reddy, A. S. N., and Rhee, J. K.,** Auxin-induced delay of abscission: the involvement of calcium ions and protein phosphorylation in bean explants, *Physiol. Plant.*, 73, 354, 1988.
51. **Osborne, D. J.,** Ethylene, in *Phytohormones and Related Compounds—A Comprehensive Treatise*, Vol. 1, Latham, D. S., Goodwin, P. B., and Higgins, T. J. V., Eds., Elsevier/North-Holland, Amsterdam, 1978, 266.
52. **Roberts, J. A. and Osborne, D. J.,** Auxin and the control of ethylene production during the development and senescence of leaves and fruits, *J. Exp. Bot.*, 32, 875, 1981.
52a. **Osborne, D. J.,** unpublished data.
53. **Sargent, J. A., Osborne, D. J., and Dunford, S. M.,** Cell separation and its hormonal control during fruit abscission in the Gramineae, *J. Exp. Bot.*, 35, 1663, 1984.
54. **Tsai, D. S., Arteca, R. N., Phillips, A. T., and Bachman, J. M.,** Characterization of ACC synthase purified from etiolated mung bean hypocotyl segments, *Plant Physiol.*, 86, 113, 1988.
55. **Arteca, R. N.,** Hormonal induction of ACC synthase, in *Biochemical and Physiological Aspects of Ethylene Production in Lower and Higher Plants*, Clijsters, H., de Proft, M., Marcelle, R., and van Poucke, M., Eds., Kluwer Academic Publishers, Dordrecht, The Netherlands, 1989, 119.
56. **Suttle, J. C.,** Cytokinin-induced ethylene biosynthesis in nonsenescing cotton leaves, *Plant Physiol.*, 82, 930, 1986.
57. **Porter, A. J. R., Borlakoglu, J. T., and John, P.,** Activity of the ethylene-forming enzyme in relation to plant cell structure and organization, *J. Plant Physiol.*, 125, 207, 1986.
58. **Guy, M. and Kende, H.,** Conversion of ACC to ethylene by isolated vacuoles of *Pisum sativum*, *Planta*, 160, 281, 1984.
59. **Erdman, H., Griesbach, R. J., Lawson, R. H., and Mattoo, A. K.,** 1-Aminocyclopropane-1-carboxylic-acid-dependent ethylene production during re-formation of vacuoles in evacuolated protoplasts of *Petunia hybrida*, *Planta*, 179, 196, 1989.
60. **Bouzayen, M., Latché, A., Alibert, G., and Pech, J. C.,** Intracellular sites of synthesis and storage of 1-(malonylamino)cyclopropane-1-carboxylic acid in *Acer pseudoplatanus* cells, *Plant Physiol.*, 88, 613, 1988.
61. **Romani, R. J., Hess, B. M., and Leslie, C. A.,** Salicylic acid inhibition of ethylene production by apple discs and other plant tissues, *J. Plant Growth Regul.*, 8, 63, 1989.
62. **Masuda, Y. and Yamamoto, R.,** Control of auxin-induced stem elongation by the epidermis, *Physiol. Plant.*, 27, 109, 1972.
63. **Tokada, I. and Imaseki, H.,** Epidermal cells do not contribute to auxin-induced ethylene production in mung bean stem sections, *Plant Cell Physiol.*, 26, 865, 1985.
64. **Rodecap, K. D. and Tingey, D. T.,** Ozone-induced ethylene release from leaf surfaces, *Plant Sci.*, 44, 73, 1986.
65. **Osborne, D. J.,** The control of ethylene in plant growth and development in *Biochemical and Physiological Aspects of Ethylene Production in Lower and Higher Plants*, Clijsters, H., de Proft, M., Marcelle, R., and van Poucke, M., Eds., Kluwer Academic Publishers, Dordrecht, The Netherlands, 1989, 1.
65a. **Osborne, D. J.,** unpublished data.
66. **Cookson, C. and Osborne, D. J.,** The stimulation of cell extension by ethylene and auxin in aquatic plants, *Planta*, 144, 39, 1978.
67. **Tittle, F. L.,** Auxin-stimulated ethylene production in fern gametophytes and sporophytes, *Physiol. Plant.*, 70, 499, 1987.
68. **Bleecker, A. B., Kenyon, W. H., Somerville, S. C., and Kende, H.,** Use of monoclonal antibodies in the purification and characterization of 1-aminocyclopropane-1-carboxylate synthase, an enzyme in ethylene biosynthesis, *Proc. Natl. Acad. Sci. U.S.A.*, 83, 7755, 1986.

69. **Mehta, A. M., Jordon, R. L., Anderson, J. D., and Mattoo, A. K.,** Identification of a unique isoform of 1-aminocyclopropane-1-carboxylic acid synthase by monoclonal antibody, *Proc. Natl. Acad. Sci. U.S.A.*, 85, 8810, 1988.
70. **Hoffman, N. E., Yang, S. F., Ichihara, A., and Sakamura, S.,** Stereospecific conversion of 1-aminocyclopropanecarboxylic acid to ethylene by plant tissues. Conversion of stereoisomers of 1-amino-2-ethylcyclopropanecarboxylic acid to 1-butene, *Plant Physiol.*, 70, 195, 1982.
71. **Sato, T. and Theologis, A.,** Cloning the mRNA encoding 1-aminocyclopropane-1-carboxylate synthase. The key enzyme for ethylene biosynthesis in plants, *Proc. Natl. Acad. Sci. U.S.A.*, 86, 6621, 1990.
72. **Baldwin, I. T. and Schultz, J. C.,** Rapid changes in tree leaf chemistry induced by damage: evidence for communication between plants, *Science*, 2211, 277, 1983.
73. **Osborne, D. J. and Mullins, M. G.,** Auxin, ethylene and kinetin as a carrier-protein model system for the polar transport of auxins in petiole segments of *Phaseolus vulgaris, New Phytol.*, 68, 977, 1969.
74. **Suttle, J. C.,** Effect of ethylene treatment on polar IAA transport net IAA uptake and specific binding of N-1-naphthylphthalamic acid in tissues and microsomes isolated from etiolated pea epicotyls, *Plant Physiol.*, 88, 795, 1988.
75. **Sisler, E. C.,** Measurement of ethylene binding in plant tissue, *Plant Physiol.*, 64, 538, 1979.
76. **Sisler, E. C. and Wood, C.,** Ethylene binding and evidence that binding in vivo and in vitro is to the physiological receptor, in *Plant Hormone Receptors*, (NATO ASI Ser., Vol. H10), Klambt, D., Ed., Springer-Verlag, Berlin, 1987, 239.
77. **Smith, A. R., Robertson, D., Sanders, I. O., Williams, R. A. N., and Hall, M. A.,** Ethylene binding sites, in *Plant Hormone Receptors*, (NATO ASI Ser., Vol. H10), Klambt, D., Ed., Springer-Verlag, Berlin, 1987, 299.
78. **Sanders, I. O., Smith, A. R., and Hall, M. A.,** The measurement of ethylene binding and metabolism in plant tissue, *Planta*, 179, 97, 1989.
79. **Jerie, P. H., Shaari, A. R., and Hall, M. A.,** Compartmentation of ethylene in developing cotyledons of *Phaseolus vulgaris* L., *Planta*, 144, 503, 1979.
80a. **Roberts, J. A.,** personal communication.
80b. **Atkins, C. G.,** personal communication.
81. **McCready, C. C. and Osborne, D. J.,** unpublished data.
82. **Osborne, D. J., Yip, W.-K., and Yang, S. F.,** unpublished data.

Note: For other important reviews on the subject readers are referred to:

Addicott, F. T., *Abscission*, University of California Press, Berkeley, CA, 1982.
Baker, N. R., Davies, W. J., and Ong, C. K., Eds., *Control of Leaf Growth*, Soc. Exp. Biol. Sem. Ser. 27, Cambridge University Press, London, 1985.
Dale, J. E., The control of leaf expansion, *Annu. Rev. Plant Physiol.*, 39, 267, 1988.

Chapter 12

ETHYLENE IN FLOWER DEVELOPMENT AND SENESCENCE

Michael S. Reid and Meng-Jen Wu

TABLE OF CONTENTS

I. INTRODUCTION

The importance of ethylene as a natural regulator of many phases of plant growth and development can be gauged from the extent of the literature reviewed in this volume. In flowers, too, ethylene has long been considered to play an important role. One of the earliest known responses of plants to ethylene was the observation of "sleepiness" of carnation flowers in greenhouses contaminated with ethylene from faulty heaters or leaky gas pipes. Effects of ethylene on flowering were also first noted in a greenhouse, in this case in greenhouse-grown pineapple plants. In these, as in other responses, ethylene was at first considered to be a plant growth regulator only; its hormonal role was not established until the development of analytical methods sufficiently sensitive to measure the minute quantities produced by plant tissues. This chapter summarizes studies that have provided much insight into the role of ethylene in flower induction, flower development, and flower senescence.

II. ETHYLENE AND FLOWER DEVELOPMENT

A. FLOWER INDUCTION

The striking effects of ethylene on flowering of bromeliads are now used commercially in the production of pineapples[1] and ornamental bromeliads popular in the potted plant trade.[2] For many years, pineapple growers dusted their plants with calcium carbide. Its hydrolysis products include sufficient ethylene to provide the desired acceleration and coordination of flowering. The availability of ethylene-releasing compounds such as 2-chloroethyl phosphonic acid has simplified practical application of this technique. Ethylene-mediated induction of flowering is also utilized in the commercial production of mangos. Growers in the Philippines and in India maintain smoky fires in mango orchards for several days during a vegetative flush, to induce good flowering. Sprouting and flowering in a number of geophytes have also been shown to be stimulated by ethylene.[3] The response is said to have been first noted by Japanese iris growers, who burned senescent foliage in their fields to reduce the carryover of foliar diseases. Bulbs lifted from fields treated in this way flowered more uniformly than those from fields that were not burned. Growers in Japan still induce flowering in iris bulbs by "smoking" them in unventilated barns.[3a] In cormous crops, such as freesia and gladiolus, ethylene or ethephon treatments result in earlier sprouting and flowering.[4] In tarzetta narcissus and iris, in which flower induction occurs after the bulbs are lifted, ethylene treatment accelerates flower induction.[5] In addition, the ethylene treatment induces flowering of small bulbs, which otherwise would not flower.

Despite these numerous examples of the spectacular effects of ethylene on the induction of flowering, there has been little study of the way in which this hormone interacts with the flowering process. Han et al.[6] used the 'Queen Fabiola' cultivar of brodiaea *(Triteleia laxa)* to study the mechanism of ethylene induction of flowering in geophytes. Small corms of this plant flowered very sparsely (Figure 1A). Treatment of dormant corms with ethylene resulted in a spectacular increase in the percentage of small corms which flowered (Figure 1A), and in the quality of the flowers in terms of numbers of flowers per inflorescence. The hypothesis that ethylene treatment resulted in enhanced mobilization of carbohydrate in the dormant organ, which in turn resulted in induction of flowering in normally vegetative bulbs, was not sustained. No differences in carbohydrate patterns in the storage tissues and the meristematic apex could be detected following ethylene treatment.[6]

Ethylene treatment resulted in a rapid increase in size and growth rate of the meristematic dome of dormant brodiaea corms (Figure 1B). In brodieae, as in other geophytes, the size of the meristematic dome is related to the propensity of a propagule to produce flowers, and to the number of flowers or florets that are produced on the resulting axis. These data suggest that the stimulation of flowering in this species may be due directly to an effect of

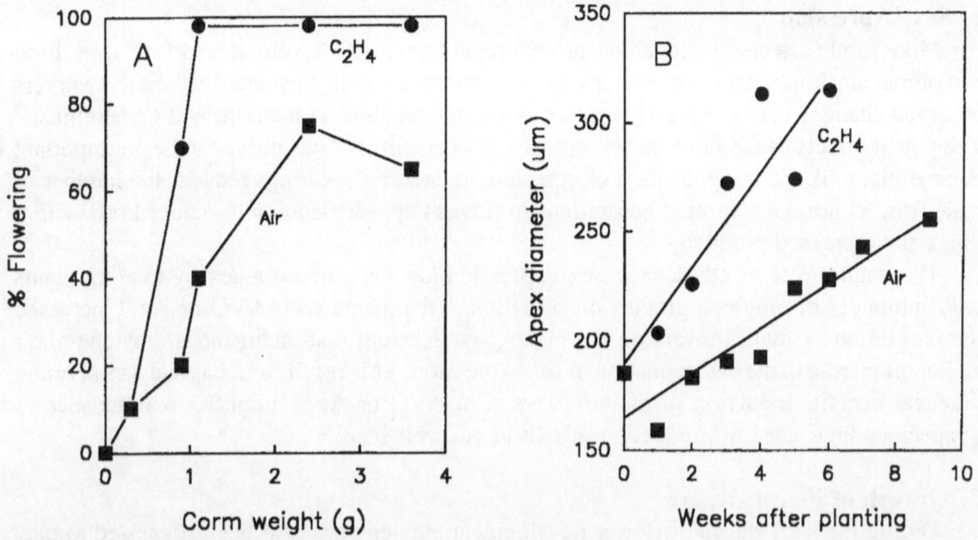

FIGURE 1. Effect of ethylene treatment (20 ppm, 7 d) of dry, dormant corms on (A) subsequent flowering and (B) diameter of the apical dome in *Triteleia laxa*. Apex diameter was determined from scanning electron micrographs of replicate corms sacrificed at intervals during normal forcing. (Data from Han, S. S. et al., *J. Am. Soc. Hortic. Sci.*, 115, 482, 1990.)

ethylene on the diameter of the apical meristem. The possibility that a similar mechanism operates in other geophytes and possibly in other plants where ethylene stimulates flowering[7] requires investigation.

In some photoperiodic plants, ethylene treatment has been shown to inhibit flowering under photoinductive conditions.[8,9] Recent studies have examined this effect using the short-day plant *Pharbitis nil*. In the 'Violet' strain of this species, exposure of young seedlings to one long night will induce complete flowering. This effect is inhibited by treatment with ethylene during the inductive dark period,[10] an inhibition that was suggested to be due to synthesis of a "flower inhibiting entity". Lay-Yee et al.[11] examined *in vitro* translation products of mRNA extracted from the cotyledons of *Pharbitis* seedlings given various photoperiodic treatments in the presence and absence of ethylene. Ethylene treatment induced many changes in the pattern of translation products, including a pronounced decrease in a polypeptide that had been shown[12] to be associated with the induced condition. These data suggest that ethylene may also act to prevent flowering at an early stage of the induction process.

B. FLOWER DEVELOPMENT

1. Bud Abortion

The stimulation of flowering in geophytes by ethylene is particularly curious, because ethylene has long been known to have deleterious effects on flower development in many bulbous crops. Inhibition of normal floral development, termed "blasting", is a common and commercially important symptom of accidental exposure of tulips and other flowering bulbs to ethylene during flower development.[13] Ethylene also appears to be the causal agent in flower blasting in plants grown under environmental stress. For example, the flower abortion commonly observed in 'Enchantment' lilies grown in low light conditions can be prevented by spraying the plants with silver thiosulfate (STS)[14] that inhibits ethylene action, or with aminoethoxyvinylglycine (AVG) that inhibits its synthesis. Further investigation of this phenomenon revealed that low light conditions resulted in greatly stimulated ethylene production during gametogenesis in the developing anthers.[15] This ethylene was presumably the cause of the subsequent blasting and abscission of the flowers.

2. Sex Expression

More subtle effects of exogenous growth regulators in the development of induced floral meristems are implicit in the changes in sex expression that they induce.[16] Early workers observed changes in sex expression in response to treatments with auxins and gibberellins.[17] The potent effects of ethylene on sex expression in cucurbits[18] has proved to be an important commercial tool. Early application of ethephon to cucurbit seedlings reduces the number of male flowers normally formed before female flowers appear, leading to reduced production times and increased production.[19]

The natural role of ethylene in sex expression has been investigated by treating plants with inhibitors of ethylene production or action. Treatments with AVG or Ag^+ increased the proportion of male flowers on the plants,[20] suggesting that endogenous ethylene plays an important role in the determination of sex expression. This result, too, has had an important practical benefit. Induction of male flowers with AVG or Ag^+ simplifies maintenance of gynoecious lines used in production of hybrid cucurbit seed.[21]

3. Growth of Floral Organs

During the later stages of flower development ethylene has also been suggested to play a beneficial role. Koning[22] studied the growth of various parts of the developing flowers of *Gaillardia,* and proposed the participation of auxin-mediated ethylene production in the rapid growth of their filaments. This hypothesis needs to be tested by exploring the effects of inhibitors of ethylene production and action on the process.

C. FLOWER OPENING

The failure of carnation flowers to open in greenhouse contaminated by leaking gas pipes or cracked heater vents, a disorder termed "sleepiness", was among the earliest observed effects of ethylene. In contrast to this common observation, Camprubi and Nichols[23] found that very low concentrations of ethylene stimulated bud opening in carnations. The facts that STS strongly inhibits the action of endogenous ethylene in triggering carnation petal senescence, but that carnation buds pretreated with STS open normally, suggest that ethylene is not a natural regulator of bud opening in carnations. Zacarias et al.[24] have shown that opening of citrus flowers is associated with the onset of ethylene biosynthesis, and is retarded, but not prevented, by application of STS. In citrus, therefore, it would appear that the action of ethylene is part of the natural opening process.

Very low concentrations of ethylene strongly affect the opening of some cultivars of rose.[25] The response of rose buds to applied ethylene (Table 1) varies by cultivar, from complete inhibition of opening ('Lovely Girl'), to accelerated opening and promotion of abscission ('Golden Fantasy'). Kinetic evaluation of the effects of different concentrations of ethylene on the inhibition of opening of 'Lovely Girl' buds indicates that the effect is half-maximal at 4 ppb, one of the most sensitive responses to ethylene known. As with carnations, STS does not affect the opening of rose flowers held in ethylene-free air, which again indicates that ethylene is not a natural regulator of rose opening. After cool storage, roses *do* produce ethylene[26] and STS or AOA (aminooxyacetic acid, which also inhibits ACC synthase) treatments then increase flower life. Another effect of ethylene on the development of floral tissues was shown by Nichols and Kofranek[27] in tulips, where ethylene inhibited growth of the scape, and prior application of STS allowed normal growth.

III. ETHYLENE AND FLOWER SENESCENCE

Senescence can be broadly defined as the combination of events that lead to the death of cells, tissues, or organs. Such a definition, applied to cut flowers, might include adverse water relations and floret abscission. Although ethylene probably plays a role in these

TABLE 1
Effect of Exposure to 0.5 ppm Ethylene for 2 d on Flower Opening of Different Rose Cultivars

Cultivar	Ethylene effect on opening
Angel	Acceleration
Bettina	Acceleration, abscission
Candia	Acceleration
Capella	Abscission
Cara Mia	Acceleration
Celica	Distortion
Cerisa	Acceleration
Chantilly Lace	Inhibition
Coed	No effect
Excitement	Acceleration
Gold Rush	No effect
Golden Fantasy	Acceleration, abscission
Golden Times	No effect
Golden Wave	Acceleration
Grace	No effect
Jack Frost	Acceleration, distortion
Lady Diana	No effect
Lavande	Acceleration, abscission
Lovely Girl	Inhibition
Paul's Pink	Acceleration
Privé	Abscission
Royalty	Distortion
Sonia	Abscission
St. Louis	Inhibition, abscission
Sterling Silver	Acceleration
Tobone	Acceleration, abscission
White Success	Acceleration

Data from Reid, M. S., Dodge, L. L., Mor, Y., and Evans, R. Y., *Acta Hortic.*, 261, 215, 1989.

processes, we will consider, in this chapter, only the stricter sense of senescence, the actual events occurring during aging and death of floral tissues.

A. EFFECTS OF APPLIED ETHYLENE

For most cut flowers, the importance of ethylene in their senescence was first examined by testing the effects of exogenous ethylene. These effects often are first noted as a practical problem resulting from environmental pollution with ethylene. Vehicle exhausts, cracked heater vents, and direct-fired greenhouse heaters are all common sources of such pollution. Some workers have used application of aqueous solutions of the ethylene-releasing compound, 2-chloroethylphosphonic acid (ethephon), to examine "ethylene" effects.[28,29] Although this technique has some practical attractions, care must always be exercised to use concentrations that will generate physiologically meaningful ethylene concentrations and to avoid the strong acid effects of this compound.[30] In this regard, it is important to note that the release of ethylene from 1 ml of 1 ppm ethephon into a liter of air results in a concentration of ethylene greater than 100 ppm! Although direct application of ethylene is the most satisfactory means of examining its effects, the practical difficulties of setting up complex dilution systems often mitigate against this approach. The simple diffusion system described by Saltveit[31] makes it easy to apply low concentrations of ethylene to cut flowers. Ethylene is allowed to diffuse from a vessel containing pure ethylene through the walls of a plastic

TABLE 2
Ethylene and Senescence of Flowers from Different Families[a]

Family	Symptoms	Sensitivity
Monocotyledonae		
Amaryllidaceae	Wilting	0—2
Iridaceae	Wilting	0—1
Liliaceae	Wilting	0—1
Orchidaceae	Wilting	3—4
Dicotyledonae		
Campanulaceae	Wilting	3—4
Caryophyllaceae	Wilting	4
Compositae	Wilting	0—1
Dipsacaceae	Wilting	2—4
Euphorbiaceae	Wilting	1
Geraniaceae	Abscission	4
Labiatae	Abscission	4
Malvaceae	Wilting	4
Primulaceae	Abscission	3—4
Ranunculacaea	Abscission	3—4
Rosaceae	Abscission	3—4
Scrophulariaceae	Abscission	2—4
Umbelliferae	Wilting	0

[a] Sensitivity from 0 (insensitive) to 4 (very sensitive).

Adapted from Woltering, E. J. and van Doom, W. G., *J. Exp. Bot.*, 208, 1605, 1988.

or rubber tube carrying the stream of treatment air. The concentration of ethylene in the treatment air is easily controlled by the length and type of tubing inside the vessel, and the flow rate of air through the tube.

In an exhaustive survey, Woltering and van Doorn[32] studied the sensitivity to ethylene of petal senescence in flowers of 93 species representing 23 families. Except for a few families (all tested members of Campanulaceae, Caryophyllaceae, and Malvaceae, and most Orchidaceae), most of those flowers in which wilting was the primary senescence symptom were not sensitive to ethylene (Table 2).[32,33] Where petal abscission was the initial symptom of senescence the flowers were mostly sensitive to exogenous ethylene (Geraniaceae, Labiateae, Ranunculaceae, Rosaceae, Scrophulareaceae). Flowers that respond to low exogenous concentrations of ethylene are probably those in which ethylene is naturally involved in senescence.

B. ETHYLENE PRODUCTION DURING FLOWER SENESCENCE

The small mass of most flowers, and the minute amounts of ethylene produced by petal tissues until the onset of the climacteric burst has meant that it was only after the application of gas chromatography to ethylene analysis that the production of ethylene by senescing flowers was established. Maxie et al.[34] demonstrated the close coincidence between the onset of the respiratory climacteric and of a burst of ethylene synthesis in cut carnations. A similar relationship has been shown in studies of ephemeral flowers such as those of *Ipomoea*,[35-37] *Tradescantia*,[38] and *Hibiscus*.[39]

The advent of more sensitive techniques of ethylene analysis such as the photoionization detector for gas chromatography and the laser photoacoustic system described by Woltering et al.[40] will permit even more detailed evaluation of the role of ethylene biosynthesis in the natural senescence of cut flowers. In orchids, for example, Woltering and Harren[41] used this extremely sensitive instrument to examine the role of ethylene in senescence of emasculated orchid flowers. The reason that removal of the pollinia in *Cymbidium* results in senescence of the flower (and blushing of the labellum) is that desiccation of the tissue of the rostrum stimulates the early production of trace amounts of ethylene. These minute quantities of ethylene are, in turn, responsible for initiating general senescence of the flower.

The pathway of ethylene biosynthesis in plants has now been almost completely described,[42] and is dealt with in detail in Chapter 1. The essential steps of the pathway have been completely identified in cut flowers. ACC content,[43] ACC synthase,[44] activity of the EFE[45] and of β-cyanoalanine synthase[46] have all been shown to rise coordinately with the onset of the ethylene climacteric burst in various flower parts. In a study of the control of ethylene biosynthesis during the climacteric rise, Whitehead et al.[45] showed that, unlike some fruit tissues (where EFE activity is always high), the activity of the EFE in senescing carnation petals rises coordinately with the increased synthesis of ACC. Adam and Mayak[47] prepared an extract from carnation petals which converted ACC to ethylene. Although this preparation had some of the characteristics of *in vivo* EFE activity, its high Km, and lack of stereospecificity[47a] indicates that it is probably a nonenzymic system, perhaps linked to lipid peroxidation rather than the *in vivo* enzyme.

The elucidation of the pathway of ethylene biosynthesis was assisted by the discovery of inhibitors of ethylene biosynthesis. These inhibitors have now become useful tools for interfering with ethylene-mediated events by inhibiting its biosynthesis. The activity of ACC synthase (normally the rate-limiting step) is markedly reduced by inhibitors of pyridoxal phosphate enzymes, such as AOA. This is also the mode of action of AVG and its analogs.[48] Application of AOA to cut carnations inhibits the rise in ethylene and respiration and extends the vase life of cut carnation flowers.[49] Since AOA only prevents ethylene biosynthesis, it has limited practical value, offering little protection in ethylene-polluted environments.[49,50]

C. ETHYLENE ACTION AND FLOWER SENESCENCE

The early onset of ethylene synthesis during senescence of carnations and the retardation of senescence by inhibitors of ethylene biosynthesis indicate the importance of ethylene in regulation of the process. Treatment of *Tradescantia* or *Ipomoea* tissues with AVG greatly reduced ethylene production but only partially delayed symptoms of senescence.[37,38] These workers concluded that ethylene, while not required for petal senescence, coordinated and accelerated the process. The availability of inhibitors of ethylene action has made it possible to reexplore the role of ethylene in the senescence of ephemeral flowers. Woodson[39] found that STS pretreatment of petals excised from *Hibiscus* buds one day before opening prevented the normal wilting of the petals observed in controls or petals treated on the day of opening. These data indicate an absolute requirement for ethylene in senescence of this ephemeral flower. The lack of effectiveness of STS in petals excised on the day of opening may well relate to their very small water uptake. In day lily, pretreatments are best applied during the period of rapid petal expansion, as the bud develops just prior to flower opening. Use of 2,5-norbornadiene (NBD), a volatile inhibitor of ethylene action, can overcome the problem posed by the need for transpiration in tissues being treated with Ag^+.[51] The life of dandelion flowers was extended when they were held in an atmosphere containing NBD.[51a] The striking effect of Ag^+ in extending the life of carnations and other ethylene-sensitive flowers is telling evidence of the central role ethylene plays in natural regulation of their senescence.

A hypothetical model for the action of ethylene in flower senescence is presented in Figure 2. This scheme suggests a membrane-based binding site that is activated or repressed by a "sensitivity factor". The ethylene molecule binds to a site where the inhibitors of ethylene action, Ag^+ and NBD, can also bind. When the binding site is sensitized and ethylene binds to it, a second message is generated which interacts with the 5′ (promotor)

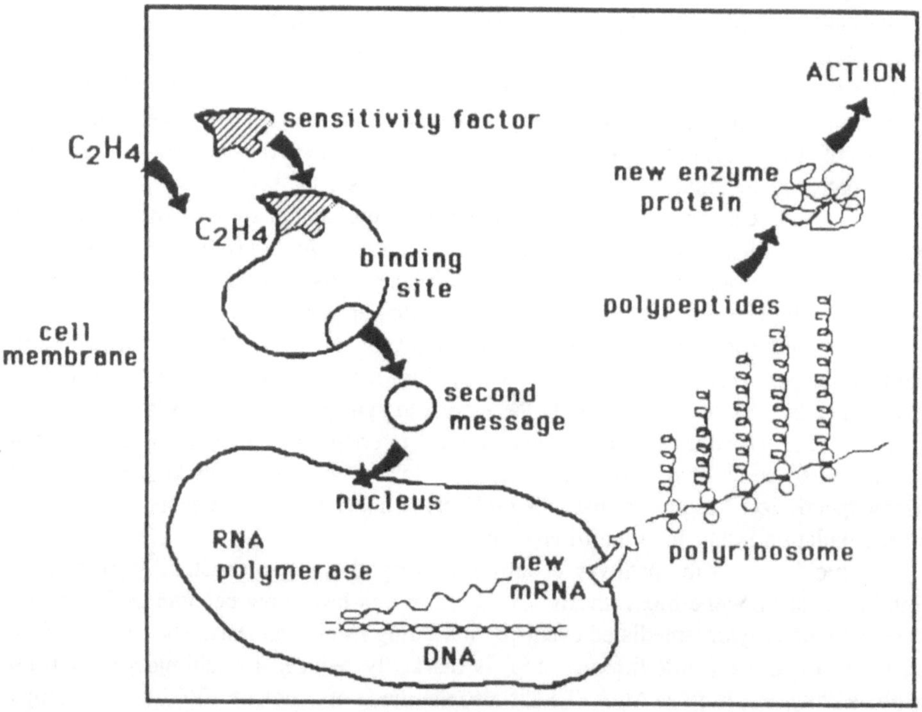

FIGURE 2. Hypothetical scheme for the action of ethylene in inducing flower senescence.

regions of genes involved in ethylene-regulated senescence, inducing transcription of the genes, and synthesis of the proteins encoded by these genes. A considerable amount of evidence has accumulated from studies with cut flowers that is consistent with this scheme.

1. The Binding Site

Using an isotopic competition technique, Sisler et al.[51] demonstrated the presence of a specific ethylene binding site in carnation petals (see Chapter 5). The application of Ag^+ or NBD to the flowers substantially reduced binding activity. Veen[52] suggested that Ag^+ competitively inhibited the ethylene-stimulated growth of carnation styles, and the inhibition of opening of roses by ethylene was prevented by Ag^+ in a competitive fashion.[25]

2. The Sensitivity Factor

As Trewavas[53] has pointed out, response to a plant hormone may involve increased biosynthesis, increased sensitivity to the hormone, or both. In aging of some cut flowers there is a clear increase in sensitivity to ethylene.[45] This effect is perhaps most striking in the ephemeral ethylene-sensitive flowers,[38,39] whose sensitivity increases very rapidly as the bud nears the day of opening. Suttle and Kende[38] used pigment efflux as a measure of senescence in *Tradescantia* petals. While ethylene treatment hastened efflux in petals excised from opening flowers, it had no effect on petals taken from buds one or two days before opening. The nature of the sensitivity factor and its site of action is unknown, although there is some evidence supporting a short-chain fatty acid as a sensitivity factor in pollinated carnations.[54] The senescence-delaying effects of other growth regulators[55] is consistent with the notion that sensitivity to ethylene is a function of the balance of growth regulators. One means by which sensitivity to ethylene might be regulated is an interaction of the sensitivity factor with the binding site itself, as suggested in Figure 2, but this is only one of a number of possibilities.

TABLE 3
Effects of Petal Removal on Growth of Ovaries in Detached Individual Flower Buds of *Triteleia laxa*

	Ovary fresh weight (mg)			
	Buds showing color		Open flowers	
	petals		petals	
	+	−	+	−
Initial weight	14.2a[a]	14.2a	16.4a	16.4a
Final weight (0% sucrose)	16.9a	29.0b	47.3c	25.1b
Final weight (2% sucrose)	31.2b	40.1c	60.8d	50.7c

Note: Flower buds were harvested at various stages of development and placed in opening solution containing either 0% or 2% sucrose with a biocide (hydroxyquinoline citrate) for 7 d. Data are the means of 10 flowers.

[a] Means separated within columns by Duncan's multiple range test (P = 0.05).

Data from Han, S. S., Halevy, A. H., and Reid, M. S., *J. Am. Soc. Hortic. Sci.*, 116, 68, 1991.

The timing of the onset of the ethylene burst and, thus, the timing of senescence, seems likely to be controlled by the sensitivity factor. By analogy with the situation in many fruits, the signal would be the endogenous production of ethylene, that would interact with the binding site and induce a response, including a stimulation of ethylene biosynthesis, once the tissue had become sensitive. Some authors have espoused the attractive notion that the timing of senescence is directed by interaction between different floral organs.[9] Mature carnation petals, however, senesce at the same time, whether attached to the flower or held individually in small vials.[56] Furthermore, removal of the gynoecium from the flower does not affect the timing of flower senescence. These data tend to indicate that the timing of petal senescence is related to the petal tissues themselves. There is, however, increasing evidence that the base of the petal plays an important role in timing.[57]

Related to the question of the influence of the ovary on petal senescence are a number of studies on source-sink relationships during carnation flower senescence. There has been some controversy as to the role of carbohydrate movement between the senescing petals and the developing ovaries.[58-60] Carbohydrates are obviously required for ovary development and it has been suggested that the ovary of C_2H_4-treated carnations grows at the expense of carbohydrate reserves in the wilting petals.[60] Removal of the petals has no effect on ovary growth in naturally senescing carnations.[59] Thus, the growth of the ovary is not dependent on reallocation of carbohydrate materials from the petals. Other researchers accelerated petal senescence and ovary growth by treating carnation flowers with ethrel.[58] Ovaries from treated flowers with petals removed weighed less than those from intact flowers, suggesting that metabolites move from the senescing petals to the developing ovary.[58]

Flowers of *Trieleia laxa* (brodiaea) provide a good system to study the role of petals in ovary growth because their single and easily detached corolla is a more uniform tissue than the petals of carnations. Study of the interaction of corolla and ovary in these flowers[61] indicates time-dependent changes in competition among the floral organs for carbohydrate (Table 3). In young brodiaea flowers, removal of the petals increased ovary growth, indicating that the growing petals were the strongest sink for carbohydrate from the stem. These data are similar to those obtained from carnation flowers pretreated with chemicals that delay flower senescence, such as STS and ethanol.[58,59] Addition of sucrose to the vase solution increased growth of both organs, which supports this sink-strength hypothesis. In contrast, the ovaries of flowers with fully developed petals grew only slightly if the petals were removed, but increased 3- to 4-fold when the petals were present, or when the stems were provided with sugar. These data are consistent with those from carnation flowers, whose senescence was enhanced by pretreatment with ethrel.[58] The changed response to petal removal as the brodiaea flower ages indicates that the ovary in the aging brodiaea flower becomes a very strong sink for carbohydrate from either the stem or the senescing petals. These results suggest that in brodiaeas and in carnations,[59,61] ovary growth is a function of the relative sink strengths of the ovaries and petals for carbohydrate from the stem and the vase solution.

3. The Second Message

Sisler's analysis of the compounds stimulating or inhibiting an ethylene response[62] suggested to him that a *trans*-effect pi-bonding scheme might be involved. He suggested that the interaction of ethylene with the binding site might release another ligand, which would be an obvious candidate for the second messenger. The nature of this compound is still unknown, but analysis of the structure of the promotor regions of senescence-specific genes in flower petals may provide important clues.

4. Messenger RNA Synthesis

Woodson and co-workers[63-65] have provided elegant evidence that aging of carnation flowers and the onset of ethylene biosynthesis are accompanied by synthesis of mRNA species not found in young flowers. They have isolated three cDNA clones from a cDNA library prepared from mRNA isolated from senescing carnation petals. The cDNAs are representative of two classes of mRNAs that increase in abundance during petal senescence. One mRNA (pSR5) is present at low levels during the early stages of development, and begins to accumulate in mature petals prior to the increase in ethylene production (Figure 3). Accumulation of this mRNA is retarded, but not prevented, by inhibiting the synthesis or action of ethylene. The other class of mRNAs are highly regulated by ethylene. They are not detectable (by Northern hybridization analysis) prior to the rise in ethylene production, and increase in abundance concomitantly with the ethylene climacteric. Furthermore, their expression is markedly inhibited by inhibition of ethylene production or action (Figure 4). Analysis of the products of these genes and examination of the control of their expression should provide important information about the way in which ethylene controls senescence.

5. Polyribosomes

The rapid rise in polyribosome profiles as petal senescence commences in carnations[66] is consistent with the observation that new genes are transcribed during this period. The *increase* in polyribosome profiles is somewhat unexpected, since total protein showed a concomitant decrease. This is presumably a reflection of the rapid synthesis of catabolic enzymes during senescence.

6. Proteins

Treatment of flower petals with inhibitors of protein synthesis substantially delays their senescence.[38,67] Increased activities in catabolic enzymes such as ribonuclease are associated with decline in macromolecules including DNA, RNA, proteins, and membrane lipids.[68] Clearly, the onset of senescence is associated with elaboration of catabolic enzymes that presumably play an important role in remobilizing the cell contents of the petal and the eventual death of the cells.

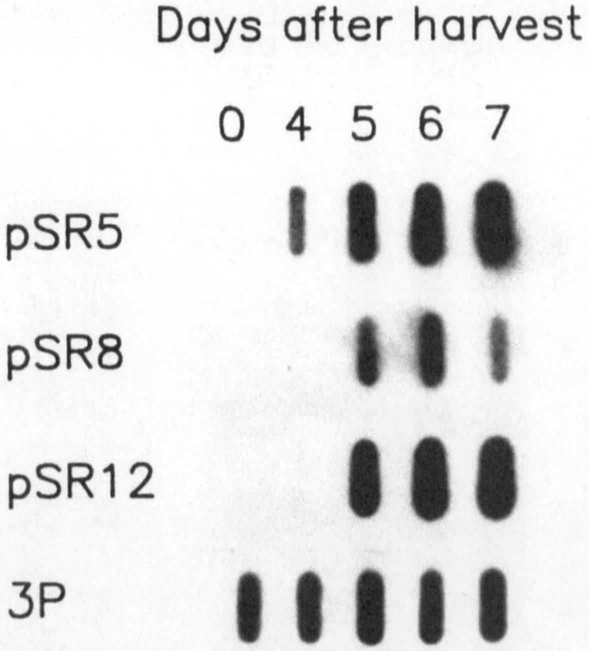

FIGURE 3. Accumulation of specific mRNAs during carnation petal senescence. Ten μg of total RNA isolated from petals at various times after harvest was slot-blotted onto nitrocellulose, and hybridized with ^{32}P-labeled cDNA clones. Petals exhibited initial morphological symptoms of senescence and increased ethylene production 5 d after harvest. Blots of pSR5, pSR8, and pSR12 were exposed to X-ray film with an intensifying screen at −70°C for 18, 36, and 18 h, respectively. (Reprinted from Woodson, W. R., Lawton, K. A., and Goldsbrough, P. B., *Acta Hortic.*, 261, 137, 1989. With permission.)

The evidence that has accumulated indicating the general applicability of the scheme depicted in Figure 2 provides substantive support for the view that the senescence of petals is a carefully regulated event, involving the synthesis of specific proteins. The hypothesis that random peroxidation of membrane lipids and consequent changes in membrane function are primary events[69,70] remains unproven.

D. INHIBITION OF ETHYLENE ACTION

The demonstration by Beyer[71] that Ag^+ was an effective and specific inhibitor of ethylene action provided a powerful tool for examining the action of ethylene in plant growth and development. The more mobile STS complex makes it possible to treat tissues systemically and has become an important commercial treatment for flowers sensitive to ethylene (Figure 5). Application of STS can be used to test the role of ethylene in natural senescence of flowers.[72] In roses, as noted above, STS has no effect on vase life unless ethylene is applied exogenously. This suggests that their normal senescence is therefore not directly linked to ethylene.

The cyclic olefin NBD also inhibits ethylene action (Figure 6), presumably by competitively inhibiting the ethylene binding site.[51] This material, being volatile, is a useful experimental tool but its foul odor and toxicity to humans make it of no practical value. Nevertheless, it points to the possibility of discovering other organic compounds that may interfere with the ethylene binding site, and hence prevent ethylene action. Such compounds

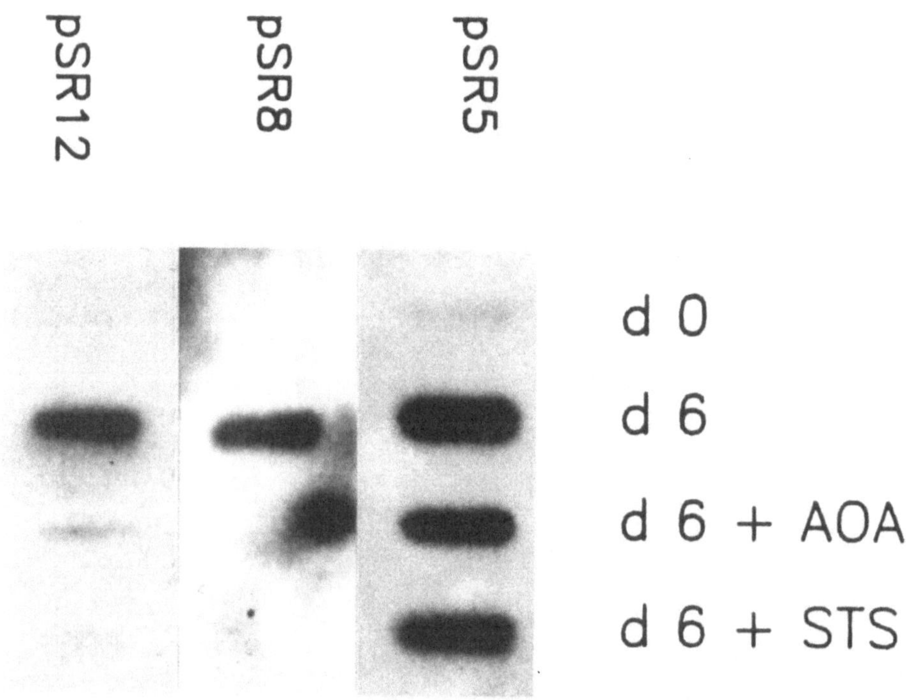

FIGURE 4. Effect of inhibitors of ethylene biosynthesis and action on the accumulation of specific mRNAs during carnation petal aging. Flowers were harvested on day 0 (d 0) and treated with either 4 m*M* AOA (continuously) or 4 mM STS (for 1 h). Ten μg of total RNA isolated from petals the day of harvest (d 0) or 6 days after harvest (d 6) was slot-blotted onto nitrocellulose and hybridized with ^{32}P-labeled cDNA clones. Blots of pSR5, pSR8, and pSR12 were exposed to X-ray film with an intensifying screen at −70°C for 18, 36, and 18 h, respectively. (Reprinted from Woodson, W. R., Lawton, K. A., and Goldsbrough, P. B., *Acta Hortic.*, 261, 137, 1989. With permission.)

are likely to be of interest in practical control of ethylene responses in flowers long before they can be applied to fruits and vegetables. NBD has been shown to reverse the early stages of ethylene-mediated petal wilting in carnation flowers.[73]

E. POLLINATION

The pollination of many flowers results in rapid wilting[74,75] or abscission of florets.[76] In ethylene-sensitive flowers, this response is associated with enhanced ethylene production.[77] In ethylene-insensitive flowers, pollination may induce ethylene sensitivity (as in cyclamen[78]) or even ethylene production (as in brodiaea[79]). Studies of ethylene production following pollination of carnations showed a progression of ethylene synthesis from the stigma and style to the ovary, the receptacle, and, finally, the petals (Figure 7). This suggests movement of a stimulus for ethylene production from the stigma following pollination. The EFE activity of the stigma was found to be more than an order of magnitude greater than that of other floral tissues,[77] and pollen of many species was shown to contain very high concentrations of ACC.[80] These observations, and the rapidity with which ethylene production increases following pollination[74,75] suggested that ACC in the pollen might be the initial pollination stimulus.[74]

The fact that application of AVG to the stigmas prevented the pollination-induced ethylene synthesis[81] showed, however, that the prerequisite ACC was synthesized by the stigma. Application of cycloheximide but not of actinomycin-D prevented the rise (Figure 8), suggesting that ACC synthase in stigmas is synthesized from preexisting mRNA.[44] Although ACC can move from the stigma to the petals,[82] the fact that application of ACC

FIGURE 5. Effect of treatment with STS (4 m*M*, 1 h) on the longevity of cut carnations.

to the corolla does not always initiate the pollination response suggests that a "sensitivity" factor must move to petals following pollination.[83] The nature of this factor remains unknown, although it has been suggested that short-chain fatty acids may be involved.[54]

F. ETHYLENE-RESISTANT MUTANTS

In tomatoes, much has been learned about the role of ethylene in fruit ripening by the use of ethylene-resistant slow-ripening mutants (*nor, rin, nr* and others — see review by Roberts et al.[84]). Similar variants have been sought in carnation,[85] and two commercial cultivars, 'Sandra', and 'Chinera', were found to have vase lives about twice that of 'White Sim'.[86] Measurement of ethylene production showed that the senescence of 'Sandra' was qualitatively different to that of 'White Sim' in that there was no petal inrolling, and no ethylene produced (Figure 9). Levels of ACC and EFE activity were also low throughout the postharvest period. 'Sandra' flowers never showed normal corolla inrolling; their senescence was marked by loss of petal color, and gradual necrosis of the petals. In contrast, the long vase life of 'Chinera' was due to a marked delay in the onset of the normal symptoms of senescence — petal inrolling, ethylene production, and increases in ACC content and EFE activity. 'Sandra' proved to be a very interesting model system. Although its flowers produce no ethylene during natural senescence, they are much more sensitive to ethylene than those of 'White Sim' (Figure 10). Furthermore, ethylene treatment induces ethylene biosynthesis in 'Sandra'. The longer vase life of this cultivar is therefore solely due to its failure to initiate autocatalytic ethylene production. 'Chinera' was considerably less sensitive to ethylene than 'White Sim' (Figure 10). This reduced sensitivity may explain its long vase life.

IV. CONCLUSIONS

The striking response of sensitive flowers to ethylene, the availability of highly sensitive equipment for measuring this gas, and the usefulness of inhibitors of ethylene synthesis and

FIGURE 6. Effect of holding in an atmosphere containing various concentrations of 2,5-norbornadiene on the quality of carnation flowers. The photograph was taken after the flowers had been held for 10 days. From left to right: 0, 500, 2000, and 4000 μl/l 2,5-norbornadiene.

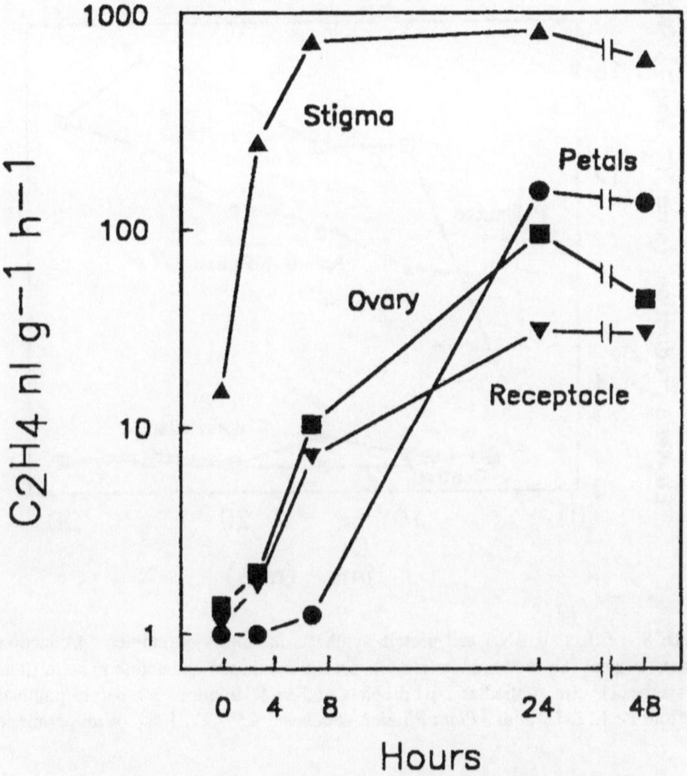

FIGURE 7. Ethylene production by parts of pollinated carnation flowers. Replicate pollinated flowers were dissected at intervals, and the ethylene production of individual tissues was immediately determined. (Data from Nichols, R. et al., *J. Plant Growth Regul.*, 2, 1, 1983. With permission.)

action, have all led postharvest researchers to concentrate on ethylene-sensitive flowers. As a result of this interest, we now have an excellent picture of many of the events occurring during ethylene-induced flower senescence. Commercially, application of inhibitors of ethylene synthesis and action have improved the quality and extended the life of a range of important cut flowers. In the near future, we may expect the development of ethylene-resistant cultivars, based on existing breeding lines, or on recent discoveries in the molecular biology of ethylene action in plants. As noted above, however, relatively few of the commercially important cut flowers have an ethylene-sensitive senescence, and these are in only a handful of families. Senescence of the petals of many cut flowers, including most of the important geophytes (such as gladiolus, iris, narcissus, and tulip) appears not to be related to ethylene. Despite the convenience of working with ethylene and its responses, it is clearly time to extend our experiments to other flowers to learn whether the sequence of events taking place in senescing petals of ethylene-sensitive flowers are in any way analogous to those occurring in flowers whose senescence is neither associated with, nor stimulated by, ethylene.

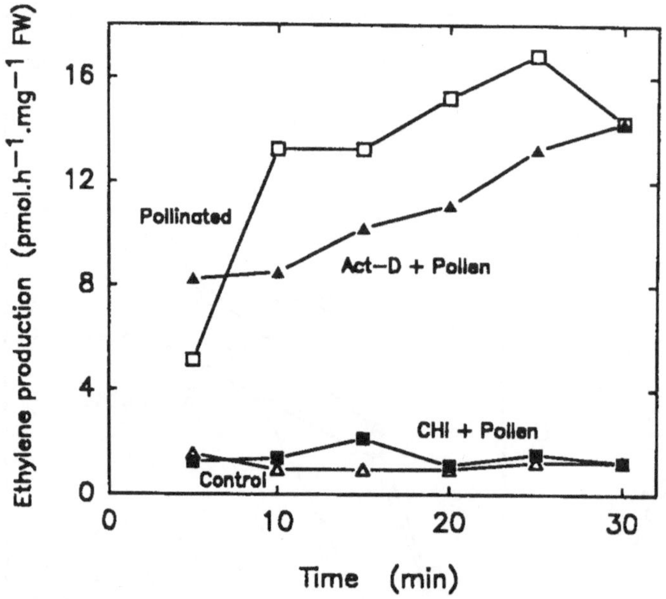

FIGURE 8. Effect of RNA and protein synthesis inhibitors on ethylene production by pollinated stigmas. Cycloheximide (inhibits protein synthesis) and actinomycin-D (inhibits RNA synthesis) were applied as 2-μl droplets of 50-μ*M* solutions 2 h before pollination. (Data from Pech, J.-C. et al., *Plant Physiol. Biochem.*, 25, 431, 1987. With permission.)

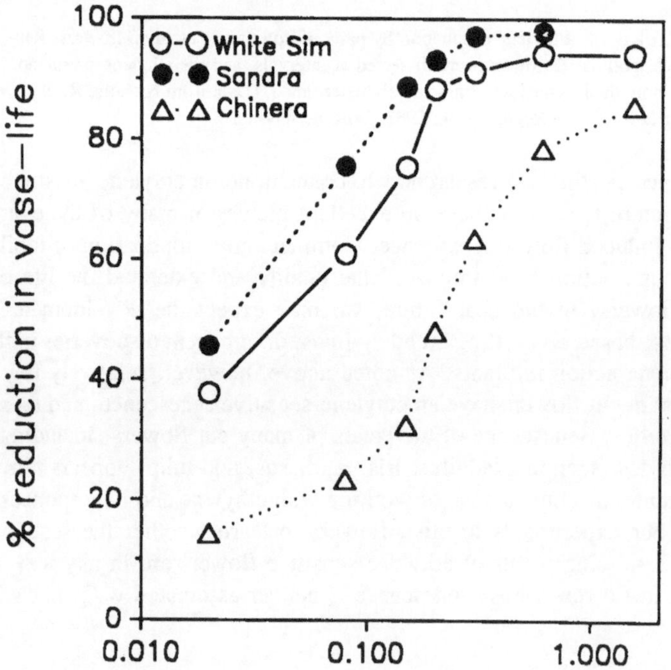

FIGURE 9. Ethylene biosynthesis by long-lived carnation cultivars. Ethylene production, ACC content, and EFE activity of replicate cut flowers of the 'White Sim', 'Sandra', and 'Chinera' cultivars of carnation were measured at intervals as previously described.[86]

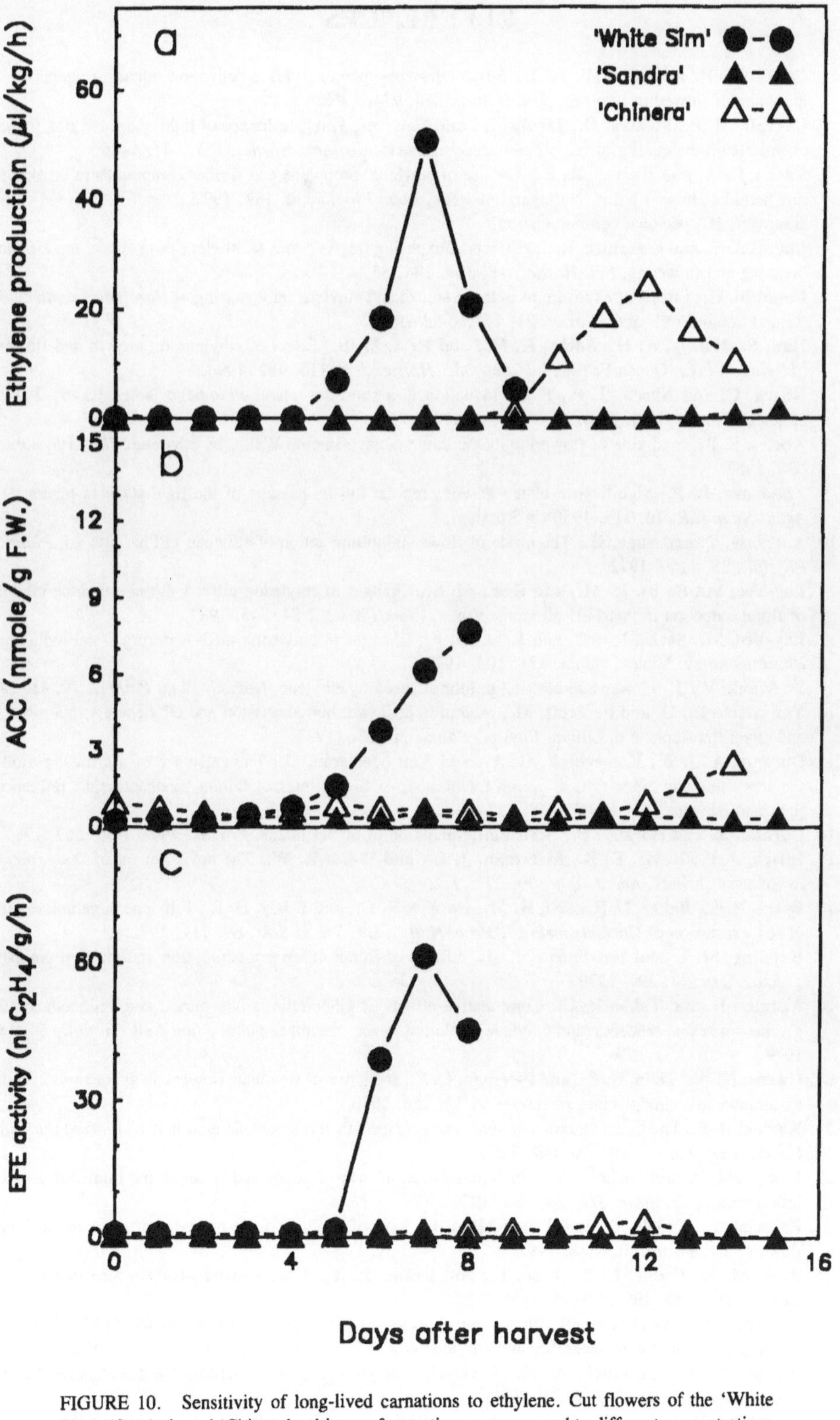

FIGURE 10. Sensitivity of long-lived carnations to ethylene. Cut flowers of the 'White Sim', 'Sandra', and 'Chinera' cultivars of carnation were exposed to different concentrations of ethylene in flowing air streams.[86]

REFERENCES

1. **Cooke, A. R. and Randall, D. I.,** 2-Haloethanephosphonic acids as ethylene releasing agents for the induction of flowering in pineapples, *Nature*, 218, 974, 1968.
2. **DeProft, M. P., Mekers, O., Jacobs, L., and DeGreef, J. A.,** Influence of light and flowering inducing chemicals on the quality of the bromeliaceae inflorescence, *Acta Hortic.*, 181, 141, 1986.
3. **Vacha, G. A. and Harvey, R. B.,** The use of ethylene, propylene and similar compounds in breaking the rest period of tubers, bulbs, cuttings, and seeds, *Plant Physiol.*, 2, 187, 1927.

3a. **Imanishi, H.,** personal communication.

4. **Imanishi, H. and Fortainer, E. J.,** Effects of exposing freesia corms to ethylene or to smoke on dormancy-breaking and flowering, *Sci. Hortic.*, 18, 381, 1982/83.
5. **Imanishi, H.,** Effects of exposure of bulbs to smoke and ethylene on flowering of *Narcissus tazetta* cultivar 'Grand Soleil d'Or', *Sci. Hortic.*, 21, 173, 1982/83.
6. **Han, S., Halevy, A. H., Sachs, R. M., and Reid, M. S.,** Effect of ethylene on growth and flowering of *Triteleia laxa* 'Queen Fabiola', *J. Am. Soc. Hortic. Sci.*, 115, 482, 1990.
7. **Nitsch, C. and Nitsch, J. P.,** Floral induction in a short-day plant, *Plumbago indica* L., by 2-chloroethanephosphonic acid, *Plant Physiol.*, 44, 1747, 1969.
8. **Abeles, F. B.,** Inhibition of flowering in Xanthium pensylvanicum Walln. by ethylene, *Plant Physiol.*, 42, 608, 1967.
9. **Zhdanova, L. P.,** Significance of the gaseous regime for the passage of the light stage in plants, *Dokl. Akad. Nauk SSR*, 70, 715, 1950 (in Russian).
10. **Amagasa, T. and Suge, H.,** The mode of flower-inhibiting action of ethylene in Pharbitis nil, *Plant Cell Physiol.*, 28, 1159, 1972.
11. **Lay-Yee, M., Sachs, R. M., and Reid, M. S.,** Changes in cotyledon mRNA during ethylene inhibition of floral induction in Pharbitis nil strain Violet, *Plant Physiol.*, 84, 545, 1987.
12. **Lay-Yee, M., Sachs, R. M., and Reid, M. S.,** Changes in cotyledon mRNA during floral induction of *Pharbitis nil* cv. Violet, *Planta*, 171, 104, 1987.
13. **De Munk, W. J.,** Flower-bud blasting in tulips caused by ethylene, *Neth. J. Plant Pathol.*, 79, 41, 1973.
14. **Van Meeteren, U. and de Proft, M.,** Inhibition of flower bud abscission and ethylene evolution by light and silver thiosulphate in *Lilium*, *Physiol. Plant.*, 56, 236, 1982.
15. **Durieux, A. J. B., Kamerbeek, G. A., and Van Meeteren, U.,** The existence of a critical period for the abscission and a non-critical period for blasting of flower-buds of *Lilium* 'Enchantment'; influence of light and ethylene, *Sci. Hortic.*, 18, 287, 1982.
16. **Durand, R. and Durand, B.,** Sexual differentiation in higher plants, *Physiol. Plant.*, 60, 267, 1984.
17. **Nitsch, J. P., Kurtz, E. B., Liverman, J. L., and Went, R. W.,** The development of sex expression in Cucurbit flowers, *Am. J. Bot.*, 39, 32, 1952.
18. **Byers, R. E., Baker, L. R., Sell, H. M., Herner, R. C., and Dilley, D. R.,** Ethylene: a natural regulator of sex expression of *Cucumis melo* L., *Proc. Natl. Acad. Sci. U.S.A.*, 69, 717, 1972.
19. **Hopping, M. E. and Hawthorne, B. T.,** Effects of Ethrel 48 on sex expression and yield in pumpkins, *J. Exp. Agric.*, 7, 399, 1979.
20. **Atsmon, D. and Tabback, C.,** Comparative effects of gibberellin, silver nitrate and aminoethoxyvinylglycine on sexual tendency and ethylene evolution in the cucumber plant, *Plant Cell Physiol.*, 20, 1547, 1979.
21. **Owens, K. W., Tolla, G. E., and Peterson, C. E.,** Induction of staminate flowers on gynoecious cucumber by aminoethoxyvinylglycine, *HortScience*, 15, 256, 1980.
22. **Koning, R. E.,** The roles of auxin, ethylene, and acid growth in filament elongation in *Gaillardia grandiflora* (Asteraceae), *Am. J. Bot.*, 70, 602, 1983.
23. **Camprubi, P. and Nichols, R.,** Ethylene-induced growth of petals and styles in the immature carnation inflorescence, *J. Hortic. Sci.*, 54, 225, 1979.
24. **Zacarias, L., Tudela, D., and Primo-Millo, E.,** Role of ethylene in the opening and senescence of citrus flowers, *Sci. Hortic.*, in press, 1991.
25. **Reid, M. S., Dodge, L. L., Mor, Y., and Evans, R. Y.,** Effects of ethylene on rose opening, *Acta Hortic.*, 261, 215, 1989.
26. **Faragher, J. D. and Mayak, S.,** Physiological responses of cut rose flowers to exposure to low temperature: changes in membrane permeability and ethylene production, *J. Exp. Bot.*, 35(156), 965, 1984.
27. **Nichols, R. and Kofranek, A. M.,** Reversal of ethylene inhibition of tulip stem elongation by silver thiosulphate, *Sci. Hortic.*, 17, 71, 1982.
28. **De Stigter, H. C. M.,** Ethephon effects in cut 'Sonia' roses after pretreatment with silver thiosulfate, *Acta Hortic.*, 113, 27, 1981.
29. **Van Staden, J.,** Cytokinins and auxins in carnation senescence as related to chemical treatments, *Acta Hortic.*, 261, 69, 1989.

30. **Reid, M. S., Paul, J. L., and Young, R. E.,** Effects of ethephon on betacyanin leakage from beet root discs, *Plant Physiol.*, 66, 1015, 1980.
31. **Saltveit, M. E.,** A simple apparatus for diluting and dispensing trace concentrations of ethylene in air, *HortScience*, 13, 249, 1978.
32. **Woltering, E. J. and van Doorn, W. G.,** Role of ethylene in senescence of petal — morphological and taxonomical relationships, *J. Exp. Bot.*, 208, 1605, 1988.
33. **Woltering, E. J.,** Ethyleengevoeligheid van zomerbloemen. Voorbehandeling voorkomt schade, *Vakbl. Bloemisterij*, 17, 34, 1984.
34. **Maxie, E. C., Farnham, D. S., Mitchell, F. G., Sommer, N. F., Parsons, R. A., Snyder, R. G., and Rae, H. L.,** Temperature and ethylene effects on cut flowers of carnation (*Dianthus caryophyllus* L.), *J. Am. Soc. Hortic. Sci.*, 98(6), 568, 1973.
35. **Baumgartner, B., Kende, H., and Matile, P.,** Ribonuclease in senescing morning glory, *Plant Physiol.*, 55, 734, 1975.
36. **Matile, Ph. and Winkenbach, F.,** Function of lysosomes and lysosomal enzymes in the senescing corolla of the morning glory *(Ipomoea purpurea)*, *J. Exp. Bot.*, 22(73), 759, 1971.
37. **Kende, H. and Hanson, A. D.,** Relationship between ethylene evolution and senescence in Morning-Glory flower tissue, *Plant Physiol.*, 57, 523, 1976.
38. **Suttle, J. C. and Kende, H.,** Ethylene action and loss of membrane integrity during petal senescence in Tradescantia, *Plant Physiol.*, 65, 1067, 1980.
39. **Woodson, W. R.,** Role of ethylene in the senescence of isolated hibiscus petals, *Plant Physiol.*, 79, 679, 1985.
40. **Woltering, E. J., Harren, F., and Bicanic, D. D.,** Laser photoacoustics: a novel method for ethylene determination in plant physiological studies, *Acta Hortic.*, 261, 201, 1989.
41. **Woltering, E. J. and Harren, F.,** Role of rostellum desiccation in emasculation-induced phenomena in orchid flowers, *J. Exp. Bot.*, 40, 907, 1989.
42. **Yang, S. F.,** Biosynthesis and action of ethylene, *HortScience*, 20, 41, 1985.
43. **Bufler, G., Mor, Y., Reid, M. S., and Yang, S. F.,** Changes in the 1-aminocyclopropane-1-carboxylic acid content of cut carnation flowers in relation to their senescence, *Planta*, 150, 439, 1980.
44. **Pech, J.-C., Latche, A., Larrigaudiere, C., and Reid, M. S.,** Control of early ethylene synthesis in pollinated petunia flowers, *Plant Physiol. Biochem.*, 25, 431, 1987.
45. **Whitehead, C. S., Halevy, A. H., and Reid, M. S.,** Control of ethylene synthesis during development and senescence of carnation petals, *J. Am. Soc. Hortic. Sci.*, 109, 473, 1984.
46. **Manning, K.,** Ethylene production and β-cyanoalanine synthase activity in carnation flowers, *Planta*, 168, 61, 1986.
47. **Adam, Z. and Mayak, S.,** Solubilization and partial purification of an enzyme converting 1-aminocyclopropane-1-carboxylic acid to ethylene in plants, *FEBS Lett.*, 172, 47, 1984.
47a. **Borochov, A. and Reid, M. S.,** unpublished observation.
48. **Baker, J. E., Wang, C. Y., Lieberman, M., and Hardenburg, R.,** Delay of senescence in carnations by a rhizobitoxine analog and sodium benzoate, *HortScience*, 12, 38, 1977.
49. **Fujino, D. W., Reid, M. S., and Yang, S. F.,** Effects of aminooxyacetic acid on postharvest characteristics of carnation, *Acta Hortic.*, 113, 59, 1981.
50. **Harkema, H. and Struijlaart, P. F.,** Effect of aminooxyacetic acid on coloration of the labellum and longevity of cut cymbidium flowers, *Acta Hortic.*, 261, 293, 1989.
51. **Sisler, E. C., Reid, M. S., and Fujino, D. W.,** Investigation of the mode of action of ethylene in carnation senescence, *Acta Hortic.*, 141, 229, 1983.
51a. **Sisler, E. C.,** unpublished observation.
52. **Veen, H.,** A theoretical model for anti-ethylene effects of silver thiosulphate and 2,5-norbornadiene, *Acta Hortic.*, 181, 129, 1986.
53. **Trewavas, A.,** How do plant growth substances work?, *Plant Cell Environ.*, 4, 203, 1981.
54. **Whitehead, C. S. and Halevy, A. H.,** Ethylene sensitivity: the role of short-chain fatty acids in pollination-induced senescence of *Petunia hybrida* flowers, *Plant Growth Regul.*, 8, 41, 1989.
55. **Mayak, S. and Dilley, D. R.,** Effect of sucrose on response of cut carnation to kinetin, ethylene and abscisic acid, *J. Am. Soc. Hortic. Sci.*, 101, 583, 1976.
56. **Mor, Y. and Reid, M. S.,** Isolated petals — a useful system for studying senescence, *Acta Hortic.*, 113, 18, 1980.
57. **Mor, Y., Halevy, A. H., Spiegelstein, H., and Mayak, S.,** The site of 1-aminocyclopropane-1-carboxylic acid synthesis in senescing carnation petals, *Physiol. Plant.*, 65, 196, 1985.
58. **Cook, E. L. and Van Staden, J.,** Senescence of cut carnation flowers: ovary development and CO_2 fixation, *Plant Growth Regul.*, 1, 221, 1983.
59. **Mor, Y., Reid, M. S., and Kofranek, A. M.,** Role of the ovary in carnation senescence, *Sci. Hortic.*, 13, 377, 1980.
60. **Nichols, R.,** Induction of flower senescence and gynoecium development in the carnation *(Dianthus caryophyllus)* by ethylene and 2-chloroethylphosphonic acid, *J. Hortic. Sci.*, 46, 323, 1971.

61. **Han, S. S., Halevy, A. H., and Reid, M. S.,** The role of ethylene and pollination in petal senescence and ovary growth of Brodiaea (*Triteleia laxa* Benth.), *J. Am. Soc. Hortic. Sci.*, 116, 68, 1991.
62. **Sisler, E. C.,** Ethylene activity of some pi-acceptor compounds, *Tob. Sci.*, 21, 43, 1978.
63. **Woodson, W. R.,** Changes in protein and mRNA populations during the senescence of carnation petals, *Physiol. Plant.*, 71, 445, 1987.
64. **Woodson, W. R., Lawton, K. A., and Goldsbrough, P. B.,** Ethylene-regulated gene expression during carnation petal senescence, *Acta Hortic.*, 261, 137, 1989.
65. **Woodson, W. R. and Lawton, K. A.,** Ethylene-induced gene expression in carnation petals. Relationship to autocatalytic ethylene production and senescence, *Plant Physiol.*, 87, 478, 1988.
66. **Bufler, G., Romani, R., and Reid, M. S.,** Polysomal population in relation to ethylene production and the senescence of cut carnation flowers and floral parts, *J. Am. Soc. Hortic. Sci.*, 108, 554, 1983.
67. **Wulster, G., Sacalis, J., and Janes, H. W.,** The effect of inhibitors of protein synthesis on ethylene induced senescence in isolated carnation petals, *J. Am. Soc. Hortic. Sci.*, 107, 112, 1982.
68. **Mayak, S. and Halevy, A. H.,** Flower senescence, in *Senescence in Plants*, Thimann, K. V., Ed., CRC Press, Boca Raton, FL, 1980, 131.
69. **Borochov, A., Halevy, A. H., and Shinitzky, M.,** Senescence and fluidity of rose petal membranes, *Plant Physiol.*, 69, 296, 1982.
70. **Paulin, A. and Droillard, M. J.,** Membrane lipids peroxidation during the senescence of cut carnation, *(Dianthus caryophyllus* L.), *Acta Hortic.*, 261, 51, 1989.
71. **Beyer, E. M., Jr.,** A potent inhibitor of ethylene action in plants, *Plant Physiol.*, 58, 268, 1976.
72. **Veen, H.,** Silver thiosulphate: an experimental tool in plant science, *Sci. Hortic.*, 20, 211, 1983.
73. **Wang, H. and Woodson, W. R.,** Reversible inhibition of ethylene action and interruption of petal senescence in carnation flowers by norbornadiene, *Plant Physiol.*, 89, 434, 1989.
74. **Whitehead, C. S., Fujino, D. W., and Reid, M. S.,** The roles of pollen ACC and pollen tube growth in ethylene production by carnations, *Acta Hortic.*, 141, 221, 1983.
75. **Whitehead, C. S., Halevy, A. H., and Reid, M. S.,** Roles of ethylene and 1-aminocyclopropane-1-carboxylic acid in pollination and wound-induced senescence of Petunia hybrida flowers, *Physiol. Plant.*, 61, 643, 1984.
76. **Stead, A. D. and Moore, K. G.,** Studies on flower longevity in Digitalis: pollination induced corolla abscision in Digitalis flowers, *Planta*, 146, 409, 1979.
77. **Nichols, R., Bufler, G., Mor, Y., Fujino, D. W., and Reid, M. S.,** Changes in ethylene production and 1-aminocyclopropane-1-carboxylic acid content of pollinated carnation flowers, *J. Plant Growth Regul.*, 2, 1, 1983.
78. **Halevy, A. H., Whitehead, C. S., and Kofranek, A. M.,** Does pollination induce corolla abscission of cyclamen flowers by promoting ethylene production?, *Plant Physiol.*, 75, 1090, 1984.
79. **Han, S. S., Halevy, A. H., and Reid, M. S.,** Postharvest handling of *Triteleia laxa*, *HortScience*, 25, 1268, 1990.
80. **Whitehead, C. S., Fujino, D. W., and Reid, M. S.,** Identification of the ethylene precursor, 1-aminocyclopropane-1-carboxylic acid (ACC) in pollen, *Sci. Hortic.*, 21, 291, 1983.
81. **Hoekstra, F. A. and Weges, R.,** Lack of control by early pistillate ethylene of the accelerated wilting of Petunia hybrida flowers, *Plant Physiol.*, 80, 403, 1986.
82. **Reid, M. S., Fujino, D. W., Hoffman, N. E., and Whitehead, C. S.,** 1-Aminocyclopropane-1-carboxylic acid (ACC) — the transmitted stimulus in pollinated flowers?, *J. Plant Growth Regul.*, 3, 189, 1984.
83. **Gilissen, L. J. W. and Hoekstra, F. A.,** Pollination-induced corolla wilting in Petunia hybrida: rapid transfer through the style of a wilting-inducing substance, *Plant Physiol.*, 75, 496, 1984.
84. **Roberts, J. A., Grierson, D., and Tucker, G. A.,** Genetic variants as aids to examine the significance of ethylene in development, in *Hormone Action in Plant Development: A Critical Appraisal*, Hoad, G. V., Lenton, J. R., Jackson, M. B., and Atkin, R. K., Eds., Butterworth, Cornwall, 107, 1987.
85. **Wu, M. J., van Doorn, W. G., Mayak, S., and Reid, M. S.,** Senescence of 'Sandra' carnations, *Acta Hortic.*, 261, 221, 1989.
86. **Wu, M. J.,** Studies of genetic and chemical control of senescence in carnation *(Dianthus caryophyllus* L.), Ph.D. thesis, University of California, Davis, 1989.

Chapter 13

ETHYLENE IN FRUIT ONTOGENY AND ABSCISSION

Colin J. Brady and Jim Speirs

TABLE OF CONTENTS

I. INTRODUCTION

That ethylene influences the development of commercial fruit is obvious from a whole range of practices within the horticultural industry throughout the world. It is apparent as ethylene-releasing chemicals are used to promote the early development of stone fruits, as thinning agents, as an aid to the harvesting of olive fruits, or to stimulate the degreening of citrus fruits. There is evidence of it, too, as ethylene is pumped into ripening rooms to give controlled ripening of bananas, tomatoes, or other fruits. In the opposite sense, the evidence is there as convertors or absorbant chemicals function to remove ethylene from fruit storage rooms, and handling procedures are arranged to minimize the exposure of fruits to ethylene.

The recognition of the fact that ethylene has a powerful influence on commercial fruits came early. The history of the early discoveries is well covered in the excellent monograph by Abeles[1] and there are numerous other reviews that deal with the physiology of fruit ripening and the role of ethylene in fruit ripening.[2-7] Research activity continues unabated and the advent of powerful molecular techniques for the study of gene action and gene products has quickened the pace of discovery and placed us within the shadow of understanding how ethylene interacts with the plant genome to influence fruit development. Some knowledge of the molecular interactions in fruit allows a more sensible comparison to be made with other tissues and across species. Thus, gradually do we learn how this small, but not chemically simple, plant regulator exerts its effect.

The ripening fruit has been a useful tissue for the study of ethylene action and gene regulation. Current methodologies most effectively study gene products that accumulate to be abundant messenger RNAs or abundant proteins. At least some of the reactions that occur during ripening represent massive changes in tissue physiology and tissue chemistry. To achieve these changes major changes in the population of RNA templates, and of enzyme catalysts, are sometimes needed. Moreover, fruit tissues can be relatively homogeneous and bulky so that there is no shortage of experimental material. Thus, the ripening fruit has attractions that are not apparent when the changes are more subtle, as in early development, or where the tissue is limited, as in abscission zones. Knowing something of the changes in gene expression that occur in ripening fruit can be a useful input to the study of ethylene action in other tissues. It can provide molecular probes that can be used where the changes are less extreme or the tissue less abundant. It can allow the definition of sequences that signify tissue, species, or developmental stage.

Of course, there are dangers in the approach now commonly taken of using probes developed in one fruit to explore gene expression in other tissues or in the fruit of an unrelated species. This is a way to emphasize the similarities while ignoring the subtle, or not so subtle, differences. There are lessons to be learnt from the similarities, especially those conserved across the boundaries of species, genus, or family. However, it cannot lightly be assumed that a conserved sequence represents a conserved function. A gene expressed in a different chemical or biochemical environment or in a cell of different form, may not have the significance imagined from its role in another tissue. There is a need, sometimes neglected, to explore cell function as well as gene expression. If conserved sequences are connected to conserved functions, then the sequence differences between tissues or species are especially fascinating. Are these the regions that are of little importance such that they can mutate without significance to cell function, or are they the clue to changes in gene regulation that contribute to tissue differentiation?

II. ETHYLENE PRODUCTION DURING ONTOGENY

There is a pattern to the trend in ethylene production rates during fruit development. There are, of course, variations from species to species; in part, this reflects how fruit

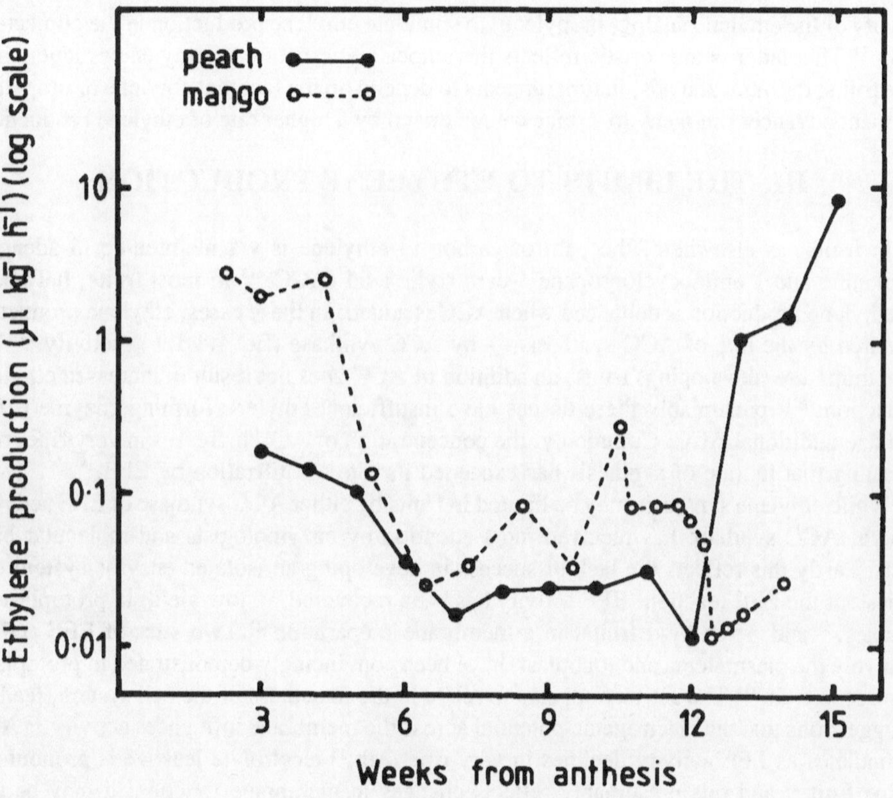

FIGURE 1. The trend in ethylene production during development of mango (dotted) and peach fruit. Ethylene measurements were made on detached fruit, shortly after detachment. For details on mango see Reference 8; the peach data are plotted from data in Reference 9, for cultivar Redhaven.

develop from different parts of the flower and these parts themselves differ in rates of ethylene production. The pattern shown in Figure 1 for mango[8] and peach[9] is typical of many fruits. Ethylene is produced at a high rate during anthesis and during early development, when the auxin content is high. A lower rate is maintained during the growth phase; typically this is reported as less than 0.1 μl kg^{-1} h^{-1} and it may be much less than this. There has been little emphasis given to species variation in this low rate of ethylene production, often because the measurements challenge instrument sensitivity. However, modern instruments will allow careful comparisons to be made and there may be physiological significance in variations in this "basal" production rate. As ripening approaches, the basal rate of ethylene production increases in those fruits classified as "climacteric". Figure 1 illustrates this for mango and peach. Mango is unusual in that the respiratory climacteric is measured two or three days before the ethylene production rate increases. In most climacteric fruits, the rise in respiration rate accompanies or briefly trails the increase in ethylene production. In typical nonclimacteric fruit, such as citrus or grapes, the low rate of ethylene production measured before ripening persists throughout fruit development.

In apple and cherry, the emergence of flower buds from dormancy is accompanied by an increase in ethylene biosynthesis.[10] This declines as the flowers develop, to increase again at petal drop (see Chapter 12 for a consideration of flower senescence). The large increase in ethylene production associated with pollination[11] is largely produced by the stigma and stylus.

Previous reviews have dealt in depth with the distinction between the climacteric and nonclimacteric fruits. It will not be pursued here except to make the points that nonclimacteric fruits are ethylene-sensitive and the two groups are most satisfactorily distinguished by the capacity of the ethylene analog, propylene, to stimulate ethylene production in the climacteric group.[12] This latter characteristic reflects the autocatalytic nature of ethylene production in these fruits; the autocatalysis, in turn, appears to depend on the fact that ethylene or propylene treatment advances ontogeny to a stage characterized by a higher rate of ethylene production.

III. THE LIMITS TO ETHYLENE PRODUCTION

In fruits, as elsewhere, the path of carbon to ethylene is via methionine, *S*-adenosyl methionine and 1-aminocyclopropane-1-carboxylic acid (ACC).[13] In most fruits, but not in all, ethylene production is enhanced when ACC is added. In these cases, ethylene production is limited by the rate of ACC synthesis — by ACC synthase (EC 4.4.1.14) activity.[14,15] In some immature (developing) fruits, an addition of ACC does not result in increased ethylene production;[16,17] presumably these tissues have insufficient ethylene-forming enzyme (EFE) to utilize additional ACC. Commonly, the concentration of ACC increases in very ripe fruit, indicating that its rate of synthesis has exceeded its rate of utilization by EFE.[18]

While ethylene synthesis may be limited in fruits by either ACC synthase or EFE activity, or both, ACC synthase has received most attention by enzymologists and molecular biologists. Partly this reflects the lack of success in developing an isolated enzyme system that carries out the EFE reaction. EFE activity has been recovered, in low yield, in protoplasts,[19] vacuoles,[20] and, from kiwi fruits, in a membrane preparation.[21] Two sites of EFE action, putatively the plasmalema and tonoplast, have been convincingly demonstrated in protoplasts from grape fruit.[21a] The activity appears to relate to the intactness of the cell system, leading to suggestions that the electrogenic potential across the membrane influences activity.[22] ACC accumulates as EFE activity declines in very ripe fruit;[18] electrolyte leakage is pronounced in ripe fruit,[24] and this presumably reflects changes in membrane function: it may be that the accumulation of ACC, the decline in EFE activity, and the loss of transmembrane potential in ripe fruit are causally related. ACC synthase is also a preferred target for research because it precedes EFE in the metabolic pathway, accumulates rapidly when ethylene synthesis is induced, and decays rapidly if the induction is blocked.[25-27] These attributes suggest that the enzyme is tightly regulated, perhaps transcriptionally. This is a circumstance in which antisense technology[28] may be useful, leading to plants genetically modified so that the fruits produce minimal amounts of ethylene while retaining ethylene sensitivity. Such fruits could be harvested and shipped in the mature, preripe condition and ripened at an appropriate time by ethylene treatment.

The information needed to explore this approach is at several levels. Is there one or more enzyme? Is one or more than one gene involved? If there is more than one gene, how homologous are the sequences, and are they expressed differentially between tissues, or do they respond to different signals?

In the absence of a direct approach to the gene, attempts have been made to purify the ACC synthase protein. From the protein sequence, nucleotide probes may be developed to explore gene expression and gene structure. Also, the purified proteins may yield antibodies that can recognize the polypeptides synthesized *in vivo,* or *in vitro* from RNA recovered from developing fruit. ACC synthase has been an elusive target and only recently have small amounts of purified protein been recovered.

Among fruits, most attempts to purify the enzyme have been from ripe tomato or squash. A number of attempts to purify the tomato fruit enzyme are summarized in Table 1. The tissue used has been the pericarp of pink or red fruit in which the level of ACC activity has been enhanced by wounding and/or LiCl, or auxin treatment. This approach carries the assumption that the induced enzyme is identical to the enzyme that is induced during ripening; since LiCl treatment[32,33,35] results in a hundredfold increase in activity, it will be the salt-induced enzyme that is purified. It is yet to be shown that the same enzyme is induced

TABLE 1
Aminocyclopropane-1-carboxylate Synthase Preparations From Wounded or LiCl-Treated Pericarp Tissue From Pink or Red Tomato Fruits

Purification ×	Specific activity nmol/h/mg	M_r	pI	Ref.
44	110	57,000[a]		29
400	391	50,000	4.5	30
900	805	67,000	7.0	31
6,500	27,500	55,000		32
6,000	500,000[b]	50,000		33
138	11,730	50,000		34
5,000	206,390	45,000	5.8	35

[a] Native enzyme, other M_r estimates are for the denatured enzyme in presence of SDS.

[b] Estimated from the initial specific activity and recovered immunoreactive protein.

during ripening. In the table, there is little correspondence in the estimates of purification or in the final specific activities. This reflects variance in the level of induction of the enzyme in the starting material, and in the retention of activity during purification. ACC synthase is not an abundant protein, being of the order of 0.0002% of the total protein in wounded, pink tomato pericarp tissue.[32,35] Nonetheless, it has been possible[35] to recover sufficient amounts of the protein for amino acid sequence analysis. This has allowed the synthesis of nucleic acid probes for the RNA and genomic sequences.[35a,39b] It has also been possible to develop monoclonal antibodies that recognize and inactivate the ACC synthase protein.[31-33,35a] With the antibodies, it has been possible to substantiate the thesis, previously precariously based on the use of metabolic inhibitors,[36,37] that there is a net synthesis of ACC synthase protein when the enzyme is induced.

It is apparent (Table 1) that there is no agreement on the size and charge properties of ACC synthase. Further detailed work is needed to characterize the structural differences that contribute to these differences. Mehta et al.[31] recovered a monoclonal antibody that recognized an enzyme of pI 7.0, but not those of pI 5.3 and 9.0, also present in the tissue extracts. This suggests that there are significant sequence differences between the isoforms and that they may be the products of different genes.

Enzymes carrying out similar reactions and responding to similar physiological signals commonly have sequence homologies.[38] Molecular probes for the protein and gene in tomato may aid the search for the enzyme and the gene in other plants. Although initial reports[39,40] suggested that the squash enzyme may differ substantially from the tomato enzyme, it appears that the squash[39a] and zucchini[39b] enzymes are similar in size and physicochemical nature to the tomato enzyme(s). The apple fruit ACC synthase differs from the tomato enzyme immunologically, but the active site sequences are similar.[39c]

IV. ETHYLENE RECEPTORS

Models of how ethylene influences fruit development invariably assume the presence of specific receptors for ethylene at the cell surface,[41] or elsewhere. Whether there is one or more receptor is not specified, but it is commonly speculated that changes in tissue sensitivity to ethylene reflect changes in the concentration or sensitivity of receptors (see Chapter 5). The nature of the receptors has been postulated by Burg and Burg,[42] and their properties have been anticipated and used for their assay.[43] Inhibitors that presumably block the receptors have been used with some success to describe ethylene effects in fruits. Despite all of this, there is no direct evidence of the presence of specific protein receptors in fruit tissue and no substantiation of the hypothesis that their presence or their properties determine the sensitivity of tissues to ethylene, and that they play a role in regulating fruit development.

Nonetheless, the evidence for ethylene receptors in other tissues is strong and the emerging genetic evidence in *Arabidopsis*[44] is of particular interest. It is reasonable to explore the hypothesis that specific ethylene receptors exist in fruit tissues and that these may be structurally related to those in vegetative tissues. The definition of the receptor molecules in fruits is such a glaring omission in our description of ethylene action in fruit development that attempts to develop receptor probes in sensitive vegetative tissues for use on fruit tissues is a matter of high priority. The questions that would follow include receptor heterogeneity, receptors and ethylene sensitivity, receptors in mutant lines, including ripening mutants, and the possibility of engineering modified receptors. There would be an opportunity to ask, too, whether the common "metabolic" responses to ethylene, for example the common increase in respiration rate, are mediated via specific receptors or via a more direct effect as on membrane systems (see also Chapter 5).

V. ETHYLENE AND EARLY DEVELOPMENT

Early development of many fruits is characterized by a high rate of ethylene evolution[9,45] (Figure 1). However, there is little information concerning the significance to early development of this ethylene. At this early stage of development the tissues are particularly heterogeneous. High levels of auxins in some tissue may result in high rates of ethylene production and low sensitivity to ethylene. Senescing flower tissues may also contribute.

Although endogenous ethylene production may not rise, the sensitivity to applied ethylene rises in a number of fruits during development. This is a matter of horticultural significance, especially in those fruits that have a lag period separating early and late stages of rapid growth.

In fig,[46] grape berry,[47] and drupe fruits[10,48] external applications of ethylene or ethylene-releasing chemicals result in an advance in fruit development and the possibility of an earlier, commercial harvest. Sensitivity to applied ethylene is related to the stage of fruit development measured as relative growth rate, and in grape[49] and peach[9] the increased sensitivity correlates with an accumulation of abscisic acid in the developing fruit. In fig, the growth response to ethylene application is accompanied by an apparent increase in the rate of RNA turnover.[50]

VI. ETHYLENE AND FRUIT ABSCISSION

It may seem anomalous that abscission is considered here, prior to fruit maturation and ripening, for the formation of abscission zones is often associated with the final stages of fruit ripening. However, abscission can occur at any time, and the abscission of young fruits at an early stage of development and the use of ethylene-releasing chemicals to provoke abscission and so reduce fruit load are important horticultural events. Early fruit drop occurs as the levels of endogenous growth regulators are changing rapidly, and complex regulatory patterns are surely involved. Interactions between auxins,[51,52] polyamines,[53] gibberellins,[54] abscisic acid,[55] and ethylene[56] appear to be involved but the evidence strongly favors the view that auxin and ethylene are the major players.[57] The interactions may operate by influencing auxin distribution and/or ethylene production, for example as abscisic acid provokes senescence and ethylene production in distal tissues.

In cotton flower buds, flowers, and young fruits,[58] the indoleacetic acid concentration is high prior to anthesis, declines at anthesis, remains low for 3 or 4 d and then increases as the young bolls grow rapidly. In contrast, ethylene production is high at anthesis and for 3 or 4 d after anthesis, and then declines rapidly.[56] The probability of abscission during early development is at a maximum 2 to 4 d after anthesis when the auxin level is low and ethylene production is high.[58]

Very young fruits generally have high rates of ethylene production.[59] During early development of 'Haas' avocado fruit, Adato and Gazit[60] recorded mean rates of ethylene production of about 50 μl/kg/h; the seed was the main site of ethylene production, and defective seeds had much higher rates of ethylene production than normal seeds. Consequently, fruit with defective seeds have a higher probability of abscission. Do defects of this type commonly contribute to the high rates of ethylene production measured in young fruits? Is ethylene production exceptionally high in nonseed fruit tissue undergoing rapid cell division? These questions await careful investigation but it is apparent that the auxin/ethylene balance is important to the continued attachment of young fruits, and there is an obvious advantage in shedding defective tissue at an early stage. Individual fruits with a higher than average auxin content will have a greater chance of avoiding abscission. These individuals may also be stronger sinks than their neighbors, so is it possible that sink strength independently influences abscission? Competition for nutrients has frequently been suggested as the mechanism controlling the early summer drop in deciduous fruit trees.[61] In prune, Weinbaum and Muraoka[62] found that girdling and defoliation treatments did not cause abscission, while the application of ethephon did. On the other hand, ethephon treatment reduced assimilate accumulation. The reduced sink strength of the fruit reflected ethephon-provoked fruit senescence, while the ethephon-provoked abscission was mediated independently of the coincident effect on assimilate movement.

There are numerous demonstrations that ethylene treatments result in fruit abscission and that applied auxins limit fruit drop. The influence of ethylene is not irreversible because removal of ethylene from the atmosphere before the abscission zone is formed halts the abscission process after it has commenced.[63] This is consistent with the demonstration that endogenous ethylene production is not always induced when ethylene-releasing chemicals induce fruit abscission.[64]

In fruit, as in leaves, abscission involves the fracture, apparently in the middle lamellar region, of a plane of cells at the abscission zone. Partial dissolution of the cell walls may occur. The forces causing the fracture are generated by the expansion of adjacent cells, in which membrane systems develop that may be involved in enzyme secretion.[65-68] The abscission zone is characterized by a reversal of normal responses in that cell expansion here is stimulated by ethylene and inhibited by auxin.

In leaf abscission zones there has been intense study of the role of cell wall hydrolases in abscission, and of ethylene in stimulating cell wall hydrolase activity (see Chapter 11). The production of wall hydrolases — and β-1,4-endoglucanhydrolase is the best-characterized response[69] — is clearly stimulated by ethylene treatment and inhibited by auxins. In bean leaves, at least, ethylene treatment activates gene expression and the increase in β-1,4-endoglucanhydrolase activity reflects enhanced synthesis of its protein. Are the molecular events involved in fruit abscission similar?

Because the physiology of leaf and fruit abscission is very similar, there is an expectation that the molecular events will be similar. Limited evidence supports this notion. Ethylene treatments enhance and auxin treatments delay or inhibit fruit abscission. An increase in β-1,4-endoglucanhydrolase and polygalacturonase activities occurs in the abscission zones of orange fruits;[70,71] the increases in activity are stimulated by ethylene treatment and inhibited by auxin treatment;[70,71] in peach fruit abscission zones, ribosomes are recruited to polysomes as abscission is induced by ethylene.[72] The evidence is that superficially similar enzymatic events occur in leaf and fruit abscission zones, and that the enzyme activities respond to similar signals. There is as yet no evidence that identical enzymes act in leaf and fruit tissues, or that the same genes are involved in the responses to the regulatory signals.

VII. ETHYLENE AND FRUIT RIPENING

It is now well established that ethylene is an important regulator of ripening events.[73] In climacteric fruit, the onset of ripening is accompanied by a respiratory rise and a rise in the rate of ethylene evolution. In some fruits, for example, cherimoya, some avocados, and mango, the rise in respiration rate clearly precedes the rise in ethylene evolution,[74] although more generally the rise in ethylene production briefly precedes or accompanies the respiratory burst. In most mature fruits, climacteric or nonclimacteric, exposed to ethylene there is an increase in respiratory activity and it is often concluded that the respiratory climacteric is a direct consequence of the increased concentrations of internal ethylene. This may be so in some circumstances, but the observations on mangos, in which the respiration rate rises several days before the increment in internal ethylene, suggests that controls independent of ethylene regulate respiration in some developing fruit.

Partial removal of ethylene, for example by hypobaric storage,[75,76] slows ripening as does inhibition of ethylene synthesis by competitive inhibitors.,[77-79] Anionic silver, an inhibitor of ethylene action, completely blocks ripening in several climacteric fruits[80,81] but has no effect on strawberry.[82] In climacteric fruit, treatment with silver halts the ripening process even after it is well advanced, indicating that ethylene has a continuing role as a regulator and does more than irreversibly initiate a process.[83] Reducing endogenous ethylene production by reducing the steady-state level of ethylene by the use of antisense gene constructs also influences the onset and progress of ripening.[83a]

Trewavas[84] has argued that the involvement of ethylene in ripening, as in senescence, is as an accelerator rather than as a switch, and that advancing senescence is characterized by an increase in ethylene sensitivity. Undoubtedly, some fruits do become more sensitive to ethylene as they mature. Cantaloupe[85] and banana[86] will ripen if exposed to sufficient ethylene at any stage — although higher concentrations are needed for younger fruit. Less sensitive fruit such as tomato and apple[87,88] will not ripen immediately in response to ethylene unless they are close to full maturity. Less mature fruit are nonetheless sensitive to ethylene, for ethylene treatment reduces the time prior to natural ripening. Liu et al.[89] illustrated one mechanism that may contribute to this when they showed that the treatment of immature tomato fruit with ethylene induced EFE. This induction removed a limit to autocatalytic ethylene production. The debate on this matter will be much advanced when reliable methods for measuring ethylene receptors in fruit are available.

The nonclimacteric fruits, in which ethylene is not produced in response to added ethylene, are nonetheless sensitive to ethylene. Exogenous ethylene causes an increase in respiration rate in many[82,90-92] and ethylene treatment accelerates peel degreening and fruit senescence in citrus.[93] In strawberry, the progress of ripening appears to be independent of ethylene in that it does not respond to added ethylene or ethylene antagonists, although it is sensitive to auxins.[82] Watermelon fruit exposed to exogenous ethylene respond by accumulating cell-wall-degrading enzymes.[94] The level of auxins in the tissue also influences the sensitivity of climacteric fruits to endogenous or added ethylene.[95,96]

A. MOLECULAR RESPONSES

In the fruit so far examined, ripening is associated with some redirection of protein synthesis. Such a redirection has been proposed for a number of years,[7] but it is only recently that methods have been available to unambiguously establish that there is an accumulation of ripening-specific proteins and that this involves changes in gene expression and in the spectrum of proteins synthesized. In tomato, *in vitro* translation of messenger RNA populations from immature, mature, and ripening fruit identifies three classes of messenger RNA.[97-99] The most abundant class remains constant throughout maturation and ripening. A second class, encoding at least eight major classes of protein, appears or is greatly increased

TABLE 2
Genes Regulated by Ethylene and/or Ripening

Fruit	mRNA size (kb)	Size (Da)	Peptide encoded enzyme/function	Ethylene regulation	Increase/decrease during ripening	Ref.
Tomato	2.1	54,000	Endopolygalacturonase	−	Increase	98, 102-105
	0.58		Proteinase inhibitor I	+	Increase	105-106
	0.88	25,000		+	Increase	105
	1.6	42,000		+	Increase	105
	1.0	27,000		+	Increase	105
	1.6	42,506	Pectin esterase	−	Decrease	107
		37,500	Ethylene associated	−	Increase	99
		35,000	Ethylene associated		Increase	108
		33,500	Ethylene associated		Increase	109
		38,000		+	Decrease	3
		190,000				
		80,000				
		57,000		+	Increase	110
		55,000				
		48,000				
		44,000				
		35,000				
		20,000				
Avocado		53,000	Cellulase		Increase	111
		40,000				
		31,000			Increase	111
		20,000				
		25,000			Decrease	111
		18,000				
		17,000				
Peach		100,000				
		85,000				
		59,000			Increase	101
		54,000				
		42,000				
		41,000				
		25,500				
		46,000				
		40,000			Decrease	101
		25,500				
		25,000				

during ripening. A third class, encoding at least six protein types, declines in abundance during maturation and ripening. In avocado, three messenger RNAs increase dramatically with the onset of ripening;[100] in peach, seven messenger RNAs increase in abundance and four decrease with ripening.[101]

A number of the mRNAs, mostly from tomato, that change in abundance during ripening, have been identified in terms of enzyme product. One encodes the tomato endopolygalacturonase enzyme (Table 2), that hydrolyzes pectin and is believed to play a role in tomato fruit softening. The mRNA for endopolygalacturonase is undetectable in mature-green fruit, detectable in fruit in which ethylene synthesis is beginning to increase at the onset of ripening, and increases to constitute some 2.3% of the total mRNA mass,[112] 3 to 4 d after the increase in ethylene production.[113] A second tomato mRNA encodes one member of a family of proteinase inhibitors[106] (Table 2), and accumulates during ripening of tomato fruit. The third identified tomato mRNA encodes pectin esterase.[107] Unlike the other two mRNAs, pectin esterase mRNA declines in abundance during tomato fruit ripening.

A fourth tomato mRNA has been assigned the identity of "ethylene-associated" on the basis of its induction in the fruit tissue coincident with the onset of autocatalytic ethylene synthesis. Variously referred to as encoding a peptide of 35 kDa,[108] 37.5 kDa,[99] or 33.5 kDa[109] (Table 2), the mRNA is encoded by a gene that is related to others in a small multigene family.[114]

One further ripening-related mRNA that has been identified encodes avocado β-1,4-endoglucanase or "cellulase". This mRNA increases in abundance during ripening of avocado fruit.

B. ETHYLENE AND TISSUE SPECIFICITY

Are such changes in mRNA abundance in the ripening fruit, as are discussed above, dependent on ethylene? The question has been approached in a number of ways. The most superficial is to observe the time course of change and to relate this to ethylene production. It may then be argued that changes that are rapid and coincident with a change in ethylene synthesis, may be ethylene dependent, and those that are delayed may be a response to other signals that are secondary to the ethylene response. A better method of asking the question is by adding ethylene, or its analog propylene, to immature fruit that fail to ripen in the presence of ethylene. Changes that are ethylene dependent, but independent of secondary events that occur during ripening, may be so detected. A comparison with other tissues will determine whether the response is tissue specific. This approach may be extended by using mutant fruit that mature, but do not ripen, in the presence of ethylene. The effect of adding inhibitors of ethylene action, silver ions or norbornadiene, can be observed. Ethylene-dependent events should not occur in the presence of effective concentrations of the inhibitors. However, on long exposure, secondary events will also be inhibited so that kinetic studies are an important aspect of inhibitor studies. Ultimately the identification of responsive concensus sequence in gene regulatory elements and the isolation of interacting regulatory molecules will allow the recognition of genes that respond to common signals. A start has been made in using these methods to identify responsive genes, particularly in tomato fruit.

Maunders et al.[115] used cDNAs to ripening-related mRNAs to study ethylene regulation of gene expression in tomato fruit. In fruit ripening on the vine, one RNA, recognized by the cDNA pTom75 (see Table 3), rose rapidly to its maximum abundance as ethylene production increased. The mRNAs complementary to pTom13 ("ethylene-associated"), pTom36, and pTom99 (Table 3), also rose rapidly as ethylene production increased, but these were at greatest abundance several days later as the fruit ripened. Messenger RNAs for pTom5, pTom6 (endopolygalacturonase), and pTom137 reached their maxima gradually, and several days after ethylene had peaked. When ripening was initiated by exposing green fruit to ethylene, RNA complementary to pTom75, pTom36, and pTom99 peaked rapidly and the others much more slowly. Of these RNAs, those complementary to pTom13, pTom75, and pTom137 also occur in senescing leaves,[116] so that only pTom5, pTom6, and pTom99 are ripening specific. Of the latter three, the pTom99-specific transcript responds most rapidly to ethylene. The presence of silver ions over 2 d reduces the level of the transcripts specific for these cDNAs. The reduction is slight for pTom36 and extreme for the three ripening-specific clones and for pTom13.[117]

Lee et al.[99] exposed immature fruit to the ethylene analog propylene,[12] and determined the effects on mRNA populations by *in vitro* translation studies. Their experiments, summarized in Figure 2, showed reversible changes in abundance of only a few ripening-related mRNAs while others significantly did not respond immediately. To investigate specifically the response to the endopolygalacturonase gene to ethylene/propylene treatment, dot-blots of the mRNA preparations were probed with cDNA encoding tomato endopolygalacturonase (Figure 2). Endopolygalacturonase mRNA was seen to increase only in fruit that had entered ripening — as indicated by autocatalytic synthesis of ethylene. In this experiment, the

TABLE 3
Ripening-Related cDNAs

Clone	Fruit	RNA size (b)	Transcript size (Da)	Ripening specific	Ethylene induction	Encodes	Ref.
pTom 6	Tomato	1,580	55,000	Yes	+/−	Endopolygalacturonase (110)	115
pTom 13	Tomato	1,400	35,000	Yes	+/−	Ethylene-associated (111)	115
pTom 36	Tomato	1,470	52,000	Yes	+/−		115
pTom 5	Tomato	1,960	48,000	Yes	+/−		115
pTom 137	Tomato	1,680	57,000	No	+/−		115
pTom 75	Tomato	1,200	28,000	No	+		115
pTom 99	Tomato	1,520	44,000	Yes	+/−		115
E 4	Tomato	880	25,000	N.t	+		105
E 8	Tomato	1,600	42,000	N.t	+		105
E 17	Tomato	580		N.t	+	Proteinase inhibitor 1	105
J 49	Tomato	1,000	27,000	N.t	+		105
E 41	Tomato	2,100		N.t	−	Endopolygalacturonase	105
22—2	Tomato	1,400		No	N.t		122
24—2	Tomato	700		Yes	N.t		122
28—8	Tomato	1,600		Yes	N.t		122
9—24	Tomato	1,900		Yes	N.t		122
C 3	Tomato	1,600		Yes	N.t	Endopolygalacturonase	122a
pAV 5	Avocado		53,000	Yes	N.t	"Cellulase"	121

Note: N.t., not tested.

[a] Specific to ripening fruit.

endopolygalacturonase gene product does not accumulate as a direct response to ethylene, yet it is activated in ripening fruit (this experiment and Reference[115]) and its activity is inhibited by inhibitors of ethylene perception.[117] It therefore appears that stimulation of the endopolygalacturonase gene requires ethylene plus another factor (or factors) evidently absent in immature fruit but present in ripening fruit.

Lincoln et al.[105] directed their studies of tomato fruit specifically at ethylene-stimulated genes. They selected four cDNAs (E 4, 8, 17 [proteinase inhibitor], and J 49) on the basis of ethylene sensitivity, and one (E 41 [endopolygalacturonase]) as ripening specific (see Table 3). In immature fruit exposed to ethylene for 8 h, the mRNAs complementary to the four "ethylene-sensitive" cDNAs increased but that specific for E 41 showed no increase (Figure 3).

In naturally developing fruit (Figure 4), two of the mRNAs, E 17 and J 49, accumulated prior to the increase in ethylene synthesis that accompanies ripening. Because the accumulation of these RNAs was also stimulated when immature fruit were exposed to ethylene, it was suggested that the expression of these genes became more sensitive to ethylene as the fruit matured. This thesis suggests that multiple regulatory elements are involved. Treatment of the fruit with norbornadiene inhibited the accumulation of each of these mRNAs. Norbornadiene also inhibited ripening so that direct and indirect effects of ethylene are not distinguished by this observation.

Lincoln and Fischer[118] utilized their ethylene-sensitive cDNAs as hydridization probes to examine nuclear transcripts in nuclear run-on experiments. In nuclei from ripening fruit, active transcription of the four genes was measured. In equivalent aged fruit of the ripening-inhibitor (*rin*) mutant, transcription of the E 4 gene was significantly suppressed, transcription of the E 8 and J 49 genes was less drastically reduced, while that of the E 17 gene was as in normal fruit. Mature fruit of the *rin* mutant do not synthesize additional ethylene with maturity as do the normal ripening fruit. Nevertheless, transcription of only some of the

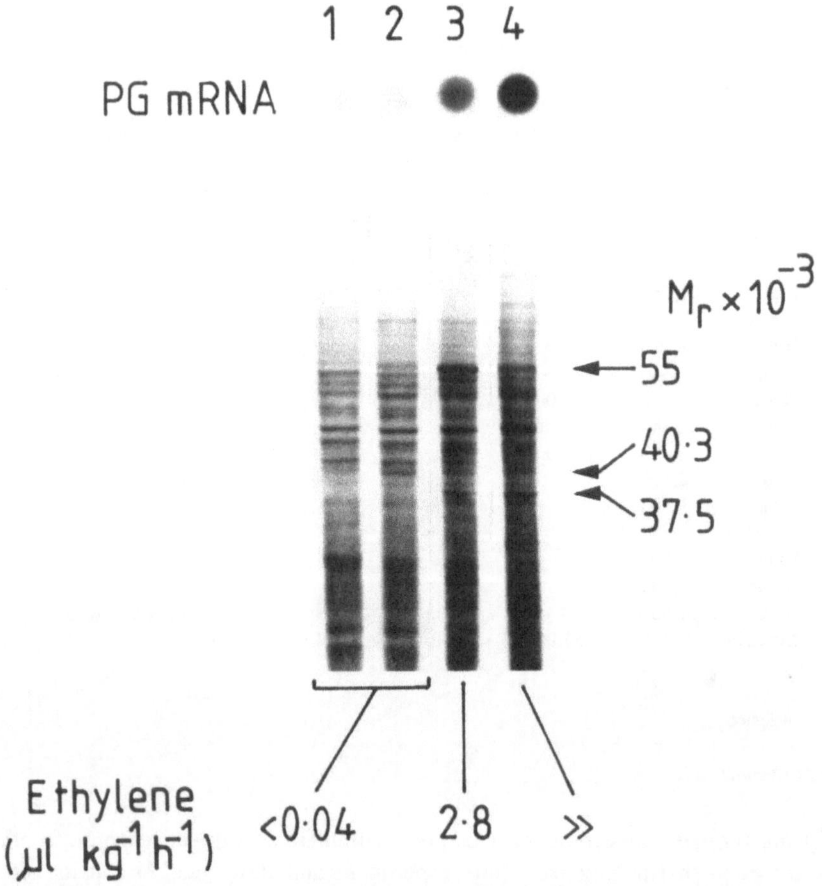

FIGURE 2. Analysis of tomato fruit mRNAs by *in vitro* translation and dot blot hybridization. Total RNA from fruit pericarp was translated in a wheat germ system and the products analyzed by fractionation on SDS gradient acrylamide gels and exposure to X-ray film. Experimental details are in Lee et al.[99] Arrows indicate major translation products, which differ between tracks. 2 μg of total RNA from each experiment were bound to a membrane and hybridized with ^{32}P-labeled cDNA to tomato endopolygalacturonase, as described in Reference 113. RNA was obtained from pericarp of immature green tomato fruit, Tracks 1 to 3, and ripe fruit, Track 4. Track 1, fruit incubated for 4 d in air; Tracks 2 and 3, fruit incubated for 4 d in 500 $\mu l\ l^{-1}$ propylene; Track 4, RNA isolated from fruit 4 d past first ethylene rise. Ethylene evolution of fruit, indicated below tracks, had a basal level of less than 0.04 $\mu l\ kg^{-1}\ h^{-1}$.

ethylene-inducible genes was inhibited in the mutant fruit. In mutant fruit, accumulation of the J 49 transcript was normal, of the E 8 transcript about one third of normal, and E 4 and E 17 transcripts did not accumulate. Each of the transcripts accumulated in ethylene-treated mutant fruit. Thus, ethylene regulated the E 4 gene at the transcription level, and had a less dramatic effect on the transcription of the E 8 gene. Regulation of the E 17 gene expression by ethylene is at a post-transcriptional step. DellaPenna et al.[119] examined the transcription of the endopolygalacturonase gene in normal and mutant tomato fruit. In normal fruit, transcription of the gene was activated early in ripening and remained active during ripening. Transcription of the endopolygalacturonase gene was impaired in mutants that have greatly reduced endopolygalacturonase mRNA and protein contents. These are important experiments in delineating the mode of ethylene action. They indicate that genes differ in the level of ethylene to which they respond[120] and that ethylene can act directly on gene activation or on a later molecular event.

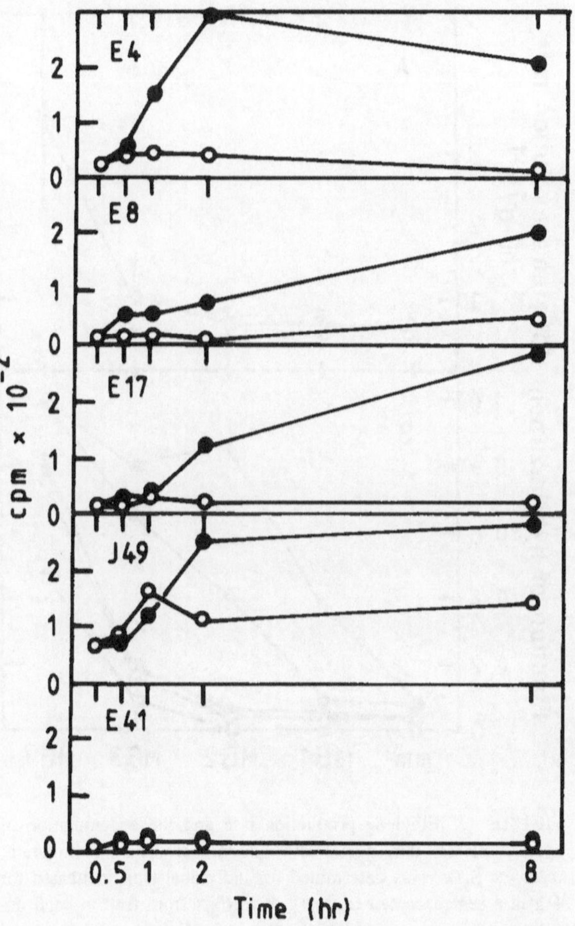

FIGURE 3. Accumulation of specific mRNAs in tomato fruit exposed to ethylene. mRNA was isolated from the pericarp of MG1 fruit treated with either ethylene (●) or air (○) for the indicated period of time, or from MG4 fruit. One microgram of each mRNA was bound to nitrocellulose and hybridized to the indicated ^{32}P-labeled cDNA probes. Hybridized dots were excised and hybridization determined by liquid scintillation spectrometry. (From Lincoln, J. E., Cordes, S., Read, E., and Fischer, R. L., *Proc. Natl. Acad. Sci. U.S.A.*, 84, 2793, 1987. With permission.)

In avocado, the mRNA for β-1,4-endoglucanase shows a 60-fold increase in abundance during ripening (cDNA pAV5, Table 3), but there is no evidence that indicates how direct a response this is to the large increase in ethylene synthesis that occurs in ripening avocado.[121]

C. *CIS*- AND *TRANS*-ACTING FACTORS

The regulation of gene transcription is generally mediated by one or more "*trans*-acting" factors — probably proteins — acting on nucleotide sequence(s) adjacent, or even internal, to the gene — the "*cis*-acting" elements — and these in turn influence the binding of RNA polymerase molecules to the gene.

A number of the genes discussed above appear to be tissue specific, to respond to ethylene induction, and to be dependent on fruit developmental stage. One would therefore expect their regulation to be mediated by more than one mechanism. Deikman and Fischer[123] and Cordes et al.[124] have examined two of the ripening-associated genes of tomato with a

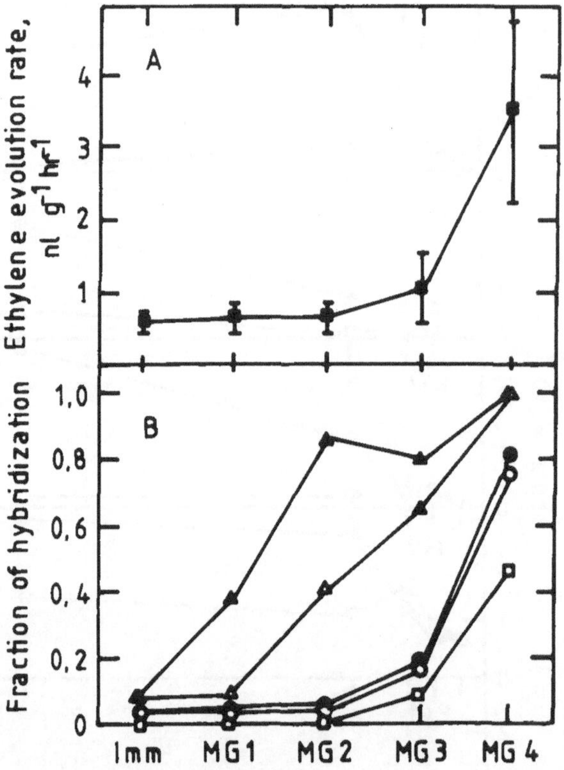

FIGURE 4. Ethylene production rate and the accumulation of specific mRNAs during fruit development. Ethylene evolution rate, mean ± S.D., was determined for individual fruit incubated for 1 h in a sealed container, 1 μg of mRNA from fruit at each developmental stage was blotted onto nitrocellulose and hybridized to the indicated ^{32}P-labeled cDNA probes. Hybridized dots were excised and hybridization determined by liquid scintillation spectrometry. (From Lincoln, J. E., Cordes, S., Read, E., and Fischer, R. L., *Proc. Natl. Acad. Sci. U.S.A.*, 84, 2793, 1987. With permission.)

view to investigating their regulatory mechanisms. In each gene, they have identified a region of sequence located 5′ to the coding sequence that is involved in regulating its transcription. In the E 8 gene the regulatory sequence is located −936 to −920, 5′ of the transcription start site, while in the E 4 gene an equivalent sequence is located −34 to −18, 5′ of the start site and overlapping the TATA box. Gel retardation and competition binding experiments indicate that the same *trans*-acting factor binds selectively to both sequences. The exact binding site was determined for each gene and a degree of homology was found, suggesting a concensus sequence for this particular *cis*-acting element, which is shown in Figure 5.

The *trans*-acting factor is a nuclear protein of as yet unknown size, and accumulates in ripening fruit, but does not accumulate significantly in unripe fruit exposed for 2 h to exogenous ethylene. This suggests that the factor is associated with fruit ripening and not with ethylene-mediated induction. Ethylene induction was seen to be associated with the formation of a second complex of nuclear protein with nucleotide sequence; however, this complex appeared to be less specific and is yet to be further investigated.

During studies on these two genes, a number of potential regulatory sequences that may interact with different nuclear proteins were observed. Hence it appears that the genes may be under several regulatory influences yet to be resolved.

```
E4   -034   ATTTCTATATAAAGAAA   -018
            TAAAGATATATTTCTTT
            ||| | |    | |||||
E8   -936   ATTCCCAACAATAGAAA   -920
            TAAGGGTTGTTATCTTT

CONSENSUS   ATT-C-A---A-AGAAA
            TAA-G-T---T-TCTTT
```

FIGURE 5. Comparison of the binding sites flanking the ethylene-sensitive E 4 and E 8 genes. DNA sequences are numbered relative to the transcription start of the E 4 and E 8 genes. (|) DNA sequences identical in E 4 and E 8; (—) DNA sequences not identical in E 4 and E 8. (Data from Reference 124.)

D. CHIMERIC GENES

The molecular trigger (or triggers) of ripening in fruit is as yet unknown, nor is it clear whether the signals are the same in all fruit. In climacteric fruit, increased synthesis of ethylene plays a critical role, as has been discussed above. Yet nonclimacteric fruit ripen in the absence of increased ethylene synthesis and immature fruit do not ripen immediately when placed in an ethylene atmosphere. Tigchelaar et al.[125] speculated that changes in the cell wall, associated with softening, may release factors coordinating ripening of the tomato fruit.

With the development of gene manipulation and plant transformation methodologies, a direct approach can be taken to examine this hypothesis. Using the ethylene-sensitive promoter region of a ripening-associated gene from tomato (the E 8 gene[123]), coupled to the coding region of the gene encoding tomato endopolygalacturonase, Giovannoni et al.[126] constructed a chimeric gene and used it to transform ripening inhibitor (*rin*) mutant tomato plants. In air, fruit of these plants behave in the same way as fruit of untransformed *rin* plants. That is, they soften little, do not accumulate lycopene, and display no respiratory climacteric or increase in ethylene biosynthesis. In an atmosphere of propylene, however, simulating an ethylene environment, the chimeric gene in the transformed fruits is activated, and both endopolygalacturonase mRNA and the active enzyme accumulate. Associated with accumulation of endopolygalacturonase enzyme, soluble uronides, indicative of endolytic hydrolysis of cell wall pectins, are released as in the normal ripening fruit. However, even with pectin hydrolysis occurring, the transformed fruit do not soften, nor do they display any of the other features of ripening, including enhanced ethylene biosynthesis. The absence of softening probably reflects the pleiotropic nature of ripening, involving numerous changes in cell processes. Alterations in the status of cell wall calcium may be such a change, and may be required before hydrolysis of the pectins will have a visible effect.

E. ETHYLENE AND RESPIRATION

In climacteric fruit, ethylene stimulates gene expression;[99,118] the ethylene-stimulated genes are also expressed during the ripening of climacteric fruits, when autocatalytic ethylene production and an increase in respiration occur. In tomato, expression of one of the genes relates to ethylene production.[99,108] What then is the relationship between gene expression and respiration?

In ripening climacteric fruits, an increase in the rate of protein synthesis accompanies the increase in respiration.[127] This has often been interpreted as implying a need for energy to drive the synthesis of proteins that catalyze ripening reactions. Several lines of evidence indicate that this is not so. Respiration and protein synthesis in fruit tissues increase in response to exposures of ethylene that are insufficient to cause ripening.[99,128] A similar increase in respiration and in protein synthesis in response to ethylene occurs in nonripening

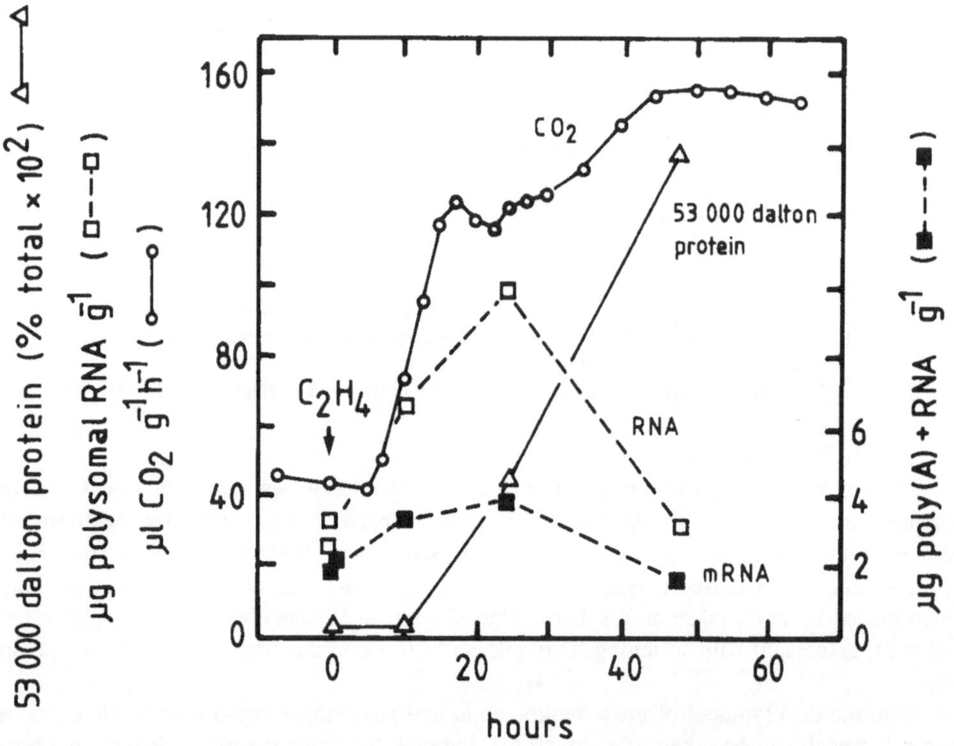

FIGURE 6. Effect of 10 µl l^{-1} ethylene treatment of avocado fruit on respiration, polysome prevalence, and poly(A)$^+$ RNA extracted from polysomes. (From Tucker, M. L. and Laties, G. G., *Plant Physiol.*, 74, 307, 1984. With permission.)

mutant genotypes[99] and in cultured fruit cells that also do not ripen.[129] The increase in protein synthesis in ripening fruit includes an increase in the turnover of many proteins,[128,111] see Figure 6, and may reflect a general tendency in plants for the rates of protein turnover to be determined by the supply of energy for maintenance.[129] There is also evidence that respiration increases in response to ethylene in fruit[130,131] or other storage organs[132] treated with cycloheximide to inhibit protein synthesis. Finally, nonclimacteric fruits, in which ripening-specific proteins are not induced by ethylene, nonetheless show an increase in respiration rate when ethylene is applied.[82]

While this evidence *in toto* establishes that the respiration response is not solely dependent on the synthesis of ethylene-dependent, ripening-specific proteins, it does not entirely divorce the respiration response from changes in the rate of protein synthesis. The observations with cycloheximide appear to do this, but the mode of action of cycloheximide in these systems has not been explored in detail, and complex interactions can occur when metabolic inhibitors are present in tissues for an extended time.[132,133] There have been suggestions in the literature for many years that ethylene may have an effect on cell membranes that influences compartmentation and metabolic regulation.[134,136] It is implied that the membrane effect is independent of protein synthesis and contemporary gene action; it may involve membrane proteins, particularly as the implied effects are subtle. However, there is no direct evidence of an interaction between ethylene and a membrane component.

The mechanisms involved in the respiratory response have been reviewed,[137] but not resolved. There is an increase in the glycolytic flux without an increase in the content of glycolytic enzymes.[138] The large increase in the pool of fructose-1,6-phosphate that occurs during the respiratory climacteric indicates that the regulation of ATP phosphofructokinase

(EC 2.7.1.11) may be involved.[139] However, the recent demonstration that the fructose-2,6-phosphate (F2,6P_2) concentration increases in ripening bananas[140] and avocados,[141] offers an alternative explanation; F2,6P_2 activates pyrophosphate:fructose-6-phosphate-1-phosphotransferase (EC 2.7.1.90) and inhibits cytoplasmic fructose-1,6-bisphosphatase (EC 3.1.3.11) favoring fructose-1,6-phosphate accumulation. Beaudry et al.,[142] also studying ripening bananas, detected a transitory rise in F2,6P_2 prior to the respiratory rise and coincident with the first rise in ethylene production. A second and much larger increase in F2,6P_2 occurred on the second day of the respiratory climacteric. Because this larger rise significantly trailed the increase in respiration, Beaudry et al. discounted the possibility that an increasing concentration of F2,6P_2 activated the glycolytic flux by activating the pyrophosphate-dependent phosphofructokinase (EC 2.7.1.90). They argued that the activation of the ATP phosphofructokinase (EC 2.7.1.11) was the significant contribution (see Reference 137), and the F2,6P_2 served to maintain a balance between glycolysis and gluconeogenesis. An increase in F2,6P_2 coincident with the respiratory increase was observed in ethylene-treated storage tissues,[143] where the authors suggested that regulation of existing enzymes by metabolites did not account for the changes in activity.

At this time, there is no convincing evidence that the increase in respiration rate in the presence of ethylene is dependent on gene activation, and the possibility of a direct effect of ethylene on a membrane component is not excluded. The complex pattern of compartmentation that occurs in plant cells makes it difficult to rationalize metabolite activation and inhibition from tissue analyses. When the receptors, that are presumably involved in ethylene-dependent gene expression, can be selectively blocked or mutated, a more precise analysis of the role of ethylene in the regulation of glycolysis may be possible.

VIII. OLIGOSACCHARINS AND THE REGULATION OF ETHYLENE SYNTHESIS

If ethylene is a signal that advances ripening, and ethylene synthesis is regulated by enzymes that are not constitutive, what then regulates the induction of these critical inductive enzymes? There is no clear answer to this but there is the suggestion that catabolites released from the cell wall are active in promoting ethylene synthesis by fruit tissues. The term oligosaccharins has been coined to describe cell wall fragments that may have a role in modifying cell function[144,145] and is applicable here. Tigchelaar et al.[125] speculated that changes in the cell wall, associated with softening, may release factors that coordinate ripening of the tomato fruit. Their suggestions included the hypothesis that polygalacturonase action may induce changes that lead to enhanced ethylene synthesis. Kinetic analyses[146,147] showed that the increase in ethylene synthesis preceded any increase in polygalacturonase protein in ripening tomato fruit, discounting the notion that products of polygalacturonase action initiated the increase in ethylene synthesis. However, cell wall fragments[148,149] were shown to stimulate ethylene synthesis in tomato and other fruit tissues. So it is possible that pectic polymers, released by polygalacturonase action, contribute to the induction of the so called system 2[12] ethylene in climacteric fruits. In this model a little ethylene provokes polygalacturonase synthesis and action, and the activity of the enzyme, in turn, provokes more ethylene synthesis. It is a model for the autocatalytic production of ethylene, but as a model it lacks detail. The induction of polygalacturonase is a relatively slow event after exposure to ethylene[102] and the release of small pectic fragments occurs only late in ripening in tomatoes.[150] The initial peak in ethylene synthesis is reached before the wall fragments appear to be available in reasonable concentration. However, the fragments that elicit ethylene are quite specific in their structures[144] and, because ripening proceeds as a wave through the fruit,[151] the local concentrations of specific elicitors is not revealed by analyses of gross segments of tissue. More refined analysis, perhaps using immunological tools, may provide the detail that is needed to substantiate the model.

A response, in terms of ethylene production, to cell wall fragments seems common to climacteric and nonclimacteric fruit.[148,152,153] This is not to suggest that each fruit is sensitive to the same elicitor molecules or that the physiological consequences of enhanced ethylene synthesis are similar in different fruits. But it does suggest that the structurally complex carbohydrate oligomers are utilized widely as metabolic signals in fruits. Cell wall breakdown, either endogenously during ripening or as a consequence of the action of invading microorganisms, will yield a range of different oligomers in different fruits so there is the potential for the development of diverse metabolic languages.

In naturally ripening tomatoes, at least two ripening-specific messenger RNAs increase before the rate of ethylene synthesis increases.[105] This implies that transcription of their genes has become sensitive to the low preclimacteric ethylene level, presumably as a response to a second effector signal. Some indication that carbohydrate oligomers may be involved in this, or in the induction of climacteric ethylene is found in the experiments of Tong and Gross.[154,155] They observed that an elicitor that provoked ethylene synthesis when infiltrated into mature-green tomato fruit was found in the pectic fraction of tomato fruit at the very earliest stages of ripening. The elicitor was not present in immature fruit, in colored (pink) fruit or in the mutant "ripening inhibited" (*rin*) fruit. *Rin* fruit produced ethylene in response to the elicitor. This transient appearance of a specific elicitor immediately prior to the ethylene climacteric suggests that this may be an important regulatory event. Changes in cell wall chemistry prior to or in the absence of polygalacturonase action have been demonstrated.[156] These changes may be linked to elicitor synthesis although it is not excluded that elicitors are generated by specific anabolic reactions.

IX. CONCLUDING REMARKS

There is but brief comment in this chapter on some of the research achievements in the study of the responses of fruits to ethylene. An attempt has been made to emphasize where assumptions have persisted without substantial evidence and where knowledge is too thin to support the fabric of a metabolic model. Some of the more spectacular recent achievements have resulted in the confirmation of the thesis that ripening is controlled by contemporary gene action and that ethylene is rather directly involved in the regulation of gene expression. It does not follow from this that the action of ethylene as a regulatory molecule is confined to gene regulation following its binding to a single receptor. The characterization of ethylene receptor(s) is an immediate challenge. Now that relevant genes can be sequenced, the definition of regulatory sequences upstream and downstream of the coding sequences will follow. It is a reasonable expectation that concensus sequences responsive to particular signals will be defined and the conservation of these between species explored. The potential of carbohydrate oligomers as information molecules is exciting and perhaps they hold some of the answers to the long-standing problem of what regulates maturity and sensitivity to ethylene. It is unlikely that changes in gene expression during fruit development are confined to the transition from nonripening to ripening, and other changes in gene expression through development will surely be defined.

The application of transformation techniques to yield specifically modified fruits is an immensely powerful tool in the study of fruit development. Of course, shrewdly used, it can also be valuable to the horticultural industry. The prospects for regulating ethylene production in a tissue-specific manner is very real but here, as in many related cases, there must be considerable uncertainty of the consequences, highlighting the limits to our understanding of the integrated physiology of plants.

REFERENCES

1. **Abeles, F. B.,** *Ethylene in Plant Biology,* Academic Press, New York, 1973.
2. **Brady, C. J.,** Fruit ripening, *Annu. Rev. Plant Physiol.,* 38, 155, 1987.
3. **Grierson, D.,** Gene expression in ripening tomato fruit, *Crit. Rev. Plant Sci.,* 3, 113, 1985.
4. **Grierson, D.,** Senescence in fruits, *HortScience,* 22, 859, 1987.
5. **Jerie, P. H., Hall, M. A., and Zeroni, M.,** Aspects of the role of ethylene in fruit ripening, *Acta Hortic.,* 80, 325, 1978.
6. **Marriott, J.,** Bananas—physiology and biochemistry of storage and ripening for optimum quality, *Crit. Rev. Food Sci. Nutr.,* 13, 41, 1980.
7. **Rhodes, M. J. C.,** Respiration and senescence of plant organs, in *The Biochemistry of Plants. A Comprehensive Treatise,* Vol. 2, Stumpf, P. K. and Conn, E. E., Eds., Academic Press, London, 1980, 419.
8. **Quintana, E. G., Nanthachai, P., Hiranpradit, H., Mendosa, D. B., and Ketsa, S.,** Changes in mango during growth and maturation, in *Mango: Fruit Development, Postharvest Physiology and Marketing in ASEAN,* Mendosa, D. B. and Wills, R. B. H., Eds., ASEAN Food Handling Bureau, Kuala Lumpur, 1984, 25.
9. **Looney, N. E., McGlasson, W. B., and Coombe, B. G.,** Control of fruit ripening in peach, *Prunus persica:* action of succinic acid, -2,2-dimethyl hydrazide and (2-chloroethyl)phosphonic acid, *Aust. J. Plant Physiol.,* 1, 77, 1974.
10. **Blanpeid, G. D.,** A study of ethylene in apple, red raspberry and cherry, *Plant Physiol.,* 49, 627, 1972.
11. **Hall, I. V. and Forsyth, F. R.,** Production of ethylene by flowers following pollination and treatments with water and auxin, *Can. J. Bot.,* 45, 1163, 1967.
12. **McMurchie, E. J., McGlasson, W. B., and Eaks, I. L.,** Treatment of fruit with propylene gives information about the biogenesis of ethylene, *Nature,* 237, 235, 1972.
13. **Yang, S. F. and Hoffman, N. E.,** Ethylene biosynthesis and its regulation in higher plants, *Annu. Rev. Plant Physiol.,* 35, 155, 1984.
14. **Terai, H. and Mizuno, S.,** Changes in 1-aminocyclopropane-1-carboxylic acid and ethylene forming enzyme activity in growing and ripening fruits of tomato and cucumber, *J. Jpn. Soc. Hortic. Sci.,* 53, 467, 1985.
15. **Brecht, J. K. and Kader, A. A.,** Regulation of ethylene production by ripening nectarine fruit *Prunus persica* as influenced by ethylene and low temperature, *J. Am. Soc. Hortic. Sci.,* 109, 596, 1984.
16. **Bufler, G.,** Ethylene-promoted conversion of 1-aminocyclopropane-1-carboxylic acid to ethylene in peel of apple at various stages of fruit development, *Plant Physiol.,* 80, 539, 1986.
17. **Liu, Y., Hoffman, N. E., and Yang, S. F.,** Promotion by ethylene of the capability to convert 1-aminocyclopropane-1-carboxylic acid to ethylene in preclimacteric tomato and cantaloupe fruits, *Plant Physiol.,* 77, 407, 1985.
18. **Hoffman, N. E. and Yang, S. F.,** Changes in 1-aminocyclopropane-1-carboxylic acid content in ripening fruits in relation to their ethylene production rates, *J. Am. Soc. Hortic. Sci.,* 105, 492, 1980.
19. **Guy, M. and Kende, H.,** Ethylene formation in *Pisum sativum* and *Vicia faba* protoplasts, *Planta,* 160, 276, 1984.
20. **Guy, M. and Kende, H.,** Conversion of 1-aminocyclopropane-1-carboxylic acid to ethylene in isolated vacuoles of *Pisum sativum* L., *Planta,* 160, 281, 1984.
21. **Mitchell, T., Porter, A. J. R., and Johns, P.,** Authentic activity of the ethylene-forming enzyme observed in membranes obtained from kiwifruit *(Actinidis deliciosa), New Phytol.,* 109, 313, 1988.

21a. **Bouzayen, M., Latche, A., and Pech, J.-C.,** Subcellular localization of the sites of conversion of 1-aminocyclopropane-1-carboxylic acid into ethylene in plant cells, *Planta,* 180, 175, 1990.

22. **John, P.,** The coupling of ethylene biosynthesis to a transmembrane, electrogenic proton flux, *FEBS Lett.,* 152, 141, 1983.
23. **Pengelly, W. L. and Su, L.-Y.,** Ethylene and plant tissue culture, in this volume, chap. 14.
24. **Brady, C. J., O'Connell, P. B. H., Smydzuk, J., and Wade, N. L.,** Permeability, sugar accumulation and respiration rate in ripening banana fruits, *Aust. J. Biol. Sci.,* 23, 1143, 1970.
25. **Biggs, M. S., Woodson, W. R., and Handa, A. K.,** Biochemical basis of high-temperature inhibition of ethylene biosynthesis in ripening tomato fruits, *Physiol. Plant.,* 72, 572, 1988.
26. **Yoshii, H. and Imaseki, H.,** Regulation of auxin-induced ethylene biosynthesis. Repression of inductive formation of 1-aminocyclopropane-1-carboxylic acid, *Plant Cell Physiol.,* 23, 639, 1982.
27. **Kende, H. and Boller, T.,** Wound ethylene and 1-aminocyclopropane-1-carboxylate synthase in ripening tomato fruit, *Planta,* 151, 476, 1981.
28. **Rothstein, S. J., De Maio, J., Strand, M., and Rice, D.,** Stable and heritable inhibition of the expression of nopaline synthase in tobacco expressing antisense RNA, *Proc. Natl. Acad. Sci. U.S.A.,* 84, 8439, 1987.
29. **Acaster, M. A. and Kende, H.,** Properties and partial purification of 1-aminocyclopropane-1-carboxylate synthase, *Plant Physiol.,* 72, 139, 1983.

30. **Privalle, L. S. and Graham, J. S.,** Radiolabeling of a wound-inducible pyridoxal phosphate utulizing enzyme: evidence for its identification as ACC synthase, *Arch. Biochem. Biophys.*, 253, 333, 1987.
31. **Mehta, A., Jordan, R. L., Anderson, J. D., and Mattoo, A. K.,** Identification of a unique isoform of 1-aminocyclopropane-1-carboxylic acid synthase by monoclonal antibody, *Proc. Natl. Acad. Sci. U.S.A.*, 85, 8810, 1988.
32. **Bleecker, A. B., Kenyon, W. H., Somerville, S. C., and Kende, H.,** Use of monoclonal antibodies in the purification and characterization of 1-aminocyclopropane-1-carboxylate synthase, an enzyme in ethylene biosynthesis, *Proc. Natl. Acad. Sci. U.S.A.*, 83, 7755, 1986.
33. **Bleecker, A. B., Robinson, G., and Kende, H.,** Studies on the regulation of 1-aminocyclopropane-1-carboxylate synthase in tomato using monoclonal antibodies, *Planta*, 173, 385, 1988.
34. **Satoh, S. and Yang, S. F.,** S-adenosylmethionine-dependent inactivation and radiolabeling of 1-aminocyclopropane-1-carboxylate synthase isolated from tomato fruits, *Plant Physiol.*, 88, 109, 1988.
35. **Van Der Straeten, D., Van Wiemeersch, L., Goodman, H. M., and Van Montagu, M.,** Purification and partial characterization of 1-aminocyclopropane-1-carboxylate synthase from tomato pericarp, *Eur. J. Biochem.*, 182, 639, 1989.

35a. **Van Der Straeten, D., Van Wiemeersch, L., Goodman, H. M., and Van Montagu, M.,** Cloning and sequence of two different cDNAs encoding 1-aminocyclopropane-1-carboxylate synthase in tomato, *Proc. Natl. Acad. Sci. U.S.A.*, 87, 4859, 1990.

36. **Boller, T. and Kende, H.,** Regulation of wound ethylene synthesis in plants, *Nature*, 286, 259, 1980.
37. **Kende, H. and Boller, T.,** Wound ethylene and 1-aminocyclopropane-1-carboxylate synthase in ripening tomato fruit, *Planta*, 151, 476, 1981.
38. **Reichard, J. K. V. and Berg, P.,** Conservation of short patches of amino acid sequence amongst proteins with a common function but evolutionary distinct origins. Implications for cloning genes and for structure-function analysis, *Nucleic Acids Res.*, 16, 9017, 1988.
39. **Nakajima, N. and Imaseki, H.,** Purification and properties of 1-aminocyclopropane-1-carboxylate synthase of mesocarp of *Cucurbita maxima* Duch fruits, *Plant Cell Physiol.*, 27, 969, 1986.

39a. **Nakajima, N., Nakagawa, N., and Imaseki, H.,** Molecular size of wound-induced 1-aminocyclopropane-1-carboxylate synthase from *Cucurbita maxima* Duch. and change translatable mRNA of the enzyme after wounding, *Plant Cell Physiol.*, 29, 989, 1990.

39b. **Sato, T. and Theologis, A.,** Cloning the mRNA encoding 1-aminocyclopropane-1-carboxylate synthase, the key enzyme for ethylene biosynthesis in plants, *Proc. Natl. Acad. Sci. U.S.A.*, 86, 6621, 1989.

39c. **Yip, W.-K., Dong, J.-G., Kenny, J. W., Thompson, G. A., and Yang, S. F.,** Characterization and sequencing of the active site of 1-aminocyclopropane-1-carboxylate synthase, *Proc. Natl. Acad. Sci. U.S.A.*, 87, 7930, 1990.

40. **Imaseki, H., Nakajima, N., and Todaka, I.,** Biosynthesis of ethylene and its regulation in plants, in *Biomechanisms Regulating Growth and Development*, Steffens, G. L. and Rumsey, J. S., Eds., Kluwer Academic Publishers, Dordrecht, The Netherlands, 1988, 205.
41. **McGlasson, W. B.,** Ethylene and fruit ripening, *HortScience*, 20, 51, 1985.
42. **Burg, S. P. and Burg, E. A.,** Ethylene action and the ripening of fruits, *Science*, 148, 1190, 1965.
43. **Sisler, E. C.,** Measurement of ethylene binding in plant tissue, *Plant Physiol.*, 64, 538, 1979.
44. **Bleecker, A. B., Estelle, M. G., Somerville, C., and Kende, H.,** Insensitivity to ethylene conferred by a dominant mutation in *Arabidopsis thaliana*, *Science*, 241, 1086, 1988.
45. **Zeroni, M., Ben-Yehoshua, S., and Galil, J.,** Relationship between ethylene and the growth of *Ficus sycomorus*, *Plant Physiol.*, 50, 378, 1972.
46. **Marei, N. and Crane, J. C.,** Growth and respiratory response of fig (*Ficus carica* L. cv. Mission) fruits to ethylene, *Plant Physiol.*, 48, 249, 1971.
47. **Hale, C. R., Coombe, B. G., and Hawker, J. S.,** Effects of ethylene and 2-chloroethylphosphonic acid on the ripening of grapes, *Plant Physiol.*, 45, 620, 1970.
48. **Looney, N. E.,** Effects of succinic acid 2,2-dimethylhydrazide, 2-chloroethylphosphonic acid, and ethylene on respiration, ethylene production and ripening of 'Redhaven' peaches, *Can. J. Plant Sci.*, 52, 73, 1972.
49. **Coombe, B. G. and Hale, C. R.,** The hormone content of ripening grape berries and the effects of growth substance treatments, *Plant Physiol.*, 51, 629, 1973.
50. **Marei, N. and Romani, R.,** Ethylene-stimulated synthesis of ribosomes, ribonucleic acid, and protein in developing fig fruits, *Plant Physiol.*, 48, 806, 1971.
51. **Rodgers, J. P.,** Cotton fruit development and abscission: variations in the levels of auxins, *Trop. Agric. (Trinidad)*, 58, 63, 1981.
52. **Guinn, G. and Brummett, D. L.,** Changes in free and conjugated indole-3-acetic acid and abscisic acid in young cotton fruits and their abscission zones in relation to fruit retention during and after moisture stress, *Plant Physiol.*, 86, 28, 1988.
53. **Kushad, M. M., Yelenosky, G., and Knight, R.,** Interrelationship of polyamine and ethylene biosynthesis during avocado fruit development and ripening, *Plant Physiol.*, 87, 463, 1988.
54. **Morgan, P. W.,** Gibberellic acid and IAA compete in ethylene-promoted abscission, *Planta*, 129, 275, 1976.

55. **Davis, L. A. and Addicott, F. T.,** Abscisic acid: correlations with abscission and with development in the cotton fruit, *Plant Physiol.*, 49, 644, 1972.
56. **Lipe, J. A. and Morgan, P. W.,** Ethylene: role in fruit abscission and dehiscence processes, *Plant Physiol.*, 50, 759, 1972.
57. **Osborne, D. J.,** Abscission, *Crit. Rev. Plant, Sci.*, 8, 103, 1989.
58. **Guinn, G. and Brummett, D. L.,** Changes in abscisic acid and indoleacetic acid before and after anthesis relative to change in abscission rates of cotton fruiting forms, *Plant Physiol.*, 87, 629, 1988.
59. **Akamine, K. E. and Goo, T.,** Respiration and ethylene production during ontogeny in fruit, *J. Am. Soc. Hortic. Sci.*, 98, 381, 1973.
60. **Adato, I. and Gazit, S.,** Role of ethylene in avocado fruit development and ripening, *J. Exp. Bot.*, 28, 636, 1977.
61. **Quinlan, J. D. and Preston, A. P.,** The influence of shoot competition on fruit retention and cropping of apple trees, *J. Hortic. Sci.*, 46, 525, 1971.
62. **Weinbaum, S. A. and Muraoka, T. T.,** Chemical thinning of prune: relation of assimilate deprivation to ethylene-mediated fruit abscission, *HortScience*, 13, 159, 1978.
63. **Holm, R. E. and Wilson, W. C.,** Ethylene and fruit loosening from combination of *Citrus* abscission chemicals, *J. Am. Soc. Hortic. Sci.*, 102, 576, 1977.
64. **Blumenfeld, A., Epstein, E., and Ben-Tal, Y.,** Ethylene treatment, *HortScience*, 13, 47, 1978.
65. **Mackenzie, K. A. D.,** The structure of the fruit of the red raspberry *Rubus idaeus* in relation to abscission, *Ann. Bot.*, 43, 355, 1979.
66. **Iwahori, S. and Van Steveninck, R. F. M.,** Ultrastructural observation of lemon fruit abscission, *Sci. Hortic.*, 4, 235, 1976.
67. **Strosser, R., Rasmusen, H. P., and Bukovac, M. J.,** Histochemical changes in the developing abscission layer in fruits of *Prunus cerasus* L., *Planta*, 86, 151, 1969.
68. **Sexton, R., Jamieson, G. G. C., and Allen, M. H. I. L.,** An ultrastructural study of abscission zone cells with special reference to the mechanism of enzyme excretion, *Protoplasma*, 91, 369, 1977.
69. **Sexton, R., Durbin, M., Lewis, L. N., and Thompson, W. N.,** The immunological localization of 9.5 cellulase in abscission zone of bean (*Phaseolus vulgaris* cv Red Kidney), *Protoplasma*, 109, 335, 1981.
70. **Greenberg, J., Goren, R., and Riov, J.,** The role of cellulase and polygalacturonase in abscission of young and mature Shamouti orange fruits, *Physiol. Plant.*, 34, 1, 1975.
71. **Goren, R. and Huberman, M.,** Effects of ethylene and 2,4-D on the activity of cellulase isoenzymes in abscission zones of the developing orange fruit, *Physiol. Plant.*, 37, 123, 1976.
72. **Rascio, N., Casadoro, G., Ramino, A., and Masia, A.,** Structural and biochemical aspects of peach fruit abscission (*Prunus persica* L. Batsch), *Planta*, 164, 1, 1985.
73. **Yang, S. F. and Hoffman, N. E.,** Ethylene biosynthesis and its regulation in higher plants, *Annu. Rev. Plant Physiol.*, 35, 155, 1984.
74. **McGlasson, W. B., Wade, N. L., and Adato, I.,** Phytohormones and fruit ripening, in *Phytohormones and Related Compounds—A Comprehensive Treatise*, Vol. 2, Letham, D. S., Goodwin, P. B., and Higgins, T. J., Eds., Elsevier/North-Holland, Amsterdam, 1978, 447.
75. **Bufler, G. and Bangerth, F.,** Effects of propylene and oxygen on the ethylene producing system of apples (*Malus sylvestris* cultivar Golden Delicious), *Plant Physiol.*, 58, 486, 1983.
76. **Bangerth, F.,** Changes in sensitivity for ethylene during storage of apple and banana fruits under hypobaric conditions, *Sci. Hortic.*, 24, 151, 1984.
77. **Bramlage, W. J., Greene, D. W., Autio, W. R., and McLaughlin, J. M.,** Effects of aminoethoxyvinylglycine on internal ethylene concentrations and storage of apples, *J. Am. Soc. Hortic. Sci.*, 105, 847, 1980.
78. **Dekazos, E. D.,** Effect of aminoethoxyvinylglycine, carboxymethylcellulose and growth regulators on longevity of fresh rabbit-eye blueberries, *Proc. Fl. State Hortic. Soc.*, 93, 145, 1980.
79. **Ness, P. J. and Romani, R. J.,** Effects of aminoethoxyvinylglycine and counter effects of ethylene on ripening of 'Bartlett' pear fruits, *Plant Physiol.*, 65, 372, 1980.
80. **Hobson, G. E., Nichols, R., Davies, J., and Atkey, P. T.,** The inhibition of tomato fruit ripening by silver, *J. Plant Physiol.*, 116, 21, 1984.
81. **Saltveit, M. E., Bradford, K. J., and Dilley, D. R.,** Silver ion inhibits ethylene synthesis and action in ripening fruits, *J. Am. Soc. Hortic. Sci.*, 103, 472, 1978.
82. **Given, N. K., Venis, M. A., and Grierson, D.,** Hormonal regulation of ripening in the strawberry, a non-climacteric fruit, *Planta*, 174, 402, 1988.
83. **Tucker, G. and Brady, C. J.,** Silver ions interrupt fruit ripening, *J. Plant Physiol.*, 127, 165, 1987.
83a. **Hamilton, A. J., Lycett, G. W., and Grierson, D.,** Antisense gene that inhibits synthesis of the hormone ethylene in transgenic plants, *Nature*, 346, 284, 1990.
84. **Trewavas, A.,** How do plant growth substances work? *Plant Cell Environ.*, 4, 203, 1981.
85. **McGlasson, W. B. and Pratt, H. K.,** Effects of ethylene on canteloupe fruits harvested at various ages, *Plant Physiol.*, 39, 120, 1964.

86. **Brady, C. J., O'Connell, P. B. H., Smydzuk, J., and Wade, N. L.**, Permeability, sugar accumulation and respiration rate in ripening banana fruits, *Aust. J. Biol. Sci.*, 23, 1143, 1970.
87. **McGlasson, W. B., Dostal, H. C., and Tigchelaar, E. C.**, Comparison of propylene-induced responses of immature fruit of normal and mutant tomatoes, *Plant Physiol.*, 85, 218, 1975.
88. **Rai, R. M., Trewari, J. D., and Pant, N.**, Physiological effect of growth regulators on ripening of cultivar early-Shanbury apples, *Prog. Hortic.*, 15, 276, 1983.
89. **Liu, Y., Hoffman, N. E., and Yang, S. F.**, Promotion by ethylene of the capability to convert 1-aminocyclopropane-1-carboxylic acid the ethylene in preclimacteric tomato and cantaloupe fruits, *Plant Physiol.*, 72, 402, 1985.
90. **Burg, S. P. and Burg, E. A.**, Role of ethylene in fruit ripening, *Plant Physiol.*, 37, 179, 1962.
91. **Hoad, G. V., Anderson, H. M., Gutteridge, C. G., and Sparks, T. R.**, Ethylene and the ripening of strawberry fruits, in Bristol University Long Ashton Res. Station Annu. Rep., 1971, 33.
92. **Reid, M. S. and Pratt, H. K.**, Ethylene and the respiration climacteric, *Nature*, 226, 976, 1970.
93. **Knee, M., Tsantili, E., and Hatfield, S. G. S.**, Promotion and inhibition by ethylene of chlorophyll degradation in orange fruits, *Ann. Appl. Biol.*, 113, 129, 1988.
94. **Elkashif, M. E. and Huber, D. J.**, Electrolyte leakage, firmness, and scanning electron microscope studies of watermelon fruit treated with ethylene, *J. Am. Soc. Hortic. Sci.*, 113, 378, 1989.
95. **Frenkel, C. and Dyck, R.**, Auxin inhibition of ripening in Bartlett pears, *Plant Physiol.*, 51, 6, 1973.
96. **Vendrell, M.**, Reversion of senescence, *Aust. J. Biol. Sci.*, 22, 601, 1969.
97. **Grierson, D., Slater, A., Speirs, J., and Tucker, G. A.**, The appearance of polygalacturonase mRNA in tomatoes: one of a series of changes in gene expression during development and ripening, *Planta*, 163, 263, 1985.
98. **Biggs, M. S., Harriman, R. W., and Handa, A. K.**, Changes in gene expression during tomato fruit ripening, *Plant Physiol.*, 81, 395, 1986.
99. **Lee, E., Speirs, J., McGlasson, W. B., and Brady, C. J.**, Messenger RNA changes in tomato fruit pericarp in response to propylene, wounding or ripening, *J. Plant Physiol.*, 129, 287, 1987.
100. **Christoffersen, R. E., Warm, E., and Laties, G. G.**, Gene expression during fruit ripening in avocado, *Planta*, 155, 52, 1982.
101. **Callahan, A., Morgens, P., and Walton, E.**, Isolation and in vitro translation of RNAs from developing peach fruit, *HortScience*, 24, 356, 1989.
102. **Sato, T., Kusaba, S., Nakagawa, H., and Ogura, N.**, Cell-free synthesis of a putative precursor of polygalacturonase in tomato fruits, *Plant Cell Physiol.*, 25, 1069, 1984.
103. **Sato, T., Kusaba, S., Nakagawa, H., and Ogura, N.**, Polygalacturonase mRNA of tomato: size and content in ripe fruit, *Plant Cell Physiol.*, 26, 211, 1985.
104. **DellaPenna, D., Alexander, D. G., and Bennett, A. B.**, Molecular cloning of tomato fruit polygalacturonase: analysis of polygalacturonase mRNA levels during ripening, *Proc. Natl. Acad. Sci. U.S.A.*, 83, 6420, 1986.
105. **Lincoln, J. E., Cordes, S., Read, E., and Fischer, R. L.**, Regulation of gene expression by ethylene during *Lycopersicum esculentum* (tomato) fruit development, *Proc. Natl. Acad. Sci. U.S.A.*, 84, 2793, 1987.
106. **Margossian, L. J., Federman, A. D., Giovannoni, J. J., and Fischer, R. L.**, Ethylene-regulated expression of a tomato fruit ripening gene encoding a proteinase inhibitor 1 with a glutamic residue at the reactive site, *Proc. Natl. Acad. Sci. U.S.A.*, 85, 8012, 1988.
107. **Ray, J., Knapp, J., Grierson, D., Bird, C., and Schuch, W.**, Identification and sequence determination of a cDNA clone for tomato pectin esterase, *Eur. J. Biochem.*, 174, 119, 1988.
108. **Smith, C. J. S., Slater, A., and Grierson, D.**, Rapid appearance of an mRNA correlated with ethylene synthesis encoding a protein of molecular weight 35000, *Planta*, 168, 94, 1986.
109. **Holdsworth, M. J., Bird, C. R., Ray, J., Schuch, W., and Grierson, D.**, Structure and expression of an ethylene-related mRNA from tomato, *Nucleic Acids Res.*, 15, 731, 1987.
110. **Grierson, D., Slater, A., Maunders, M., Crooks, P., Tucker, G. A., Schuch, W., and Edwards, K.**, Regulation of the expression of tomato fruit ripening genes: the involvement of ethylene, in *Ethylene and Plant Development*, Roberts, J. and Tucker, G. A., Eds., Butterworths, Oxford, 1985, 147.
111. **Tucker, M. L. and Laties, G. G.**, Interrelationship of gene expression, polysome prevalence, and respiration during ripening of ethylene and/or cyanide-treated avocado fruit, *Plant Physiol.*, 74, 307, 1984.
112. **DellaPenna, D., Kates, D. S., and Bennett, A. B.**, Polygalacturonase gene expression in Rutgers, *rin*, *nor*, and *Nr* tomato fruit, *Plant Physiol.*, 85, 502, 1987.
113. **Speirs, J., Lee, E., Brady, C. J., Robertson, J., and McGlasson, W. B.**, Endopolygalacturonase: messenger RNA, enzyme and softening in the ripening fruit of a range of tomato genotypes, *J. Plant Physiol.*, 13, 567, 1989.
114. **Holdsworth, H. J., Schuch, W., and Grierson, D.**, Organisation and expression of a wound/ripening-related small multigene family from tomato, *Plant Mol. Biol.*, 11, 81, 1988.

115. **Maunders, M. J., Holdsworth, M. J., Slater, A., Knapp, J. E., Bird, C. R., Schuch, W., and Grierson, D.,** Ethylene stimulates the accumulation of ripening-related mRNAs in tomato, *Plant Cell Environ.*, 10, 177, 1987.
116. **Davies, K. M. and Grierson, D.,** Identification of cDNA clones for tomato (*Lycopersicon esculentum* Mill) mRNAs that accumulate during fruit ripening and leaf senescence in response to ethylene, *Planta,* 179, 73, 1989.
117. **Davies, K. M., Hobson, G. E., and Grierson, D.,** Silver ions inhibit the ethylene-stimulated production of ripening-related mRNAs in tomato, *Plant Cell Environ.*, 11, 729, 1988.
118. **Lincoln, J. E. and Fischer, R. L.,** Regulation of gene expression by ethylene in wild-type and *rin* tomato (*Lycopersicum esculentum*) fruit, *Plant Physiol.*, 88, 370, 1988.
119. **DellaPenna, D., Lincoln, J. E., Fischer, R. L., and Bennett, A. B.,** Transcriptional analysis of polygalacturonase and other ripening-induced genes in normal and mutant tomato fruit, *Plant Physiol.*, 90, 1372, 1989.
120. **Lincoln, J. E. and Fischer, R. L.,** Diverse mechanisms for the regulation of ethylene-inducible gene expression, *Mol. Gen. Genet.*, 212, 71, 1988.
121. **Christoffersen, R. E., Tucker, M. L., and Laties, G. G.,** Cellulase gene expression in ripening avocado fruit. The accumulation of cellulase mRNA and protein as demonstrated by hybridization and immunodetection, *Plant Mol. Biol.*, 3, 385, 1984.
122. **Mansson, P. E., Hsu, D., and Stalker, D.,** Characterization of fruit specific cDNAs from tomato, *Mol. Gen. Genet.*, 200, 356, 1985.
.22a. **Sheahy, R. E., Pearson, J., Brady, C. J., and Hiatt, W. R.,** Molecular characterization of tomato fruit polygalacturonase, *Mol. Gen. Genet.*, 208, 30, 1987.
123. **Deikman, J. and Fischer, R. L.,** Interaction of a DNA binding factor with the 5'-flanking region of an ethylene-responsive fruit ripening gene from tomato, *EMBO J.*, 7, 3315, 1988.
124. **Cordes, S., Deikman, J., Margossian, L. J., and Fischer, R. L.,** Interaction of a developmentally regulated DNA-binding factor with sites flanking two different fruit-ripening genes from tomato, *The Plant Cell,* 1, 1025, 1989.
125. **Tigchelaar, E. C., McGlasson, W. B., and Buescher, R. W.,** Genetic regulation of tomato fruit ripening, *HortScience,* 13, 508, 1978.
126. **Giovannoni, J. J., DellaPenna, D., Bennett, A., and Fischer, R. L.,** Expression of a chimaeric polygalacturonase gene in transgenic *rin* (ripening inhibitor) tomato fruit results in polyuronide degradation but not fruit softening, *The Plant Cell,* 1, 53, 1989.
127. **Richmond, A. and Biale, J. B.,** Protein and nucleic acid metabolism in fruits. I. Studies of amino acid incorporation during the climacteric rise of avocado, *Plant Physiol.*, 41, 1247, 1966.
128. **Brady, C. J. and O'Connell, P. B. H.,** On the significance of increased protein synthesis in ripening banana fruits, *Aust. J. Plant Physiol.*, 3, 301, 1976.
129. **Brady, C. J. and Romani, R. J.,** Respiration and protein synthesis in nongrowing cultured pear fruit cells in response to ethylene and modified atmospheres. A model system for fruits postharvest, *Plant Physiol.*, 87, 571, 1988.
130. **Frenkel, C., Klein, I., and Dilley, D. R.,** Protein synthesis in relation to ripening in pome fruits, *Plant Physiol.*, 43, 1146, 1968.
131. **McGlasson, W. B., Palmer, J. K., Vendrell, M., and Brady, C. J.,** Metabolic studies with banana fruit slices. I. Changes in the incorporation of ^{14}C-labelled compounds in response to cutting, *Aust. J. Biol. Sci.*, 24, 7, 1971.
132. **Kahl, G. and Laties, G. G.,** Ethylene-induced respiration in thin slices of carrot root, *J. Plant Physiol.*, 134, 496, 1989.
133. **Brady, C. J.,** Nucleic acid and protein synthesis, in *Senescence and Aging in Plants,* Nooden, L. D. and Leopold, A. C., Eds., Academic Press, New York, 1988, 147.
134. **Solomos, T. and Laties, G. G.,** Cellular organization and fruit ripening, *Nature,* 245, 390, 1973.
135. **Jones, R. L.,** Ethylene enhanced release of α-amylase from barley aleurone cells, *Plant Physiol.*, 43, 442, 1968.
136. **Abeles, F. B., Leather, G. R., Forrence, L. E., and Craker, L. E.,** Abscission regulation of senescence, protein synthesis and enzyme secretion by ethylene, *HortScience,* 6, 371, 1971.
137. **Solomos, T.,** Respiration and energy metabolism in senescing plant tissues, in *Post-Harvest Physiology and Crop Improvement,* Lieberman, M., Ed., Plenum Press, New York, 1983, 61.
138. **Nair, P. M. and Darak, B. G.,** Identification of multiple forms of phosphofructokinase in ripening dwarf Cavendish banana, *Phytochemistry,* 20, 605, 1981.
139. **Salminen, S. O. and Young, R. E.,** The control properties of phosphofructokinase in relation to the respiratory climacteric in banana fruit, *Plant Physiol.*, 55, 54, 1975.
140. **Mertens, E., Marcellin, P., Van Schaftingen, E., and Hers, H. G.,** Effect of ethylene treatment on the concentration of fructose-2,6-bisphosphate and on the activity of phosphofructokinase-2/fructose-2,6-bisphosphatase in banana, *Eur. J. Biochem.*, 167, 579, 1987.

141. **Bennett, A. B., Smith, G. M., and Nichols, B. G.,** Regulation of climacteric respiration in ripening avocado fruit, *Plant Physiol.*, 83, 973, 1987.
142. **Beaudry, R. M., Paz, N., Black, C. C., and Kays, S. J.,** Banana ripening, implications of changes in internal ethylene and CO_2 concentrations, pulp fructose 2,6-bisphosphate concentration, and activity of some glycolytic enzymes, *Plant Physiol.*, 85, 277, 1987.
143. **Stitt, M., Cseke, C., and Buchanan, B.,** Ethylene-induced increases in fructose-2,6-bisphosphate in storage tissues, *Plant Physiol.*, 80, 246, 1986.
144. **McNeil, M., Darvill, A. G., Fry, S. C., and Albersheim, P.,** Structure and function of the primary cell walls in plants, *Annu. Rev. Biochem.*, 53, 625, 1984.
145. **Ecker, J. R. and Davis, R. W.,** Plant defense genes are regulated by ethylene, *Proc. Natl. Acad. Sci. U.S.A.*, 86, 5202, 1987.
146. **Brady, C. J., MacAlpine, G., McGlasson, W. B., and Ueda, Y.,** Polygalacturonase in tomato fruits and the induction of ripening, *Aust. J. Plant Physiol.*, 9, 171, 1982.
147. **Grierson, D. and Tucker, G. A.,** Timing of ethylene and polygalacturonase synthesis in relation to the control of tomato fruit ripening, *Planta*, 157, 174, 1983.
148. **Brecht, J. K. and Huber, D. J.,** Stimulation of tomato fruit ripening by tomato cell-wall fragments, *HortScience*, 21, 319, 1986.
149. **Tong, C. B., Labavitch, J. M., and Yang, S. F.,** The induction of ethylene production from pear cell culture by cell wall fragments, *Plant Physiol.*, 81, 929, 1986.
150. **Huber, D. J.,** Polyuronide degradation and hemicellulose modification in ripening tomato fruit, *J. Am. Soc. Hortic. Sci.*, 108, 405, 1983.
151. **Brecht, J. K.,** Locular gel formation in developing tomato fruit and initiation of ethylene production, *HortScience*, 22, 476, 1987.
152. **Baldwin, E. A. and Pressey, R.,** Tomato polygalacturonase elicits ethylene production in tomato fruit, *J. Am. Soc. Hortic. Sci.*, 113, 92, 1988.
153. **Baldwin, E. A. and Biggs, R. H.,** Cell wall lyzing enzymes and products of cell wall digestion elicit ethylene in citrus, *Physiol. Plant.*, 73, 58, 1988.
154. **Tong, C. and Gross, K.,** Purification of an ethylene elicitor from tomato fruit cell walls, *Plant Physiol.*, 89(Suppl.), 99, 1989.
155. **Gross, K. C.,** Cell wall dynamics, in *Biotechnology and Food Quality*, Kung, S. D., Bills, D. D., and Quatrano, R., Eds., Butterworths, Boston, 1988, 143.
156. **Kim, J. and Gross, K. C.,** Early changes in ethylene, ACC and cell wall composition of tomato locule and pericarp tissue during initial ripening, *Plant Physiol.*, 83, 113, 1987.

Chapter 14

ETHYLENE AND PLANT TISSUE CULTURE

William L. Pengelly and Ling-Yuan Su

TABLE OF CONTENTS

I. INTRODUCTION

Plant culture provides experimental systems well suited to the study of plant growth and development. The relatively simple nutritional requirements of plants permit cells to be grown on comparatively inexpensive and completely defined culture media. It is now possible to regenerate fertile plants from single-cell isolates of many plant species. The ability to control growth and differentiation by modifying culture conditions, particularly by changing the chemical constituents of the culture medium, has obvious advantages for investigations of developmental regulation. The more practical problem of plant propagation in culture is also an area of increasing interest, particularly with respect to plant production from mutant or transgenic cells.

Plant hormones have long been a focus of plant culture. Growth of cultured plant cells most often shows an absolute requirement for hormones, and modification of hormone levels in the medium has been the primary technique for manipulating cell differentiation and morphogenesis. Auxins and cytokinins are two classes of plant hormones receiving the greatest attention with respect to plant culture. In most cases, the maintenance of proliferating callus, the induction of root or shoot organogenesis, and the initiation of embryogenic cultures involve some regime of auxin and/or cytokinin feeding. Ethylene has received less attention, but the fact that ethylene production is regulated by auxins and cytokinins suggests that ethylene might mediate many auxin and cytokinin responses.[1] This chapter reviews what is known about ethylene production in culture, and assesses the possible role of ethylene in regulating growth and development *in vitro*.

Since ethylene is a gas with limited solubility under physiological conditions, many studies of ethylene production and action have been carried out *in vitro* using tissue slices or organ segments placed in vessels that are sealed to contain ethylene. Thus, plant culture methods, which by necessity employ culture vessels such as flasks or tubes, are readily amenable to investigations of ethylene.

II. ETHYLENE PRODUCTION IN CULTURED PLANT CELLS

Ethylene production appears to be as universal in cultured cells as it is in intact plants. Several studies have shown that cultures of a variety of monocotyledonous and dicotyledonous plants produce ethylene.[2-6] Rohwer and Bopp[7] reported that ethylene was produced by cultured moss protonema (*Funaria hygrometrica*). Ethylene is produced by suspension cultures,[2,5-7] callus grown on solid medium,[3,4,8] and protoplast preparations.[9-12]

As described in Chapter 1, higher plants synthesize ethylene from L-methionine via *S*-adenosylmethionine (SAM) and 1-aminocyclopropane-1-carboxylic acid (ACC).[13] The final step in ethylene biosynthesis is the oxidation of ACC by the ethylene-forming enzyme (EFE).[14] Other than oxidation, the only known fate of ACC is its conjugation with malonyl-CoA to form *N*-malonyl-ACC (MACC).[15,16]

There is good reason to believe that cultured cells produce ethylene by the same pathway used in the plant. Lieberman et al.[17] showed with radiotracer studies that apple (*Malus domestica*) suspension cultures converted L-methionine to ethylene. There are also numerous examples showing stimulation of ethylene biosynthesis in culture with ACC feeding (see below), and the occurrence of both ACC and MACC in tobacco (*Nicotiana tabacum*) callus has been demonstrated by mass spectrometry.[18] Ethylene biosynthesis inhibitors believed to act at the level of EFE,[13] such as Co^{2+},[19,20] uncouplers,[20] and free radical scavengers,[17,20] are effective in blocking ethylene production in culture.

The enzymatic nature of the ethylene-forming reaction was demonstrated using the 2-ethyl derivative of ACC, 1-amino-2-ethylcyclopropane-1-carboxylic acid (AEC).[13] Of the four possible stereoisomers, only one, (1R,2S)-AEC, proved to be an effective substrate and a competitive inhibitor of ACC oxidation. Guy and Kende[21] showed that over 80% of the ethylene produced by *Pisum sativum* protoplasts originated in the vacuole. Using AEC isomers they demonstrated that ethylene production by isolated vacuoles showed the same stereospecificity as plant tissues.

Although the bona fide pathway for ethylene biosynthesis appears to operate in cultured cells, it is by no means clear that this is the only route by which ethylene is synthesized. Other nonspecific, ACC-oxidizing systems may also be involved, a possibility perhaps made more probable by the fact that cultured cells often contain much higher ACC levels than the vegetative plant.[22] A possible route of nonspecific ethylene formation involves lipoxygenase that may oxidize ACC during leaf senescence.[23] Cultured tobacco cells produce ethane (an indicator of lipid peroxidation) when grown on 2,4-dichlorophenoxyacetic acid (2,4-D) concentrations greater than 10 mg l^{-1}, or in response to NaCl stress.[24] In this case, increased ethane production was correlated with a rise in ethylene output, suggesting a connection between the two.

In most higher plants EFE activity is constitutively expressed, and ethylene production is normally limited by ACC biosynthesis.[13] This situation can be quite different in cultured cells where EFE capacity is often the factor limiting ethylene production. In some cases this may result from the saturation of EFE by the very high ACC levels that occur in cultured tissues.[25] However, there are other examples where tissues contain low ACC levels and continue to produce small amounts of ethylene even when treated with hormones or ACC.[25] Thus, ethylene production rates in cultured cells may be determined by either ACC synthesis or oxidation, and in some cases it is the oxidation of ACC which limits ethylene synthesis.

Ethylene production in batch cultures is almost always biphasic. Within the first few hours after transfer there is occasionally a burst of ethylene production that quickly dissipates. This is particularly evident in primary explants[26] and is likely to be a wound effect. Independent of whether there is an initial burst of wound ethylene, there is another rise in ethylene production that lasts several days or weeks.[6,8,17,20,24-29] The time of maximum ethylene production can be quite variable, occurring as early as a day or two after transfer, but always within the first two weeks of culture. The timing, intensity, and shape of this peak of ethylene production can be affected by a variety of factors, many of which are common constituents of culture media.

III. FACTORS AFFECTING ETHYLENE BIOSYNTHESIS IN CELL CULTURE

A. PLANT HORMONES

1. Auxins

It is well known that auxins promote ethylene production in plants by stimulating the biosynthesis of ACC.[13] Initially, it was not clear whether auxins were also effective with cultured cells. Whereas auxin promoted ethylene production in cultures of some species such as sycamore[6] and rue (*Ruta graveolens*),[29] other species such as horseradish and sweet clover showed no auxin stimulation.[2] Rose suspensions showed only marginal increases in ethylene output with auxin treatment.[29] However, species that responded poorly to auxins produced substantial quantities of ethylene on hormone-free medium,[2,29] suggesting that ACC was not limiting. Studies of cultured tissues of three solanaceous species yielded several examples where auxin would not promote ethylene formation, but in every case auxin markedly increased the ACC content of the tissue.[25] Thus, the failure of auxins to promote ethylene production in some tissues can be attributed to insufficient EFE capacity of the tissue, and not to a lack of auxin-induced ACC synthesis.

The importance of EFE as a determinant of ethylene production rate was clearly shown by Puschmann and Romani[28] in studies of cultured pear cells. Transfer of auxin-starved cells to an aging medium containing auxin resulted in a burst of ethylene production that reached a maximum level 12 h after transfer. Thereafter, ethylene production declined rapidly even though ACC levels continued to increase for another 12 h. By 72 h ethylene production was barely detectable, despite the fact that ACC levels at this time were comparable to those at 12 h when ethylene production was maximal. Addition of ACC to cultures at different times after transfer confirmed that the decline in ethylene production rate resulted from a loss of EFE activity.

Plant cells transformed by *Agrobacterium tumefaciens* or *A. rhizogenes* show markedly elevated rates of ethylene biosynthesis.[30] This effect requires the tumorigenic plasmids harbored by the bacteria[30] and is determined specifically by *tms* genes,[25,31] which code for auxin biosynthesis.[32] Cells transformed by wild-type octopine strains of *Agrobacterium* may contain hundredfold higher ACC levels than cells transformed by *tms*-mutant strains.[25] Endogenous indole-3-acetic acid (IAA) levels in *Agrobacterium*-transformed tobacco cells typically show a biphasic pattern during the culture period where auxin increases during the first two weeks after subculture and then declines progressively for the remainder of the culture period.[33,34] Tobacco tumor cells show a similar rise and fall in ACC content and ethylene production,[25] suggesting a close relationship between auxin level and ethylene biosynthesis. Indeed, one tobacco teratoma line (TA66C3-78) failed to follow this pattern and showed a relatively constant IAA level[35] and ethylene production rate[25] over a three-week culture period.

There is also evidence that ethylene production in nontransformed, hormone-dependent cultures is related to the auxin content of the cells. In continuous (turbidostat) cultures of *Acer pseudoplatanus*, King[36] showed that a tenfold reduction in the 2,4-D concentration of the input medium resulted in a marked reduction in the 2,4-D content of the cells and a tenfold lower rate of ethylene production. In batch cultures of carrot (*Daucus carota*), Bender and Neumann[26] found that the decline in ethylene production after 6 d in culture could be prevented if the medium was supplemented with IAA and *myo*-inositol at days 7, 13, and 16. Unfortunately, they did not test the addition of IAA alone, and the contribution of *myo*-inositol to this effect is not known (see Section III.B.). Lee and co-workers[37-39] found that glyphosate treatment increased the rate of IAA metabolism in tobacco callus and lowered free IAA levels in the tissue. The effect of glyphosate on IAA levels paralleled a dose-dependent inhibition of ethylene production. Glyphosate did not inhibit the basal rate of ethylene production in tissues grown without IAA, showing that the herbicide did not directly affect the ethylene-forming reaction.

Auxin stimulation of ethylene production in culture is dose dependent and biphasic. Increasing auxin concentrations promotes ethylene biosynthesis to a maximum rate. However, higher auxin doses become increasingly toxic and are progressively inhibitory to ethylene biosynthesis.[20,24,25,27,28] In aged (auxin-starved) pear cells, Puschmann and Romani[28] showed that inhibition of ethylene production by high IAA levels was transient and was eventually overcome so that the peak of maximum ethylene production was shifted to a later time after auxin addition. In other cases no shift in the time of maximal ethylene production was noted,[20,24,25,27] but a common feature of high auxin treatments is a slower and more protracted decline in ethylene production rate.

Inhibition of ethylene production by excessive auxin concentrations is likely due to inhibition of EFE rather than to a lack of ACC. Addition of NAA (0.1 m*M*) to the growth medium of tobacco cells transformed by *A. tumefacients* strain A6 inhibited the ethylene production rate by 36% but increased the ACC content of the cells fourfold.[25] How ethylene formation is inhibited by auxin is not known, but the effect may be a secondary one arising from the severe and often lethal tissue damage that excessive auxin concentrations can cause.

A variety of natural and synthetic auxins promote ethylene production in cultured cells. Table 1 shows the effect of several auxins on ACC and MACC contents of two tobacco

TABLE 1
The Effect of Various Auxins on ACC and MACC Contents of Tobacco Crown-Gall Tumor Lines TA6-5 and TA66C3-78 After 5 and 21 Days in Culture

Auxin[a] (0.1 m*M*)	Days in culture	ACC (nmol g^{-1} F. wt) TA6-5	ACC (nmol g^{-1} F. wt) TA66C3-78	MACC (nmol g^{-1} F. wt) TA6-5	MACC (nmol g^{-1} F. wt) TA66C3-78
None	5	577 ± 126[b]	1.7 ± 0.6	171 ± 38	18 ± 0.5
	21	122 ± 33	3.2 ± 0.5	136 ± 18	57 ± 14
IAA	5	356 ± 113	8.9 ± 1.9	259 ± 87	121 ± 32
	21	1120 ± 366	10 ± 7.2	314 (2)	185 ± 68
IPA	5	872 ± 97	261 ± 93	180 ± 49	210 (2)
	21	345 ± 127	14 ± 2.7	262 ± 65	48 (2)
IBA	5	789 ± 171	146 ± 41	202 ± 25	128 ± 43
	21	1510 ± 180	1150 ± 151	351 ± 69	562 ± 85
PAA	5	2680 ± 340	28 ± 7.8	130 ± 26	43 ± 19
	21	2917 ± 738	22 ± 3.8	393 ± 60	216 ± 29
NAA	5	1890 ± 32	2960 ± 570	129 ± 12	469 ± 120
	21	3930 ± 987	6610 ± 39	449 ± 10	1959 ± 490
2,4-D	5	834 ± 400	1820 ± 195	300 ± 91	192 ± 26
	21	3920 ± 680	10200 ± 1320	500 ± 46	1094 ± 64

[a] Auxins used were: indole-3-acetic acid (IAA), indole-3-propionic acid (IPA), indole-3-butyric acid (IBA), phenylacetic acid (PAA), α-naphthaleneacetic acid (NAA), and 2,4-dichlorophenoxyacetic acid (2,4-D).

[b] Mean values for ACC and MACC expressed ± sem (n = 3, except where noted).

crown-gall tumor lines, one line (TA6-5) transformed by the virulent A6 strain of *A. tumefaciens*, and the second (TA66C3-78) transformed by the *tms*-mutant A66 strain. On hormone-free medium, TA6-5 contained more ACC and MACC than TA66C3-78, due to differences in auxin production encoded by *tms* genes. Addition of auxin to the growth medium increased ACC levels in the tumor tissues, and the amount of this increase varied markedly with the auxin used. With indole-3-propionic acid (IPA), stimulation of ACC accumulation was transient, and high levels of ACC observed at 5 days declined markedly by day 21. With other, perhaps more persistent auxins such as NAA and 2,4-D, the cells continued to accumulate large amounts of ACC through three weeks of culture. Interestingly, IAA proved to be a relatively weak inducer of ACC accumulation, particularly in TA66C3-78, despite the fact that IAA was nearly as effective as NAA at inhibiting the growth of this tissue. Thus, the activity of auxins with respect to ACC synthesis cannot always be correlated with their activity with respect to growth.

Another interesting result in Table 1 is the unusual behavior of phenylacetic acid (PAA). Although a weak inducer of ACC in TA66C3-78, PAA was as potent as NAA or 2,4-D in stimulating ACC accumulation in TA6-5. Phenylacetic acid is generally considered to be a weak auxin and would not support rapid growth of nontransformed tobacco callus when applied over a wide concentration range (Figure 1a). Similarly, PAA did not significantly affect the growth of TA66C3-78, but was as potent as NAA at inhibiting the growth of the auxin-sensitive line TA6-5 (Figure 1b). Thus, TA6-5 provided an example where PAA showed unusually potent auxin activity that was not found with other indole auxins. Cell lines like these that exhibit stable differences in their response to auxins can be used to investigate basic questions concerning auxin sensitivity and the role of PAA and other naturally occurring auxins in plants.

In summary, auxins are potent inducers of ACC biosynthesis in cultured cells. The extent and time course of ACC accumulation depends on the inherent activity and persistence

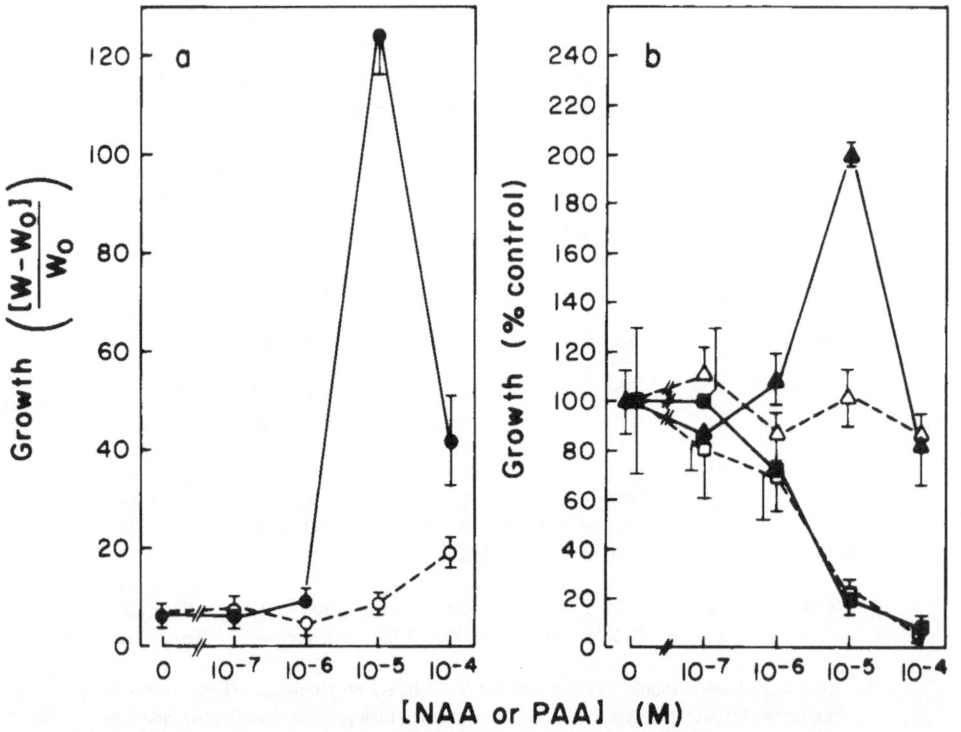

FIGURE 1. The effect of NAA (solid symbols) and PAA (open symbols) on the growth of (a) tobacco pith callus, and (b) tobacco crown-gall tumor lines TA6-5 (squares) and TA66C3-78 (triangles). Tobacco pith was cultured with 1.4 μM kinetin in all cases; tumor lines were cultured without cytokinin supplements. Tissue was grown on 10 ml MS medium[40] at 27°C with continuous fluorescent light (7 to 9 W m^{-2} photosynthetically active radiation).

of the auxin used, the plant species examined, and the physiological or developmental state of the cells. Since the response of cells to auxins can be highly variable, examination of auxin-induced ACC accumulation at a fixed time may fail to give an accurate, quantitative estimate of auxin induction. Moreover, since ethylene synthesis in culture is often limited at the ethylene-forming step, measurement of ethylene output alone will often fail to detect large increases in ACC content.

2. Cytokinins

Cytokinins may promote[26] or inhibit[17] ethylene production in cultured cells. In dose-response experiments, Gamborg and LaRue[29] found the effect of kinetin on ethylene production in rose cells to be inversely related to growth, whereas in lettuce pith[41] and carrot root cultures[26] addition of cytokinin promoted both growth and ethylene output. Lieberman et al.[17] found that inhibition of ethylene biosynthesis by ^{6}N-isopentenyladenosine (i^{6}Ado) depended upon both the concentration and time in culture. After 2 days of subculture, 10 μM i^{6}Ado was much less inhibitory to ethylene production than 100 μM i^{6}Ado, but after 6 days the lower dose was more inhibitory than the higher dose.

Variation in the response of cells to cytokinins may be explained in part by their interaction with other factors, particularly auxins. Cytokinins and auxins can act synergistically in promoting ethylene production in culture. Lettuce pith explants treated with IAA and kinetin evolved much more ethylene than explants treated with either IAA or kinetin alone.[41] Such synergism is also observed in tobacco pith explants (Table 2). In the presence of NAA and 2 μM kinetin, ethylene production was greater than the sum of ethylene

TABLE 2
The Effect of Kinetin, NAA, and ACC on Ethylene Production in Unorganized Tobacco Pith Callus on the Fifth Day After Subculture

		Ethylene production (nmol h^{-1} g^{-1} F. wt)		
		Kinetin (μM)		
NAA (μM)	ACC (μM)	0	2	20
—	—	0.06 ± 0.01[a]	0.10 ± 0.04	0.12 ± 0.02
1	—	0.09 ± 0.01	0.34 ± 0.05	0.30 ± 0.06
10	—	0.21 ± 0.03	0.97 ± 0.15	0.35 ± 0.07
100	—	0.47 ± 0.06	1.03 ± 0.15	0.69 ± 0.07
—	10	0.06 ± 0.02	0.17 ± 0.06	0.06 ± 0.01
—	100	0.09 ± 0.03	0.29 ± 0.04	0.15 ± 0.04
—	1000	0.11 ± 0.02	0.25 ± 0.08	0.16 ± 0.04

Note: Pith callus was grown on 10 ml Murashige and Skoog (MS) medium[40] at 27°C with continuous fluorescent light (7 to 9 W m^{-2} photosynthetically active radiation).

[a] Mean ethylene values expressed ± sem (n = 5).

production rates obtained with NAA or kinetin alone. Increasing kinetin concentration to 20 μM caused a significant reduction in ethylene production, but only at the higher NAA doses. Addition of 2 μM kinetin caused a 2- to 5-fold increase in ethylene production over that obtained with NAA alone, as well as a 3-fold increase in ethylene production in response to ACC feeding (Table 2), suggesting that kinetin may promote EFE activity.

One interesting and consistent feature of hormone synergism in tobacco callus is that ethylene production with auxin and cytokinin treatment was always much greater than could be obtained with ACC treatment, even when ACC was supplied at a high and presumably saturating concentration (Table 1). Table 3 shows that feeding ACC in combination with NAA and kinetin can significantly inhibit ethylene production. This finding is worrisome because ACC feeding is commonly used to quantify EFE activity in plants. The possible inhibition of ethylene formation by applied ACC must be taken into account when ACC feeding is used to quantify EFE activity *in vivo*.

3. Abscisic Acid and Gibberellins

The two plant hormones abscisic acid (ABA) and gibberellins (GAs) have received relatively little attention with regard to ethylene biosynthesis in culture. In plant tissues, ABA may either promote or inhibit ethylene production.[42] Similarly in culture, ABA slightly promoted ethylene production in rose suspensions but inhibited ethylene production in *Ruta* cells.[29] In apple suspensions, ABA slightly promoted ethylene production.[17]

In higher plants, GAs may promote ethylene biosynthesis.[43] Induction of seed germination in *Amaranthus caudatus* by GA_3 was shown to depend on ethylene production.[44] Induction of ethylene synthesis by GAs might be mediated by auxin, since GAs have been shown to promote IAA synthesis in pea seedlings.[45,46] However, GA_3 inhibited ethylene production in rose suspensions,[29] whereas it could promote or inhibit ethylene production in apple suspensions depending on the GA_3 concentration used and time in culture.[17] In crown-gall teratoma tissues of tobacco, GA_3 markedly stimulated growth but did not significantly increase either ACC, MACC, or IAA levels in the tissue.[47] Huxter et al.[48] also found that GA_3 did not affect ethylene production in tobacco callus. In general, the effect of either GA_3 or ABA on ethylene biosynthesis in culture appears to be modest compared to that of auxins and cytokinins.

TABLE 3
Inhibition of Hormone-Induced Ethylene Biosynthesis in Tobacco Pith Callus by ACC

Growth media supplements[a]			Ethylene production
Kinetin	NAA	ACC	(nmol h^{-1} g^{-1} F. wt)
+	−	−	0.07 ± 0.01[b]
+	+	−	1.38 ± 0.25
+	−	+	0.21 ± 0.04
+	+	+	0.55 ± 0.09

Note: Pith callus was grown on 10 ml MS medium[40] at 27°C with continuous fluorescent light (7 to 9 W m^{-2} photosynthetically active radiation).

[a] Concentrations of supplements were: kinetin, 1.4 μ*M*; NAA, 10 μ*M*; ACC, 1 m*M*
[b] Mean ethylene values expressed ± sem (n = 5)

B. CARBOHYDRATES

Several studies[49-54] have shown that carbohydrates stimulate ethylene production in plant tissues. A survey of carbohydrates revealed that several mono-, di-, and trisaccharides were capable of stimulating ethylene production in tobacco leaf discs.[51] Carbohydrate stimulation of ethylene synthesis is not believed to be related to osmotic effects since the response shows considerable specificity for the carbohydrate tested. The most striking example of this is the marked stimulation of ethylene synthesis in tobacco leaf discs by D-glucose, while L-glucose is without effect.[51] Species or tissue specificity is also indicated by the fact that carbohydrate specificities are different for tobacco leaf discs[51] and tomato fruit.[52] The effect of carbohydrates on ethylene biosynthesis is complex and involves stimulation of both ACC synthesis and EFE activity.[50,52,53] For a more complete discussion of the role of carbohydrates in ethylene production, the reader is referred to Chapter 2.

Sucrose and glucose are the most common carbon sources in plant culture, although a variety of other carbohydrates have been used.[55] The effect of carbohydrates on ethylene production in cultured tissues has not been systematically studied, but results with tissue explants cited above strongly indicate that carbohydrates in culture media would exert a strong, promotive effect on ethylene production. Dose-response studies in leaf discs showed that mannitol,[50] sucrose, and glucose[51] stimulated ethylene production at concentrations as low as 1 m*M*, with maximum stimulation occurring at 100 m*M*. Sucrose concentrations in culture media are typically 2 to 4% (60 to 120 m*M*) and thus are in the range of maximum stimulation found for leaf discs. Moreover, sugar alcohols such as mannitol and sorbitol are often used at higher concentrations (0.45 to 0.8 *M*) for osmotic stabilization of protoplasts,[56] and carbon sources such as sucrose may be increased to concentrations as high as 12% to induce differentiation.[57] How these manipulations of carbohydrate affect ethylene production is not known.

C. LIGHT, CO_2, AND TISSUE MORPHOLOGY

Light may either promote or inhibit ethylene biosynthesis in plants.[13] The factor that appears to be most important in determining the type of light effect is CO_2. In ambient air, light inhibits ethylene formation,[58-60] whereas light promotes ethylene production in CO_2-enriched atmospheres.[58] Variation in response to light is also observed in culture. Huxter et al.[48] found that light markedly inhibited ethylene production in tobacco pith callus. LaRue and Gamborg[5] found little or no effect of light on ethylene production in *Ruta* and sweet clover suspensions, but marked light stimulation of ethylene production in rose suspensions. The role of CO_2 in mediating the light effect in culture is unclear since the aforementioned experiments were all carried out in ambient air. Understanding the variable effects of light on ethylene production in cultured cells requires further discussion of the interaction of light and CO_2. Here the reader is referred to Chapter 2.

Light inhibition of ethylene biosynthesis can be of two types. The first type, found in etiolated pea[61] and bean,[62] was shown to be mediated by phytochrome; red light inhibited ethylene biosynthesis, and this effect was reversed by far-red light. This phenomenon may be related to phytochrome-mediated reduction in free IAA levels that has been shown to occur in etiolated maize seedlings.[63] However, studies by Jiao et al.[64] with etiolated wheat seedlings indicated that light actually promoted ACC synthesis and EFE activity. They found instead that light inhibition of ethylene production was caused by a marked stimulation of ACC malonylation leading to a drastic reduction in free ACC levels. Interestingly, no significant difference in extractable ACC-malonyltransferase activity could be detected in light- and dark-grown plants.

A second type of light inhibition was found in green leaf tissue.[58-60] In this case, inhibition of ethylene production was not mediated by phytochrome but depended instead on photosynthesis.[60] Light strongly inhibited the conversion of ACC to ethylene,[58-60] but did not affect ACC levels.[58] Kao and Yang[58] showed that light inhibition of ethylene could be reversed by CO_2 and suggested that light acted via photosynthesis by reducing CO_2 levels in the tissue. Interestingly, ethylene production by green leaf segments in the presence of exogenous ACC (1 m*M*) and CO_2 (3%) was greater in the light than in the dark.[58] Thus, while light may inhibit ethylene production by lowering CO_2 levels, it also appears to increase EFE capacity when sufficient CO_2 is present. Therefore, the effect of light on the ethylene pathway cannot be explained solely by fluctuations in CO_2 concentration.

The effect of light on EFE capacity may be related to tissue differentiation. In *Pharbitis nil*, de Laat et al.[60] showed that ethylene production from exogenous ACC (1 m*M*) was greater in cotyledons from light-grown plants than in cotyledons from dark-grown plants when tissues were incubated under dark conditions where photosynthesis would not occur. The capacity of light-grown cotyledons to show this enhanced EFE capacity was abolished if plants were raised with the chlorophyll synthesis inhibitor SAN 9789. Thus, it appears that EFE capacity is related in some way to chloroplast function.

Variation in EFE capacity with greening is also evident in cultured cells. Crown-gall teratoma cells of tobacco and tomato, which formed green, leafy shoots, had a much higher EFE capacity than unorganized, nongreeen tumors of the same species.[25] Reversibility of this phenomenon is shown in Table 4. Unorganized (O^-) tobacco pith callus maintained on NAA and kinetin at concentrations optimal for growth produced less ethylene in response to NAA or ACC than did shoot-forming callus (O^+) maintained on a higher kinetin concentration and without NAA. When O^+ tissue was returned to optimal growth media, the morphology again became unorganized ($O^+ \rightarrow O^-$) and ethylene-forming capacity returned to the lower level. Similar results were obtained for the tobacco teratoma line TA66C3-78. When this tissue was maintained on growth medium containing 10 μM NAA, the tissue lost its chlorophyll, became unorganized, and showed a lower EFE capacity than the shooty teratoma (Table 4).

Although shoot-forming tobacco tissues generally show a higher EFE capacity than unorganized tissues, some unorganized tissues may retain a high EFE capacity.[25] This is illustrated by two unorganized variants isolated from the teratoma line TA66C3-78. The two variants no longer formed shoots, lacked chlorophyll, and proliferated in culture as friable tissues.[35] One of the variant lines, TA66-D1, showed a marked reduction in EFE capacity; whereas the second line, TA66-D2, showed an EFE capacity comparable to the shooty teratoma (Table 5). Thus, the presence of shoots and intense greening of the tissue are not

TABLE 4
The Effect of Tissue Morphology on Ethylene-Forming Capacity in Cultured Pith Callus and Crown-Gall Tumor Tissue of Tobacco

Tissue[a]	Morphology	Test medium[b] NAA	Kinetin	ACC	Ethylene production (nmol h^{-1} g^{-1} F. wt)
O^- Pith	Unorganized	−	+	−	0.03 ± 0.01[c]
		+	+	−	1.74 ± 0.25
		−	+	+	0.41 ± 0.04
O^+ Pith	Shooty, green	−	+	−	0.12 ± 0.03
		+	+	−	2.37 ± 0.33
		−	+	+	1.87 ± 0.29
$O^+ \rightarrow O^-$ Pith	Unorganized	−	+	−	0.03 ± 0.01
		+	+	−	1.61 ± 0.25
		−	+	+	0.60 ± 0.12
TA66C3-78	Shooty, green	−	−	−	0.17 ± 0.02
		+	−	−	3.21 ± 0.67
		−	−	+	2.84 ± 0.52
TA66C3-78 (O^-)	Unorganized	−	−	−	0.08 ± 0.01
		+	−	−	1.17 ± 0.10
		−	−	+	0.48 ± 0.17

Note: Tissue was grown on 10 ml MS medium[40] at 27°C with continunous fluorescent light (7 to 9 W m^{-2} photosynthetically active radiation).

[a] O^- pith maintained with 2 μ*M* kinetin and 10 μ*M* NAA; O^+ pith maintained with 20 μ*M* kinetin; $O^+ \rightarrow O^-$ pith maintained same as O^- pith; TA66C3-78 maintained without hormones; TA66C3-78(O^-) maintained with 10 μ*M* NAA.
[b] NAA provided at 10 μ*M*, kinetin at 2 μ*M*, and ACC at 1 m*M*.
[c] Mean values expressed ± sem (n = 5)

TABLE 5
The Effect of ACC on Ethylene Production in Crown-Gall Teratoma Line TA66C3-78 and Two Unorganized Variants Isolated From It

Tissue line	Treatment[a]	Ethylene production[b] nmol/h/g F. wt	nmol/h/100 mg dry wt
TA66C3-78	Basal medium	0.09	0.09
	1 m*M* ACC	8.82	9.22
TA66-D1	Basal medium	0.04	0.10
	1 m*M* ACC	0.57	1.25
TA66-D2	Basal medium	0.11	0.19
	1 m*M* ACC	5.37	9.12

[a] Tissue was grown on 10 ml MS medium[40] at 27°C with continuous fluorescent light (7 to 9 W m^{-1} photosynthetically active radiation). Tumor lines were grown without auxin or cytokinin supplements.
[b] Ethylene production measured on day 4 of culture.

required for high EFE activity. Cell lines such as these may be useful systems for studying factors that control EFE activity *in vivo*.

D. INORGANIC NUTRIENTS

Several inorganic ions have been shown to affect ethylene biosynthesis in plants. Effects of various ions include both promotion and inhibition of ethylene biosynthesis as summarized

TABLE 6
The Effect of Inorganic Ions on Ethylene Biosynthesis in Plants

Ion	Effect on ethylene production[a]	Ref.
Ag^{+}	+/−	65,65a
Ba^{2+}	−	66
Ca^{2+}	+/−	66—70
Cd^{2+}	+	71
Co^{2+}	−	72,73
Cu^{2+}	+	66,74,75
Fe^{2+}	+	66,76,77
Hg^{2+}	−	66
Li^{+}	+	78
Mg^{2+}	−	66
NH_4^{+}	+	78a,78b
Ni^{2+}	−	66,72
HPO_4^{2-}	−	79,80
Sn^{2+}	−	66
Sr^{2+}	+	66
Zn^{2+}	−	66

[a] +, Promotion, −, Inhibition, and +/−, both promotion and inhibition in different instances.

in Table 6. Many ions having strong effects on ethylene production such as Ag^{+}, Cd^{2+}, Hg^{2+}, Li^{+}, Ni^{2+}, and Sr^{2+} are not normally added to culture media. Other effective ions such as Co^{2+}, (0 to 0.4 μM), Cu^{2+} (0.04 to 0.8 μM), Fe^{2+} (3 to 100 μM), and Zn^{2+} (1 to 35 μM) are added at concentrations (given in parentheses) too low to significantly affect ethylene production. On the other hand, Ca^{2+} (1 to 4 m*M*), Mg^{2+} (0.3 to 3 m*M*), NH_4^{+} (0 to 20 m*M*), and HPO_4^{2-} (0.1 to 4 m*M*) are often provided at levels that are at or near the concentrations used to demonstrate altered ethylene production rates. Of these, Ca^{2+} is of particular interest because it shows a strong synergistic promotion of ethylene production with cytokinins.[66-68] There is increasing evidence that Ca^{2+} is a second messenger in auxin responses,[81] suggesting that Ca^{2+} may play a role in synergistic induction of ethylene by auxin and cytokinin.

Ammonium ions applied at 40 m*M* increased ethylene production in tomato plants relative to plants treated with an equimolar concentration of NO_3^{-}.[78a,78b] Ammonium-treated plants showed leaf epinasty and lower levels of N, S, K, Ca, and Mg than nitrate-treated plants. Deficiencies of K, Ca, and Mg promoted ethylene evolution.[78b] Ammonium-induced leaf epinasty and ethylene evolution could be blocked with addition of KCl in amounts equimolar to NH_4^{+}.[78a] In culture, NH_4^{+} is often added at concentrations as high as 20 m*M*, but NO_3^{-} is always supplied at higher concentrations. How ethylene production in culture might be affected by ionic interactions and nutrient deficiencies that may develop at later times in the culture period has not been investigated.

In summary, several components of tissue culture media including hormones, carbon source, and inorganic nutrients, as well as culture conditions such as light can affect ethylene biosynthesis. The composition of culture media and the culture conditions are often changed for the purpose of modifying tissue growth and development. The question that arises is whether ethylene plays a role in mediating these developmental changes.

IV. ETHYLENE AND THE DEVELOPMENT OF CULTURED CELLS

A. GROWTH

Hsu and Stewart[82] found that the ethylene-releasing agent, 2-chloroethyl phosphonic acid (CEPA), induced callus formation in cultured cotton ovules. Callus formed at the micropylar end in both fertilized and unfertilized ovules. Addition of ABA or IAA inhibited CEPA-induced callus formation, whereas kinetin and particularly GA_3 promoted it.[83] Callus formation occurred with GA_3 alone, but the effect with CEPA was strongly synergistic. Robinson et al.[84] found that ACC induced callus formation in sunflower shoot tip cultures. However, this callus did not continue to proliferate in subculture with ACC and benzyladenine (BA), whereas callus initiated on NAA and BA could be maintained in subculture with NAA and BA. Thus, ethylene could not replace auxin as a growth factor, and it appears that induction of callus by ethylene depends on the presence of other growth factors in the primary explant.

An early question that was posed was whether ethylene was required for growth or simply a byproduct of growth processes.[8] Some studies[5,6] have shown a close correspondence between the time course of ethylene production and the growth of cultured cells, whereas others[17,20] have found that ethylene production peaks well before any decline in growth rate. With the use of the ACC synthase inhibitor, aminoethoxyvinylglycine (AVG), Gude and van der Plas,[85] using callus from potato tubers, and Sauerbrey et al.,[20] using sunflower suspension cultures, showed that ethylene production could be inhibited without affecting growth. Moreover, rapidly growing *N. glutinosa* crown-gall tumors have been isolated, which produce very little ethylene in culture.[25] The growth of tobacco crown-gall tumors and pith callus can be markedly stimulated with GA_3 without affecting ethylene biosynthesis.[47] Thus, ethylene does not appear to be required for growth, at least in a dose-dependent manner. It is not known if a low, threshold level of ethylene is essential for growth.

In soybean suspension cultures, Constabel et al.[86] showed that periodic flushes with ethylene at intervals roughly equal to the cell-cycle time induced mitotic synchrony. The amplitude of the peaks in mitotic index could be increased if the ethylene flushes were followed immediately by flushes of CO_2, which alone did not induce synchrony but increased growth rate. Comparison of mitotic phasing relative to ethylene treatment suggested that ethylene caused cells to arrest in the G2 and S phases of the cell cycle.

Whether growth is promoted or inhibited by ethylene can depend on the culture conditions. In tobacco callus, Ricci[87] found that ethylene treatment promoted growth if tissues were cultured without hormones or with auxin and cytokinin together. However, if tissues were cultured with auxin alone, then ethylene inhibited growth. Burg and Burg[88] proposed that growth inhibition at higher auxin doses was the consequence of increased ethylene biosynthesis. However, auxin has been shown to inhibit the growth of tobacco crown-gall cells without an increase in ethylene production;[25] ethylene actually promoted the growth of cells cultured with a supraoptimal auxin concentration.[87] Tobacco crown-gall cells that were shown to vary in sensitivity to auxin were relatively insensitive to CEPA applied over a wide range of concentration.[89] On the other hand, tumor cells that were more sensitive to auxin were also more sensitive to exogenous ACC and accumulated higher levels of ACC in response to auxin than cells that were more resistant to auxin. These distinct responses were found to result from differences in the capacity of the cells to form MACC.[89] Thus, accumulation of ACC may account in part for the toxic effects of high auxin concentrations, and the capacity of cells to malonylate ACC may, in turn, play a role in auxin sensitivity.

B. CYTODIFFERENTIATION

Cultured cells provide useful models for cytodifferentiation because cells are removed from the complex interactions existing between cells and tissues within the plant and are exposed instead to the controlled experimental conditions of culture. Although it has been shown that cultured cells from many plant species are totipotent, and hence can form all cell types of the plant, this fact has only been demonstrated by the regeneration of organs or embryos, which again introduce the complex cellular interactions inherent in ontogeny. Direct differentiation of cultured cells has been limited to tracheary and sieve elements,[90-94] and of the two, tracheary elements have received greater attention. The advantage of tracheary elements is that the characteristic annular or reticulate secondary cell walls are readily identified and stable to acid maceration techniques used for quantification. The disadvantage of tracheary elements is that the secondary wall deposition represents the final stage of differentiation, and it is not possible at the present time to deduce whether inductive conditions affect early events in determination or late events such as lignification.

All of the major plant hormones[90-94] and cyclic AMP[95] have been implicated in the control of tracheary differentiation. Auxin, cytokinin, GA, ethylene, and cAMP have been shown to promote tracheary element formation, whereas ABA was found to inhibit differentiation.[96,97] Auxin appears to be the most important factor for induction, although several species require both auxin and cytokinin for tracheary element differentiation.

Much of the evidence implicating ethylene in tracheary element differentiation has been reviewed by Roberts and Miller.[98] The most compelling evidence includes studies with ethylene precursors and inhibitors. Thus, auxin and cytokinin induction of differentiation in lettuce pith explants was promoted by methionine, SAM, ACC, and CEPA, and was inhibited by AVG, Co^{2+}, and Ag^{+}.[19] Moreover, inhibition of differentiation by AVG, Co^{2+}, and Ag^{+} was reversed by ACC, CEPA, and methionine, respectively.[19] Subsequent studies showed that ACC or CEPA would partially substitute for cytokinin in the lettuce system that did not show tracheary element differentiation with auxin, cytokinin, ACC, or CEPA supplied separately.[41] The synergism between auxin and ethylene could not be explained by ethylene concentration, since tissue treated with CEPA alone or IAA plus CEPA evolved similar amounts of ethylene. Thus, auxin and ethylene may be required at different stages of tracheary element development.

There is also evidence that ethylene is inhibitory to tracheary element differentiation.[99-101] Dose-response studies showed that tracheary element differentiation promoted by ethylene precursors[19] or ethylene directly[101] was biphasic, being enhanced at lower doses and inhibited at higher doses. It is not yet known what specific role ethylene plays in tracheary element development. It has been suggested that ethylene may be involved in lignification.[102]

C. MORPHOGENESIS

Skoog and Miller[103] first demonstrated that development of roots or shoots in culture could be controlled by the ratio of auxin and cytokinin in the growth medium. Such manipulations of auxin and cytokinin affect ethylene biosynthesis, and it has been shown that tobacco cells cultured under shoot-forming conditions produce less ACC[104] and ethylene[48] than cells cultured under conditions that promote unorganized growth. Ethylene inhibited greening in spinach suspensions,[105] bud initiation on *Begonia* stem segments,[106] and root formation in cultures of carrot,[107] barley,[108] and tomato.[109] In addition, Ag^{+} promoted shoot formation in tomato,[109] wheat, and *Nicotiana plumbaginifolia*.[110] Shoot formation in tomato leaf discs was also promoted by absorption of ethylene with $HgClO_4$ traps.[109] Katoh et al.[108] showed that, in barley, inhibition of ACC synthesis with aminooxyacetic acid (AOA) or inhibition of ACC oxidation with aminoisobutyric acid (AIB) reduced ethylene production and promoted root formation, whereas inhibition of SAM decarboxylation by methylglyoxal-bis(guanylhydrazone) (MGBG) strongly promoted ethylene production and completely blocked root formation.

While there is considerable evidence that ethylene promotes unorganized growth, there is also evidence that ethylene stimulates organogenesis. Thus, ethylene promoted bud formation in endive,[111,112] rice,[113] *Lilium*,[114] *Pinus radiata*,[115] and *Digitalis obscura*.[116] Ethylene promoted embryogenesis in potato[117] and induced apogamous bud formation in *Pteridium* gametophytes.[119] In *Digitalis*, ethylene promoted root formation in the absence of other growth regulators, but promoted bud formation when applied with IAA and kinetin.[116] Treatment with CEPA enhanced sprouting in potato tubers,[119] and induced floral meristems in *Plumbago indica* explants.[120] The latter case is of interest since the tissue culture work predated the discovery that ethylene induced flowering in *Plumbago* plants.[121] Thus, it is likely that floral induction by ethylene *in vitro* is limited to those few species that are similarly responsive at the whole-plant level.

Conflicting results regarding the role of ethylene in morphogenesis may be accounted for in part by ethylene dose. In many cases where low ethylene doses were found to promote organogenesis, higher doses proved to be inhibitory.[113,115,116] Huxter et al.[48] also found that ethylene can have different effects depending on the time of application. Bud formation in tobacco callus was inhibited by CEPA applied during the first 5 days of culture, but promoted by CEPA applied between 5 and 10 days. Thus, ethylene may not induce buds directly, but rather it may stimulate the development of cells already committed to bud formation.

Promotion of organogenesis by ethylene has been found to be enhanced by CO_2, which alone has been shown to promote root[122] and shoot[123] formation. In rice callus, however, CO_2 alone had no effect on morphogenesis but did enhance ethylene-induced shoot formation.[113] Elimination of both CO_2 and ethylene with gas traps inhibited bud formation in *P. radiata* cotyledons.[115] Removing CO_2 with KOH traps inhibited shoot formation in tobacco callus, but the effect was transient and by day 17 of culture there was no difference in the number of shoot primordia in control and KOH-treated cultures.[48] In a more thorough survey of volatile emissions by plant cultures, Thomas and Murashige[3,4] could find no relationship between ethylene or CO_2 evolution and organogenesis. However, they did find that unorganized cultures of several species produced significant quantities of ethanol, which could not be detected in any shoot-forming cultures. Moreover, they found an inverse relationship between ethanol production and embryogenesis in *Daucus carota* and *Phoenix dactylifera*, and Tisseral and Murashig[124] showed that ethanol could reversibly inhibit somatic embryogenesis in citrus.

The results described above suggest that morphogenesis is associated with changes in respiratory metabolism. In *D. carota*, embryogenesis has been correlated with increased cyanide-resistant respiration,[125] and ethylene is well known to stimulate the cyanide-resistant (alternate) pathway.[126,127] It is of interest that somatic embryogenesis in *Medicago sativa* was inhibited by AVG, AOA, 2,4-dinitrophenol, and salicylic acid at concentrations that did not affect ethylene production or growth.[128] Pirrung and Brauman[129] recently proposed that regulation of the alternate pathway by ethylene may lead to changes in gene expression that account for the hormonal effects of ethylene. Further work is required to determine whether altered respiration does in fact mediate the development changes induced by ethylene.

V. CONCLUSIONS

Ethylene biosynthesis is universal in cultured plant cells and, where studied, occurs by the same biosynthetic pathway as in the intact plant. Ethylene production is affected by a variety of factors including hormones, carbohydrates, inorganic nutrients, and light. Because each of these factors can be precisely controlled, cultured cells provide useful systems for studying the regulation of ethylene production. Of particular interest is the fact that cultured cells can vary markedly in EFE activity, which often limits ethylene production in culture.

It has been possible to obtain variant cell lines of common genetic origin that vary markedly in EFE capacity (cf. Table 5), and such systems should prove useful for the investigation of factors that regulate EFE activity *in vivo*.

Ethylene can affect the growth and development of cultured cells. The action of ethylene on growth and differentiation is highly variable, and it is not yet clear why ethylene promotes growth and differentiation in some cases and inhibits them in others. Thus, the utilization of ethylene as a tool to promote growth or differentiation in culture remains for the present a largely empirical enterprise. Nevertheless, it is clear that while ethylene may be a second messenger for auxins, it cannot replace auxin as a growth factor. The apparent requirement for ethylene to interact with other factors such as hormones and with respiratory metabolism may account for the variable and often disparate effects of ethylene on cultured cells. Cell culture systems that give definitive responses to ethylene should prove useful in elucidating the mode of action of ethylene in plants.

ACKNOWLEDGMENTS

Support for work in the authors' laboratory from the National Science Foundation (DMB-8417087) and the Competitive Research Grants Office of the U.S. Department of Agriculture (86-CRCR-1-2150) is gratefully acknowledged.

REFERENCES

1. **Abeles, F. B.,** *Ethylene in Plant Biology*, Academic Press, New York, 1973, 1.
2. **Gamborg, O. L. and LaRue, T. A. G.,** Ethylene produced by plant cells in suspension cultures, *Nature*, 220, 604, 1968.
3. **Thomas, D. des S. and Murashige, T.,** Volatile emissions of plant tissue cultures. I. Identification of the major component, *In Vitro*, 15, 654, 1979.
4. **Thomas, D. des S. and Murashige, T.,** Volatile emissions of plant tissue cultures. II. Effects of the auxin 2,4-D on production of volatiles in callus cultures, *In Vitro*, 15, 659, 1979.
5. **LaRue, T. A. G. and Gamborg, O. L.,** Ethylene production by plant cell cultures. Variations in poduction during growing cycle and in different plant species, *Plant Physiol.*, 48, 394, 1971.
6. **MacKenzie, I. A. and Street, H. E.,** Studies on the growth in culture of plant cells. VIII. The production of ethylene by suspension cultures of *Acer pseudoplatanus*, L., *J. Exp. Bot.*, 21, 824, 1970.
7. **Rohwer, F. and Bopp, M.,** Ethylene synthesis in moss protonema, *J. Plant Physiol.*, 117, 331, 1985.
8. **Huxter, T. J., Reid, D. M., and Thorpe, T. A.,** Ethylene production by tobacco (*Nicotiana tabacum*) callus, *Physiol. Plant.*, 46, 374, 1979.
9. **Mattoo, A. K. and Lieberman, M.,** Localization of the ethylene-synthesizing system in apple tissue, *Plant Physiol.*, 60, 794, 1977.
10. **Anderson, J. D., Mattoo, A. K., and Lieberman, M.,** Induction of ethylene biosynthesis in tobacco leaf discs by cell wall digesting enzymes, *Biochem. Biophys. Res. Commun.*, 107, 588, 1982.
11. **Konze, J. R., Jones, J. F., Boller, T., and Kende, H.,** Effect of 1-aminocyclopropane-1-carboxylic acid on the production of ethylene in senescing flowers of *Ipomoea tricolor* Cav., *Plant Physiol.*, 66, 566, 1980.
12. **Guy, M. and Kende, H.,** Ethylene formation in *Pisum sativum* and *Vicia faba* protoplasts, *Planta*, 160, 276, 1984.
13. **Yang, S. F. and Hoffman, N. E.,** Ethylene biosynthesis and its regulation in higher plants, *Annu. Rev. Plant Physiol.*, 35, 155, 1984.
14. **Hoffman, N. E., Yang, S. F., Ichihara, A., and Sakamura, S.,** Stereospecific conversion of 1-aminocyclopropanecarboxylic acid to ethylene by plant tissues. Conversion of stereoisomers of 1-amino-2-ethylcyclopropanecarboxylic acid to 1-butene, *Plant Physiol.*, 70, 195, 1982.
15. **Amrhein, N., Schneebeck, D., Skorupka, H., Tophof, S., and Stöckigt, J.,** Identification of a major metabolite of the ethylene precursor 1-aminocyclopropane-1-carboxylic acid in higher plants, *Naturwissenschaften*, 68, 619, 1981.

16. **Hoffman, N. E., Yang, S. F., and McKeon, T.**, Identification of 1-(malonylamino)cyclopropane-1-carboxylic acid as a major conjugate of 1-aminocyclopropane-1-carboxylic acid, an ethylene precursor in higher plants, *Biochem. Biophys. Res. Commun.*, 104, 765, 1982.
17. **Lieberman, M., Wang, S. Y., and Owens, L. D.**, Ethylene production by callus and suspension cells from cortex tissue of postclimacteric apples, *Plant Physiol.*, 63, 811, 1979.
18. **Amrhein, N., Dorzok, U., Kionka, C., Kondziolka, U., Skorupka, H., and Tophofs, S.**, The biochemistry and physiology of 1-aminocyclopropane-1-carboxylic acid conjugation, in *Ethylene Biochemical, Physiological and Applied Aspects*, Fuchs, Y. and Chalutz, E., Eds., Martinus Nijhoff, Dr. W. Junk, Boston, 1984, chap. 2.
19. **Miller, A. R. and Robert, L. W.**, Ethylene biosynthesis and xylogenesis in *Lactuca* pith explants cultured *in vitro* in the presence of auxin and cytokinin: the effect of ethylene precursors and inhibitors, *J. Exp. Bot.*, 35, 691, 1984.
20. **Sauerbrey, E., Grossmann, K., and Jung, J.**, Is ethylene involved in the regulation of growth of sunflower cell suspension cultures?, *J. Plant Physiol.*, 127, 471, 1987.
21. **Guy, M. and Kende, H.**, Conversion of 1-aminocyclopropane-1-carboxylic acid to ethylene by isolated vacuoles of *Pisum sativum* L., *Planta*, 160, 281, 1984.
22. **Amrhein, N., Breuing, F., Eberle, J., Skorupka, H., and Tophof, S.**, The metabolism of 1-aminocyclopropane-1-carboxylic acid, in *Plant Growth Substances 1982*, Wareing, P. F., Ed., Academic Press, London, 1982, 249.
23. **Bousquet, J. and Thimann, K. V.**, Lipid peroxidation forms ethylene from 1-aminocyclopropane-1-carboxylic acid and may operate in leaf senescence, *Proc. Natl. Acad. Sci. U.S.A.*, 81, 1724, 1984.
24. **Garcia, F. G. and Einset, J. W.**, Ethylene and ethane production in 2,4-D treated and salt treated tobacco tissue cultures, *Ann. Bot.*, 51, 287, 1983.
25. **Miller, A. R. and Pengelly, W. L.**, Ethylene production by shoot-forming and unorganized crown-gall tumor tissues of *Nicotiana* and *Lycopersicon* cultured in vitro, *Planta*, 161, 418, 1984.
26. **Bender, L. and Neumann, K. H.**, Ethylene production and growth of carrot root tissue cultures (*Daucus carota* L.) as influenced by β-indole-3-acetic acid and kinetin, *Z. Pflanzenphysiol.*, 88, 227, 1978.
27. **Köves, E. and Szabó, M.**, Ethylene production in habituated and auxin-requiring tobacco callus cultures. Does ethylene play a role in the habituation?, *Physiol. Plant.*, 69, 351, 1987.
28. **Puschmann, R. and Romani, R.**, Ethylene production by auxin-deprived, suspension-cultured pear fruit cells in response to auxins, stress, or precursor, *Plant Physiol.*, 73, 1013, 1983.
29. **Gamborg, O. L. and LaRue, T. A. G.**, Ethylene production by plant cell cultures. The effect of auxins, abscisic acid, and kinetin on ethylene production in suspension cultures of rose and *Ruta* cells, *Plant Physiol.*, 48, 399, 1971.
30. **Canfield, M. L. and Moore, L. W.**, Production of ethylene by *Daucus carota* inoculated with *Agrobacterium tumefaciens* and *Agrobacterium rhizogenes*, *Z. Pflanzenphysiol.*, 112, 471, 1983.
31. **Goodman, T. C., Montoya, A. L., Williams, S., and Chilton, M. D.**, Sustained ethylene production in *Agrobacterium*-transformed carrot disks caused by expression of the T-DNA *tms* gene products, *J. Bacteriol.*, 167, 387, 1986.
32. **Thomashow, M., Hugly, S., Buchholz, W., and Thomashow, L.**, Molecular basis for the auxin-independent phenotype of crown gall tumor tissues, *Science*, 231, 616, 1986.
33. **Pengelly, W. L. and Meins, F., Jr.**, The relationship of indole-3-acetic acid content and growth of crown-gall tumor tissues of tobacco in culture, *Differentiation*, 21, 27, 1982.
34. **Pengelly, W. L. and Meins, F., Jr.**, Growth, auxin requirement, and indole-3-acetic acid content of cultured crown-gall and habituated tissues of tobacco, *Differentiation*, 25, 101, 1983.
35. **Pengelly, W. L., Vijayaraghavan, S. J., and Sciaky, D.**, Neoplastic progression in crown gall in tobacco without elevated auxin levels, *Planta*, 169, 454, 1986.
36. **King, P. J.**, Studies on the growth in culture of plant cells. XX. Utilization of 2,4-dichlorophenoxyacetic acid by steady-state cell cultures of *Acer pseudoplatanus* L., *J. Exp. Bot.*, 27, 1053, 1976.
37. **Lee, T. T.**, Mode of action of glyphosate in relation to metabolism of indole-3-acetic acid, *Physiol. Plant.*, 54, 289, 1982.
38. **Lee, T. T. and Dumas, T.**, Effect of glyphosate on ethylene production in tobacco callus, *Plant Physiol.*, 72, 855, 1983.
39. **Lee, T. T., Dumas, T., and Jevnikar, J. J.**, Comparison of the effects of glyphosate and related compounds on indole-3-acetic acid metabolism and ethylene production in tobacco callus, *Pestic. Biochem. Physiol.*, 20, 354, 1983.
40. **Murashige, T. and Skoog, F.**, A revised medium for rapid growth and bioassays with tobacco tissue cultures, *Physiol. Plant.*, 15, 473, 1962.
41. **Miller, A. R., Pengelly, W. L., and Roberts, L. W.**, Induction of xylem differentiation in *Lactuca* by ethylene, *Plant Physiol.*, 75, 1165, 1984.
42. **McGlasson, W. B., Wade, N. L., and Adato, I.**, Phytohormones and fruit ripening, in *Phytohormones and Related Compounds: A Comprehensive Treatise*, Vol. 2, Letham, D. S., Goodwin, P. B., and Higgins, T. J. V., Eds., Elsevier/North-Holland, Amsterdam, 1978, 447.

43. **Garcia-Martinez, J. L., Ben-Shalom, N., and Rappaport, L.,** Gibberellin-induced ethylene production and its effect on cowpea epicotyl elongation, *Plant Growth Regul.*, 2, 209, 1984.
44. **Kepczynski, J.,** Ethylene-dependent action of gibberellin in seed germination of *Amaranthus caudatus*, *Physiol. Plant.*, 67, 584, 1986.
45. **Law, D. M. and Hamilton, R. H.,** Effects of gibberellic acid on endogenous indole-3-acetic acid and indoleacetyl aspartic acid levels in a dwarf pea, *Plant Physiol.*, 75, 255, 1984.
46. **Law, D. M.,** Gibberellin-enhanced indole-3-acetic acid biosynthesis: D-tryptophan as the precursor of indole-3-acetic acid, *Physiol. Plant.*, 70, 626, 1987.
47. **Pengelly, W. L. and Su, L. Y.,** Host compensation for mutant auxin oncogenes of *Agrobacterium*, in *Physiology and Biochemistry of Auxins in Plants*, Kutáček, M., Bandurski, R. S., and Krekule, J., Eds., Academia, Prague, 1988, 401.
48. **Huxter, T. J., Thorpe, T. A., and Reid, D. M.,** Shoot initiation in light- and dark-grown tobacco callus: the role of ethylene, *Physiol. Plant.*, 53, 319, 1981.
49. **Colclasure, G. C. and Yopp, J. H.,** Galactose-induced ethylene evolution in mung bean hypocotyls: a possible mechanism for galactose retardation of plant growth, *Physiol. Plant.*, 37, 298, 1976.
50. **Riov, J. and Yang, S. F.,** Stimulation of ethylene production in citrus leaf discs by mannitol, *Plant Physiol.*, 70, 142, 1982.
51. **Meir, S., Philosoph-Hadas, S., Epstein, E., and Aharoni, N.,** Carbohydrates stimulate ethylene production in tobacco leaf discs. I. Interaction with auxin and the relation to auxin metabolism, *Plant Physiol.*, 78, 131, 1985.
52. **Kim, J., Gross, K. C., and Solomos, T.,** Characterization of the stimulation of ethylene production by galactose in tomato (*Lycopersicon esculentum* Mill.) fruit, *Plant Physiol.*, 85, 804, 1987.
53. **Philosoph-Hadas, S., Meir, S., and Aharoni, N.,** Carbohydrates stimulate ethylene production in tobacco leaf discs. II. Sites of stimulation in the ethylene biosynthesis pathway, *Plant Physiol.*, 78, 139, 1985.
54. **Philosoph-Hadas, S. and Aharoni, N.,** Galactose inhibits the conversion of 1-aminocyclopropane-1-carboxylic acid to ethylene in aged tobacco leaf discs, *Plant Physiol.*, 83, 8, 1987.
55. **Maretzki, A., Thom, M., and Nickell, L. G.,** Utilization and metabolism of carbohydrates in cell and callus cultures, in *Tissue Culture and Plant Science 1974*, Street, H. E., Ed., Academic Press, London, 1974, chap. 14.
56. **Evans, P. K. and Cocking, E. C.,** Isolated plant protoplasts, in *Plant Tissue and Cell Culture*, 2nd ed., Street, H. E., Ed., University of California Press, Berkeley, 1977, 103.
57. **Collins, G. B. and Grosser, J. W.,** Culture of embryos, in *Cell Culture and Somatic Cell Genetics of Plants*, Vol. 1, Vasil, I. K., Ed., Academic Press, Orlando, 1984, chap. 30.
58. **Kao, C. H. and Yang, S. F.,** Light inhibition of the conversion of 1-aminocyclopropane-1-carboxylic acid to ethylene in leaves is mediated through carbon dioxide, *Planta*, 155, 261, 1982.
59. **Gepstein, S. and Thimann, K. V.,** The effect of light on the production of ethylene from 1-aminocyclopropane-1-carboxylic acid by leaves, *Planta*, 149, 196, 1980.
60. **de Laat, A. M. M., Brandenburg, D. C. C., and van Loon, L. C.,** The modulation of the conversion of 1-aminocyclopropane-1-carboxylic acid to ethylene by light, *Planta*, 153, 193, 1981.
61. **Goeschl, J. D., Pratt, H. K., and Bonner, B. A.,** An effect of light on the production of ethylene and the growth of the plumular portion of etiolated pea seedlings, *Plant Physiol.*, 42, 1077, 1967.
62. **Kang, B. G. and Ray, P. M.,** Ethylene and carbon dioxide as mediators in the response of the bean hypocotyl hook to light and auxins, *Planta*, 87, 206, 1969.
63. **Iino, M.,** Inhibitory action of red light on the growth of the maize mesocotyl: evaluation of the auxin hypothesis, *Planta*, 156, 388, 1982.
64. **Jiao, X. Z., Yip, W. K., and Yang, S. F.,** The effect of light and phytochrome on 1-aminocyclopropane-1-carboxylic acid metabolism in etiolated wheat seedling leaves, *Plant Physiol.*, 85, 643, 1987.
65. **Aharoni, N.,** Effect of silver ions and ethylene on auxin metabolism and auxin-induced ethylene production in tobacco leaf discs, *Physiol. Plant.*, 63, 438, 1985.
65a. **Lis, E. K., Kwakkenbos, A. A. M., and Veen, H.,** Differential effects of silver salts in apple tissue, *Plant Sci. Lett.*, 33, 1, 1984.
66. **Lau, O. L. and Yang, S. F.,** Stimulation of ethylene production in the mung bean hypocotyls by cupric ion, calcium ion, and kinetin, *Plant Physiol.*, 57, 88, 1976.
66a. **Lieberman, M.,** Biosynthesis and action of ethylene, *Annu. Rev. Plant Physiol.*, 30, 533, 1979.
67. **Lau, O. L. and Yang, S. F.,** Interaction of kinetin and calcium in relation to their effect on stimulation of ethylene production, *Plant Physiol.*, 55, 738, 1975.
68. **Lau, O. L., John, W. W., and Yang, S. F.,** Effect of different cytokinins on ethylene production by mung bean hypocotyls in the presence of indole-3-acetic acid or calcium ion, *Physiol. Plant.*, 39, 1, 1977.
69. **Ferguson, I. B.,** Calcium stimulation of ethylene production induced by 1-aminocyclopropane-1-carboxylic acid and indole-3-acetic acid, *J. Plant Growth Regul.*, 2, 205, 1983.
70. **Evensen, K. B.,** Calcium effects on ethylene and ethane production and 1-aminocyclopropane-1-carboxylic acid content in potato disks, *Physiol. Plant.*, 60, 125, 1984.

71. **Fuhrer, J.**, Ethylene biosynthesis and cadmium toxicity in leaf tissue of beans (*Phaseolus vulgaris* L.), *Plant Physiol.*, 70, 162, 1982.
72. **Lau, O. L. and Yang, S. F.**, Inhibition of ethylene production by cobaltous ion, *Plant Physiol.*, 58, 114, 1976.
73. **Yu, Y. B. and Yang, S. F.**, Auxin-induced ethylene production and its inhibition by aminoethoxyvinylglycine and cobalt ion, *Plant Physiol.*, 64, 1074, 1979.
74. **Abeles, A. L. and Abeles, F. B.**, Biochemical pathway for stress-induced ethylene, *Plant Physiol.*, 50, 496, 1972.
75. **Mattoo, A. K., Baker, J. E., and Moline, H. E.**, Induction by copper ions of ethylene production in *Spriodela oligorrhiza:* evidence for a pathway independent of 1-aminocyclopropane-1-carboxylic acid, *J. Plant Physiol.*, 123, 193, 1986.
76. **Cooper, W. C., Rasmussen, G. K., Rogers, B. J., Reece, P. C., and Henry, W. H.**, Control of abscission in agricultural crops and its physiological basis, *Plant Physiol.*, 43, 1560, 1968.
77. **Ben-Yehoshua, S. and Biggs, R. H.**, Effects of iron and copper ions in promotion of selective abscission and ethylene production by citrus fruit and the inactivation of indoleacetic acid, *Plant Physiol.*, 45, 604, 1970.
78. **Boller, T.**, Superinduction of ACC synthase in tomato pericarp by lithium ions, in *Ethylene Biochemical, Physiological and Applied Aspects,* Fuchs, Y. and Chalutz, E., Eds., Martinus Nijhoff, Dr W. Junk, Boston, 1984, 87.
78a. **Corey, K. A., Barker, A. V., and Craker, L. E.**, Ethylene evolution by tomato plants under stress of ammonium toxicity, *HortScience,* 22, 471, 1987.
78b. **Barker, A. V. and Corey, K. A.**, Ethylene evolution by tomato plants under nutrient stress, *HortScience,* 23, 202, 1988.
79. **Chalutz, E., Mattoo, A. K., and Fuchs, Y.**, Biosynthesis of ethylene: the effect of phosphate, *Plant Cell Environ.*, 3, 349, 1980.
80. **Fuchs, Y., Mattoo, A. K., Chalutz, A., and Rot, I.**, Biosynthesis of ethylene in higher plants: the metabolic site of inhibition by phosphate, *Plant Cell Environ.*, 4, 291, 1981.
81. **Brummell, D. A. and Hall, J. L.**, Rapid cellular responses to auxin and the regulation of growth, *Plant Cell Environ.*, 10, 523, 1987.
82. **Hsu, C. L. and Stewart, J. M.**, Callus induction by (2-chloroethyl) phosphonic acid on cultured cotton ovules, *Physiol. Plant.*, 36, 150, 1976.
83. **Stewart, J. M. and Hsu, C. L.**, Influence of phytohormones on the response of cultured cotton ovules to (2-chloroethyl)phosphonic acid, *Physiol. Plant.*, 39, 79, 1977.
84. **Robinson, K. E. P., Adams, D. O., and Lee, R. Y.**, Differential physiological and morphological responses of inbred lines to the ethylene precursor 1-aminocyclopropane-1-carboxylic acid by cultured *Helianthus annuus* (sunflower) shoot tips, *Plant Cell Rep.*, 6, 405, 1987.
85. **Gude, H. and van der Plas, L. H. W.**, Endogenous ethylene formation and the development of the alternative pathway in potato tuber discs, *Physiol. Plant.*, 65, 57, 1985.
86. **Constabel, F., Kurz, W. G. W., Chatson, K. B., and Kirkpatrick, J. W.**, Partial synchrony in soybean cell suspension cultures induced by ethylene, *Exp. Cell Res.*, 105, 263, 1977.
87. **Ricci, R.**, Ethylene effects on tissue cultures of *Nicotiana tabacum, Experientia,* 15, 597, 1972.
88. **Burg, S. P. and Burg, E. A.**, The interaction between auxin and ethylene and its role in plant growth, *Proc. Natl. Acad. Sci. U.S.A.*, 55, 262, 1966.
89. **Pengelly, W. L., Su, L. Y., and Campell, B. C.**, The regulation of tumor morphology in crown gall, in *Plant Growth Substances 1988,* Pharis, R. P. and Rood, S. B., Eds., Springer-Verlag, Berlin, 1990, 503.
90. **Bornman, C. H.**, Cytodifferentiation in tissue culture, in *Tissue Culture and Plant Science 1974,* Street, H. E., Ed., Academic Press, London, 1974, chap. 3.
91. **Roberts, L. W.**, *Cytodifferentiation in Plants: Xylogenesis as a Model System,* Cambridge University Press, London, 1976, 1.
92. **Reinert, J., Bajaj, Y. P. S., and Zbell, B.**, Aspects of organization—organogenesis, embryogenesis, cytodifferentiation, in *Plant Tissue and Cell Culture,* 2nd ed., Street, H. E., Ed., University of California Press, Berkeley, 1977, chap. 12.
93. **Phillips, R.**, Cytodifferentiation, *Int. Rev. Cytol. Suppl.*, 11A, 55, 1980.
94. **Fukuda, H. and Komamine, A.**, Cytodifferentiation, in *Cell Culture and Somatic Cell Genetics of Plants,* Vol. 2, Vasil, I. K., Ed., Academic Press, Orlando, 1984, 149.
95. **Basile, D. V., Wood, H. N., and Braun, A. C.**, Programming of cells for death under defined experimental conditions: relevance to the tumor problem, *Proc. Natl. Acad. Sci. U.S.A.*, 70, 3055, 1973.
96. **Minocha, S. C. and Halperin, W.**, Hormones and metabolites which control tracheid differentiation, with or without concomitant effects on growth, in cultured tuber tissue of *Helianthus tuberosus L.*, *Planta,* 116, 319, 1974.

97. **Haddon, L. E. and Northcote, D. H.,** The influence of gibberellic acid and abscisic acid on cell and tissue differentiation of bean callus, *J. Cell Sci.*, 20, 47, 1976.
98. **Roberts, L. W. and Miller, A. R.,** Is ethylene involved in xylem differentiation?, in *Vistas in Plant Sciences,* Vol. 6, Varghese, T. M., Ed., Today and Tomorrow Printers and Publishers, New Dehli, 1983, 1.
99. **Bradley, M. V. and Dahmen, W. J.,** Cytohistological effects of ethylene, 2,4-D, kinetin and carbon dioxide on peach mesocarp callus cultured in vitro, *Phytomorphology,* 21, 154, 1971.
100. **Apelbaum, A., Fisher, J. B., and Burg, S. P.,** Effect of ethylene on cellular differentiation in etiolated pea seedlings, *Am. J. Bot.*, 59, 697, 1972.
101. **Zobel, R. W. and Roberts, L. W.,** Effects of low concentrations of ethylene on cell division and cytodifferentiation in lettuce pith explants, *Can. J. Bot.*, 56, 987, 1978.
102. **Miller, A. R., Crawford, D. L., and Roberts, L. W.,** Lignification and xylogenesis in *Lactuca* pith explants cultured *in vitro* in the presence of auxin and cytokinin: a role for endogenous ethylene, *J. Exp. Bot.*, 36, 110, 1985.
103. **Skoog, F. and Miller, C. O.,** Chemical regulation of growth and organ formation in plant tissues cultured in vitro, *Symp. Soc. Exp. Biol.*, 11, 118, 1957.
104. **Grady, K. L. and Bassham, J. A.,** 1-Aminocyclopropane-1-carboxylic acid concentrations in shoot-forming and non-shoot-forming tobacco callus cultures, *Plant Physiol.*, 70, 919, 1982.
105. **Dalton, C. C. and Street, H. E.,** The role of the gas phase in the greening and growth of illuminated cell suspension cultures of spinach (*Spinacia oleracea* L.), *In Vitro,* 12, 485, 1976.
106. **Ringe, F.,** Förderung der Zellproliferation unter der Einwirkung von 2-Chloräthylphosphorsaure bei Explanten von *Begonia x richmondensis* in steriler Kultur, *Z. Pflanzenphysiol.*, 67, 45, 1972.
107. **Wochok, Z. S. and Wetherell, D. F.,** Suppression of organized growth in cultured wild carrot tissue by 2-chloroethylphosphonic acid, *Plant Cell Physiol.*, 12, 771, 1971.
108. **Katoh, Y., Hasegawa, T., Suzuki, T., and Fujii, T.,** Changes in 1-aminocyclopropane-1-carboxylic acid content and ethylene production in Hiproly barley callus during differentiation, *Agric. Biol. Chem.*, 51, 2185, 1987.
109. **Coleman, W. K., Huxter, T. J., Reid, D. M., and Thorpe, T. A.,** Ethylene as an endogenous inhibitor of root regeneration in tomato leaf discs cultured in vitro, *Physiol. Plant.*, 48, 519, 1980.
110. **Purnhauser, L., Medgyesy, P., Czakó, M., Dix, P. J., and Márton, L.,** Stimulation of shoot regeneration in *Triticum aestivum* and *Nicotiana plumbaginifolia* Viv. tissue cultures using the ethylene inhibitor $AgNO_3$, *Plant Cell Rep.*, 6, 1, 1987.
111. **Bouriquet, R.,** Action de l'éthylène sur le bourgeonnement et la floraison in vitro de fragments de racines d'endive, *C. R. Acad. Sci. (Paris) Ser. D.*, 275, 33, 1972.
112. **Lefebvre, R.,** Effets comparés d'um traitement anaérobie et de l'éthylène sur le bourgeonnement et la synthèse des pigments foliaires, *C. R. Acad. Sci. (Paris) Ser. D.*, 275, 193, 1972.
113. **Cornejo-Martin, M. J., Mingo-Castel, A. M., and Primo-Millo, E.,** Organ redifferentiation in rice callus: effects of C_2H_4, CO_2 and cytokinins, *Z. Pflanzenphysiol.*, 94, 117, 1979.
114. **Van Aartrijk, J., Blom-Barnhoorn, G. J., and Bruinsma, J.,** Adventitious bud formation from bulb-scale explants of *Lilium speciosum* Thunb. in vitro: effects of aminoethoxyvinylglycine, 1-aminocyclopropane-1-carboxylic acid, and ethylene, *J. Plant Physiol.*, 117, 401, 1985.
115. **Kumar, P. P., Reid, D. M., and Thorpe, T. A.,** The role of ethylene and carbon dioxide in differentiation of shoot buds in excised cotyledons of *Pinus radiata* in vitro, *Physiol. Plant.*, 69, 244, 1987.
116. **Pérez-Bermúdez, P., Cornejo, M. J., and Segura, J.,** A morphogenetic role for ethylene in hypocotyl cultures of *Digitalis obscura* L., *Plant Cell Rep.*, 4, 188, 1985.
117. **Reynolds, T. L.,** A possible role for ethylene during IAA-induced pollen embryogenesis in anther cultures of *Solanum carolinense* L., *Am. J. Bot.*, 74, 967, 1987.
118. **Elmore, H. W. and Whittier, D. P.,** The involvement of ethylene and sucrose in the inductive and developmental phases of apogamous bud formation in *Pteridium* gametophytes, *Can. J. Bot.*, 53, 375, 1975.
119. **Rekha, M. N. S., Rama, M. V., Narasimham, P.,** Synergistic action of gibberellin and ethrel on the inducement of sprouting in potatoes, *J. Food Sci. Technol.*, 20, 120, 1983.
120. **Nitsch, C.,** Induction *in vitro* de la floraison chez une plante de jours courts: *Plumbago indica* L., *Ann. Sci. Natl. Bot. 12e Ser.*, 9, 1, 1968.
121. **Nitsch, C. and Nitsch, J. P.,** Floral induction in a short-day plant, *Plumbago indica* L., by 2-chloroethanephosphonic acid, *Plant Physiol.*, 44, 1747, 1969.
122. **Molnar, J. M. and Cumming, W. A.,** Effect of carbon dioxide on propagation of softwood, conifer and herbaceous cuttings, *Can. J. Plant Sci.*, 48, 595, 1968.
123. **Howard, J. A., Mitchell, K. L., and Paterson, K. E.,** The effect of exogenous CO_2 on shoot initiation in tobacco cultures, *Plant Sci. Lett.*, 31, 261, 1983.
124. **Tisserat, B. H. and Murashige, T.,** Probable identity of substances in citrus that repress asexual embryogenesis, *In Vitro,* 13, 785, 1977.

125. **Kessell, R. H. J., Goodwin, C., Philp, J., and Fowler, M. W.,** The relationship between dissolved oxygen concentration, ATP and embryogenesis in carrot (*Daucus carota*) tissue cultures, *Plant Sci. Lett.*, 10, 265, 1977.
126. **Solomos, T. and Laties, G. G.,** Effects of cyanide and ethylene on the respiration of cyanide-sensitive and cyanide-resistant plant tissues, *Plant Physiol.*, 58, 47, 1976.
127. **Solomos, T. and Laties, G. G.,** Induction by ethylene of cyanide-resistant respiration, *Biochem. Biophys. Res. Commun.*, 70, 663, 1976.
128. **Meijer, E. G. M. and Brown, D. C. W.,** Inhibition of somatic embryogenesis in tissue cultures of *Medicago sativa* by aminoethoxyvinylglycine, amino-oxyacetic acid, 2,4-dinitrophenol and salicylic acid at concentrations which do not inhibit ethylene biosynthesis and growth, *J. Ex. Bot.*, 39, 263, 1988.
129. **Pirrung, M. C. and Brauman, J. I.,** Involvement of cyanide in the regulation of ethylene biosynthesis, *Plant Physiol. Biochem.*, 25, 55, 1987.

Chapter 15

MICROBIAL ETHYLENE PRODUCTION

Hideo Fukuda and Takahira Ogawa

TABLE OF CONTENTS

I. INTRODUCTION

A series of hydrocarbons having two carbon atoms (C_2-hydrocarbons), such as ethane and ethylene, are important starting materials for petrochemicals including plastics and synthetic fibers. Among them, ethylene, which is one of the most important chemicals for the industries, is mainly produced from natural gas and/or cracking of crude oil. From the viewpoint of the global shortage of fossil resources some day in the future, it is desirable to develop a process for the microbial production of such C_2-hydrocarbons from various renewable resources and waste materials.[1-15]

Ethylene as a plant hormone is produced by various plants. Also, since 1940,[16,17] there have been many reports on the production of ethylene by aerobic heterotrophs. Furthermore, it has been reported that traces of C_2-hydrocarbons were found under anaerobic conditions in sewage or in enrichment cultures of certain methanogenic bacteria.[18-21]

In this chapter, we focus on the microbial production of ethylene, especially the "ethylene-forming enzyme" of microbes, and present our view on the future prospects of microbial ethylene production.

II. REVIEW OF MICROBIAL ETHYLENE PRODUCTION

A. BACTERIA

Freebairn and Buddenhagen reported[22] that a pathogenic bacterium, *Pseudomonas solanacearum* produced ethylene from peptone, glucuronic, glutamic, or fumaric acid as a substrate. This was the first report of an axenic bacterial ethylene production. Subsequently, Primrose[24,25] and Primrose and Dilworth[23] showed that many soil bacteria had an ethylene-producing ability and demonstrated *Escherichia coli* as a producer of ethylene using methionine as a substrate.[25,26] Ince and Knowles[27,28] have developed an ethylene-forming system using cell-free extracts from *E. coli*. Ethylene production by *Pseudomonas syringae* pv. *phaseolicola,*[29,30] and *P. syringae* pv. *glycinea*[31] has also been reported.

B. YEASTS

Except for reports on *Saccharomyces cerevisiae*[32] and *Cryptococcus albidus*,[11,13] no others exist on the production of ethylene by yeasts.

C. FUNGI

Production of ethylene by fungi was first recognized in 1940 with *Penicillium digitatum*,[16,17] the green mold of citrus fruit, and then with *Blastomyces dermatitidis*.[33] Ilag and Curtis[34] reported that 58 of 228 species of fungi examined produced ethylene and they concluded that ethylene is a common metabolic product of fungi.

D. ACTINOMYCES

Ilag and Curtis[34] examined 20 unidentified streptomyces routinely isolated from soil and showed the presence of ethylene in the atmosphere surrounding some of the culture. However, ethylene was verified in only one sample.

E. DISTRIBUTION OF ETHYLENE-PRODUCING MICROBES

The appearance rate of ethylene-producing microbes, i.e., the ratio of strains that produced ethylene to strains of microorganisms tested, was examined by Fukuda et al.[1] Ethylene was produced by 7 out of 34 strains in bacteria, 13 out of 45 strains in yeast, 27 out of 52 strains in fungi, and 2 out of 35 strains in actinomycetes. Overall, the appearance rate of ethylene producers was about 30% (49 strains out of the total 166 strains used).

TABLE 1
The Typical Biosynthetic Pathways of Ethylene of Higher Plants and Microbes

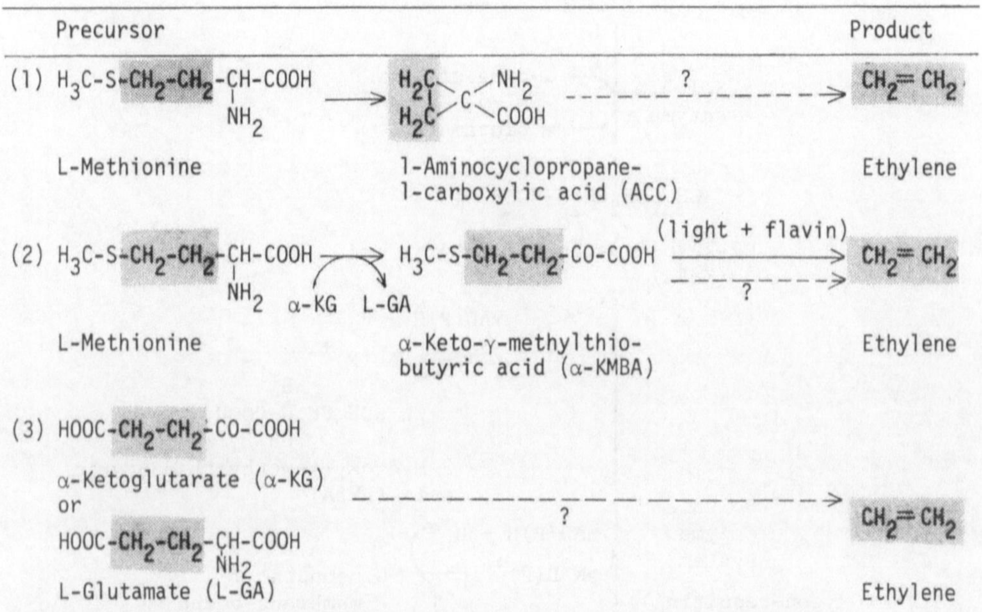

F. REGULATION OF ETHYLENE PRODUCTION BY *PENICILLIUM DIGITATUM*

Chalutz et al.[35] demonstrated that Pi is a potent inhibitor of ethylene production by shake cultures of *P. digitatum*. Mattoo et al.[36] followed the characterization of the phosphate-mediated control on ethylene production by *P. digitatum*. They suggested that phosphate level regulates ethylene production by this fungus and that the regulation involves a phosphorylation or dephosphorylation reaction of some enzyme system associated with ethylene production.

III. BIOSYNTHETIC PATHWAYS OF ETHYLENE IN MICROBES

A. TYPICAL PATHWAYS

During the last decade, microbial biosynthetic pathways of ethylene have gradually become clearer. The typical biosynthetic pathways of ethylene in microbes reported up to 1980 are summarized in Table 1. These are compared to the biosynthetic pathway of ethylene in higher plants.[37] Although ethylene-forming enzyme was not elucidated at that time, two precursors were known for microbial ethylene production. One is L-methionine[24,38] and the other class belongs to TCA cycle-related compounds.[39-42] Hereafter, we refer to the former as methionine-dependent ethylene-forming microbes and the latter as α-ketoglutarate-dependent ethylene-forming microbes. It was reported that *P. digitatum* is capable of producing ethylene from different substrates, depending on the type of medium and conditions of culture under which the fungus is grown. These substrates included acrylate,[39,40] isocitrate,[41] α-ketoglutarate and/or glutamate,[42] and even methionine.[43] The ethylene production in *P. digitatum* was reviewed in detail by Lieberman in 1979.[44]

B. METHIONINE-DEPENDENT ETHYLENE-FORMING MICROBES

1. Biosynthesis of Ethylene by *Escherichia coli*

Primrose proposed[26] that 2-keto-4-methylthiobutyric acid (KMBA) is an intermediate in

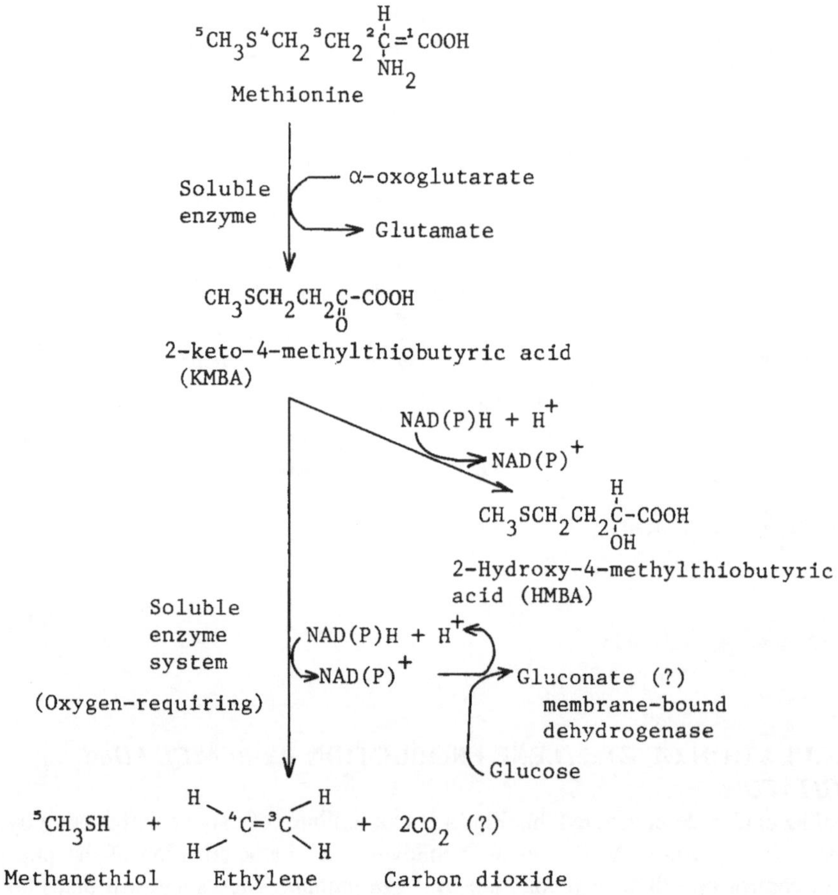

FIGURE 1. Pathway of the formation of ethylene by *Escherichia coli* strain B SPAO. (Data from Ince, J. E. and Knowles, C. H., *Arch. Microbiol.*, 46, 151, 1986.)

ethylene formation and suggested two possible mechanisms for its conversion to ethylene: (1) by the action of light and excreted flavins, and (2) by an enzymatic process, possibly via a peroxidase. Billington at el.[38] described methods for identifying KMBA and showed that KMBA is a common metabolic product of microorganisms such as *Escherichia, Pseudomonas, Bacillus, Acinetobacter, Aeromonas, Rhizobium,* and *Corynebacterium* when grown in the presence of methionine. Ince and Knowles reported[27,28] ethylene formation from methionine by the cell-free extracts of *E. coli.* KMBA was converted to ethylene, methanethiol, and probably carbon dioxide by a soluble enzyme system requiring the presence of NAD(P)H, Fe^{3+} chelated to EDTA, and oxygen. Their proposed pathway for ethylene formation from methionine is shown in Figure 1. They suggested that ethylene formation by other bacteria might generally occur via the route seen in *E. coli.* However, the ethylene-forming enzyme was not isolated.

2. Biosynthesis of Ethylene by *Cryptococcus albidus*

Fukuda at el.[11,13] selected *Cryptococcus albidus* IFO 0939 as one of the typical methionine-dependent ethylene-forming microbes and constructed a cell-free ethylene-forming system.[11] This system resembled the Ince and Knowles's cell-free system. The standard reaction mixture consisted of 3 m*M* KMBA, 0.5 m*M* NADH, 0.4 m*M* Fe(III) EDTA, and cell extract in 20 m*M* potassium phosphate buffer, pH 6.5. Fukuda et al.[13] purified the ethylene-forming enzyme 1,587-fold from the cell-free extract to a specific activity of

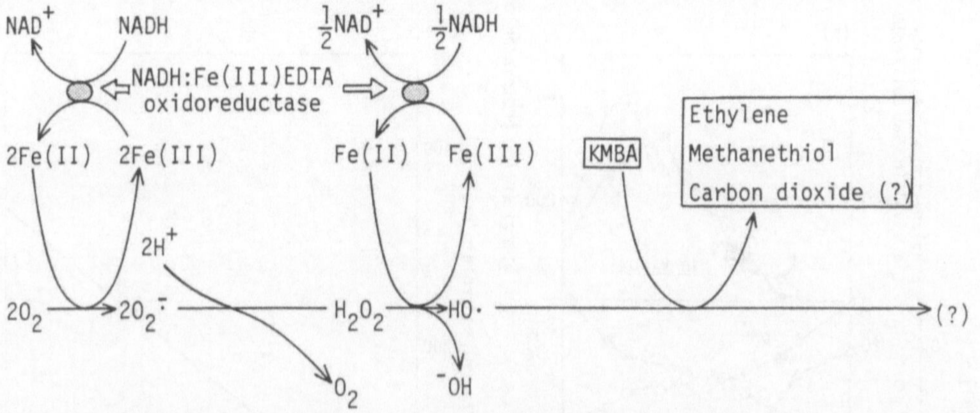

FIGURE 2. Hypothetical mechanism for the formation of ethylene from KMBA by *Cryptococcus albidus* IFO 0939. (Data from Fukuda, H. et al., *FEMS Microbiol. Lett.*, 60, 107, 1989.)

8,250 μl mg protein^{-1} hr^{-1} with a yield of 1.7%. The homogeneity of the enzyme was confirmed by polyacrylamide gel electrophoresis. The ethylene-forming enzyme migrated as a single protein band in 7.5% polyacrylamide gel electrophoresis and SDS-polyacrylamide gel electrophoresis. The molecular weight of the ethylene-forming enzyme was estimated to be 56,000 by gel filtration and 62,000 by SDS-polyacrylamide gel electrophoresis. These results demonstrate that the ethylene-forming enzyme from *C. albidus* IFO 0939 is a monomeric protein. The properties of the purified ethylene-forming enzyme were also studied by these authors.[13] They concluded that this ethylene-forming enzyme is an NADH:Fe(III)EDTA oxidoreductase, which reduces 2 mol of Fe(III)EDTA with 1 mol of NADH to give 2 mol of Fe(II)EDTA and 1 mol of NAD^+ under anaerobic conditions at pH 6.5. From these results, Fukuda and Ogawa proposed a mechanism for ethylene formation from methionine, shown in Figure 2. The first step involves an NADH-dependent reduction of Fe(III) to Fe(II) catalyzed by an NADH:Fe(III)EDTA oxidoreductase. Oxidation of Fe(II)EDTA by molecular oxygen (Equation 1) yields superoxide radical anion ($O_2^{\cdot-}$), which can undergo the dismutation reaction (Equation 2) to form hydrogen peroxide. Hydrogen peroxide in turn can react with Fe^{2+} via the Fenton reaction (Equation 3), generating the hydroxyl radical ($OH^{\cdot}$).

$$Fe^{2+} + O_2 \rightarrow Fe^{3+} + O_2^{\cdot-} \quad (1)$$

$$2O_2^{\cdot-} + 2H^+ \rightarrow H_2O_2 + O_2 \quad (2)$$

$$Fe^{2+} + H_2O_2 \rightarrow Fe^{3+} + OH^{\cdot} + OH^- \quad (3)$$

It had previously been established that the hydroxyl radicals can serve as oxidizing agents in the oxidation of KMBA to ethylene.[45,46] Although some enzymes that catalyze Fe(III) to Fe(II) are known,[47] all the enzymes reported are membrane bound and of the oxidase type. Since the purified ethylene-forming enzyme by *C. albidus* IFO 0939 is neither membrane bound nor an oxidase, it seems to be a new type. The NADH:Fe(III) oxidoreductase may act as an Fe(III)-Fe(II) recycling enzyme *in vivo*. It should be noted that ethylene production requires the existence of a specific transaminase that catalyzes the formation of KMBA from L-methionine. It will be of interest to know whether or not an NADH:Fe(III)EDTA oxidoreductase is present in microbes that do not produce ethylene even after the addition of L-methionine.

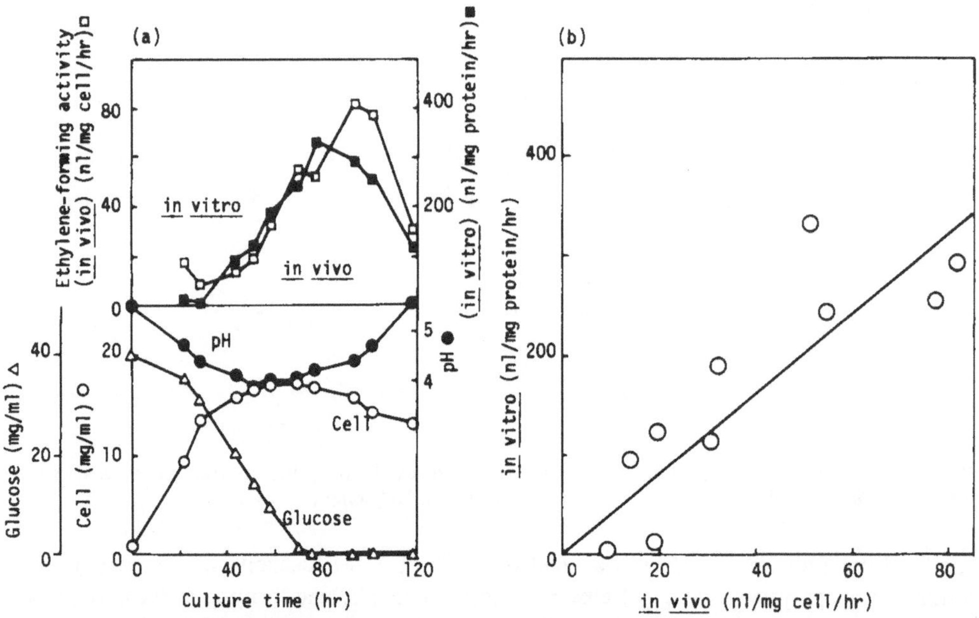

FIGURE 3. Time courses of ethylene-forming activities of living cell and cell-free system: (○) cell concentration, (△) glucose concentration, (●) pH, (□) ethylene-forming activity of the living cell system, (■) ethylene-forming activity of the cell-free system. (Data from Fukuda H., Fujii, T., and Ogawa, T., *Agric. Biol. Chem.*, 50, 977, 1986.)

C. α-KETOGLUTARATE-DEPENDENT ETHYLENE-FORMING MICROBES

1. Biosynthesis of Ethylene by *Penicillium digitatum*

P. digitatum, the green mold of citrus fruit, produces large quantities of ethylene. Chou and Yang elegantly demonstrated[42] that the carbon unit of ethylene originated from carbons 3 and 4 of glutamate or corresponding α-ketoglutarate in ^{14}C-labeling experiments. In 1986, Fukuda et al.[5] reported a partial purification of an ethylene-forming enzyme isolated from *P. digitatum* IFO 9372 by means of DEAE-Sepharose CL-6B ion exchange chromatography. Using this system as a tool for elucidation of the biochemical basis of ethylene formation, the immediate precursor of ethylene was found to be α-ketoglutarate. Interestingly, the enzyme system required L-arginine and the ferrous ion in reduced form (with reducing agents such as dithiothreitol [DTT]) and oxygen as essential factors. A comparison of the time courses of ethylene-forming activity of the cell-free system and that of the living system (*in vivo*) of the fungus is shown in Figure 3. Clearly, the two time courses are similar. Fukuda et al.[12] therefore concluded that this cell-free ethylene-forming system operates in living cells. This ethylene-forming enzyme has been purified from *P. digitatum* IFO 9372. The purified enzyme was confirmed to be homogenous on polyacrylamide disc gel electrophoresis, and the molecular weight of the enzyme was estimated to be about 42,000 by gel filtration and SDS-gel electrophoresis, suggesting that the enzyme activity is associated with a single protein. The specific substrate and cofactor of the purified enzyme were α-ketoglutarate and L-arginine, respectively.

The effects of a number of inhibitors and agents on the ethylene-forming activity were examined.[12] The enzyme was strongly inhibited by chelating reagents, EDTA and Tiron. This suggested that a certain Fe(II) complex may be required for the enzyme activity. Divalent transition metals, Co(II), Cu(II), and Mn(II) inhibited the ethylene-forming activity to a certain extent, probably by competing with Fe(II) in the formation of such a complex. Some SH-groups in the enzyme were suggested to play an important role in the enzyme activity, as the enzyme was inhibited by 5,5′-dithio-bis(2-nitrobenzoate) (DTNB). Superoxide and hydrogen peroxide did not seem to be involved in the enzymatic reaction since superoxide dismutase and catalase failed to suppress the reaction. The involvement of the hydroxyl radical also seems unlikely, since mannitol and Na-benzoate exerted little inhibition. These results imply that the activation of molecular oxygen occurs via direct coordination of Fe(II), which may be loosely bound to the enzyme through SH-groups. Other free radical inhibitors, such as propyl gallate, diazabicyclooctane, and hydroquinone, suppressed the reaction considerably. These effects, however, may not necessarily be attributable to the radical scavenging properties of these agents. They may instead act as chelating agents for Fe(II), and thus suppress the reaction in a similar manner to EDTA.

Figure 4 shows a proposed mechanism for ethylene formation by *P. digitatum*. Fe(II) can have six coordination numbers. Three of them may combine with the ethylene-forming enzyme, 2 with α-ketoglutarate, and 1 with oxygen. L-Arginine may act in the reaction between Fe(II) and enzyme protein. The intermediate (see Figure 4 [3]) is very unstable, especially –O–O– bonds, and, as a result, ethylene and carbon dioxide are released from α-ketoglutarate. The intermediate (4) produces 2 mol of carbon dioxide and then returns to the original state (1). Consequently, one mol of ethylene and three mol of carbon dioxide are produced from one mol of α-ketoglutarate as shown in the figure. It was observed that the molar ratio between ethylene formation and CO_2 release was 1:3.6 using [U-^{14}C] α-ketoglutarate (unpublished data).

It is interesting to compare the properties of this ethylene-forming enzyme with those of the *in vivo* ethylene-forming system. Both systems require oxygen and are similarly inhibited by a variety of inhibitors and other agents except that the former specifically requires L-arginine. The reason why the ethylene-forming enzyme of higher plants has not been isolated may be the lack of knowledge of an indispensable cofactor such as L-arginine.

The ethylene-forming enzyme of *P. digitatum* has similar properties to those of an α-ketoglutarate-dependent dioxygenase[48] with respect to the requirement for α-ketoglutarate, Fe(II), oxygen, and a reducing agent such as DTT and/or ascorbate. Table 2 shows the similarity between the ethylene-forming enzyme of *P. digitatum* and α-ketoglutarate-dependent dioxygenase. The dioxygenase needs another substrate, viz. propyl-glycyl peptide, beside α-ketoglutarate. Therefore, the ethylene-forming enzyme of *P. digitatum* is a previously undescribed enzyme.

2. Biosynthesis of Ethylene by *Penicillium cyclopium*

In order to survey whether or not there are microbes other than *P. digitatum* that produce ethylene via α-ketoglutarate, a screening of α-ketoglutarate-dependent ethylene-forming fungi was conducted. Ethylene production was examined in 70 strains from IFO and 23 fungi isolated from soil. Table 3 shows the results of the *in vitro* ethylene-forming activity. Beside *P. digitatum*, one strain of *P. cyclopium* (KIT 0229), which was newly isolated from soil, was shown to exhibit the highest α-ketoglutarate-dependent ethylene-forming activity. The ethylene-forming enzyme from *P. cyclopium* KIT 0229 was partially purified and its properties were shown to be similar to those of *P. digitatum*.

3. Biosynthesis of Ethylene by *Pseudomonas syringae*

Goto et al.[29] and Goto and Hyodo[30] studied the production of ethylene from many pathovars of plant pathogenic bacteria. A significant amount of ethylene was produced by *Pseudomonas solanacearum*, *P. syringae* pv. *phaseolicola* (Kudzu strains isolated from *Pueraria lobata*), and *Erwinia rhapontici*. Among these strains, the Kudzu strain of *P. syringae* pv. *phaseolicola* was selected and its *in vivo* ethylene production was studied. They constructed a cell-free ethylene-forming system, in which α-ketoglutarate served the substrate and Fe^{2+}, DTT, histidine, and oxygen were essential for the ethylene formation. It should

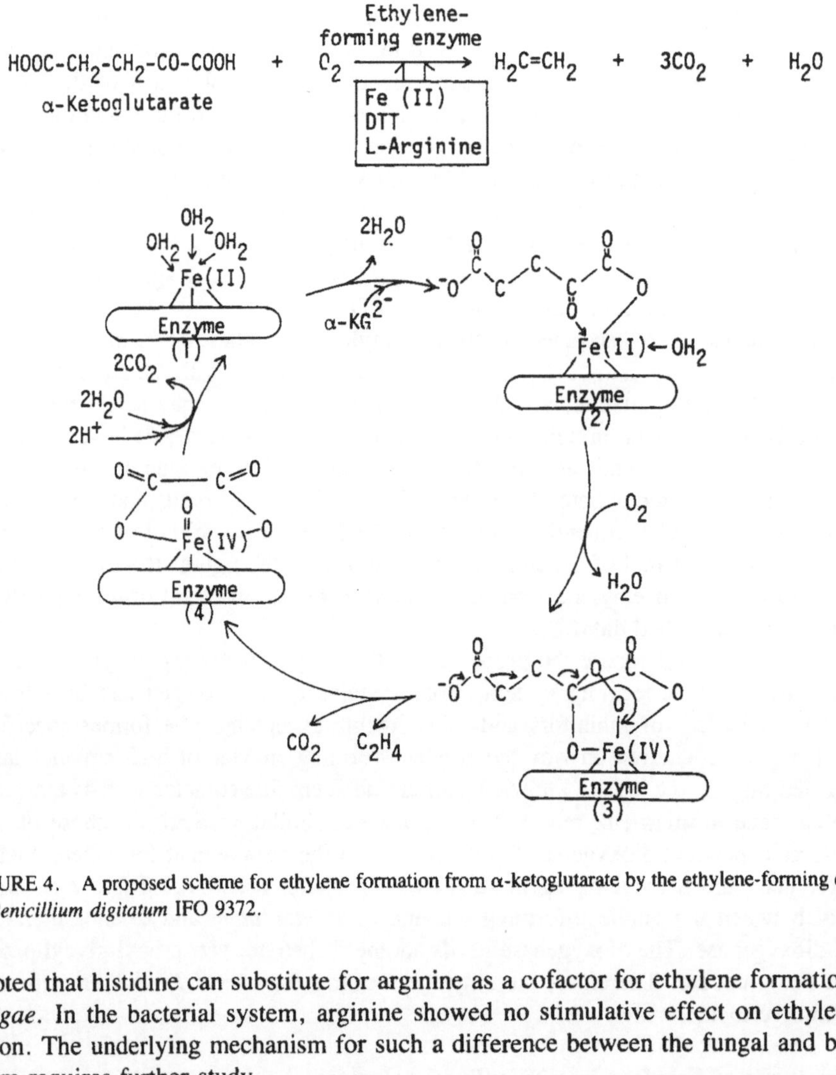

FIGURE 4. A proposed scheme for ethylene formation from α-ketoglutarate by the ethylene-forming enzyme of *Penicillium digitatum* IFO 9372.

be noted that histidine can substitute for arginine as a cofactor for ethylene formation in *P. syringae*. In the bacterial system, arginine showed no stimulative effect on ethylene production. The underlying mechanism for such a difference between the fungal and bacterial system requires further study.

IV. SIGNIFICANCE OF MICROBIAL ETHYLENE IN PLANT DEVELOPMENT

The possible involvement of ethylene gas in pathogenesis has been suggested and extensively studied[49] (see Chapter 16). These studies have shown that low concentrations of ethylene cause changes in host metabolism, such as increased enzyme activity and formation of antifungal compounds associated with plant resistance.[50-53] On the other hand, ethylene is known to cause some disease symptoms[54] and to accelerate senescence,[55] and thus may be important in the predisposition of fruits and leaves to diseases. Chalutz[56] studied the role of ethylene in the pathogenicity of *P. digitatum* on citrus fruits such as oranges, grapefruits, and lemons with the normal ethylene-producing fungus strain and the non-ethylene-producing one. He found that the incubation time of the disease of citrus fruits caused by the ethylene-producing and nonproducing strains was similar. Based on these findings, it was concluded that ethylene has no clear role in the pathogenicity of *P. digitatum*. Therefore, there is, as yet, no conclusive evidence that ethylene directly affects host-parasite relationships.

TABLE 2
Comparison of Ethylene-Forming Enzyme and α-Ketoglutarate-Dependent Dioxygenase

Properties	Ethylene-forming enzyme (*P. digitatum*)	Prolyl hydroxylase	Lysyl hydroxylase
Purification	100 units[a]/OD_{280}	(1) 86—100 units/mg (from skin of newborn rat) (2) 540 units/mg (from chick embryo)	60 units/mg
Absorption spectrum	Simple protein	Simple protein	—
Molecular weight	Approx. 40,000 (monomer)	(1) 130,000 (two of 65,000 subunits) (2) 230,000 (tetramer)	550,000 200,000
Km			
Fe^{++}	0.005 m*M*	0.001 m*M* (rat) 0.005 m*M* (chick)	0.001 m*M*
α-KG	0.1 m*M*	0.01 m*M* (rat) 0.01 m*M* (chick)	0.05 m*M*
Substrate	0.1 m*M*	0.1 m*M* (rat) 0.006 m*M* (chick)	—
Oxygen	2.8 vol%	2.6—6.0 vol% (chick)	—
Reducing agent	DTT	Ascorbate	Ascorbate
Product of α-KG	C_2H_4, CO_2	Succinate, CO_2	Succinate, CO_2
Product of substrate	?	*trans*-4-Hydroxyproline	5-Hydroxylysine

[a] A unit of activity: the amount of enzyme which catalyzes the oxidation of 1 nmol of substrate per minute.

Data from Hayaishi, O., Nozaki, M., and Abbot, M. T., *The Enzyme*, Vol. 12, 3rd ed., Academic Press, New York, 1975, 152.

TABLE 3
Ethylene-Forming Activities of Various Strains Using Cell-Free Extracts (*in vitro*)

Strain		Ethylene-forming activity (nl/mg protein/hr)			
		None	+ α-KG	+ α-KG + Arg	α-KG Type
Cunninghamella elegans	IFO 4441	1.2	1.0	1.0	No
Chaetomium globosum	IFO 6347	2.5	8.5	15.3	Yes
Phycomyces nitens	IFO 9422	2.0	10.5	23.6	Yes
Penicillium cyclopium	KIT 0024	2.9	7.8	16.7	Yes
NI[a]	KIT 0127	5.0	25.5	54.2	Yes
Penicillium digitatum	KIT 0175	11.9	124.3	273.8	Yes
NI[a]	KIT 0180	0.3	0.1	0.3	No
Penicillium cyclopium	KIT 0229	11.2	117.8	247.8	Yes
Penicillium cyclopium	KIT 0400	10.5	48.7	234.5	Yes
Penicillium digitatum	IFO 9372	12.2	63.2	285.7	Yes

[a] NI, not identified.

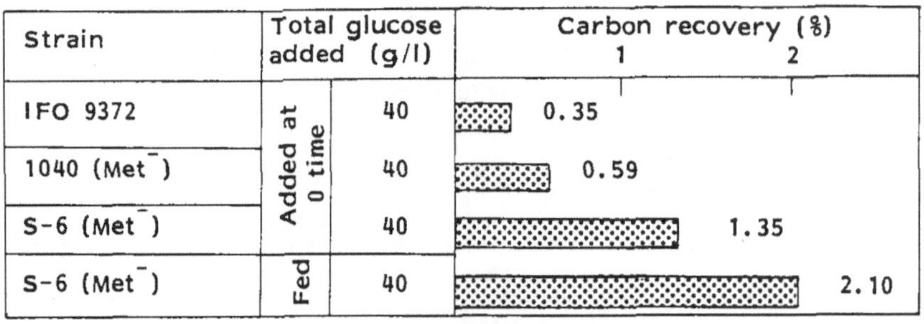

FIGURE 5. Recovery of carbon in ethylene produced by the three different variants of *Penicillium digitatum* IFO 9372. The recovery of carbon in ethylene was defined as the conversion % of carbon from glucose added to ethylene produced. (Data from Fukuda, H., Fujii, T., and Ogawa, T., *Biotechnol. Bioeng.*, 31, 620, 1988.)

V. BIOTECHNOLOGICAL EXPLOITATION OF MICROBIAL ETHYLENE PRODUCTION

A. BREEDING FOR ETHYLENE PRODUCTION IN *PENICILLIUM DIGITATUM*

There are three factors that may regulate the rate of ethylene production in *P. digitatum:* (1) α-ketoglutarate concentration, (2) L-arginine concentration, and (3) ethylene-forming enzyme. If the synthetic (or anabolic) reactions proceed more slowly in a mutant than in a parent strain, an increase of carbon flux from α-ketoglutarate to ethylene might be expected. Fukuda et al.[9] attempted to control the amount of carbon consumed by formation of daughter cells and respiration. In order to achieve this objective, an attempt was made to obtain mutants that form smaller colonies on agar plates than the parent. A methionine auxotroph, *P. digitatum* IFO 9372-1040 (Met^-) [hereafter 1040 (Met^-)] was induced by NTG (*N*-methyl-*N'*-nitro-*N*-nitrosoguanidine) treatment of the wild strain IFO 9372. Forty-six growth-suppressed mutants, which formed phenotypically smaller colonies on agar plates, were isolated from 1040 (Met^-). *P. digitatum* IFO 9372-1040-S-6 [hereafter S-6 (Met^-)] was the highest producer of ethylene among the strains tested. Figure 5 summarizes carbon recovery data under different culture conditions for the three strains of *P. digitatum* under investigation. The conversion of carbon from glucose to ethylene by S-6 (Met^-) was 4 and 2.3 times higher than that by IFO 9372 and 1040 (Met^-), respectively. If the S-6 (Met^-) strain was batch-fed with glucose in culture, the extent of conversion of carbon to ethylene increased to 2.1%.

B. ETHYLENE PRODUCTION FROM RENEWABLE RESOURCES BY *P. DIGITATUM*

Chalutz et al.[57] reported ethylene production by *P. digitatum*-infected citrus fruit peel. Fukuda et al. (unpublished data) observed cellulase activity in *P. digitatum* IFO 9372 and investigated ethylene production from the homogenate of recycled paper and unshiu fruit peel by *P. digitatum* IFO 9372. Figure 6 shows the time courses of ethylene production with renewable resources. The carbon recovery of ethylene from carbon source 10 g/l recycled paper and 40 g/l of the unshiu peel were 1.10% and 0.48%, respectively.

VI. CONCLUDING REMARKS

Two microbial ethylene-forming enzymes have been identified and purified to electrophoretic homogeneity state. However, there are still many problems that remain to be elucidated.

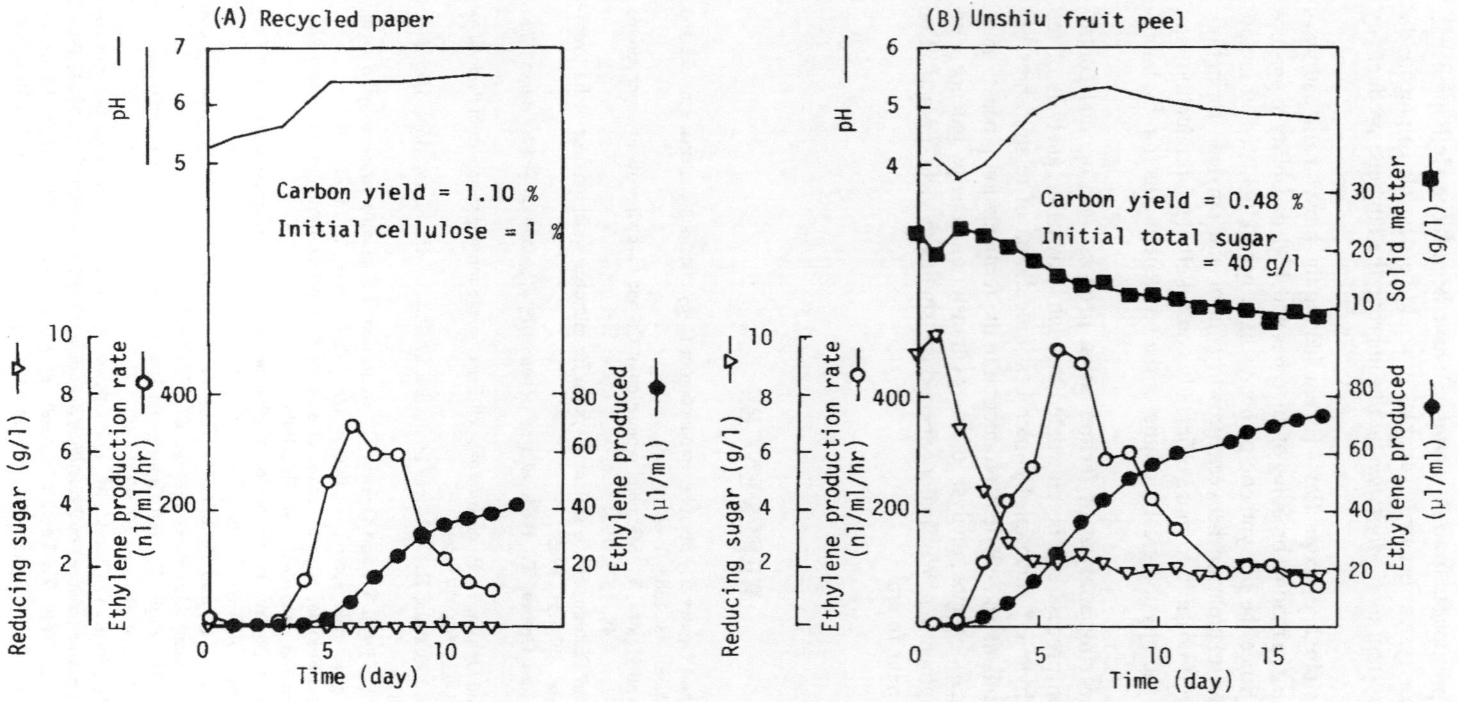

FIGURE 6. Ethylene production from *Penicillium digitatum* IFO 9372 using recycled paper or unshiu fruit peel (homogenate + 0.5 g/l $NaNO_3$) as carbon source.

1. Improvement in genetic breeding of ethylene-producing strains—By fully utilizing genetic engineering and other techniques, it should be possible to obtain strains with a production capacity 10 to 100 times higher than that of producers obtained thus far. It is worth mentioning that the protoplasts from *P. digitatum*[58] have been successfully prepared. If the fungus plasmid-vector system became available, it would be possible to prepare the protoplast transformation system of *P. digitatum* and to improve the ethylene productivity of this fungus.

2. Elucidation of physiological properties—By elucidating the biosynthetic pathways of ethylene in microorganisms, a rational breeding of microbes with high capability to produce ethylene will be possible. Since the biosynthetic pathways have been elucidated, it is now possible to characterize the relationship between growth control and ethylene formation. Future studies should make it possible to more efficiently control ethylene production, and should also help to clarify the physiological meaning of the ethylene production by microorganisms.

3. Biochemical engineering analysis of fermentation process—An optimal control method of the fermentation process can be elucidated through a dynamic analysis of the fermentation process of ethylene production by microbes, elucidation of relationships between growth of microorganisms and ethylene production in the fermentation process, analysis of mass transfer and mass balance, analysis of energy transfer and energy balance, and analysis of biochemical engineering. We finally expect these studies to yield a practical or feasible technology in the near future.

REFERENCES

1. **Fukuda, H., Fujii, T., and Ogawa T.,** Microbial production of C_2-hydrocarbons, ethane, ethylene and acetylene, *Agric. Biol. Chem.*, 48, 1363, 1984.
2. **Fukuda, H., Fujii, T., and Ogawa, T.,** Microbial production of C_3- and C_4-hydrocarbons under aerobic conditions, *Agric. Biol. Chem.*, 48, 1679, 1984.
3. **Fujii, T., Ogawa, T., and Fukuda, H.,** A screening system for microbes which produce olefin hydrocarbons, *Agric. Biol. Chem.*, 49, 651, 1985.
4. **Fukuda, H., Fujii, T., and Ogawa, T.,** Production of isobutene by *Rhodotorula* yeasts, *Agric. Biol. Chem.*, 49, 1541, 1985.
5. **Fukuda, H., Fujii, T., and Ogawa, T.,** Preparation of a cell-free ethylene-forming system from *Penicillium digitatum, Agric. Biol. Chem.*, 50, 977, 1986.
6. **Fujii, T., Ogawa, T., and Fukuda, H.,** Isobutene production by *Rhodotorula minuta, Appl. Microbiol. Biotechnol.*, 25, 430, 1987.
7. **Fukuda, T., Kawaoka, Y., Fujii, T., and Ogawa, T.,** Production of a gaseous saturated-hydrocarbon mixture by *Rhizopus japonicus* under aerobic conditions, *Agric. Biol. Chem.*, 51, 1529, 1987.
8. **Fujii, T., Ogawa, T., and Fukuda, H.,** Preparation of a cell-free, isobutene-forming system from *Rhodotorula minuta, Appl. Environ. Microbiol.*, 54, 583, 1988.
9. **Fukuda, H., Fujii, T., and Ogawa, T.,** Production of ethylene by a growth-suppressed mutant of *Penicillium digitatum, Biotechnol. Bioeng.*, 31, 620, 1988.
10. **Fujii, T., Ohga, Y., Ogawa, T., and Fukuda, H.,** The role of L-phenylalanine on the production of isobutene by *Rhodotorula minuta, J. Ferment. Bioeng.*, 67, 115, 1989.
11. **Fukuda, H., Takahashi, M., Fujii, T., and Ogawa, T.,** Ethylene production from L-methionine by *Cryptococcus albidus, J. Ferment. Bioeng.*, 67, 173, 1989.
12. **Fukuda, H., Kitajima, H., Fujii, T., Tazaki, M., and Ogawa, T.,** Purification and some properties of a novel ethylene-forming enzyme produced by *Penicillium digitatum, FEMS Microbiol. Lett.*, 59, 1, 1989.
13. **Fukuda, T., Takahashi, M., Fujii, T., Tazaki, M., and Ogawa, T.,** An NADH:Fe(III)EDTA oxidoreductase from *Cryptococcus albidus:* an enzyme involved in ethylene production *in vivo?, FEMS Microbiol. Lett.*, 60,107, 1989.
14. **Fujii, T., Ogawa, T., and Fukuda, H.,** A membrane-bound isobutene-forming enzyme from *Rhodotorula minuta, FEMS Microbiol. Lett.*, 58, 199, 1989.
15. **Fujii, T., Ogawa, T., and Fukuda, H.,** The relationship between isobutene-forming activity and a cytochrome P-450 in *Rhodotorula minuta, J. Ferment. Bioeng.*, 68, 174, 1989.

16. **Biale, J. B.,** Effect of emanations from several species of fungi on respiration and colour development of citrus fruits, *Science*, 91, 458, 1940.
17. **Miller, E. V., Winston, J. R., and Fisher, D. F.,** Production of epinasty by emanations from normal and decaying citrus fruits and from *Penicillium digitatum, J. Agric. Res.*, 60, 269, 1940.
18. **Davis, J. B. and Squires, R. M.,** Detection of microbially produced gaseous hydrocarbons other than methane, *Science*, 119, 381, 1954.
19. **Smith, K. A. and Russell, R S.,** Occurrence of ethylene, and its significance, in anaerobic soil, *Nature*, 222, 769, 1969.
20. **Oremland, R. S.,** Microbial formation of ethane in anoxic estuarine sediments, *Appl. Environ. Microbiol.*, 42, 122, 1981.
21. **Gollakota, K. G. and Jayalakshmi, B.,** Biogas (natural gas?) production by anaerobic digestion of oil cake by a mixed culture isolated from cow dung, *Biochem. Biophys. Res. Commun.*, 110, 32, 1983.
22. **Freebairn, H. T. and Buddenhagen, I. W.,** Ethylene production by *Pseudomonas solanacearum, Nature*, 202, 313, 1964.
23. **Primrose, S. B. and Dilworth, M. J.,** Ethylene production by bacteria, *J. Gen. Microbiol.*, 93, 177, 1976.
24. **Primrose, S. B.,** Formation of ethylene by *Escherichia coli, J. Gen. Microbiol.*, 95, 159, 1976.
25. **Primrose, S. B.,** Ethylene-forming bacteria from soil and water, *J. Gen. Microbiol.*, 97, 343, 1976.
26. **Primrose, S. B.,** Evaluation of the role of methional, 2-keto-4-methylthiobutyric acid and peroxidase in ethylene formation by *Escherichia coli, J. Gen. Microbiol.*, 98, 519, 1977.
27. **Ince, J. E. and Knowles, C. J.,** Ethylene formation by cultures of *Escherichia coli, Arch. Microbiol.*, 141, 209, 1985.
28. **Ince, J. E. and Knowles, C. J.,** Ethylene formation by cell-free extracts of *Escherichia coli, Arch. Microbiol.*, 146, 151, 1986.
29. **Goto, M. Y., Ishida, Y., Takikawa, Y., and Hyodo, H.,** Ethylene production by the Kudzu strains of *Pseudomonas syringae* pv. *phaseolicola* causing halo blight in Pusraria Lobata (Wild) Ohwi, *Plant Cell Physiol.*, 26, 141, 1985.
30. **Goto, M. and Hyodo, H.,** Ethylene production by cell-free extract of the Kudzu strains of *Pseudomonas syringae* pv. *phaseolicola, Plant Cell Physiol.*, 28, 405, 1987.
31. **Sato, M., Urushizaki, S., Nishiyama, K., Sakai, F., and Ohta, Y.,** Efficient production of ethylene by *Pseudomonas syringae* pv *glycinea* which causes halo blight in soybeans, *Agric. Biol. Chem.*, 51, 1177, 1987.
32. **Thomas, K. C. and Spencer, M.,** L-Methionine as an ethylene precursor in *Saccharomyces cerevisiae, Can. J. Microbiol.*, 23, 1669, 1977.
33. **Nickerson, W. J.,** Ethylene as a metabolic product of the pathogenic fungus *Blastomyces dermatitidis, Arch. Biochem.*, 17, 225, 1948.
34. **Ilag, L. and Curtis, R.,** Production of ethylene by fungi, *Science*, 159, 1357, 1968.
35. **Chalutz, E., Mattoo, A. K., Anderson, J. D., and Lieberman, M.,** Regulation of ethylene production by phosphate in *Penicillium digitatum, Plant Cell Physiol.*, 19, 189, 1978.
36. **Mattoo, A. K., Chalutz, E., Anderson, J. D., and Lieberman, M.,** Characterization of the phosphate-mediated control of ethylene production by *Penicillium digitatum, Plant Physiol.*, 64, 55, 1979.
37. **Adams, D. O. and Yang, S. F.,** Ethylene biosynthesis: identification of 1-aminocyclopropane-1-carboxylic acid as an intermediate in the conversion of methionine to ethylene, *Proc. Natl. Acad. Sci. U.S.A.*, 76, 170, 1979.
38. **Billington, D. C., Golding, B. T., and Primrose, S. B.,** Biosynthesis of ethylene from methionine, *Biochem. J.*, 182, 827, 1979.
39. **Wang, C. H., Persyn, A., and Krackov, J.,** Role of the Krebs cycle in ethylene biosynthesis, *Nature*, 195, 1306, 1962.
40. **Jacobsen, D. W. and Wang, C. H.,** The biogenesis of ethylene in *Penicillium digitatum, Plant Physiol.*, 43, 1959, 1968.
41. **Ketring, D. L., Young, R. E., and Biale, J. B.,** Effect of monofluoroacetate on *Penicillium digitatum* metabolism and on ethylene biosynthesis, *Plant Cell Physiol.*, 9, 617, 1968.
42. **Chou, T. W. and Yang, S. F.,** The biogenesis of ethylene in *Penicillium digitatum, Arch. Biochem. Biophys.*, 157, 73, 1973.
43. **Chalutz, E., Lieberman, M., and Sisler, H. D.,** Methionine-induced ethylene production by *Penicillium digitatum, Plant Physiol.*, 60, 402, 1977.
44. **Lieberman, M.,** Biosynthesis and action of ethylene, *Annu. Rev. Plant. Physiol.*, 30, 533, 1979.
45. **Diguiseppi, J. and Fridovich, I.,** Ethylene from 2-keto-4-thiomethyl butyric acid: the Haber-Weiss reaction, *Arch. Biochem. Biophys.*, 205, 323, 1980.
46. **Tauber, A. and Babior, B. M.,** Evidence for hydroxyl radical production by human neutrophils, *J. Clin. Invest.*, 60, 374, 1980.

47. **Halliwell, B. and Gutteridge, J. M. C.,** Role of iron in oxygen radical reactions, in *Methods in Enzymology*, Vol. 105, Packer, L., Ed., Academic Press, New York, 1984, 47.
48. **Hayaishi, O., Nozaki, M., and Abbott, M.,** α-Ketoglutarate dioxygenases, in *The Enzymes*, Vol. 12, 3rd ed., Boyer, P. D., Ed., Academic Press, New York, 1975, 151.
49. **Pegg, C. F.,** The involvement of ethylene in plant pathogenesis, in *Encyclopedia of Plant Physiology*, New Ser., *Physiological Plant Pathology*, Vol. 4, Springer-Verlag, Berlin, 1976, 582.
50. **Chalutz, E.,** Ethylene-induced phenylalanine ammonia lyase activity in carrot roots, *Plant Physiol.*, 51, 1033, 1973.
51. **Chalutz, E. and Stahmann, M. A.,** Induction of pisatin by ethylene, *Phytopathol.*, 59, 1972, 1969.
52. **Chalutz, E., Devay, J. E., and Maxie, E. C.,** Ethylene-induced isocoumarin formation in carrot root tissue, *Plant Physiol.*, 44, 235, 1969.
53. **Stahmann, M. A., Clare, B. G., and Woodbury, W.,** Increased disease resistance and enzyme activity induced by ethylene and ethylene production by black rot infected sweet potato tissue, *Plant Physiol.*, 41, 1505, 1966.
54. **Diamond, A. E. and Waggoner, P. E.,** The cause of epinastic symptoms in Fusarium wilt of tomatoes, *Phytopathol.*, 43, 663, 1953.
55. **Abeles, F. B.,** *Ethylene in Plant Biology*, Academic Press, London, 1973, 302.
56. **Chalutz, E.,** No role for ethylene in the pathogenicity of *Penicillium digitatum*, *Physiol. Plant Pathol.*, 14, 259, 1979.
57. **Chalutz, E., Kapulnik, E., and Chet, I.,** Fermentative production of ethylene by *Penicillium digitatum* from citrus fruit peel, *Eur. J. Appl. Microbiol. Biotechnol.*, 18, 293, 1983.
58. **Hou, H. H. and Jong, S. C.,** Protoplast formation from mycelia of *Penicillium digitatum* by cell wall-lytic enzymes of *Trichoderma harzianum*, *J. Ferment. Technol.*, 63, 189, 1985.

Chapter 16

ETHYLENE IN PATHOGENESIS AND DISEASE RESISTANCE

Thomas Boller

TABLE OF CONTENTS

I. INTRODUCTION

Plant tissues produce increased amounts of ethylene, so called stress ethylene, in response to various stress factors, as discussed in Chapter 3 and in other reviews.[2,3] This chapter deals with the generation of stress ethylene and its significance in one specific instance, in plant pathogenesis. A separate treatment of ethylene with regard to plant-pathogen interactions has also been covered in earlier reviews[1,4-7] and may be justified for two reasons. First, both plants and pathogens contribute to the production of ethylene in diseased plants, a factor that renders the analyses and interpretations difficult. Second, the possible functions of stress ethylene have been studied particularly extensively in relation to pathogenesis and disease resistance. However, concentration on a single factor is also somewhat dangerous because it tends to restrict the view to direct cause-effect relationships between pathogens, ethylene, and defense responses. Broader concepts, such as the possibility that ethylene is a tonic factor stimulating the "awareness" of plant cells to their environment,[1] recede into the background. In any case, when the author attempts to keep the field of vision open, it is the reader who must look around and use it.

This chapter starts with a discussion of plant ethylene production in response to pathogens (Section II). The studies in this field demonstrate that enhanced ethylene production is an early, active response of plants to the perception of a pathogen attack. This leads to the question whether pathogen-induced stress ethylene has an adaptive function as a signal to induce defense reactions against pathogens. Many of the developmental and biochemical alterations in plant tissues occurring in response to a pathogen attack are believed to represent defense reactions. Some but not all of these defense reactions can also be induced by exogenous ethylene, indicating that ethylene has a potential to act as a (partial) signal for defense (Section III). The role of ethylene in resistance or susceptibility of a given plant to a given pathogen is a different matter; ethylene can increase or reduce disease incidence, depending on the plant-pathogen interaction (Section IV). Although this may be confusing at first glance, it is not too unexpected since well-adapted pathogens may overcome the defense reactions of their hosts or even use them to their advantage. The fact that some pathogens produce their own ethylene could possibly be explained in this way as well (Section V). The chapter ends by summarizing the interesting implications and the many open questions with regard to ethylene in pathogenesis (Section VI).

II. ETHYLENE PRODUCTION BY PLANTS IN REACTION TO PATHOGENS

The enhanced formation of ethylene is an early biochemical event in many plant-pathogen interactions.[1-7] In virus-infected plants, this clearly represents ''stress ethylene'' produced by the plant; viruses lack the capacity to produce ethylene. In bacterial or fungal diseases, the situation is more complicated because some pathogens have a capacity to produce ethylene themselves. Nevertheless, most of the ethylene synthesized in bacterial or fungal diseases is generally also of plant origin. This may be concluded indirectly from experiments in which plants are treated with bacterial or fungal components, so called elicitors, which induce both defense responses and stress ethylene formation. As summarized below for the various organisms causing disease as well as for elicitors, in all cases investigated ethylene is generated through the ''normal'' biosynthetic pathway, i.e., via 1-aminocyclopropane-1-carboxylic acid (ACC). Wherever this has been studied, the induction of ACC synthase is a key event in the formation of stress ethylene. Thus pathogen-induced stress ethylene production by the plant is under strict enzymatic control at the level of ACC synthase, as was first found for mechanically wounded plants[8] and later for many other situations.[3] Generally, ethylene biosynthesis is induced in cases where a plant actively ''recognizes'' a pathogen attack.[1]

A. VIRUSES

Virus-induced ethylene production was first reported in 1951 for several host-virus interactions causing local necrotic lesions[9] and has been studied extensively in tobacco plants infected with tobacco mosaic virus (TMV).[9-15] Tobacco plants without the *N* gene are susceptible to multiplication and systemic spread of TMV and do not form ethylene upon infection.[10-12] Tobacco cultivars which carry the *N* gene are resistant to the systemic spread of TMV and react hypersensitively to the virus, i.e., they form small local necrotic flecks at the infection sites. These plants show a surge of ethylene production about 48 h after mechanical infection with a virus suspension, just before the first symptoms of the hypersensitive response are seen.[12,13] De Laat et al.[12] found that the ethylene produced in tobacco plants reacting hypersensitively to TMV was derived from methionine and that ACC accumulated at the time of increased ethylene production.[13] They were unable to detect ACC synthase *in vitro* but devised an assay for ACC synthase activity *in vivo* by measuring ACC production in living leaves under a nitrogen atmosphere, which blocked conversion of ACC to ethylene. ACC accumulation in this assay was strongly increased in the course of the hypersensitive reaction.[13] Although measurement of ACC synthase *in vivo* is problematic in hindsight, since it does not address the possibility of ACC conjugation, the whole study clearly established increased ACC formation as a prime cause of increased ethylene production in the hypersensitive reaction.

Is stress ethylene production in the hypersensitive response an early reaction preceding necrosis, or is it induced in cells adjacent to the first dying cells as a type of wound ethylene comparable to mechanically induced wound ethylene? The temporal sequence of reactions is difficult to establish in the course of a normal virus infection because the first necrotic lesions appear stochastically and unpredictably in the tissue. To study the temporal sequence in more detail, De Laat and Van Loon[14,15] exploited the temperature sensitivity of the *N* gene: Samsum *NN* plants infected with TMV at 30°C show systemic spread and the usual mosaic symptoms of susceptible plants. When such plants were shifted back to 20°C, they showed a rapid development of necrotic lesions over all infected plant parts.[14] In this situation, ethylene production had already started to increase after 2 h and clearly preceded the appearance of necrotic lesions, which became visible after about 8 h.[14] When plants were inoculated at 20°C and shifted to 30°C at different times later, it was found that both the induction of ethylene biosynthesis and the formation of lesions were initiated between 15 and 18 h after inoculation,[15] indicating a close relation between lesion formation and ethylene formation.

B. BACTERIA

1. Stimulation of Plant Ethylene Production by Bacteria

An early report of ethylene production in a bacterial disease, banana fruit rot, described a case where the bacterium *Pseudomonas solanacearum* was capable of producing ethylene in pure culture.[16] Subsequently, it was found that bacteria can induce ethylene production in the plant. *Erwinia carotovora,* a pathogen that does not produce ethylene,[17] and a cell-free culture fluid from this bacterium containing pectolytic enzymes caused marked ethylene production in cauliflower florets.[18] Ethylene appeared to be derived from methionine, indicating that the "normal" biosynthetic pathway for ethylene was induced also in this case.[18] However, it remains to be seen at what step ethylene biosynthesis is induced.

Carrot discs infected with *Agrobacterium tumefaciens* or *Agrobacterium rhizogenes* produce high levels of ethylene.[19] Enhanced ethylene production appears to be a secondary result of auxin synthesis in the plant tissue due to transformation of the plants with the *tms* genes contained in the *Agrobacterium* T-DNA.[20,21]

2. Bacterial Inhibitors of Plant Ethylene Biosynthesis

In view of the possible function of ethylene as a signal inducing defense reactions, it is interesting that a potent inhibitor of plant ethylene biosynthesis, rhizobitoxine, has been isolated from plant-inhabiting bacteria.[22-25] This chlorosis-inducing toxin, first discovered in a *Bradyrhizobium* strain isolated from soybean,[22] was initially found to be an irreversible inhibitor of β-cystathionase, an enzyme of the methionine biosynthetic pathway.[26,27] Later, at low concentrations, it was shown to inhibit ethylene biosynthesis from methionine.[28] An analog of the compound, aminoethoxyvinylglycine (AVG), has been shown to be a competitive inhibitor of ACC synthase.[29] Although the function of rhizobitoxine in *Bradyrhizobium* is not known it is interesting that ethylene has been found to reduce nodulation in some Fabaceae.[30-32] Direct tests for a nodulation-promoting function of rhizobitoxine have yielded conflicting results. An early report that rhizobitoxine-producing strains showed enhanced colonization and nodulation on certain soybean cultivars[33] could not be verified on closer scrutiny.[34,35] However, the rhizobitoxine analog, AVG, increased nodule formation on *Vicia sativa*[36] and on *Medicago sativa*[37] by *Rhizobium leguminosarum* and *Rh. meliloti,* respectively, two *Rhizobium* species not known to produce rhizobitoxine.[34] It is interesting that rhizobitoxine also occurs in a plant pathogenic bacterium, *Pseudomonas andropogonis.*[24,25] It might be speculated that inhibition of ethylene biosynthesis is the important function of the toxin for pathogens as well as for symbionts, while its chlorosis-inducing effect, which is not related to inhbition of ethylene biosynthesis,[28] is a mere side effect.

C. FUNGI

In 1950, Williamson[38] described ethylene as a metabolic product of plants infected with various fungal diseases. Only one of the pathogens he investigated, *Penicillium digitatum*, produced ethylene in pure culture. He concluded that fungal infection generally caused increased ethylene production by the plant. Later studies reached similar conclusions.[39]

Induction of ethylene formation in the course of infection can be measured so easily and rapidly that ethylene evolution of germinating, infected barley grains has been suggested as a simple screening test to evaluate the efficiency of fungicides.[40]

The induction of ethylene formation in pea pods by compatible and incompatible pathogens could be blocked with AVG, indicating that pathogen-induced ethylene biosynthesis depended on ACC synthase.[41] In a more recent study, the pathway of ethylene biosynthesis was examined in tomato leaves infected by *Phytophthora infestans*.[42] Infected leaves reacted hypersensitively and started to produce large amounts of ethylene just before the necrotic lesions became visible.[42] In this system, it was possible to measure ACC synthase *in vitro*. Induction of this enzyme appeared to be the prime factor in the induction of ethylene formation, although the ethylene-forming enzyme (measured *in vivo*) was also increased about threefold.[42]

D. ELICITORS

To circumvent the conceptual problem that a pathogen itself may contribute to the ethylene generated in the diseased plant, induction of ethylene biosynthesis has been examined in plant tissues treated with cell-free components of pathogens, so called elicitors.

Paradies et al.[43] have found that hot water extracts from germ tube cell walls of bean rust are highly active as inducers of ethylene biosynthesis in bean plants. Induction of ethylene biosynthesis was also tested in elicitor-treated parsley cell suspension cultures.[44] Although ethylene biosynthesis was already high in the untreated controls, a fungal elicitor increased the rate of ethylene production after a lag of about 4 h, apparently by stimulating the activity of EFE. ACC synthase was induced much more rapidly, increasing tenfold within 40 min.[44] Ethylene was also induced by glycopeptides derived from *Colletotrichum lagenarium* in intact tissue of melon plants.[45]

The fungal enzymes used to digest cell walls for the preparation of plant protoplasts frequently induce ethylene production.[46-49] It has been demonstrated that the products released from plant cell walls by such enzymes can induce ethylene production in pear cells[50] and in citrus fruit.[51]

Microbial toxins may also induce ethylene biosynthesis. For example, oat cultivars carrying the resistance gene Pc-2 against *Puccinia coronata* are susceptible to the pathogen *Helminthosporium victoriae* and to its toxin, victorin. Victorin causes a rapid increase in ethylene biosynthesis in susceptible but not in resistant oat lines.[52]

III. INDUCTION OF PLANT DEFENSE RESPONSES BY ETHYLENE

As discussed in the previous section, stress ethylene production is an early reaction of plants to recognition of a pathogen attack. Does pathogen-induced stress ethylene play a role in the induction of defense reactions against pathogens? Stahmann et al.[53] were the first to suggest this possibility after their finding that ethylene treatment induces resistance to black rot in sweet potato tissue, accompanied by the induction of enzyme activities possibly related to defense. In this section, evidence is discussed that **exogenously applied ethylene** induces biochemical and structural changes that are thought to play a role in defense against pathogens. It also deals with the more difficult question whether **endogenously produced stress ethylene** is **necessary** as a "second messenger"[54] for the observed induction of these defense reactions in plant-pathogen interactions.

A. BIOCHEMICAL DEFENSE REACTIONS

Plants react to pathogen attack by an array of biochemical changes that are thought to represent defenses against pathogens.[55] For a number of these defense reactions, ethylene has been implicated in the induction process.[4-7,56,57] These are discussed below.

1. Induction of Pathogenesis-Related Proteins and Antifungal Hydrolases

Pathogenesis-related proteins (PR) are a group of proteins accumulating in large amounts in response to infection.[57,58] The PRs of tobacco (cultivars Samsun and Xanthi) have been studied most extensively, particularly in the hypersensitive response against TMV. The best-characterized tobacco PRs are acidic proteins present in the extracellular space, selectively extractable at low pH.[57,58] They can be arranged into five families on the basis of serology and sequence comparisons, and most of them have basic homologs.[58] In tobacco leaves, the PRs are not only induced by infections but also by various chemicals, including 2-chloro-ethylphosphonic acid (ethephon), an ethylene-releasing compound.[59,60] In tomato leaves, a related group of PRs is induced by ethephon[61-63] as well as by viroid[61] or fungal[63] infection. Since ethylene is produced in large amounts in the hypersensitive response of tobacco, Van Loon[60] proposed that ethylene is a natural signal for the induction of PRs. It will be interesting to examine whether exogenous ethylene (rather than ethephon) induces the PRs, and whether endogenously produced ethylene is actually necessary for induction of PRs in response to infection. In pepper leaves, PRs appear to be induced by TMV both in the presence and absence of induced ethylene synthesis.[64]

As reviewed elsewhere,[58,65] several of the long-studied acidic PRs have chitinase or β-1,3-glucanase activities. These two enzymes have been studied extensively in their own right, particularly with regard to ethylene and pathogens, and are discussed in more detail below.

As first described by Abeles and co-workers[66,67] and reviewed elsewhere,[65] chitinase and β-1,3-glucanase are induced by ethylene in various plant tissues. In bean and tobacco, the best-studied models, the main ethylene-induced chitinases and β-1,3-glucanases are basic homologs of the corresponding PRs.[58,65] In bean leaves, the activities are mainly localized in the vacuoles.[68,69] Since chitin and β-1,3-glucans are important elements of fungal cell walls, the two enzymes that hydrolyze them have been hypothesized to function in the defense of plants against fungi.[7,66-72] There are three lines of indirect evidence for this hypothesis.[70-72] First, chitinase and β-1,3-glucanase are strongly induced in the course of pathogenesis[41,73-76] and upon treatments with elicitors.[41,44,77] Second, chitinase and β-1,3-glucanase degrade isolated fungal cell walls.[78] Third, some pathogenic fungi possess proteins that inhibit plant β-1,3-glucanase.[79] Furthermore, there is now direct evidence for antifungal activity of the enzymes. Plant chitinase has been shown to inhibit the growth of some fast-growing fungi.[80-82] In addition, many other fungi which are not inhibited by chitinase alone, are strongly inhibited by combinations of chitinase and β-1,3-glucanase.[83] Taken together, these data indicate that the two hydrolases have a strong antifungal potential.

a. Effects of Exogenous Ethylene

Chitinase and β-1,3-glucanase activities are strongly increased in various plant tissues upon treatment with ethylene for about 24 h.[84] In bean leaves, ethylene induces only chitinase and β-1,3-glucanase among a number of hydrolytic activities investigated, indicating that the effect is quite specific.[84] Broglie et al.[85] isolated cDNA as well as genomic clones for chitinase from bean and tomato and obtained evidence that ethylene regulated the level of chitinase mRNA. They introduced one of the bean chitinase genes, the CHB5 gene, into tobacco and found that ethylene strongly induced the bean gene in the transgenic tobacco plants.[86] A chimeric gene consisting of 1.6 kb of 5′-flanking DNA derived from the bean chitinase gene fused to the coding sequence of β-glucuronidase was also induced by ethylene, indicating that the promoter region is sufficient for ethylene-regulated expression.[86] Comparison of the induction kinetics of chitinase and β-1,3-glucanase in ethylene-treated bean leaves demonstrated a closely parallel increase of the enzymatic activities as well as of the levels of translatable mRNA.[87] Upon withdrawal of ethylene, a similar parallel decay of mRNA activities was observed with a half-life of about 10 h, whereas enzyme activities continued to increase for about 20 h and then reached a plateau.[87] These data indicate that chitinase and β-1,3-glucanase are coordinately regulated at the level of mRNA.

As discussed elsewhere in detail,[65] the two enzymes are coordinately regulated also by other factors. Cytokinin and auxin suppress these enzyme activities in cell cultures, while an increase occurs in the course of maturation and senescence.

b. Role of Endogenous Ethylene

Mauch et al.[41] have studied induction of the two enzymes by infection or by elicitor treatment in the absence or presence of AVG. AVG reduced ethylene production below the level of uninfected controls but did not interfere with the induction of chitinase and β-1,3-glucanase in response to infection or elicitors. In contrast, the effect of auxin, which also induced ethylene production and the activities of the hydrolases, could be fully reverted by AVG. These results indicate that ethylene has a role as a second messenger when auxin is the stimulus but it is a symptom without a signal function in elicitor treatments. A study of induction of chitinase in melon plants by elicitor yielded similar results,[88] although the authors noted primarily an inhibitory effect (about 30 to 40%) of AVG on chitinase induction. The authors interpreted this as a demonstration of involvement of the endogenous ethylene, overlooking their own finding that AVG inhibited chitinase induction by exogenous ethylene to a similar extent (about 20 to 30%).[88] The latter result would commonly be taken as a sign that AVG had unspecific inhibitory side effects on chitinase induction.

In cucumber,[89] an acidic, extracellular[90] chitinase is induced systemically after local necrotic infections. Ethylene also induces chitinase activity in cucumber leaves,[89] but there appeared to be no ACC transport out of the infected leaves and no enhanced ethylene production in the noninfected leaves, indicating that ethylene did not contribute to systemic induction of chitinase in cucumber.[91]

An interesting result was obtained with tobacco callus cultures in which β-1,3-glucanase is induced after calluses are transferred to media without auxin or cytokinin.[92] It was observed that ethylene did not induce β-1,3-glucanase as long as auxin and cytokinin were present. However, endogenous ethylene production appeared to be necessary for the induction of the enzyme in media without auxin or cytokinin, since cobalt and norbornadiene, inhibitors of ethylene biosynthesis and action respectively, inhibited the induction of β-1,3-glucanase.[92]

In bean suspension cultures, the chitinase gene was induced within 15 min by a fungal elicitor.[93] Since induction of ethylene biosynthesis requires at least 15 min in the fastest responding system, chitinase gene activation by elicitor is almost certainly not caused by ethylene in this system.

2. Induction of Enzymes of the Phenylpropanoid Pathway

The key enzymes of flavonoid biosynthesis, phenylalanine ammonia-lyase (PAL), 4-coumarate:CoA ligase (4-CL), and chalcone synthase (CHS), have become paradigms of enzymes involved in defense[55,56,94] since they are strongly induced by elicitors and since many well investigated phytoalexins of the Fabaceae are isoflavonoids derived from this pathway. In addition, other products of PAL and 4-CL, including polyphenols and lignin, are thought to play a role in the defense response.

a. Effects of Exogenous Ethylene

PAL has been found to be induced by ethylene in a number of tissues, including citrus fruit peel,[95,96] pea,[97] and carrot.[98] In carrot roots, RNA blot hybridization analysis revealed marked increases in mRNAs of PAL, 4-CL, and CHS in response to elicitors.[56] This is somewhat surprising since the known phytoalexins from carrot are not derived from this pathway. Ethephon and ACC were found to induce PAL in rice leaves.[99] In lettuce, ethylene causes a metabolic disorder called russet spotting, characterized by brown, lignified flecks and an induction of PAL activity.[100,101] The process resembles a hypersensitive reaction and may be considered an idling defense reaction. Interestingly, auxin counteracts this ethylene effect.[101] Ethylene also stimulates lignification[102] along with PAL[103] in Swede roots. It should be noted that PAL is frequently **not** induced by ethylene in plant cells in which PAL is strongly induced by pathogenesis or elicitors.[44]

b. Role of Endogenous Ethylene

Several studies have examined the possible involvement of ethylene in the induction of PAL. In parsley cells, in which PAL can be induced by elicitor but not by exogenous ethylene, AVG partially reduced elicitor-stimulated ethylene production and elicitor-induced PAL activity, suggesting a link between induction of ACC synthase and PAL.[44] ACC but not ethylene reversed the effect of AVG on PAL activity. It was suggested that functioning of the ACC pathway is important for PAL induction in this system.[44] Subsequently, experiments with elicitor-treated bean leaves yielded similar results,[104] indicating that even though ethylene is not sufficient to induce PAL, it enhances induction by elicitors. In lettuce, a tissue in which PAL is strongly induced by exogenous ethylene,[101] wounding caused increases in ethylene production and PAL activity.[105] Treatment with AVG completely prevented wound-induced ethylene production but did not affect induction of PAL, indicating that in this tissue ethylene is **sufficient** for PAL induction but not **necessary** for its induction by wounding.[105]

In grapefruit flavedo, where PAL can readily be induced by exogenous ethylene,[95,96] the ethylene-producing pathogen *Penicillium digitatum* did not cause induction of PAL.[96] Interestingly, infection even suppressed induction of PAL by exogenous ethylene, indicating that the fungus is able to prevent the defense response of the host.[96]

Haga et al.[99] attempted to determine the role of endogenous ethylene for PAL induction in rice leaves infected with *Pyricularia oryzae*. Unfortunately, they chose rather unspecific inhibitors to reduce endogenous ethylene production, namely, the radical scavenger tiron and aminooxyacetic acid (an inhibitor of both ethylene biosynthesis and PAL). Although these compounds partially inhibited PAL induction, further work is needed to ascertain the contribution of endogenous ethylene to PAL induction in response to infection in rice.

3. Accumulation of Phytoalexins

A particularly well-studied defense reaction of plants is the accumulation of antibiotic secondary compounds, so called phytoalexins.[55,106] Accumulation of phytoalexins can often be induced not only by living pathogens but also by elicitors (see Section II above).

a. Effects of Exogenous Ethylene

Generally, exogenous ethylene is not an inducer of phytoalexin accumulation. Even in tissues where application of elicitors causes a strong stress ethylene production, exogenously applied ethylene does not induce phytoalexin accumulation.[107] In cases where phytoalexins have been found to accumulate in response to ethylene, the amounts observed were much smaller than in infected or elicitor-treated plants.[108,109]

An interesting exception to this rule is the induction of the phytoalexin 6-methoxymellein in carrot.[110-113] Here, ethylene induces the accumulation of large quantities of this bitter fungitoxic isocoumarin.

A different effect of ethylene has been reported for potato tuber slices pretreated with the ethylene-releasing compound ethephon[114] and ethylene,[115] which formed phytoalexins more rapidly than control slices when challenged with an elicitor. Thus, ethylene appeared to sensitize the tissue to react more rapidly to fungal elicitors. Such a sensitizing effect would be of considerable practical interest but the observations mentioned appear to represent a special, perhaps exceptional, case. In our own experience, ethylene has a sensitizing effect only in tubers stored for a long time (>4 months); tubers freshly harvested or stored for two months are less sensitive to elicitors when treated with ethylene.[7]

b. Role of Endogenous Ethylene

In an early study, Paradies et al.[107] showed that AVG, a specific inhibitor of ethylene biosynthesis, did not reduce the phytoalexin response of infected or elicitor-treated soybean plants. Hence, they concluded that ethylene was an indicator but not an inducer of phytoalexin biosynthesis. It should be noted that in this system, ethylene was not expected to be an inducer because exogenously applied ethylene failed to induce phytoalexin production in soybean. However, ethylene could have enhanced the response.

The situation is different in the case of carrot, where exogenous ethylene can induce phytoalexin production.[110-113] In carrot discs infected with *Botrytis cinerea,* treatments with AVG suppressed phytoalexin accumulation and reduced resistance, indicating that endogenous stress ethylene was involved in phytoalexin accumulation.[116]

4. Cell Wall Alterations

Localized lignification[55,117] and apposition of new cell wall material[55] are typical of the defense responses against various pathogens. These structural reinforcements of the wall may hinder penetration by fungi as well as spread of viruses.[55,117] Among the plant genes activated in the course of a pathogen attack are the hydroxyproline-rich glycoproteins (HPRGP), also called extensins. Although accumulation of HPRGP is often considered a defense reaction, their function in defense is not clear.[118]

a. Effects of Exogenous Ethylene

As already mentioned, lignification may be stimulated by ethylene in certain tissues, e.g., in Swede roots.[102,103] Induction of PAL[56] and of peroxidase[53,101] may be important for enhanced lignification. In a study on protoplast preparations from oat leaves, it was found that peeling of the abaxial epidermis induced the development of resistance to cellulolytic digestion of mesophyll cell walls,[119] a process involving lignification.[120] Exogenous ethylene partially induced the development of resistance.[119]

HPRGP are also induced in response to treatments with exogenous ethylene,[121] a process shown to involve gene activation in carrot discs.[56]

b. Role of Endogenous Ethylene

The role of endogenous ethylene was tested in the response of oat leaves to wounding.[119] The development of resistance to cellulase was inhibited by silver thiosulfate, an inhibitor of ethylene action, and by cobalt ions and AVG, both inhibitors of ethylene biosynthesis.[119] The lignification response observed within 3 to 4 h of peeling did not occur in the presence of Ag^+, Co^{2+}, or AVG.[120] The inhibitory effect of AVG on the development of resistance was reversed by the addition of ACC.[119] These results indicate that stress ethylene is necessary for lignification and the development of resistance to cellulase in response to wounding. It remains to be seen if ethylene is important also in lignification induced by pathogens or elicitors.[117]

Stermer and Hammerschmidt[122] found that cell walls of heat-shocked cucumber seedlings became more resistant to degradation, an effect that did not appear to depend on lignification, and that ethylene biosynthesis and extensin accumulation were induced in parallel. It is not yet clear whether the endogenously produced ethylene has anything to do with lignification in this system.

Roby et al.,[123] investigating whether ethylene is a signal inducing accumulation of the hydroxyproline-rich glycoproteins, assert that ethylene has this function since AVG partially inhibited HPRGP induction. However, a closer look at their data shows that AVG inhibited HPRGP induction to a similar extent in the presence of exogenous ethylene, indicating that AVG had unspecific inhibitory side effects.[123a]

5. Induction of Proteinase Inhibitors

Proteinase inhibitors are induced locally and systemically in wounded tomato and potato plants. They inhibit animal and microbial but not plant proteinases and are thought to represent primarily a defense against insects.[124] Protease inhibitors are induced by elicitors in melon seedlings[125] and in suspension-cultured tobacco cells.[126] In both tissues, an increase in ethylene biosynthesis precedes the induction of protease-inhibitor activity.

a. Effects of Exogenous Ethylene

Proteinase-inhibiting activity could be induced in melon tissue by treatments with ACC.[125] Since ACC also caused enhanced ethylene production, this effect was suggested to reflect an induction by ethylene.

b. Role of Endogenous Ethylene

Roby et al.[125] reported an increase in proteinase-inhibitor activity in infected and elicitor-treated melon seedlings. They concluded that endogenous ethylene was involved in this effect from the fact that AVG inhibited accumulation of this proteinase-inhibiting activity by about 19%. The authors did not test whether this small AVG effect could be reversed by ACC or ethylene, and in the absence of such a test the conclusion of the authors requires reexamination.

B. DEVELOPMENTAL AND STRUCTURAL CHANGES

As described elsewhere in this volume, ethylene affects plant structure and development in a variety of ways. Many of these well-known ethylene effects, for example, epinasty or stunted growth, also frequently occur as symptoms of diseases but very probably have no direct significance for the plant's defense against pathogens. Ethylene may also be a factor in specific aspects of disease development, as for example in the reduction of stomatal aperture in rust-infected leaves of *Vicia faba*.[127]

Of particular interest are ethylene-related developmental changes like abscission, wound healing, or occlusion of vascular systems, which may be considered as attempts by the plant to eliminate, localize, or isolate an infection site. These responses are discussed in more detail below.

1. Abscission

Although abscission occurs normally at specific stages of development,[128,129] it is sometimes induced prematurely in response to pathogens[130-133] or parasites.[134] Functionally, pathogen-induced abscission can be viewed as defense by elimination of infected parts. An interesting example is the "shot hole" disease of cherry leaves where formation of an abscission zone is induced in a circle around an infection site.[135]

a. Effects of Exogenous Ethylene

Exogenously applied ethylene provokes premature abscission of leaves, fruits, or flowers in many model systems.[128,129] Induction of cellulase plays a key role in ethylene-induced abscission.[128,129] In the abscission zones, antifungal hydrolases like chitinase are induced in addition to cellulase.[136] One possible explanation for this effect is that the scars remaining after abscission need special biochemical protection against pathogens.

b. Role of Endogenous Ethylene

In citrus leaves infected with *Xanthomonas citri*[130] and in pepper leaves infected with *Xanthomonas campestris,*[131] ethylene production (apparently of plant origin and due to tissue damage) was correlated with abscission. The abscission of young cotton flower buds in response to a sucking insect, the cotton fleahopper, has also been ascribed to stress ethylene.[134] The insect appears to stimulate ethylene biosynthesis in two ways, first by the feeding wound and by the salivary cell-wall-degrading enzymes, and second by inoculation of the tissue with pathogens.[134]

The production of ethylene has been compared with the induction of defoliation in two pathogen systems with fungal pathogens.[132,133] In peanut leaflets infected with *Cercospora arachidicola,* abscission was correlated with the ability of the plant genotypes to produce pathogen-induced ethylene. Ag^+ prevented pathogen-induced abscission, indicating that endogenous ethylene was involved in the process.[132] In cotton plants infected with various strains of *Verticillium dahliae,* abscission similarly correlated with ethylene production in response to the pathogen.[133]

2. Gummosis and Occlusion of Vascular Systems

Formation of vascular gels or of "gums" can be observed as a local response to infection.[137] Functionally, gummosis resembles abscission, isolating infection sites and preventing the invasion of pathogens into the plant.[137] An analogous process of defense by isolation is the so-called compartmentalization in the sapwood of trees. It has been proposed that ethylene may have a role in this compartmentalization process.[138]

a. Effects of Exogenous Ethylene

Ethylene promotes gum formation in the gum ducts of some woody plants, including *Prunus* species.[137] Ethylene also induces the formation of vascular gels in castor bean.[139]

b. Role of Endogenous Ethylene

The significance of pathogen-induced ethylene production for the formation of vascular gels has been investigated.[139] Aminooxyacetic acid (AOA), a rather unspecific inhibitor of ethylene biosynthesis, prevented pathogen-induced formation of vascular gels and exogenous ethylene overcame the AOA effect, indicating that ethylene was involved in the response.

IV. ROLE OF ETHYLENE IN RESISTANCE AND SUSCEPTIBILITY

A. ETHYLENE AND RESISTANCE

If ethylene plays a role in disease resistance, a simple way to protect plants might be to treat them with ethylene. Indeed, Stahmann et al.[53] have found that sweet potato tissue became resistant to *Ceratocystis fimbriata* after an exposure to 8 ppm ethylene for two days. However, a later study could not confirm these results,[140] indicating that effects on disease incidence of whole-plant exposures to ethylene should be viewed with caution.

This is also evident in work on post-harvest diseases of fruit. At first glance, ethylene treatments should be expected to promote ripening-related processes in fruit tissue and thus make the fruit more "digestable", not only to frugivores but also to pathogens. Furthermore, ethylene treatments have been reported to stimulate disease development when inoculation immediately followed the ethylene treatment, for example in green tangerines infected with *Colletotrichum gloeosporioides,*[141] or in strawberries exposed to *Botrytis cinerea.*[142] On the other hand, ethylene treatments were found to increase resistance of degreened, orange tangerine fruits to *Colletotrichum gloesporioides*[143] and of navel oranges to *Penicillium italicum.*[144]

Melon plants treated for 7 d with 500 ppm ethylene were more resistant to *Colletotrichum lagenarium* than untreated controls.[122] This result should be viewed with caution because of the long treatment with very high levels of ethylene.

An interesting result was obtained with carrot roots exposed to ethylene.[113] As expected, the phytoalexin 6-methoxymellein accumulated in ethylene-treated carrot. However, there was no difference between control and ethylene-treated carrot in their resistance to *Botrytis cinerea.*[113] Nevertheless, endogenous ethylene production appeared to be necessary for the induction of resistance by elicitors, indicating that ethylene is responsible for part but not all of the defense reactions observed in response to *Botrytis cinerea.*[116]

In plant-pathogen interactions ethylene is released locally in the immediate surroundings of a pathogen. Whole-plant treatments with ethylene represent a quite different situation. Van Loon[59] has attempted to mimic the natural situation by pricking tobacco leaves with ethephon. This treatment caused induction of PR proteins and local necroses similar to the hypersensitive response against viruses. Treated leaves became more resistant against subsequent virus infection. This experiment is an interesting attempt to experimentally impose an ethylene regime similar to the one found in diseased plants and to establish that ethylene is *sufficient* for an increase in resistance. On the other hand, experiments with inhibitors of ethylene biosynthesis can be used to test whether ethylene is *necessary* for increased resistance. In soybean leaves, for example, inhibitors of ethylene biosynthesis completely prevented ethylene production in response to tobacco necrosis virus but did not affect the number or size of necrotic lesions, indicating that ethylene is not necessary for the hypersensitive response in this tissue.[145]

B. ETHYLENE AND SUSCEPTIBILITY

Ethylene may not only be involved in the induction of defense reactions, but it may also increase disease severity in specific plant-pathogen interactions. In fact, some of the ethylene effects summarized above as "defense reactions" are frequently considered as grave symptoms of disease. For example, premature abscission of cotton buds is usually seen by man as a symptom of disease resulting in heavy economic losses in the crop,[134] although it may also be a defense response of the plant preventing greater loss by sacrificing the diseased part.

In early work, ethylene was found to increase disease severity in *Verticillium* wilt of tomato.[146] Exogenously applied ethylene was found to promote disease severity of grey mould (*Botrytis cinerea*) on rose and carnation flowers[147] and on detached leaves of tomato, pepper, bean, and cucumber.[148] Prevention of synthesis or action of endogenous ethylene by spraying the tissues with AVG or silver thiosulfate significantly decreased disease incidence, indicating that the plant ethylene production contributed to disease severity.[147,148] In this context, it is interesting that ethylene promotes the germination of several pathogenic fungi, including *Botrytis cinerea.*[149] Ethylene might be used as a cue for the presence of suitable host tissue by fungi like *Botrytis* colonizing flowers and fruits.

It has also been reported that exogenously applied ethylene can break natural resistance against a pathogen. Wheat leaves carrying the Sr6 resistance gene against wheat stem rust became susceptible to the rust after exposure to ethylene.[150]

V. ETHYLENE PRODUCTION BY PLANT PATHOGENS AND SYMBIONTS

Many fungi[151-153] and bacteria[154,155] form ethylene when supplied with methionine. However, methionine usually occurs at very low levels in the natural environment. Therefore, methionine-dependent ethylene biosynthesis is most probably irrelevant in natural plant-pathogen interactions. In the absence of methionine, most bacteria and fungi produce little or no ethylene. On the other hand, a small number of pathogens form very large amounts of ethylene even on a (see also Chapter 15) methionine-free medium. These are discussed below.

A. BACTERIAL PATHOGENS

1. Ethylene Production in Pure Culture

Strains of *Pseudomonas solanacearum* causing rot of banana fruit[16] were long considered to be the only bacteria to produce substantial amounts of ethylene independent of methionine. More recently, a strain of *Pseudomonas syringae* pv. *phaseolicola*, pathogenic to the Japanese weed *Pueraria*, Kudzu, was found to form large amounts of ethylene.[156] Ethylene production was also found in cell-free bacterial extracts with 2-oxoglutarate as a substrate.[157]

2. Ethylene Biosynthesis in the Host Plant

In the case of *Pseudomonas solanacearum*, the relative contributions of host and pathogen to ethylene production in the diseased tissue remain unclear.[16,18] In the case of *Pueraria* infected with *Pseudomonas syringae*, it is likely that the bacterium contributes the largest part to the ethylene formed in diseased tissue.[156]

3. Possible Functions of Ethylene Production

If the pathogenic bacteria in unripe fruit synthesize ethylene, ripening may be induced with a concomitant improvement of the nutrient supply.[16] However, the function of bacterial ethylene production in the *Pueraria-Pseudomonas syringae* interaction is not known.[156,157]

B. FUNGAL PATHOGENS

1. Ethylene Production in Pure Culture

Early reports described production of ethylene by various fungi.[38,158,159] *Penicillium digitatum*, a common mold on citrus fruit, produces particularly large amounts of ethylene and therefore has been studied extensively.[151,160-162] *Penicillium digitatum* forms ethylene from methionine when supplied with large amounts (7.5 m*M*) of this amino acid.[160] However, tracer studies in methionine-free media indicated that glutamic acid or 2-oxoglutarate, but not methionine, are precursors of ethylene in media lacking methionine.[161] A cell-free extract was isolated, which converted 2-oxoglutaric acid into ethylene.[162]

Strains of *Fusarium oxysporum* f.sp. *tulipae*, a tulip pathogen, form equally large amounts of ethylene on liquid media.[163,164] Among 19 species and formae speciales of *Fusarium* tested, only the tulip pathogen produced more than 4 μg ethylene per g dry weight during a 15-day growth period; all 7 isolates of the latter produced between 3 and 20 mg ethylene per g dry weight under the same conditions,[163] indicating that high ethylene production is an exceptional but constant trait of the isolates of *Fusarium oxysporum* from tulip.

Ethylene production was also described in the fungus *Bipolaris sorokiniana*.[165] As in other cases, ethylene production was observed only when the fungus was fed with methionine. A closer analysis appeared to indicate ACC as an intermediate of ethylene biosynthesis in this fungus.[165] This unexpected result warrants further examination and rigorous identification of ACC as a fungal product.

2. Ethylene Production in the Host Plant

Achilea et al.[166,167] succeeded in distinguishing between host and pathogen ethylene biosynthesis in citrus fruits infected by *Penicillium digitatum*, making use of the fact that the two organisms employ different biosynthetic pathways.

In the case of the interaction of *Poa pratensis* with *Bipolaris sorokiniana*, two peaks of ethylene production were observed, accompanied by substantial but erratic increases in ACC.[168] Because it is suspected that this pathogen also produces ACC,[165] it remains unclear whether the ethylene comes from the plant or from the fungus.

3. Possible Functions of Ethylene Production

Chalutz[169] studied the biological significance of ethylene production by *Penicillium digitatum*, obtaining mutants that were unable to produce ethylene. These mutants were equally pathogenic on citrus fruits, indicating that ethylene production was not a virulence factor for the pathogen.[169]

In tulip bulbs, there are hints that exogenously applied ethylene causes disappearance of an antibiotic substance, tulipalin.[170] On this basis, it was proposed that the tulip pathogen *Fusarium oxysporum* f.sp. *tulipae* produces ethylene as a pathogenicity factor, eliminating one of the resistance factors of the host plant.[164,170] Unfortunately, this interesting system seems not to have been pursued further and studies with mutants are not available.

A more recent study deals with a different group of ethylene-producing fungi isolated from tulip bulbs.[171] Here, both the ethylene-producing strains and the non-ethylene-producing strains of *Penicillium* grew similarly well on agar but the ethylene-producing strains appeared to be more virulent on tulip bulbs, particularly in competition studies with the non-ethylene-producing strains.[171] This result clearly indicates that ethylene production provides a selective advantage for a tulip bulb pathogen.

C. FUNGAL SYMBIONTS

Ethylene biosynthesis also has sometimes been mentioned as a factor in plant symbioses, particularly in the ectomycorrhizal symbiosis of tree roots with fungi.

1. Ethylene Production in Pure Culture

Ectomycorrhizal fungi produce some ethylene in pure culture, but only when the media are supplemented with 1 to 10 m*M* methionine.[172,173] It is unlikely that the fungi receive methionine at this high level in symbiosis with roots, although root exudates may contain some methionine.[172] Ectomycorrhizal fungi cannot convert ACC into ethylene.[173]

2. Ethylene Production in the Host Plant

Ethylene was produced by aseptically grown Douglas fir seedlings inoculated with mycorrhizal fungi and appearance of ethylene coincided with the formation of mycorrhizal structures.[172] It was not clear whether the ethylene originated from the plant or from the fungus. Seedlings inoculated with a pathogen, *Fusarium oxysporum* f.sp. *pini*, produced more ethylene sooner than the mycorrhizal seedlings, indicating a differential role for ethylene in symbiotic and pathogenic fungus-host interactions.[172]

3. Possible Functions of Ethylene Production

It has been speculated that ethylene is involved in the characteristic dichotomous branching of mycorrhizal roots.[174-176] Ethylene treatments did not affect mycorrhiza formation directly[174,177] but it is still possible that ethylene production is necessary for the establishment of a vital symbiosis. In support of this hypothesis, silver thiosulfate, an inhibitor of ethylene action, was found to decrease mycorrhiza formation in *Pinus mugo* roots.[177]

Note that in work with other mutualistic symbioses of plants, exogenous ethylene appeared to have a negative influence. Application of the ethylene-releasing compound ethephon to the plant roots reduced nodulation of *Pisum sativum*[30] and *Medicago sativa*[32] and suppressed the establishment of a vesicular-arbuscular mycorrhiza in *Medicago sativa* and *Triticum vulgare*.[32]

VI. CONCLUSIONS

It has become a well-established fact that in most plant-pathogen interactions the enhancement of ethylene production is primarily due to an induction of the "normal" pathway of ethylene biosynthesis in the plant. This indicates that ethylene biosynthesis in response to infection is a well-regulated process. Furthermore, it has become clear that treatments with exogenous ethylene frequently induce biochemical and structural changes that have a potential to restrain, isolate, or eliminate a pathogen. Nevertheless, the significance of ethylene in plant-pathogen interactions is far from clear. There are two main questions open: first, for a defense response known to be influenced by **exogenous** ethylene, is the ethylene produced **endogenously** in response to a pathogen attack necessary and sufficient to induce the response? Second, in a given plant-pathogen interaction, is ethylene contributing to the resistance?

During the last couple of years, these questions have been addressed primarily with inhibitor studies. Such studies, like all inhibitor studies, are somewhat problematic, since the inhibitors may not have access to the intended target, and because they may have unspecific side effects. Silver thiosulfate has frequently been used as an inhibitor of ethylene action. However, as a noncompetitive inhibitor, it has the disadvantage that unspecific side effects cannot be controlled. Norbornadiene, a competitive inhibitor of ethylene action, is more useful because it can be checked whether its effects are reversed by high levels of ethylene. AVG and Co^{2+} have also frequently been used as inhibitors of ethylene biosynthesis. Their effectiveness and specificity of action can and should always be checked as well. If they fail to inhibit a process in question, it should be ascertained that ethylene production is indeed inhibited. On the other hand, if they are inhibitory it is equally necessary to ascertain that ACC (in the case of AVG) or ethylene (in the case of AVG or cobalt) can reverse the inhibition. The results of the inhibitor studies discussed above lead to the conclusion that increased formation of endogenous ethylene production frequently may not be **necessary** for the induction of defense responses, even if exogenous ethylene is **sufficient** to induce them. However, there often remains some doubt about the interpretations.

Many new possibilities for experiments are now emerging with the availability of ethylene mutants and the tools of molecular biology. An *Arabidopsis* mutant has been obtained that is resistant to exogenous ethylene.[178] Such a mutant will be ideal to test the significance of ethylene in plant-pathogen interactions. As genes of the defense response and of ethylene biosynthesis become known, it should also become possible to obtain plants with constitutively induced or repressed defense-related genes or ethylene biosynthesis, making use of genetic transformation with either sense or antisense genes under the control of a constitutive promoter. Such transformants will be similarly useful to examine hypotheses about the role of ethylene in pathogenesis.

However, it should be noted that the ethylene response and the defense reactions are frequently induced in a specific temporal and spatial pattern just around an invading pathogen, and that the pathogen may continue to grow during this process. This results in a highly dynamic interaction which is difficult to dissect by constitutive overexpression or underexpression of genes. Even with the new tools of molecular biology at hand it will be difficult to fully understand the intricacies of the complex interactions between plants and their pathogens and the role of ethylene therein.

ACKNOWLEDGMENTS

I wish to thank Dr. P. King and Dr. J. Felix for stimulating discussions and helpful suggestions.

REFERENCES

1. **Boller, T.**, Ethylene and plant-pathogen interactions, in *Polyamines and Ethylene: Biochemistry, Physiology, and Interactions*, Flores, H. E., Arteca, R. N., and Shannon, J. C., Eds., American Society of Plant Physiologists, Rockville, MD, 1990, 138.
2. **Yang, S. F. and Pratt, H. K.**, The physiology of ethylene in wounded plant tissues, in *Biochemistry of Wounded Plant Tissues*, Kahl, G., Ed., Walter de Gruyter, Berlin, 1978, 595.
3. **Yang, S. F. and Hoffmann, N. E.**, Ethylene biosynthesis and its regulation in higher plants, *Annu. Rev. Plant Physiol.*, 35, 155, 1984.
4. **Hislop, E. C., Hoad, G. V., and Archer, S. A.**, The involvement of ethylene in plant diseases, in *Fungal Pathogenicity and the Plant's Response*, Byrde, W. and Cutting, C. V., Eds., Academic Press, New York, 1973, 87.
5. **Archer, S. A. and Hislop, E. C.**, Ethylene in host-pathogen relationships, *Ann. Appl. Biol.*, 81, 121, 1975.
6. **Pegg, G. F.**, The involvement of ethylene in plant pathogenesis, *Encycl. Plant Physiol. New Ser.*, 4, 582, 1976.
7. **Boller, T.**, Ethylene-induced biochemical defenses against pathogens, in *Plant Growth Substances 1982*, Wareing, P. F., Ed., Academic Press, London, 1982, 303.
8. **Boller, T. and Kende, H.**, Regulation of wound ethylene synthesis in plants, *Nature*, 286, 259, 1980.
9. **Ross, A. F. and Williamson, C. E.**, Physiologically active emanations from virus-infected plants, *Phytopathology*, 41, 431, 1951.
10. **Balazs, E., Gaborjany, R., Toth, A., and Kiraly, Z.**, Ethylene production in Xanthi tobacco after systemic and local virus infections, *Acta Phytopathol. Acad. Sci. Hung.*, 4, 355, 1969.
11. **Nakagaki, Y., Hirai, T., and Stahmann, M. A.**, Ethylene production by detached leaves infected with tobacco mosaic virus, *Virology*, 40, 1, 1970.
12. **De Laat, A. M. M., Van Loon, L. C., and Vonk, C. R.**, Regulation of ethylene biosynthesis in virus-infected tobacco leaves. I. Determination of the role of methionine as the precursor of ethylene, *Plant Physiol.*, 68, 256, 1981.
13. **De Laat, A. M. M. and Van Loon, L. C.**, Regulation of ethylene biosynthesis in virus-infected tobacco leaves. II. Time course of levels of intermediates and in vivo conversion rates, *Plant Physiol.*, 69, 240, 1982.
14. **De Laat, A. M. M. and Van Loon, L. C.**, The relationship between stimulated ethylene production and symptom expression in virus-infected tobacco leaves, *Physiol. Plant Pathol.*, 22, 261, 1983.
15. **De Laat, A. M. M. and Van Loon, L. C.**, Effects of temperature, light and leaf age on ethylene production and symptom expression in virus-infected tobacco leaves, *Physiol. Plant Pathol.*, 22, 275, 1983.
16. **Freebairn, H. T. and Buddenhagen, I. W.**, Ethylene production by *Pseudomonas solanacearum*, *Nature*, 202, 313, 1964.
17. **Lund, B. M. and Mapson, L. W.**, Stimulation by *Erwinia carotovora* of the synthesis of ethylene in cauliflower tissue, *Biochem. J.*, 119, 251, 1970.
18. **Lund, B. M.**, The effect of certain bacteria on ethylene production by plant tissue, in *Fungal Pathogenicity and the Plant's Response*, Byrde, W. and Cutting, C. V., Eds., Academic Press, New York, 1973, 69.
19. **Canfield, M. L. and Moore, L. W.**, Production of ethylene by *Daucus carota* inoculated with *Agrobacterium tumefaciens* and *Agrobacterium rhizogenes*, *Z. Pflanzenphysiol.*, 112, 471, 1983.
20. **Miller, R. A. and Pengelly, W. L.**, Ethylene production by shoot-forming and unorganized crown-gall tumor tissues of *Nicotiana* and *Lycopersicon* cultured *in vitro*, *Planta*, 161, 418, 1984.
21. **Goodman, T. C., Montoya, A. L., Williams, S., and Chilton, M. D.**, Sustained ethylene production in *Agrobacterium*-transformed carrot discs caused by expression of the T-DNA *tms* gene products, *J. Bacteriol.*, 167, 387, 1986.
22. **Owens, L. D. and Wright, D. A.**, Rhizobial-induced chlorosis in soybeans: isolation, production of nodules, and varietal specificity of the toxin, *Plant Physiol.*, 40, 927, 1965.
23. **Minamisawa, K.**, Comparison of extracellular polysaccharide composition, rhizobitoxine production, and hydrogenase phenotype among various strains of *Bradyrhizobium japonicum*, *Plant Cell Physiol.*, 30, 877, 1989.
24. **Mitchell, R. E., Frey, E. J., and Benn, M. H.**, Rhizobitoxine and L-*threo*-hydroxythreonine production by the plant pathogen, *Pseudomonas andropogonis*, *Phytochemistry*, 25, 2711, 1986.
25. **Mitchell, R. E. and Frey, E. J.**, Rhizobitoxine and hydroxythreonine production by *Pseudomonas andropogonis* strains, and their implications to plant disease, *Physiol. Mol. Plant Pathol.*, 32, 335, 1988.
26. **Owens, L. D., Guggenheim, S., and Hilton, J.**, Rhizobium-synthesized phytotoxin: an inhibitor of β-cystathionase in *Salmonella typhimurium*, *Biochim. Biophys. Acta*, 158, 219, 1968.
27. **Giovanelli, J., Owens, L. D., and Mudd, S. H.**, Mechanism of inhibition of spinach β-cystathionase by rhizobitoxine, *Biochim. Biophys. Acta*, 227, 671, 1971.

28. **Owens, L. D., Lieberman, M., and Kunishi, A.**, Inhibition of ethylene production by rhizobitoxine, *Plant Physiol.*, 48, 1, 1971.
29. **Boller, T., Herner, R., and Kende, H.**, Assay for and enzymatic formation of an ethylene precursor, 1-aminocyclopropane-1-carboxylic acid, *Planta*, 145, 293, 1979.
30. **Drennan, D. S. F. and Norton, C.**, The effects of ethrel on nodulation in *Pisum sativum* L., *Plant Soil*, 36, 53, 1972.
31. **Goodlass, G. and Smith, K. A.**, Effects of ethylene on root extension and nodulation of pea (*Pisum sativum* L.) and white clover (*Trifolium repens* L.), *Plant Soil*, 51, 387, 1979.
32. **Azcon-Aguilar, C., Rodriguez-Navarro, D. N., and Barea, J. M.**, Effects of ethrel on the formation and responses to VA mycorrhiza in *Medicago and Triticum*, *Plant Soil*, 60, 461, 1981.
33. **Devine, T. E. and Weber, D. F.**, Genetic specificity of nodulation, *Euphytica*, 26, 527, 1977.
34. **La Favre, J. S. and Eaglesham, A. R. J.**, Rhizobitoxine: a phytotoxin of unknown function which is commonly produced by bradyrhizobia, *Plant Soil*, 92, 443, 1986.
35. **La Favre, J. S., La Favre, A. K., and Eaglesham, A. R. J.**, Rhizobitoxine production by and nodulation characteristics of colony-type derivatives of *Bradyrhizobium japonicum* USDA 76, *Can. J. Microbiol.*, 34, 1017, 1988.
36. **Zaat, S. A. J., Van Brussel, A. A. N., Tak, T., Lugtenberg, B. J. J., and Kijne, J. W.**, The ethylene-inhibitor aminoethoxyvinylglycine restores normal nodulation by *Rhizobium leguminosarum* biovar. *viciae* by suppressing the "thick and short roots" phenotype, *Planta*, 177, 141, 1989.
37. **Peters, N. K. and Crist-Estes, D. K.**, Nodule formation is stimulated by the ethylene inhibitor aminoethoxyvinylglycine, *Plant Physiol.*, 91, 690, 1989.
38. **Williamson, C. E.**, Ethylene, a metabolic product of diseased and injured plants, *Phytopathology*, 40, 205, 1950.
39. **Montalbini, P. and Elstner, E. F.**, Ethylene evolution by rust-infected, detached bean (*Phaseolus vulgaris* L.) leaves susceptible and hypersensitive to *Uromyces phaseoli* (Pers.) Wint., *Planta*, 135, 301, 1977.
40. **Walther, H. F., Hoffmann, G. M., and Elstner, E. F.**, Ethylene formation by germinating, *Drechslera graminea*-infected barley (*Hordeum sativum*) grains: a simple test for fungicides, *Planta*, 151, 251, 1981.
41. **Mauch, F., Hadwiger, L. A., and Boller, T.**, Ethylene: symptom, not signal for the induction of chitinase and β-1,3-glucanase in pea pods by pathogens and elicitors, *Plant Physiol.*, 76, 607, 1984.
42. **Spanu, P. and Boller, T.**, Ethylene biosynthesis in tomato plants infected by *Phytophtora infestans*, *J. Plant Physiol.*, 134, 533, 1989.
43. **Paradies, I., Hümme, B., Hoppe, H. H., Heitefuss, R., and Elstner, E. F.**, Induction of ethylene formation in bean (*Phaseolus vulgaris*) hypocotyl segments by preparations isolated from germ tube cell walls of *Uromyces phaseoli*, *Planta*, 146, 193, 1979.
44. **Chappell, J., Hahlbrock, K., and Boller, T.**, Rapid induction of ethylene biosynthesis in cultured parsley cells by fungal elicitor and its relationship to the induction of phenylalanine ammonia-lyase, *Planta*, 161, 475, 1984.
45. **Toppan, A. and Esquerré-Tugayé, M.-T.**, Cell surfaces in plant-microorganism interactions. IV. Fungal glycopeptides which elicit the synthesis of ethylene in plants, *Plant Physiol.*, 75, 1133, 1984.
46. **Anderson, J. D., Mattoo, A. K., and Lieberman, M.**, Induction of ethylene biosynthesis in tobacco leaf discs by cell wall digesting enzymes, *Biochem. Biophys. Res. Commun.*, 107, 588, 1982.
47. **Chalutz, E., Mattoo, A. K., Solomos, T., and Anderson, J. D.**, Enhancement by ethylene of Cellulysin induced ethylene production by tobacco leaf discs, *Plant Physiol.*, 74, 99, 1984.
48. **Fuchs, Y. and Anderson, J. D.**, Purification and characterization of ethylene inducing proteins from Cellulysin, *Plant Physiol.*, 84, 732, 1987.
49. **Fuchs, Y., Saxena, A., Gamble, H. R., and Anderson, J. D.**, Ethylene biosynthesis-inducing protein from Cellulysin is an endoxylanase, *Plant Physiol.*, 89, 138, 1989.
50. **Tong, C. B., Labavitch, J. M., and Yang, S. F.**, The induction of ethylene production from pear cell culture by cell wall fragments, *Plant Physiol.*, 81, 929, 1986.
51. **Baldwin, E. A. and Biggs, R. H.**, Cell-wall lysing enzymes and products of cell-wall digestion elicit ethylene in citrus, *Physiol. Plant.*, 73, 58, 1988.
52. **Wheeler, H. and Elbel, E.**, Time-course and antioxidant inhibition of ethylene production by victorin-treated oat leaves, *Phytopathology*, 69, 32, 1979.
53. **Stahmann, M. A., Clare, B. G., and Woodbury, W.**, Increased disease resistance and enzyme activity induced by ethylene production by black rot-infected sweet potato tissue, *Plant Physiol.*, 41, 1505, 1966.
54. **Boller, T.**, Primary signals and second messengers in the reaction of plants to pathogens, in *Second Messengers in Plant Growth and Development*, Boss, W. F. and Morré, D. J., Eds., Alan R. Liss, New York, 1989, 227.
55. **Bell, A. A.**, Biochemical mechanisms of disease resistance, *Annu. Rev. Plant Physiol.*, 32, 21, 1981.
56. **Ecker, J. R. and Davis, R. W.**, Plant defense genes are regulated by ethylene, *Proc. Natl. Acad. Sci. U.S.A.*, 84, 5202, 1987.

57. **Van Loon, L. C.**, Pathogenesis-related proteins, *Plant Mol. Biol.*, 4, 111, 1985.
58. **Carr, J. P. and Klessig, D. F.**, The pathogenesis-related proteins of plants, in *Genetic Engineering, Principles and Methods*, Vol. 11, Setlow, J. K., Ed., Plenum Press, New York, 1989, 65.
59. **Van Loon, L. C.**, Induction by 2-chloroethylphosphonic acid of viral-like lesions, associated proteins, and systemic resistance in tobacco, *Virology*, 80, 417, 1977.
60. **Van Loon, L. C.**, The induction of pathogenesis-related proteins by pathogens and specific chemicals, *Neth. J. Plant Pathol.*, 89, 265, 1983.
61. **Vera, P. and Conejero, V.**, The induction and accumulation of the pathogenesis-related P69 proteinase in tomato during citrus exocortis viroid infection and in response to chemical treatments, *Physiol. Mol. Plant Pathol.*, 34, 323, 1989.
62. **Vera, P. and Conejero, V.**, Effect of ethephon on protein degradation and the accumulation of "pathogenesis-related" (PR) proteins in tomato leaf discs, *Plant Physiol.*, 92, 227, 1989.
63. **Christ, U. and Mösinger, E.**, Pathogenesis-related proteins of tomato. I. Induction by *Phytophthora infestans* and other biotic and abiotic inducers and correlations with resistance, *Physiol. Mol. Plant Pathol.*, 35, 53, 1989.
64. **Tobias, I., Fraser, R. S. S., and Gerwitz, A.**, The gene-for-gene relationship between *Capsicum annuum* L. and tobacco mosaic virus: effects on virus multiplication, ethylene synthesis and accumulation of pathogenesis-related proteins, *Physiol. Mol. Plant Pathol.*, 35, 271, 1989.
65. **Boller, T.**, Ethylene and the regulation of antifungal hydrolases in plants, in *Oxford Surveys of Plant Molecular and Cell Biology*, Vol. 5, Miflin, B. J., Ed., Oxford University Press, Oxford, 1988, 145.
66. **Abeles, F. B. and Forrence, L. E.**, Temporal and hormonal control of β-1,3-glucanase in *Phaseolus vulgaris* L., *Plant Physiol.*, 45, 395, 1970.
67. **Abeles, F. B., Bosshart, R. P., Forrence, L. E., and Habig, W. H.**, Preparation and purification of glucanase and chitinase from bean leaves, *Plant Physiol.*, 47, 129, 1971.
68. **Boller, T. and Vögeli, U.**, Vacuolar localization of ethylene-induced chitinase in bean leaves, *Plant Physiol.*, 74, 442, 1984.
69. **Mauch, F. and Staehelin, L. A.**, Functional implications of the subcellular localization of ethylene-induced chitinase and β-1,3-glucanase in bean leaves, *Plant Cell*, 1, 447, 1989.
70. **Pegg, G. F.**, Glucanohydrolases of higher plants: a possible defence mechanism against parasitic fungi, in *Cell Wall Biochemistry Related to Specificity in Host-Plant Pathogen Interactions*, Solheim, B. and Raa, J., Eds., Universitatsforlaget, Tromso, 1977, 305.
71. **Boller, T.**, Induction of hydrolases as a defense reaction against pathogens, in *Cellular and Molecular Biology of Plant Stress*, Key, J. L. and Kosuge, T., Eds., Alan R. Liss, New York, 1985, 247.
72. **Boller, T.**, Hydrolytic enzymes in plant disease resistance, in *Plant-Microbe Interactions*, Vol. 2, Kosuge, T. and Nester, E. W., Eds., Macmillan, New York, 1987, 385.
73. **Pegg, G. F. and Young, D. H.**, Changes in glycosidase activity and their relationship to fungal colonization during infection of tomato by *Verticillium albo-atrum*, *Physiol. Plant Pathol.*, 19, 371, 1981.
74. **Mauch, F., Hadwiger, L. A., and Boller, T.**, Antifungal hydrolases in pea tissue. I. Purification and characterization of two chitinases and two β-1,3-glucanases differentially regulated during development and in response to fungal infection, *Plant Physiol.*, 87, 325, 1988.
75. **Vögeli-Lange, R., Hansen-Gehri, A., Boller, T., and Meins, F.**, Induction of the defense related glucanohydrolases, β-1,3-glucanase and chitinase, by tobacco mosaic virus infection of tobacco leaves, *Plant Sci.*, 54, 171, 1988.
76. **Kombrink, E., Schröder, M., and Hahlbrock, K.**, Several "pathogenesis-related" proteins in potato are 1,3-β-glucanases and chitinases, *Proc. Natl. Acad. Sci. U.S.A.*, 85, 782, 1988.
77. **Kombrink, E. and Hahlbrock, K.**, Responses of cultured parsley cells to elicitors from phytopathogenic fungi. Timing and dose dependency of elicitor-induced reactions, *Plant Physiol.*, 81, 216, 1986.
78. **Young, D. H. and Pegg, G. F.**, The action of tomato and *Verticillium albo-atrum* glycosidases on the hyphal wall of *Verticillium albo-atrum*, *Physiol. Plant Pathol.*, 21, 411, 1982.
79. **Albersheim, P. and Valent, B. S.**, Host-pathogen interactions. VII. Plant pathogens secrete proteins which inhibit enzymes of the host capable of attacking the pathogen, *Plant Physiol.*, 53, 684, 1974.
80. **Schlumbaum, A., Mauch, F., Vögeli, U., and Boller, T.**, Plant chitinases are potent inhibitors of fungal growth, *Nature*, 324, 365, 1986.
81. **Roberts, W. K. and Selitrennikoff, C. P.**, Plant and bacterial chitinases differ in antifungal activity, *J. Gen. Microbiol.*, 134, 169, 1988.
82. **Broekaert, W. F., van Parijs, J., Allen, A. K., and Peumans, W.**, Comparison of some molecular, enzymatic and antifungal properties of chitinases from thorn-apple, tobacco and wheat, *Physiol. Mol. Plant Pathol.*, 33, 319, 1988.
83. **Mauch, F., Mauch-Mani, B., and Boller, T.**, Antifungal hydrolases in pea tissue. II. Inhibition of fungal growth by combinations of chitinase and β-1,3-glucanase, *Plant Physiol.*, 88, 936, 1988.
84. **Boller, T., Gehri, A., Mauch, F., and Vögeli, U.**, Chitinase in bean leaves: induction by ethylene, purification, properties, and possible function, *Planta*, 157, 22, 1983.

85. **Broglie, K. E., Gaynor, J. J., and Broglie, R. M.,** Ethylene-regulated gene expression: molecular cloning of the genes encoding an endochitinase from *Phaseolus vulgaris, Proc. Natl. Acad. Sci. U.S.A.*, 83, 6820, 1986.
86. **Broglie, K. E., Biddle, P., Cressman, R., and Broglie, R. M.,** Functional analysis of DNA sequences responsible for ethylene regulation of a bean chitinase gene in transgenic tobacco, *Plant Cell*, 1, 599, 1989.
87. **Vögeli, U., Meins, F., and Boller, T.,** Co-ordinated regulation of chitinase and β-1,3-glucanase in bean leaves, *Planta*, 174, 364, 1988.
88. **Roby, D., Toppan, A., and Esquerré-Tugayé, M.-T.,** Cell surfaces in plant-microorganism interactions. VI. Elicitors of ethylene from *Colletotrichum lagenarium* trigger chitinase activity in melon plants, *Plant Physiol.*, 81, 228, 1986.
89. **Métraux, J. P. and Boller, T.,** Local and systemic induction of chitinase in cucumber plants in response to viral, bacterial and fungal infections, *Physiol. Mol. Plant Pathol.*, 28, 161, 1986.
90. **Boller, T. and Métraux, J. P.,** Extracellular localization of chitinase in cucumber, *Physiol. Mol. Plant Pathol.*, 33, 11, 1988.
91. **Gehri, A. and Boller, T.,** unpublished work.
92. **Felix, G. and Meins, F.,** Ethylene regulation of β-1,3-glucanase in tobacco, *Planta*, 172, 386, 1987.
93. **Hedrick, S. A., Bell, J. N., Boller, T., and Lamb, C. J.,** Chitinase cDNA cloning and mRNA induction by fungal elicitor, wounding, and infection, *Plant Physiol.*, 86, 182, 1988.
94. **Lawton, M. A. and Lamb, C. J.,** Transcriptional activation of plant defense genes by fungal elicitor, wounding and infection, *Mol. Cell. Biol.*, 7, 335, 1987.
95. **Riov, J., Monselise, S. P., and Kahan, R. S.,** Ethylene-controlled induction of phenylalanine ammonia-lyase in citrus fruit peel, *Plant Physiol.*, 44, 631, 1969.
96. **Lisker, N., Cohen, L., Chalutz, E., and Fuchs, Y.,** Fungal infections suppress ethylene-induced phenylalanine ammonia-lyase in grapefruits, *Physiol. Plant Pathol.*, 22, 331, 1983.
97. **Hyodo, H. and Yang, S. F.,** Ethylene-induced synthesis of phenylalanine ammonia-lyase in pea seedlings, *Plant Physiol.* 47, 765, 1971.
98. **Chalutz, E.,** Ethylene-induced phenylalanine ammonia-lyase activity in carrot roots, *Plant Physiol.*, 51, 1033, 1973.
99. **Haga, M., Haruyama, T., Kano, H., Sekizawa, Y., Urushikazi, S., and Matsumoto, K.,** Dependence on ethylene of the induction of phenylalanine ammonia-lyase activity in rice leaf infected with blast fungus, *Agric. Biol. Chem.*, 52, 943, 1988.
100. **Hyodo, H., Kuroda, H., and Yang, S. F.,** Induction of phenylalanine ammonia-lyase and increase in phenolics in lettuce leaves in relation to the development of russet spotting, *Plant Physiol.*, 62, 31, 1978.
101. **Ke, D. and Saltveit, M. E.,** Plant hormone interaction and phenolic metabolism in the regulation of russet spotting in iceberg lettuce, *Plant Physiol.* 88, 1136, 1988.
102. **Rhodes, M. J. C. and Wooltorton, L. S. C.,** Stimulation of phenolic acid and lignin biosynthesis in swede root tissue by ethylene, *Phytochemistry*, 12, 107, 1973.
103. **Rhodes, M. J. C. and Wooltorton, L. S. C.,** The biosynthesis of phenolic compounds in wounded plant storage tissue, in *Biochemistry of Wounded Plant Tissues*, Kahl, G., Ed., Walter de Gruyter, Berlin, 1978, 243.
104. **Hughes, R. K. and Dickerson, A. G.,** The effect of ethylene on phenylalanine ammonia lyase (PAL) induction by a fungal elicitor in *Phaseolus vulgaris, Physiol. Mol. Plant Pathol.*, 34, 361, 1989.
105. **Ke, D. and Saltveit, M. E.,** Wound-induced ethylene production, phenolic metabolism and susceptibility to russet spotting in iceberg lettuce, *Physiol. Plant*, 76, 412, 1989.
106. **Ebel, J.,** Phytoalexin synthesis: the biochemical analysis of the induction process, *Annu. Rev. Phytopathol.*, 24, 235, 1986.
107. **Paradies, I., Konze, J. R., Elstner, E. F., and Paxton, J.,** Ethylene: indicator but not inducer of phytoalexin synthesis in soybean, *Plant Physiol.*, 66, 1106, 1980.
108. **Chalutz, E. and Stahmann, M. A.,** Induction of pisatin by ethylene, *Phytopathology*, 59, 1972, 1969.
109. **Uegaki, R., Fujimori, T., Kaneko, H., Kubo, S., and Kato, K.,** Phytuberin and phytuberol, sesquiterpenes from *Nicotiana tabacum* treated with ethrel, *Phytochemistry*, 19, 1543, 1980.
110. **Chalutz, E., DeVay, J. E., and Maxie, E. C.,** Ethylene-induced isocoumarin formation in carrot root tissue, *Plant Physiol.*, 44, 235, 1969.
111. **Coxon, D. T., Curtis, R. F., Price, K. R., and Levett, G.,** Abnormal metabolites produced by *Daucus carota* roots stored under conditions of stress, *Phytochemistry*, 12, 1881, 1983.
112. **Jaworski, J. G. and Kuc, J.,** Effect of ethrel and *Ceratocystis fimbriata* on the synthesis of fatty acids and 6-methoxymellein in carrot root, *Plant Physiol.*, 53, 331, 1974.
113. **Hoffman, R. M. and Heale, J. B.,** 6-Methoxymellein accumulation and induced resistance to *Botrytis cinerea* Pers. ex Pers., in carrot slices treated with phytotoxic agents and ethylene, *Physiol. Mol. Plant Pathol.*, 30, 67, 1987.
114. **Henfling, J. W. D. M., Lisker, N., and Kuc, J.,** Effect of ethylene on phytuberin and phytuberol accumulation in potato tuber slices, *Phytopathology*, 68, 857, 1978.

115. **Alves, L. M., Heisler, E. G., Kissinger, J. C., Patterson, J. M., and Kalan, E. B.**, Effects of controlled atmospheres on production of sesquiterpenoid stress metabolites by white potato tuber. Possible involvement of cyanide-resistant respiration, *Plant Physiol.*, 63, 359, 1979.
116. **Hoffman, R., Roebroeck, E., and Heale, J. B.**, Effects of ethylene biosynthesis in carrot root slices on 6-methoxymellein accumulation and resistance to *Botrytis cinerea, Physiol. Plant.*, 73, 71, 1988.
117. **Vance, C. P., Kirk, T. K., and Sherwood, R. T.**, Lignification as a mechanism of disease resistance, *Annu. Rev. Phytopathol.*, 18, 259, 1980.
118. **Cassab, G. I. and Varner, J. E.**, Cell wall proteins, *Annu. Rev. Plant Physiol. Plant Mol. Biol.*, 39, 321, 1988.
119. **Geballe, G. T. and Galston, A. W.**, Ethylene as an effector of wound-induced resistance to cellulase in oat leaves, *Plant Physiol.*, 70, 788, 1982.
120. **Geballe, G. T. and Galston, A. W.**, Wound-induced lignin formation and resistance to cellulase in oat leaves, *Phytopathology*, 73, 619, 1983.
121. **Esquerré-Tugayé, M.-T., Lafitte, C., Mazau, D., Toppan, A., and Touzé, A.**, Cell surfaces in plant-microorganism interactions. II. Evidence for the accumulation of hydroxyproline-rich glycoproteins in the cell walls of diseased plants as a defense mechanism, *Plant Physiol.*, 64, 320, 1979.
122. **Stermer, B. A. and Hammerschmidt, R.**, Association of heat shock induced resistance to disease with increased accumulation of insoluble extensin and ethylene synthesis, *Physiol. Mol. Plant Pathol.*, 31, 453, 1987.
123. **Roby, D., Toppan, A., and Esquerré-Tugayé, M.-T.**, Cell surfaces in plant microorganism interactions. V. Elicitors of fungal and of plant origin trigger the synthesis of ethylene and of cell wall hydroxyproline-rich glycoprotein in plants, *Plant Physiol.*, 77, 700, 1985.
123a. **Mattoo, A. K., Anderson, J. D., Chalutz, E., and Lieberman, M.**, Influence of enol ether amino acids, inhibitors of ethylene biosynthesis, on aminoacyl transfer RNA synthetases and protein synthesis, *Plant Physiol.*, 64, 289, 1979.
124. **Ryan C. A.**, Oligosaccharide signalling in plants, *Annu. Rev. Cell Biol.*, 3, 295, 1987.
125. **Roby, D., Toppan, A., and Esquerré-Tugayé, M.-T.**, Cell surfaces in plant micro-organism interactions. VIII. Increased proteinase inhibitor activity in melon plants in response to infection by *Colletotrichum lagenarium, Physiol. Mol. Plant Pathol.*, 30, 453, 1987.
126. **Rickauer, M., Fournier, J., and Esquerré-Tugayé, M.-T.**, Induction of proteinase inhibitors in tobacco cell suspension culture by elicitors of *Phytophthora parasitica* var. *nicotiana, Plant Physiol.*, 90, 1065, 1989.
127. **Tissera, P. and Ayres, P. G.**, Endogenous ethylene affects the behaviour of stomata in epidermis isolated from rust infected faba bean (*Vicia faba* L.), *New Phytol.*, 104, 53, 1986.
128. **Sexton, R. and Roberts, J. A.**, Cell biology of abscission, *Annu. Rev. Plant Physiol.*, 33, 133, 1982.
129. **Osborne, D. J.**, Abscission, *Crit. Rev. Plant Sci.*, 8, 103, 1989.
130. **Goto, M., Yaguchi, Y., and Hyodo, H.**, Ethylene production in citrus leaves infected with *Xanthomonas citri* and its relation to defoliation, *Physiol. Plant Pathol.*, 16, 343, 1980.
131. **Ben-David, A., Bashan, Y., and Okon, Y.**, Ethylene production in pepper (*Capsicum annuum*) leaves infected with *Xanthomonas campestris* pv. *vesicatoria, Physiol. Mol. Plant Pathol.*, 29, 305, 1986.
132. **Ketring, D. L. and Melouk, H. A.**, Ethylene production and leaflet abscission of three peanut genotypes infected with *Cercospora arachidicola* Hori, *Plant Physiol.*, 69, 789, 1982.
133. **Tzeng, D. D. and DeVay, J. E.**, Physiological response of *Gossypium hirsutum* L. to infection by defoliating and non-defoliating pathotypes of *Verticillium dahliae* Kleb., *Physiol. Plant Pathol.*, 26, 57, 1985.
134. **Martin, W. R., Morgan, P. W., Sterling, W. L., and Kenerley, C. M.**, Cotton fleahopper and associated microorganisms as components in the production of stress ethylene by cotton, *Plant Physiol.*, 87, 280, 1988.
135. **Samuel, G.**, On the shot-hole disease caused by *Clasterosporium carpophilum* and on the "shot-hole" effect, *Ann. Bot.*, 41, 375, 1927.
136. **Gomez Lim, M. A., Kelly, P., Sexton, R., and Trewavas, A. J.**, Identification of chitinase mRNA in abscission zones from bean (*Phaseolus vulgaris* Red Kidney) during ethylene-induced abscission, *Plant Cell Environ.*, 10, 741, 1987.
137. **Olien, W. C. and Bukovac, M. J.**, Ethephon-induced gummosis in sour cherry, *Plant Physiol.*, 70, 547, 1982.
138. **Shain, L. and Miller, J. B.**, Ethylene production by excised sapwood of clonal eastern cottonwood and the compartmentalization and closure of seasonal woods, *Phytopathology*, 78, 1261, 1988.
139. **Vandermolen, G. E., Labavitch, J. M., Strand, L. L., and DeVay, J. E.**, Pathogen-induced vascular gels: ethylene as a host intermediate, *Physiol. Plant.*, 59, 573, 1983.
140. **Chalutz, E. and DeVay, J. E.**, Production of ethylene *in vitro* and *in vivo* by *Ceratocystis fimbriata* in relation to disease development, *Phytopathology*, 59, 750, 1969.
141. **Brown, G. E. and Barmore, C. R.**, The effect of ethylene on susceptibility of Robinson tangerines to anthracnose, *Phytopathology*, 67, 120, 1977.

142. **El-Kazzaz, M. K., Sommer, N. F., and Fortlage, R. J.,** Effect of controlled atmospheres on postharvest decay and quality of fresh strawberries, *Phytopathology*, 73, 282, 1983.
143. **Brown, G. E.,** Hypersensitive response of orange-colored robinson tangerines to *Colletotrichum gloeosporioides* after ethylene treatment, *Phytopathology*, 68, 700, 1978.
144. **El-Kazzaz, M. K., Sommer, N. F., and Kader, A. A.,** Ethylene effects on *in vitro* and *in vivo* growth of certain postharvest fruit-infecting fungi, *Phytopathology*, 73, 998, 1983.
145. **Pennazio, S. and Roggero, P.,** Ethylene biosynthesis in soybean plants during the hypersensitive reaction to tobacco necrosis virus, *Physiol. Mol. Plant Pathol.*, 36, 121, 1990.
146. **Cronshaw, D. K. and Pegg, G. F.,** Ethylene as a toxin synergist in *Verticillium* wilt of tomato, *Physiol. Plant Pathol.*, 9, 33, 1976.
147. **Elad, Y.,** Involvement of ethylene in the disease caused by *Botrytis cinerea* on rose and carnation flowers and the possibility of control, *Ann. Appl. Biol.*, 113, 589, 1988.
148. **Elad, Y.,** Production of ethylene in tissues of tomato, pepper, French-bean and cucumber in response to infection by *Botrytis cinerea*, *Physiol. Mol. Plant Pathol.*, 36, 277, 1990.
149. **Kepczynska, E.,** Ethylene requirement during germination of *Botrytis cinerea*, *Physiol. Plant.*, 77, 369, 1989.
150. **Daly, J. M., Seevers, P. M., and Ludden, P.,** Studies on wheat stem rust resistance controlled at the Sr-6 locus. II. Ethylene and disease reaction, *Phytopathology*, 60, 1648, 1970.
151. **Lieberman, M.,** Biosynthesis and action of ethylene, *Annu. Rev. Plant Physiol.*, 30, 533, 1979.
152. **Arshad, M. and Frankenberger, W. T., Jr.,** Biosynthesis of ethylene by *Acremonium falciforme*, *Soil Biol. Biochem.*, 21, 633, 1989.
153. **Wilkes, J., Dale, G. T., and Old, K. M.,** Production of ethylene by *Endothia gyrosa* and *Cytospora eucalypticola* and its possible relationship to kino vein formation in *Eucalyptus maculata*, *Physiol. Mol. Plant Pathol.*, 34, 171, 1989.
154. **Primrose, S. B.,** Ethylene-forming bacteria from soil and water, *J. Gen. Microbiol.*, 97, 343, 1976.
155. **Shipston, N. and Bunch, A. W.,** The physiology of L-methionine catabolism to the secondary metabolite ethylene by *Escherichia coli*, *J. Gen. Microbiol.*, 135, 1489, 1989.
156. **Goto, M., Ishida, Y., Tachikawa, Y., and Hyodo, H.,** Ethylene production by the Kudzu strains of *Pseudomonas syringae* pv. *phaseolicola* causing halo blight in *Pueraria lobata* (Wild) Ohwi, *Plant Cell Physiol.*, 26, 141, 1985.
157. **Goto, M. and Hyodo, H.,** Ethylene production by cell-free extract of the Kudzu strain of *Pseudomonas syringae* pv. *phaseolicola*, *Plant Cell Physiol.*, 28, 405, 1987.
158. **Ilag, L. and Curtis, R. W.,** Production of ethylene by fungi, *Science*, 159, 1357, 1968.
159. **Sharma, A., Padwal-Desai, S. R., and Nadkarni, G. B.,** Possible implications of reciprocity between ethylene and aflatoxin biogenesis in *Aspergillus flavus* and *Aspergillus parasiticus*, *Appl. Environ. Microbiol.*, 49, 79, 1985.
160. **Chalutz, E., Lieberman, M., and Sisler, H. D.,** Methionine-induced ethylene production by *Penicillium digitatum*, *Plant Physiol.*, 60, 402, 1977.
161. **Chou, T. W. and Yang, S. F.,** The biogenesis of ethylene in *Penicillium digitatum*, *Arch. Biochem. Biophys.*, 157, 73, 1973.
162. **Fukuda, H., Fujii, T., and Ogawa, T.,** Preparation of a cell-free ethylene-forming system from *Penicillium digitatum*, *Agric. Biol. Chem.*, 50, 977, 1986.
163. **Swart, A. and Kamerbeck, G. A.,** Different ethylene production *in vitro* by several species and formae speciales of *Fusarium*, *Neth. J. Plant. Pathol.*, 82, 81, 1976.
164. **Swart, A. and Kamerbeck, G. A.,** Ethylene production and mycelium growth of the tulip strain of *Fusarium oxysporum* as influenced by shaking of and oxygen supply to the medium, *Physiol. Plant Pathol.*, 39, 38, 1977.
165. **Coleman, L. W. and Hodges, C. F.,** The effect of methionine on ethylene and 1-aminocyclopropane-1-carboxylic acid production by *Bipolaris sorokiniana*, *Phytopathology*, 76, 851, 1986.
166. **Achilea, O., Chalutz, E., Fuchs, Y., and Rot, I.,** Ethylene biosynthesis and related physiological changes in *Penicillium ditigatum*-infected grapefruit, *(Citrus paradis)*, *Physiol. Plant Pathol.*, 26, 125, 1985.
167. **Achilea, O., Fuchs, Y., Chalutz, E., and Rot, I.,** The contribution of host and pathogen to ethylene biosynthesis in *Penicillium digitatum*-infected citrus fruit, *Physiol. Plant Pathol.*, 27, 55, 1985.
168. **Coleman, L. W. and Hodges, C. F.,** Ethylene biosynthesis in *Poa pratensis* leaves in response to injury or infection by *Bipolaris sorokiniana*, *Phytopathology*, 77, 1280, 1987.
169. **Chalutz, E.,** No rôle for ethylene in the pathogenicity of *Penicillium digitatum*, *Physiol. Plant Pathol.*, 14, 259, 1979.
170. **Beijersbergen, J. C. and Bergman, B. H. H.,** The influence of ethylene on the possible resistance mechanism of the tulip (*Tulipa* spp.) against *Fusarium oxysporum*, *Acta Bot. Neerl.*, 22, 172, 1973.
171. **Prince, T., Stephens, C. T., and Herner, R. C.,** Pathogenicity, fungicide resistance, and ethylene production of *Penicillium* spp. isolated from tulip bulbs, *Phytopathology*, 78, 682, 1988.

172. **Graham, J. H. and Lindermann, R. G.,** Ethylene production by ectomycorrhizal fungi, *Fusarium oxysporum* f.sp. *pini*, and by aseptically synthesized ectomycorrhizae and Fusarium-infected Douglas fir roots, *Can. J. Microbiol.*, 26, 1340, 1980.
173. **Rupp, L. A., DeVries, H. E., II, and Mudge, K. W.,** Effect of aminocyclopropane carboxylic acid and aminoethoxyvinylglycine on ethylene production by ectomycorrhizal fungi, *Can. J. Bot.*, 67, 483, 1989.
174. **Graham, J. H. and Lindermann, R. G.,** Effect of ethylene on root growth, ectomycorrhiza formation and *Fusarium* infection of Douglas fir, *Can. J. Bot.*, 59, 149, 1981.
175. **Wilson, E. R. L. and Field, R. J.,** Dichotomous branching in lateral roots of pine: the effect of 2-chloroethylphosphonic acid on seedlings of *Pinus radiata* D. Don, *New Phytol.*, 98, 465, 1984.
176. **Rupp, L. A. and Mudge, K. W.,** Ethephon and auxin induce mycorrhiza-like changes in the morphology of root organ cultures of Mugo pine, *Physiol. Plant.*, 64, 316, 1985.
177. **Rupp, L. A., Mudge, K. W., and Negm, F. B.,** Involvement of ethylene in ectomycorrhiza formation and dichotomous branching of roots in mugo pine seedlings, *Can. J. Bot.*, 67, 477, 1989.
178. **Bleecker, A. B., Estelle, M. A., Somerville, C., and Kende, H.,** Insensitivity to ethylene conferred by a dominant mutation in *Arabidopsis thaliana, Science,* 241, 1086, 1988.

Chapter 17

ETHYLENE AND AGRICULTURE*

Klaus Lürssen

TABLE OF CONTENTS

* Dedicated to Professor Wilhelm Menke on the occasion of his 80th birthday.

I. INTRODUCTION

As early as the end of the 19th Century, the physiological effect of ethylene on plants was put to practical use by farmers on the Azores. While experimenting with fumes to control insects, a farmer observed an early and uniform flowering of pineapple plants after treatment. The use of fumes to stimulate pineapple flowering soon became a common practice on the Azores.[1] Today substances that modify the development and physiology of crops or ornamental plants are called plant growth regulators. Thus, ethylene may be regarded as the first plant growth regulator used in agriculture, although its nature as the active ingredient in fumes was not known at that time, nor was its role as a plant hormone established.

Ethylene as a gas would, of course, be difficult to handle in agriculture. As such this hormone did not play a major role for practical purposes until it became available as a liquid ethylene-releasing compound (ethephon) at the end of the 1960s.[2] Today, this compound is still one of the major plant growth regulators on the market with many different uses.

II. ETHYLENE-RELEASING COMPOUNDS AS A BASIS FOR PRACTICAL USES IN AGRICULTURE

Apart from limited uses of ethylene in greenhouses and fruit storage facilities, an application of the gas itself in agriculture is hardly possible. At the end of the 1960s a commercial product, 2-chloroethanephosphonic acid (common name ethephon) became available as a liquid formulation.[2,3] This compound is stable below pH 3.5 but decomposes in more alkaline aqueous solutions, releasing ethylene.[3] The higher the pH the faster the rate of ethylene release.[4] Several other compounds with similar modes of action were introduced later (for summary see Reference 5). The mode of decomposition in aqueous solution for some of these products is as follows:

Ethephon (Amchem/GAF):

$$Cl{-}CH_2{-}CH_2{-}PO_3H_2 + H_2O \rightarrow Cl^- + CH_2{=}CH_2 + H_3PO_4$$

Etacelasil (Ciba-Geigy):

$$Cl{-}CH_2{-}CH_2{-}Si(O{-}CH_2{-}CH_2{-}O{-}CH_3)_3 + H_2O \rightarrow$$

$$Cl^- + CH_2{=}CH_2 + SiO_2 + 3HO{-}CH_2{-}CH_2{-}O{-}CH_3$$

HOL 1302 (Bayer AG):

$$Cl{-}CH_2{-}CH_2{-}SO{-}OH + H_2O \rightarrow Cl^- + CH_2{=}CH_2 + H_2SO_3$$

Another plant growth regulator, 2-hydroxyethylhydrazine, also causes typical ethylene reactions in plants. However, this compound does not release ethylene in aqueous solutions but may be converted to ethylene within the plant.[6,7] At the end of the 1970s, 1-aminocyclopropane-1-carboxylic acid (ACC) and its derivatives were found to exert ethylene-like reactions in plants in a manner suitable for agricultural uses.[8] ACC was soon identified as a previously unknown intermediate in ethylene biosynthesis.[9,10]

It is remarkable that the different ethylene-releasing compounds reveal quite different spectra of usefulness in agriculture. Of course, all plant responses are still typical reactions to ethylene. Factors influencing the kind of response are: the penetration of the ethylene-releasing compound into the plant and possibly into different cell compartments, the speed and duration of ethylene release, the mode of ethylene generation, and the sensitivity of the plants due to their developmental stage and the plant species. The total amount of ethylene released seems to play only a minor role.[11-13]

Of all the ethylene-releasing compounds ethephon has the broadest spectrum of usefulness in agriculture. This certainly is one reason that it is today the only compound of this group available on the market, being also one of the major plant growth regulators with respect to sales volume. Ethephon has also been a very suitable tool for investigations of the different effects of the plant hormone ethylene.

III. PHYSIOLOGICAL EFFECTS OF ETHYLENE USEFUL IN AGRICULTURE AND HORTICULTURE

The plant hormone ethylene strongly influences nearly every developmental stage in plant growth, from germination to fruit ripening and senescence. It is obvious that a compound with so many different effects may be useful in many ways to modify plant growth and development and adjust crop plants to the requirements of the farmer. However, accompanying the beneficial effects are some adverse effects that limit its practical usefulness. Furthermore, effects clearly achieved under experimental conditions may not always be achieved under conditions of practical application.

A. ABSCISSION

The involvement of ethylene in the abscission processes is well documented and the mode of action is relatively well understood[14,15] (see also Chapters 11, 12, and 13). Enhancement of abscission is of interest in horticulture and agriculture in several ways. Controlled abscission of flowers or young fruit, referred to in horticulture as thinning, controlled abscission or loosening of ripe fruits for facilitation of mechanical harvesting, and controlled shedding of leaves prior to normal leaf drop in several crops for different purposes are desirable effects. In principle, all these effects can be achieved by ethylene. Accordingly, much work was done in this field when suitable ethylene-releasing compounds became available. Several ethylene-releasing compounds or intermediates in ethylene biosynthesis have been tested for these purposes, indicating a differential response of fruits and leaves to different compounds.[11-13]

In fruit trees, especially in apple, the seeds produce hormones that are assumed to reduce the number of blossoms in the following season. Alternating high and low yields in succeeding seasons result. A high number of fruits that exceeds the nutritional capacity of the tree yields smaller fruits of lower market value. Therefore, thinning is needed and routinely practiced in fruit production. A further aspect is the prevention of heavy fruit-set in young trees, to avoid disturbances in vegetative growth.[16] The effect of ethephon on thinning was investigated in apple,[17] plum,[18] peaches,[19] and coffee,[20] showing principally the usefulness of ethylene-releasing compounds for this purpose. However, acceptance by the growers seems to be limited, as the performance is rather variable and extremely dependent on an exact timing of the application with respect to the developmental stage of the young fruit. It should be noted that, at the time of application of ethylene-releasing compounds for fruit thinning, the leaves of the trees are still young and do not readily respond with the formation of an abscission layer. The major concern for application is a possible overreaction of the trees resulting in the loss of the crop.

The situation is different in the case of induction of fruit loosening to facilitate mechanical harvesting. Both the ripe fruits and the old leaves respond to ethylene treatment with the formation of an abscission layer. The possibility of leaf drop seems to be the major limitation for a widespread use of ethylene-releasing compounds for fruit loosening. The advantage of fruit loosening would be the reduction in shaking force and hence a reduction in injury to the tree and the root system. The usefulness of ethephon and other ethylene-releasing compounds has been investigated and established in many different fruit species, for example in apple,[21] cherry,[22-24] plum,[23,25] olive,[26-29] and grape.[30] In coffee, the simultaneous presence of fruits and flowers of different developmental stages complicates the situation and an unwanted thinning of young fruits and blossoms may occur.[31,32] Citrus fruit loosening is also mediated by ethylene; however, due to phytotoxic effects of ethephon, other chemicals like cycloheximide that induce ethylene biosynthesis are preferable for this crop.[33,34] Surprisingly, in oil palms an opposite effect of ethephon, namely the retardation of fruit abscission, has been reported.[35]

Defoliation is another candidate for the use of ethylene-generating compounds. However, for the most important market, preharvest defoliation in cotton, the use of ethylene-releasing compounds is more expensive and inferior to other commercial products like tribufos or thidiazuron. The action of the latter compound is at least partly mediated by a stimulation of endogenous ethylene biosynthesis.[36]

The rate of survival of trees and shrubs is much higher if the leaves are removed prior to transplantation or storage. Mechanical removal of the leaves causes damage to twigs, bark, and buds. Thus, tree nurseries are eager to obtain an effective and safe chemical for defoliation, which does not interfere with winter hardiness or sprouting in the next season. Furthermore, in some areas, defoliation of fruit trees after the harvest seems to be of interest to facilitate an early pruning of the trees.[37] Many different chemicals have been tested for defoliation of nursery stocks, including inorganic salts that enhance ethylene biosynthesis and ethylene-releasing compounds.[38,39] One problem is the species-specific response to the different chemicals. Therefore, much field research work has to be done to enable safe recommendations for a chemical treatment.

B. GERMINATION, SPROUTING, APICAL DOMINANCE, AND DORMANCY

Depending on the plant species, ethylene can either inhibit or enhance germination of seed[40-44] (see Chapter 8). Inhibition of germination could principally be used for weed control but ethylene only retards the germination and does not kill the weed seeds. Furthermore, the inhibitory effect is limited to only a few species.

There is more interest in enhancement of germination for weed control. It is often a problem that weeds germinate over a prolonged time period during the growing season, resulting in poor control by commercial herbicides. In such situations it is necessary to make sequential treatments with herbicides for sufficient weed control. Furthermore, viable but dormant weed seeds remain in the soil for many years until germination, making weed control more difficult. Several investigations were made in an attempt to get a simultaneous germination of weeds using ethylene with the aim of facilitating weed control.[40,45] A special situation occurs with the root parasite *Striga lutea* that attacks corn and other grasses. Normally, the seeds will germinate only in the presence of a host plant. However, aged but dormant seeds can be forced to germinate by ethylene and they will eventually die in the absence of host roots.[46,47]

Dormant buds or buds suppressed to sprout by apical dominance or rhizomes or aerial shoots of *Agropyron repens* make control of this troublesome weed very difficult, as many herbicides do not affect these buds. Ethephon treatment can overcome bud dormancy or growth suppression.[48,49]

Though many of the investigations to improve weed control with ethylene by stimulation of germination or by breaking dormancy revealed promising results, the method has not been accepted in agricultural practice to date.

Terpal, a plant growth regulator containing ethephon, was shown to increase tillering in barley.[50] Release from apical dominance by ethylene is probably involved in this phenomenon. Enhancement of sprouting of corms and an increased production of cormels is reported in gladiolus.[51-53]

C. FEMALE FLOWER INDUCTION

Sex determination in monoecious plants is influenced by gibberellins and ethylene. High levels of gibberellins induce male flowers, while high levels of ethylene induce female flowers[54-57] (see Chapter 12). Likewise, in the dioecious plant *Cannabis sativa* female flower induction by ethylene on genetically male plants has been reported.[58] ACC, the immediate precursor of ethylene, may be involved in female flower determination.[59] Ethephon can be successfully applied in the Cucurbitaceae for female flower differentiation. During normal development, the plants initially produce male and later female flowers. Ethylene treatment causes an earlier development of female flowers and, hence, the possibility of an earlier harvest[60,61] or of a once-over mechanical harvest in pickling cucumbers.[62] The use of ethephon to produce hybrids in Cucurbitaceae has also been discussed.[63]

D. FLOWERING

As mentioned in the introductory section, by the end of the last century induction of flowering was the first effect of ethylene used in practice. Rodriguez[64] identified ethylene as the active ingredient in the smoke used. When ethephon became available, its usefulness in this respect was shown.[2,3] Mainly Bromeliaceae can be induced to flower by ethylene treatment. Of practical importance is its use in pineapple.[65-67] Without chemical stimulation, field-grown pineapple plants do not flower synchronously, making crop management difficult. Treatment with an ethylene-releasing compound such as ethephon stimulates uniform flower development. Thus, the time of harvest and the amount of fruit can be exactly planned under tropical growing conditions. Interestingly, in the Ivory Coast, loading space on ships for transportation of fresh pineapple fruits to Europe is ordered when flowering is stimulated by ethephon in the field — about 3 to 4 months prior to harvest. Exact timing of the harvest is accomplished by a second treatment with ethephon to enhance coloration of the fruits about a week before harvest (see also Section H, on "Ripening").

Besides pineapple, there are many other species of the Bromeliaceae (i.e., *Aechmea, Guzmannia, Neoregelia, Tillandsia,* and *Vriesea*) that are used as ornamental plants. In these species, ethephon is also widely used to stimulate flowering and thus, to enable sales of flowering plants at any time of the year.[68]

E. GROWTH INHIBITION

In growing tissues, ethylene causes a modification of cell dimensions. The cells in treated tissue are inhibited in logitudinal growth but may achieve a larger diameter (Chapters 9 and 10). Ethylene-treated plants are thus stunted compared to untreated ones. Growth inhibition is of great interest in agriculture and horticulture. Accordingly, there have been many investigations conducted with ethylene-releasing compounds to find any practical usefulness in this field.

The largest market today for growth inhibitors is the cereal market in western Europe. Good climatic growing conditions with adequate rainfall and very intensive crop management practices, including the application of high amounts of nitrogen fertilizer and fungicides, have led to the highest average yields worldwide. Under these growing conditions, the cereal crop tends to lodge during heavy rainfall (i.e., the culms bend down to the ground or may even break). As a result, normal development of the cereals is disrupted. Mechanical harvesting is more difficult, especially if certain weeds have grown in and over the lodged cereals. Yield and crop quality may be reduced and fungal growth may be stimulated. Additional drying of the harvested grains may be necessary, resulting in additional costs for the farmer. Under extreme circumstances, the crop may be completely destroyed, especially in rye and barley. Thus, the application of an antilodging material is more or less essential. In wheat, chlormequat-chloride is an excellent material that practically prevents lodging. It inhibits elongation growth and thickens and stiffens the wheat culms. In principle, ethylene-releasing compounds like ethephon can be used for the same purpose in cereals[2] but their effectiveness in wheat is inferior to that of chlormequat-chloride,[69] especially with regard to yield.[70] In contrast to chlormequat-chloride, ethephon has no soil activity and seems to be superior under dry growing conditions. Chlormequat-chloride is less effective in other cereals but ethylene, applied as ethephon, reduces elongation growth and lodging substantially in barley, oats, and rye.[70-72] Application of ethephon in combination with other growth inhibitors, for example chlormequat-chloride, increases the effectiveness of the treatment in rye and barley.[73-76]

Early experiments with ethephon in oilseed rape showed a growth promotion rather than a growth inhibition.[77] Recent efforts to inhibit growth and prevent lodging of oilseed rape with ethephon or combinations of ethephon with other plant growth regulators have not been very successful. Ethephon treatments are obviously much less effective in this case than treatments with some new inhibitors of gibberellin biosynthesis. Other possible uses of ethylene-releasing compounds for growth inhibition are of minor importance in agriculture and horticulture. Under certain growing conditions, grape vines may be successfully treated with ethephon to reduce vegetative growth and increase fruit quality.[78-80]

F. LATEX STIMULATION IN *HEVEA*

One of the most important applications of ethephon is its use in stimulating latex flow in rubber trees. Many other ethylene-releasing stimulants have been tested, but in this application also ethephon is the best performing chemical. ACC turned out to be much less effective.[81] The normal harvest procedure prior to the use of stimulants was a spiral 360° cut into the bark of the rubber tree. The dripping latex was collected in vials for about 2 days, after which the latex flow stops. If ethephon is applied into the groove or is painted on the bark near the cut, the latex flow increases, depending on the concentration used and persists for a longer time. Excessive use of ethephon at high concentrations may increase the yield by several hundred percent but this treatment shortens the lifetime of the trees considerably. A well-established practice on rubber tree plantations today is the use of a relatively low ethephon concentration leading to yield increases of up to 50%. This stimulation does not adversely affect the trees provided they are sufficiently fertilized. Yield increases, however, is only one of the beneficial effects of the treatment. Other advantages are the need to cut only a 90° spiral for tapping and still get the full yield. This means (together with the longer duration of the latex flow) a considerable saving of labor. A further consequence is less damage to the bark by the tapping procedure. Most of the research for the development of latex stimulants was done at the Rubber Research Institute of Malaysia.[82-84]

Studies on the mode of action of latex stimulation by ethylene revealed a probable activation of proton pump ATPases both on the plasmalemma and on the lutoidic tonoplast leading to a rise in pH of the cytosol.[85-88] The pH changes lead to an activation of latex biosynthesis. Enhanced polyribosome formation, protein synthesis, and sucrose transport into latex vessels have been reported.[89-93]

G. MALE STERILITY

Male sterility is an important prerequisite for hybrid seed production. Great efforts have been undertaken to find chemicals that induce male sterility. Ethephon is one of the chemicals discussed in the literature as a possible candidate.[94,95] An additional mitosis induced by ethephon is responsible for the effect.[94] However, time of application seems to be extremely critical for the desired effect and reproducibility under field conditions is not good enough for a practical application.

H. RIPENING

Enhancement of fruit ripening by ethylene is one of the most widely investigated effects of the hormone[96,97] (see Chapter 13). In the past, ethylene gas was used to enhance ripening of banana fruits in storage rooms prior to marketing. This practice is no longer necessary as hands of bananas are now sealed in plastic bags and stored in temperature-controlled rooms. The temperature regime determines the onset of ethylene biosynthesis and, thus, of the ripening process. In modern controlled-atmosphere fruit storage, knowledge about the antagonism between carbon dioxide and ethylene[98,99] is used to control the ripening process.

In agriculture and horticulture, preharvest application of ethylene gas is not practical. Ethephon has become a valuable tool to control ripening in different crops. It should be pointed out that enhancement of ripening generally can only be achieved when the fruit reach a certain stage of maturity, often referred to, in tomato for example, as the "mature-green stage". Economically, the most important use is in processing tomato in the U.S. The crop is harvested once-over mechanically. Without the use of a growth regulator to enhance ripening, only about 80 to 85% of the crop would be in the mature stage at harvest time, compared to nearly 100% after ethephon treatment.[100,101] The quality of fruit induced to ripen by ethephon is not different from that of naturally ripened fruit.[102] In pineapple, ethephon is also widely used to stimulate coloration of the crop; fruit color being a characteristic which affects the market value. Unfortunately, other ripening parameters are not favorably influenced by the treatment and the correct time of application is very essential to obtain good fruit quality.[103,104] Citrus also responds with an enhancement of color formation but not with changes in other ripening parameters.[105] Increased levels of anthocyanin are reported in cranberry after preharvest ethephon application.[106] Strawberries do not respond favorably to ethylene treatment.[107] Stimulation of ripening without adverse effects can be achieved in many other vegetable crops or fruit such as pepper,[108,109] melon,[110] blueberry,[11,112] fig,[113,114] papaya,[115] grape,[116,117] peach,[118] plum, pear,[119] and apple.[120] To prevent a possible early fruit drop caused by ethylene, the simultaneous application of phenoxy acids is recommended in apple. In grape and apple, an earlier time of application during fruit development may also enhance ripening.[121] There have been many investigations on the use of ethephon in coffee. However, its usefulness could not be established convincingly, as the fruiting behavior of coffee differs from that of the other crops mentioned. There are, simultaneously, reproductive organs of different developmental stages on the plant. Thus, ethylene treatment may interfere with earlier developmental stages, adversely resulting in low-quality coffee berries or yield reductions.[122-124]

There are other useful effects of ethylene that would be called "enhancement of ripening" in an agricultural sense, but they must be regarded differently in a physiological sense. The ripening of tobacco leaves is an example.[125-127] Tobacco plants "ripen" beginning at the lower leaves. Hence harvesting starts at the lower leaves and continues with 5 to 7 separate mechanical harvests until the top leaves are mature. The use of ethephon reduces the number of leaf pickings to about 3. Ethephon can be used in sugarcane for another quite different "ripening effect". Juice purity and sugar content are increased in treated plants.[128] A further "ripening" effect can be achieved in cotton. Ethylene treatment results in an improved boll opening and, thus, increases recoverable fiber yield. Furthermore, ethylene treatment a few weeks before harvest preconditions the leaves of cotton for better response to defoliants that are applied later.[129,130] The application of ethephon to achieve boll opening in cotton is the most recent significant contribution to the use of the hormone in agricultural practice. A comparable effect of ethephon is useful for the harvest of nuts. The compound accelerates hull dehiscence, which allows an earlier harvest at a time when kernel quality is higher than later at normal harvest time. Furthermore, a once-over mechanical harvest may be possible.[131,132] A limitation for the application of ethephon is the possibility of defoliation of the trees. Studies have been undertaken to overcome this problem by adding calcium to the spraying solution.[133]

IV. OUTLOOK

Most of the major agricultural uses of ethylene are realized today with the use of the ethylene-releasing compound ethephon. According to the experience with other ethylene-releasing compounds and with the intermediate of the ethylene biosynthesis, ACC, and its derivatives, it is unlikely that other compounds with these modes of action would broaden the spectrum of possible uses over that of ethephon. Furthermore, the cost of development of plant growth regulators today is of an order of magnitude that makes the development of a second compound for a small market, compared to the market size of other pesticides, very unattractive. It seems that there is little more effort in the agrochemicals industry to develop new compounds with this mode of action.

The situation could change, however, if a new type of ethylene-active compound is discovered. It is essential for the effect of ethylene that ethylene receptors are present and active in the plant. If the response of a plant in relation to its developmental stage is mainly dependent on the availability of receptor sites then new types of ethylene-active compounds could be envisaged that would regulate the synthesis of ethylene receptors. Such hypothetical compounds could considerably broaden the spectrum of uses of ethylene.

REFERENCES

1. **Grabham, M.,** Pines in the Azores, *Gardener's Chronicle 1903,* 36.
2. **Sterry, J. R.,** Ethrel, an ethylene-evolving plant growth regulator, *Meded. Rijksfac. Landbouwwet. Gent.,* 34, 462, 1969.
3. **Cooke, A. R. and Randall, D. I.,** 2-Haloethanephosphonic acids as ethylene releasing agents for the induction of flowering in pineapple, *Nature,* 218, 974, 1968.
4. **Warner, H. L. and Leopord, A. C.,** Ethylene evolution from 2-chloroethylphosphonic acid, *Plant Physiology,* 44, 156, 1969.
5. **Draber, W.,** Natürliche und synthetische Wachstumsregulatoren, in *Chemie der Pflanzenschutz- und Schädlingsbekämpfungsmittel,* Band 4, Wegler, R., Ed., Springer-Verlag, Berlin, 1977, 1.
6. **Block, M. J. and Young, D. C.,** Conversion of β-hydroxyethylhydrazine to ethylene, *Nature (London) New Biol.,* 231, 288, 1971.
7. **Dollwet, H. H. A. and Kumamoto, J.,** The conversion of 2-hydroxyethylhydrazine to ethylene, *Plant Physiol.,* 49, 696, 1972.
8. **Schröder, R. and Lürssen, K.,** Mittel zur Regulierung des Pflanzenwachstums, *Dtsch. Offenlegungsschrift Nr.,* 28(24), 517, 1978.
9. **Lürssen, K., Nauman, K., and Schröder, R.,** 1-Aminocyclopropane-1-carboxylic acid—an intermediate of the ethylene biosynthesis in higher plants, *Z. Pflanzenphysiol.,* 92, 285, 1979.
10. **Adams, D. O. and Yang, S. F.,** Ethylene biosynthesis: identification of 1-aminocyclopropane-1-carboxylic acid as an intermediate in the conversion of methionine to ethylene, *Proc. Natl. Acad. Sci. U.S.A.,* 76, 170, 1979.
11. **Lürssen, K.,** Manipulation of crop growth by ethylene and some implications of the mode of generation, in *Chemical Manipulation of Crop Growth and Development,* McLaren J. S., Ed., Butterworth Scientific, London, 1982, 67.
12. **Konze, J. and Lürssen, K.,** Unterschiedliche Wirkungsspektren von Ethylen-Wachstumsregulatoren: Welche Rolle spielt die freigesetzte Ethylenmenge?, in *Regulation des Phytohormongehaltes,* Bangerth, F., Ed., Hohenheimer Arbeiten Band 129, Verlag Eugen Ulmer, Stuttgart, 1983, 145.
13. **Lürssen, K. and Konze, J.,** Relationship between ethylene production and plant growth after application of ethylene releasing plant growth regulators, in *Ethylene and Plant Development,* Roberts, J. A. and Tucker, G. A., Eds., Butterworths, London, 1985, 363.
14. **Sexton, R., Lewis, L. N., Trewavas, A. J., and Kelly, P.,** Ethylene and abscission, in *Ethylene and Plant Development,* Roberts, J. A. and Tucker, G. A., Eds., Butterworths, London, 1985, 173.
15. **Osborne, D. J., McManus, M. T., and Webb , J.,** Target cells for ethylene action, in *Ethylene and Plant Development,* Roberts, J. A. and Tucker, G. A., Eds., Butterworths, London, 1985, 197.
16. **Knight, J. N. and Jackson, E.,** Apple thinning, Rep. East Malling Res. Station for 1979, 1980, 227.

17. **Knight, J. N.**, Chemical thinning of apple cultivar, Laxton's Superb, *J. Hortic. Sci.*, 53, 63, 1978.
18. **Schumacher, R. and Fankhauser, F.**, Fruchtausdünnung von Fellenbergzwetschgen (Prunus domestica) mit Ethrel, *Schweiz. Z. Obst. Weinbau*, 108, 74, 1972.
19. **Blake, J. A., Biggs, R. H., and Buchanan, D. W.**, Post-bloom thinning of Florida peaches with 2-chloroethylphosphonic acid, *Proc. FL. State Hortic, Soc.*, 81, 257, 1969.
20. **Adenikinju, S. A.**, A preliminary study of the influence of chemical sprays on fruit set and abscission in Robusta coffee (Coffea canephora Pierre ex Froehner), *Turrialba*, 25, 414, 1975.
21. **Schumacher, R., Daepp, H. U., and Fankhauser, F.**, Ethrel zur Erleichterung der Ernte von Mostäpfeln, Schweiz. Z. *Obst. Weinbau*, 105, 617, 1969.
22. **Liebster, G.**, Untersuchungen zur Schüttelfähigkeit von Süβ- und Sauerkirschen, *Der Erwerbsobstbau*, 12, 126, 1970.
23. **Schumacher, R. and Fankhauser, F.**, Ethrel zur Erleichterung der Ernte bei Steinobst, *Schweiz. Z. Obst. Weinbau*, 105, 596, 1969.
24. **Stösser, R.**, Die Induktion eines Trenngewebes bei Früchten von Prunus avium L. durch 2-Chloräthylphosphonsäure, *Planta*, 90, 299, 1970.
25. **Al-Jaru, S. and Stösser, R.**, Der Enfluß von Ethrel und Cycloheximide auf die Trenngewebeausbildung und die Reduktion der Haltekräfte bei Zwetschgen und Mirabellen, *Der Erwerbsobstbau*, 15, 171, 1973.
26. **Hartmann, H. T., Heslop, A. J., and Whisler, J.**, Chemical induction of fruit abscission in olives, *Calif. Agric.*, July 1968, 14.
27. **Ben-Tal, Y. and Lavee, S.**, Ethylene influence on leaf and fruit detachment in 'Manzanillo' olive trees, *Sci. Hortic.*, 4, 337, 1976.
28. **Hartmann, H. T., Reed, W., Whisler, J., and Opitz, K. W.**, Mechanical harvesting of olives, Calif. *Agirc.*, June 1975, 4.
29. **Forlani, M., Pugliano, G., and Rotundo, A.**, La racolta meccanica delle olive: l'impiego di tre "etilen-promotori" sulla cv. "Carolea", *Ann. Fac. Sci. Agrar. Univ. Napoli in Portici Ser. IV*, 10, 3, 1976.
30. **Eynard, I.**, Effects of preharvest application of TH 6241 and CEPA on Vitis vinifera, *Vitis*, 13, 303, 1975.
31. **Browning, G. and Cannell, M. G. R.**, Use of 2-chloroethane phosphoric acid to promote the abscission and ripening of fruit of Coffea arabica L., *J. Hort. Sci.*, 45, 223, 1970.
32. **Oyebade, T.**, Influence of pre-harvest sprays of Ethrel on ripening and abscission of coffee berries, *Turrialba*, 26, 86, 1976.
33. **Wardowski, W. and Wilson, W.**, Progress in abscission of oranges, *Citrus Ind.*, December 1970, 20.
34. **Cooper, W. C. and Henry, W. H.**, Abscission chemical in relation to citrus fruit harvest, *J. Agric. Food Chem.*, 19, 559, 1971.
35. **Chan, K. W., Corley, R. H. V., and Seth, A. K.**, Effects of growth regulators on fruit abscission in oil palm, Elaeis guineensis, *Ann. Appl. Biol.*, 71, 243, 1972.
36. **Suttle, J. C.**, Involvement of ethylene in the action of cotton defoliant thidiazuron, *Plant Physiol.*, 78, 272, 1985.
37. **Gerdts, M. H., Obenhauf, G. L., LaRue, J. H., and Leavitt, G. M.**, Chemical defoliation of fruit trees, *Calif. Agric.*, April 1977, 19.
38. **Insley, H. and Boswell, R. C.**, The enhancement of the chemical defoliation of amenity tree nursery stock, Betula pendula Roth, Alnus incana (L.) Moench, Carpinus betulus L. and Platanus x hispanica Muenchh., by ethephon pretreatment, *J. Hortic. Sci.*, 55, 119, 1980.
39. **Knight, J. N.**, Chemical defoliation of nursery stock. I. Initial experiments with fruit tree material, *J. Hortic. Sci.*, 54, 229, 1979.
40. **Taylorson, R. B.**, Response of weed seeds to ethylene and related hydrocarbons, *Weed Sci.*, 27, 7, 1979.
41. **Toole, V. K., Bailey, W. K., and Toole, E. H.**, Factors influencing dormancy of peanut seeds, *Plant Physiol.*, 39, 822, 1964.
42. **Ketring, D. L. and Morgan, P. W.**, Ethylene as a component of the emanations from germinating peanut seeds and its effect on dormant Virginia-type seeds, *Plant Physiol.*, 44, 326, 1969.
43. **Ketring, D. L. and Morgan P. W.**, Physiology of oil seeds. I. Regulation of dormancy in Virginia-type peanut seeds, *Plant Physiol.*, 45, 268, 1970.
44. **Takayanagi, K. and Harrington, J. F.**, Enhancement of germination rate of aged seeds by ethylene, *Plant Physiol.*, 47, 521, 1971.
45. **Chancellor, R. J., Parker, C., and Teferedegn, T.**, Stimulation of dormant weed seed germination by 2-chloroethylphosphonic acid,. *Pestic. Sci.*, 2, 35, 1971.
46. **Egley, G. H. and Dale, J. E.**, Ethylene, 2-chloroethylphosphonic acid, and witchweed germination, *Weed Sci.*, 18, 586, 1970.
47. **Eplee, R. E. and Langston, M. A.**, Developments in the control of Striga in the USA, *Pesticide Articles and New Summaries*, 22, 61, 1976.
48. **Chancellor, R. J.**, The effects of 2-chloroethylphosphonic acid and chlorflurecol-methyl upon the sprouting of Agropyron repens (L.) Beauv. rhizomes, Proc. 10th. Br. Weed Control Conf., Brighton, November 16-19, 1970, 254.

49. **Caseley, J.**, Effect of 2-chloroethylphosphonic acid on the morphology and apical dominance of Agropyron repens, *Pestic. Sci.*, 1, 114, 1970.
50. **Woodward,, E. J. and Marshall, C.**, Effects of seed treatment with a plant grown regulator on growth and tillering in spring barley (Hordeum distichum) cv. Triumph, *Ann. Appl. Biol.*, 110, 629, 1987.
51. **Halevy, A. H., Shilo, R., and Simchon, S.**, Effect of 2-chloroethanephosphonic acid (Ethrel) on health, dormancy, and flower and corm yield of gladioli, *J. Hortic. Sci.*, 45, 427, 1970.
52. **Simchon, S., Silberstein, Y., Halevy, A. H., and Henis, Y.**, The mode of action of ethephon in increasing health of gladiolus corms *J. Hortic. Sci.*, 47, 369, 1972.
53. **Ginzburg, C.**, Studies on the role of ethylene in gladiolus cormel germination, *Plant Sci. Lett.*, 2, 133, 1974.
54. **Iwahori, S., Lyons, J. M., and Sims, W. L.**, Induced femaleness in cucumber by 2-chloroethanephosphonic acid, *Nature* 222, 271, 1969.
55. **Splitstoesser, W.**, Effects of 2-chloroethylphosphonic acid and gibberellic acid on sex expression and growth of pumpkins, *Physiol. Plant.*, 23, 762, 1970.
56. **Iwahori, S., Lyons J. M., and Smith, O. E.**, Sex expression in cucumber plants as affected by 2-chloroethylphosphonic acid, ethylene, and growth regulators, *Plant Physiol.*, 46, 412, 1970.
57. **Rudich, J., Halevy, A. H., and Kedar, N.**, Ethylene evolution from cucumber plants as related to sex expression, *Plant Physiol.*, 49, 998, 1972.
58. **Mohan Ram, H. Y. and Jaiswal, V. S.**, Induction of female flowers on male plants of Cannabis sativa L. by 2-chloroethanephosphonic acid, *Experientia*, 26, 214, 1970.
59. **Hume, R. J. and Lovell, P. H.**, Role of 1-aminocyclopropyl-1-carboxylic acid in the control of female flowering in Cucurbita pepo, *Physiol. Plant.*, 59, 324, 1983.
60. **Cantliffe, D. J. and Woods, F. E.**, Promotion of early yields in summer squash by application of ethephon, *Proc. Fl. State Hortic. Soc.*, 91, 261, 1978.
61. **Tompkins, D. R. and Smay, S. E.**, Influence of ethephon on early yields of yellow summer straightneck squash, *Arkansas Farm Res.*, Sept.-Oct., 20., 1971.
62. **Tompkins, D. R. and Shulteis, D. T.**, Using growth regulating substances to adapt pickling cucumbers to mechanical harvsting, *Arkansas Farm Res.*, Nov.-Dec., 19, 1970.
63. **Robinson, R. W., Wilczynski, H., de la Guardia, M. D., Shannon, S., and Dennis, F. G.**, Chemical regulation of fruit ripening and sex expression, *N.Y. Food Life Sci. Bull.*, 3, 10, 1070.
64. **Rodriquez, A. G.**, Influence of smoke on the fruiting of pineapple (Ananas sativus Shult.), *J. Dept. Agric. P.R.*, 16, 5, 1932.
65. **Py, C. and Guyot, A.**, La floraison controlee de l'ananas par l'ethrel, nouveau regulateur de croissance. I, *Fruits*, 25, 253, 1970.
66. **Guyot, A. and Py, C.**, La floraison controlee de l'ananas par l'ethrel, nouveau regulateur de croissance. II, *Fruits*, 25, 341, 1970.
67. **Guyot, A. and Py, C.**, La floraison controlee de l'ananas par l'ethrel nouveau regulateur de croissance (fin), *Fruits*, 25, 427, 1970.
68. **Zimmer, K.**, Über die Beeinflussung der Blütenbildung bei Bromeliaceae und ihre praktische Anwendung, *Die Gartenbauwissenschaft*, 33, 415, 1968.
69. **Karchi, Z.**, Effect of Ethrel (2-chloroethane phosphonic acid) as compared to that of CCC on height and grain yield of spring wheat, *Isr. J. Agric. Res.*, 19, 199, 1969.
70. **Wünsche, U.**, Influence of growth retarding substances on cereals. III. Field experiments with CCC, DMC and CEPA on wheat, rye barley and oats, *Z. Acker Pflanzenbau*, 145, 238, 1977.
71. **Raafat, A. and Kühn, H.**, Halmverkürzungen bei Winter- und Sommerroggen durch Ethrel (2-Chloräthylphosphonsäure), *Z. Pflanzenernähr. Bodenk.*, 138, 361, 1975.
72. **Szöke, T. G., Anthonissen, J., and Vergracht, J.**, A new ethephon based antilodging product for barley, *Meded. Rijksfac. Landbouwwet. Gent.*, 44, 843, 1979.
73. **Kühn, H., and Linser, H.**, Kombinierte Anwendung von Chlorcholinchlorid und Ethephon zur Halmverkürzung bei Roggen (Gefäßversuche), Z. *Pflanzenernähr. Bodenkd.*, 140, 233, 1977.
74. **Schott, P. E.**, BAS 079 03 W - ein neuer Wachstumsregulator zur Verbesserung der Standfestigkeit bei Roggen, *Mitt. Biol. Bundesanstalt Land Forstwirtsch.*, 178, 160, 1977.
75. **Kühn, H., Schuster, W., and Linser, H.**, Starke Halmverkürzung bei Winterroggen durch kombinierte Anwendung von CCC und Ethephon im Feldversuch, *Z. Acker Pflanzenbau*, 145 22, 1977.
76. **Scheffer, K.**, Einfluß eines Wachstumsregulators auf Wachstum und Ertrag verschiedener Braugerstensorten, *Kali-Briefe (Büntehof)*, 15,, 449, 1981.
77. **Fabry, A. and Sazimova, E.**, Application of growth substances of the hydrazine group of compounds (CMH, IMH) and of Ethrel to the vegetative stage of winter rape (Brassica napus var. arvensis f. biennis), *Rostl. Vyroba*, 18, 617, 1972.
78. **Lavee, S., Erez, A., and Shulman, Y.**, Control of vegetative growth of grape vines (Vitis vinifera) with chloroethylphosphonic acid (ethephon) and other growth inhibitors, *Vitis*, 16, 89, 1977.

79. **Hartmair, V. and Hepp, E.,** Über den Einfluß von Ethrel auf das Längenwachstum der Rebe (Vitis vinifera L.), *Mitt. Klosterneuburg,* 24, 85, 1974.
80. **Weaver, R. J. and Pool, R. M.,** Effect of Ethrel, abscissic acid, and a morphactin on flower and berry abscission and shoot growth in Vitis vinifera, *J. Am. Soc. Hortic. Sci.,* 94, 474, 1969.
81. **Sivakamuran, S. and Young, W. M.,** Effects of 1-aminocyclopropane-1-carboxylic acid and amino-ethoxyvinylglycine on ethylene formation and latex flow in Hevea brasiliensis, *J. Rubber Res. Inst. Malaysi,* 33, 9, 1985.
82. **Abraham, P. D., Wycherly, P. R., and Pakianathan, S. W.,** Stimulation of latex flow in Hevea brasiliensis by 4-amino-3,4,5-trichloropicolinic acid and 2-chloroethane-phosphonic acid, *J. Rubber Res. Inst. Malays,* 20, 291, 1968.
83. **Pakianathan, S. W.,** The search for new stimulants, *Plant. Bull. Rubber Res. Inst. Malays.,* 111, 351, 1970.
84. **Abraham, P. D.,** Field trials with Ethrel, *Plant Bull. Rubber Res. Inst. Malays.,* 111, 366, 1970.
85. **Lacrotte, R., van de Sype, H., and Chretsin, H.,** Sugar loading and metabolism in laticiferous cells. Effect of ethephon stimulation, in *Proc. Symp. "Exploitation, Physiology and Improvement of Hevea",* IRCA-GERDAT, Paris, 1984, 155.
86. **Lacrotte, R., van de Sype, H., and Chrestin, H.,** Influence de l'ethylene sur l'utilisation du saccharose exogene par les laticiferes d'Hevea brasiliensis: proposition d'un mecanisme d'action, *Physiol. Veg.,* 23, 187, 1985.
87. **Chrestin, H., Gidrol, X., Marin, B., Jacob, J. L., and d'Auzac, J.,** Role of the lutoidic tonoplast in the control of the cytosolic homeostasis within the laticiferous cells of Hevea, *Z. Pflanzenphysiol.,* 114, 269, 1984.
88. **Gidrol, X., Chrstin, H., Mounoury, G., and d'Auzac, J.,** Early activation by ethylene of the tonoplast proton-pumping ATPase in the latex from Hevea brasiliensis, *Plant Physiol.,* 86, 899, 1988.
89. **Coupé, M., Lambert, C., Primot, L., and d'Auzac, J.,** Cinetique d'action de l'acide 2-chloro ethyl-phophonique sur les polyribosome du latex d'Hevea brasiliensis, *Phytochemistry,* 16, 1133, 1977.
90. **Coupé, M., Lambert, C., and d'Auzac, J.,** Etude comparative des polyribosomes du latex d'Hevea sous l'action de l'Ethrel et d'autres produits augmentant l'ecoulement du latex, *Physiol. Veg.,* 14, 391, 1976.
91. **Low, F. C. and Yeang, H. H.,** Effects of ethephon stimulation on latex invertase in Hevea, *J. Rubber Res. Inst. Malays.,* 33, 37, 1985.
92. **Tupy, J.,** Ribosomal and polyadenylated RNA content of rubber tree latex association with sucrose level and latex pH, *Plant Sci. (Shannon),* 55, 137, 1988.
93. **Tupy, J.,** Some aspects of sucrose transport and utilization in latex producing bark of Hevea brasiliensis, *Biol. Plant.,* 27, 51, 1985.
94. **Bennett, M. D. and Hughes, W. G.,** Additional mitosis in wheat pollen induced by Ethrel, *Nature,* 240, 566, 1972.
95. **Hughes, W. G., Bennett, M. D., Bodden, J. J., and Galanopoulou, S.,** Effects of time of application of Ethrel on male sterility and ear emergence in wheat Triticum aestivum, *Ann. Appl. Biol.,* 76, 243, 1974.
96. **Burg, S. P. and Burg, E. A.,** Ethylene action and the ripening of fruits, *Science,* 148, 1190, 1965.
97. **Mapson, L. W.,** Biosynthese von Äthylen und der Reifeprozeß von Früchten, *Endeavour,* 29, 29, 1970.
98. **Burg, S. P. and Burg, E. A.,** Interaction of ethylene, oxygen and carbon dioxide in the control of fruit ripening, *Qual. Plant. Mater. Veg.,* 19, 185, 1969.
99. **Murata, T. and Minamide, T.,** Studies on organic acid metabolism and ethylene production during controlled atmosphere storage of apples (Mallus pumila Miller, cv. Rolls), *Plant Cell Physiol.,* 11, 857, 1970.
100. **Sims, W. L.,** Effects of Ethrel on fruit ripening of tomatoes—greenhouse, field and postharvest trials, *Calif. Agric.,* July 1969, 12.
101. **Rabinowitch, H. D., Rudich, J., and Kedar, N.,** The effect of Ethrel on ripening of tomato and melon fruits, *Isr. J. Agric. Res.,* 20, 47, 1970.
102. **Iwahori, S. and Lyons, J. M.,** Accelerating tomato fruit maturity with Ethrel, *Calif. Agric.,* June 1969, 17.
103. **Poignant, A.,** La maturation controlée de l'ananas. II. L'Ethrel et son action au cours des phases ascendante et descendante de la maturité, *Fruits,* 26, 23, 1971.
104. **Crochon, M., Tisseau, R., Teisson, C., and Huet, R.,** Effet d'une application d'Ethrel avant la récolte sur la qualité gustative des ananas de Cote d'Ivoire, *Fruits,* 36, 409, 1981.
105. **Fishler, M. and Monselise, S. P.,** The use of ethephon (2-chloroethylphosphonic acid) to promote color development of Shamouti orange fruit, *Isr. J. Agric. Res.,* 21, 67, 1971.
106. **Devlin, R. M. and Demoranville, I. E.,** Influence of 2-chloroethylphosphonic acid on anthocyanin formation, size, and yield in Vaccinium macrocarpon cv. Early Black, *Physiol. Plant.,* 23, 1139, 1970.
107. **Nestler, V.,** Zur Wirkung von Ethephon auf das Reifen von Erdbeeren, *Arch. Gartenbau,* 26, 99, 1978.
108. **Sims, W. L., Collins, H. B., and Gledhill, B. L.,** Ethrel effects on fruit ripening of peppers, *Calif. Agric.,* February 1970, 4.

109. **Sims, W. L., Ririe, D., Brendler, R. A., Snyder, M. J., Wright, D. N., Schweers, V. H., and Osterli, P. P.**, Factors affecting ethephon as an aid in fruit ripening peppers, *Calif. Agric.*, June 1974, 3.
110. **Tyler, K., May, D., and Miller, R.**, Ethrel sprays reduce number of pickings in hand-harvested cantaloupes, *Calif. Agric.*, April 1970, 6.
111. **Lipe, J. A.**, Effects of daminozide, ethephon, and dikegulac on yield, harvest distribution, and ripening of blackberries, *HortScience*, 15, 585, 1980.
112. **Dekazos, E. D.**, Maturity and quality responses of 'Tifblue' rabbiteye blueberries to SADH and ethephon, *Proc. Fl. State Hortic. Soc.*, 91, 168, 1978.
113. **Crane, J. C., Marei, N., and Nelson, M. M.**, Ethrel speeds growth and maturity of figs, *Calif. Agric.*, March 1970, 8.
114. **Gerdts, M. and Obenauf, G.**, Effects of prevarvest application of ethephon on maturation and quality of Calmyrna figs, *Calif. Agric.*, May 1972, 8.
115. **Shanmugavelu, K. G., Chittiraichelvan, R., and Rao, V. N. M.**, Effect of ethephon latex stimulant on papaya, *J. Hortic. Sci.*, 51, 425, 1976.
116. **Jensen, F. L., Kissler, J. J., Peacock, W. L., and Leavitt, G. M.**, Effect of ethephon on color and fruit characteristics of 'Tokay' and 'Emperor' table grapes, *Am. J. Enol. Vitic.*, 26, 79, 1975.
117. **Chakrawar, V. R. and Rane, D. A.**, Effect of Ethrel (2-chloroethyl phosphonic acid) on uneven ripening and berry characteristics of Gulabi and Banglore purple grapes, *Vitis*, 16, 97, 1977.
118. **Martin, G. F., Buchanan, D. W., and Biggs, R. H.**, Peach fruit maturity as influenced by growth regulators, *Proc. Fl. State Hortic. Soc.*, 82, 260, 1969.
119. **Nicotra, A., Manzo, P., and Damiano, C.**, Pere "Coscia" più belle e precoci con l'Ethrel, *Inf. Agrar.*, 32, 22929, 1976.
120. **McBride, R. L. and Faragher, J. D.**, The sensory evaluation of Jonathan and Delicious apples treated with the growth regulator ethephon, *J. Sci. Food Agric.*, 29, 465, 1978.
121. **Hale, C. R., Coombe, B. G., and Hawker, J. S.**, Effects of ethylene and 2-chloroethylphosphonic acid on the ripening of grapes, *Plant Physiol.*, 45, 620, 1970.
122. **Söndahl, M. R., Teixeira, A. A., Fazuoli, L. C., and Monaco, L. C.**, Efeito do etileno sobre o tipo qualidade da bibida de café, *Turrialba*, 24, 17, 1974.
123. **Clowes, M. St. J.**, The effects of Ethrel on ripening Coffea arabica L. fruits at different stages of maturity, *Rhod. J. Agric. Res.*, 15, 79, 1977.
124. **Opile, W. R. and Browning, G.**, Regulated ripening of Coffea arabica L. in Kenya: studies on the use of 2-chloroethyl phosphonic acid, *Kenya Coffee*, 42, 17, 1977.
125. **Sisler, E. C. and Pian, A.**, Effect of ethylene and cyclic olefins on tobacco leaves, *Tob. Sci.*, May 11, 68, 1973.
126. **Long, R. C., Weybrew, J. A., Woltz, W. G., and Dunn, C. A.**, Effects of 2-chloroethylphosphonic acid on the development and maturation of flue-cured tobacco, *Tob. Sci.*, June 7, 73, 1974.
127. **Domir S. C. and Foy, C. L.,**, Effect of ethephon on ripening, curing, and chemical constituents of flue-cured tobacco, *Tob. Sci.*, October 15, 158, 1976.
128. **Anon.**, A chemical ripener for sugarcane, *Afr. Sugar. J.*, May 1976; 205.
129. **Dunster, K. W., Dunlap, R. L., and Gonzales, F. J.**, Influence of Ethrel plant growth regulator on boll opening and defoliation of western cotton, Proc. Plant Growth Regul. Working Group (7th Ann. Meet. July 13-17, 1980), 15.
130. **Cothren, J. T.**, Accelerated boll opening and increased yield in cotton with an ethylene-stimulating compound, *Plant Growth Regul. Bull.*, 9, 7, 1981.
131. **Olson, W. H., Sibbett, G. S., Carnill, G. L., and Martin, G. C.**, Lower ethephon rates effective in walnut harvest, *Calif. Agric.*, July 1977, 6.
132. **Sibbett, G. S., Martin, G. C., and Draper, T. M.**, Effects of prolonged drying and harvest delay following ethephon on walnut kernel quality, *Calif. Agric.*, June 1978, 12.
133. **Martin, G. C., Campbell, R. C., and Carlson, R. M.**, Effect of calcium on offsetting defoliation induced by ethephon in pecan, *J. Am. Soc. Hortic. Sci.*, 105, 34, 1980.

INDEX

A

B

F

G

M

N

O

P

R

S

T

U

V

W

X

Y

Z